Philip J. Potter was formerly Head Professor of Mechanical Engineering at Alabama Polytechnic Institute. He previously taught at Swarthmore College, Bucknell University, the University of Kansas, and the University of North Dakota. Professor Potter is co-author (with Edward O. Jones, Jr., and Floyd S. Smith) of *Slide Rule Problems—With Operational Instructions,* published by The Ronald Press Company.

POWER PLANT THEORY AND DESIGN

PHILIP J. POTTER

HEAD PROFESSOR
DEPARTMENT OF MECHANICAL ENGINEERING
ALABAMA POLYTECHNIC INSTITUTE

Second Edition
of
STEAM POWER PLANTS

JOHN WILEY & SONS

NEW YORK · CHICHESTER · BRISBANE · TORONTO

ISBN 0 471 06689-3

Library of Congress Catalog Card Number: 58-5856

PRINTED IN THE UNITED STATES OF AMERICA

10 9 8 7 6 5 4

To
N. U. P.

PREFACE

This book presents the fundamentals of engineering, as applied to power plant design, which are essential to a thorough college training in mechanical engineering. As in the first edition, principal attention is given to large central power station design because such stations include all the problems of method and equipment that would be involved, though often less rigorously, in the design of industrial plants and smaller power plants. There is ample material for a two-semester course on the junior or senior level. By making suitable omissions, the text may be adapted for shorter courses without loss of continuity. It is hoped that engineers in industry will continue to find the text useful as a reference.

Whenever possible, the textual presentation has been improved, and full recognition has been given to the advances in the science of power plant design. The book's emphasis on engineering economy is now reflected even in the first chapter, which introduces the economic selection of equipment and the concept of the plant heat rate. Other equipment problems, including those types where economic considerations outweigh the importance of efficiency only, have been added throughout the book. Its scope has been broadened by four chapters dealing with diesel plants, gas turbine plants, hydroelectric plants, and nuclear plants—an addition of material which has made it desirable to change the book's title.

The problems and examples have been taken from practical situations whenever that has been possible. The Appendix retains the two problems in plant layout which are useful for class purposes. Some tabulations of equipment dimensions have been included in the Appendix to provide the necessary information for problems involving plant layout.

The author expresses his thanks to the teachers and students whose suggestions and corrections have been helpful in preparing the manuscript for the second edition. Acknowledgment has been given, throughout the chapters, to the manufacturers and engineering societies who have so generously contributed pictures, drawings of their equipment, and technical data. R. H. Annin, of General Electric Company, and H. A. Mayo, Jr., of S. Morgan Smith Company, were particularly helpful in reviewing portions of the manuscript. The author is also indebted to Professor Hugh S. Miles, Jr., of Virginia Polytechnic Institute, who read the complete proof and offered many excellent suggestions.

Philip J. Potter

Auburn, Alabama
January, 1959

CONTENTS

CHAPTER   PAGE

1 INTRODUCTION 3

Magnitude of the Power Industry, 3; An Industrial Plant Cycle, 4;
Central Station, 7; Package Plants and Semi-outdoor Plants, 10;
Small Versus Large Plants, 11; Other Types of Plants, 12; Plant
Performance, 14; Economic Selection of Equipment, 15

2 FLOW OF FLUIDS 25

General, 25; Steady-Flow General Energy Equation, 26; Application of the GE and Bernoulli Equations, 29; Pipe, 31; Joints and
Fittings, 35; Valves, 42; Viscosity, 47; Units of Viscosity, 50; Laminar and Turbulent Flow, 52; Turbulent Flow, 54; Friction Factors, 58; Equivalent Lengths, 68; Determination of Pipe Size, 69;
Flow of Air Under High Pressure, 73; Flow of Air Under Low Pressure, 75

3 PUMPS 90

General, 90; Hydraulic Definitions, 90; Pump Horsepower, 92; De-
scriptions of Reciprocating Pumps, 94; Calculations of Cylinder
Sizes of Direct-acting Pumps, 98; Rotary Pumps, 101; Centrifugal
Pumps, 102; Characteristics of Centrifugal Pumps, 105; Impeller
Contours, 110; Specific Speed, 113; Descriptions of Centrifugal
Pumps, 114; System Head Curves, 124; Variable-Speed Calcula-
tions for Centrifugal Pumps, 126; Variable-Speed Drives, 128; Eco-
nomics of Variable-Speed Drives, 131

4 THEORY OF HEAT TRANSFER 139

Introduction, 139; Conduction, 139; Convection, 144; Combined
Convection and Conduction, 146; Mean Temperature Difference,
149; Scale, 152; Condensing and Boiling, 153; Radiation, 154;
Insulation Materials, 155; Pipe Insulation, 161

5 FUELS AND COMBUSTION 169

Introduction, 169; Origin of Coal, 169; Proximate Analysis, 170;
Ultimate Analysis, 171; Basis of Reporting Analysis, 171; Heating
Values of Fuels, 174; Classification of Coals, 181; Oil, 186; Gaseous
Fuels, 188; Combustion, 188; Flue-Gas Analysis, 193; Dry Flue

CHAPTER   PAGE

Gases from Actual Combustion, 194; Incomplete Combustion, 197;
Air Actually Used during Combustion, 198; Boiler Heat Balance,
198

6 STEAM GENERATORS 208

General, 208; Fire-Tube Boilers, 208; Water-Tube Boilers—Circulation, 212; Water-Tube Boilers—Descriptions, 216; The Steam
Drum, 233; Superheaters and Reheaters, 235; Superheat and Reheat Temperature Control, 240; Secondary Surface, 246; Forced-
Circulation Boilers, 251; Chemical Cleaning, 254; Boiler Ratings
and Performance, 255; *Boiler Design:* General, 257; Furnace Calculations, 258; Convection-Surface Calculations, 263; Secondary
Surface, 270

7 STEAM GENERATOR AUXILIARIES 275

Fuel-burning Equipment: Pulverizers, 275; Stokers, 288; Oil, Gas,
and Coal Burners, 294; *Draft Equipment:* Chimneys, 302; Mechanical Draft, 308; Fan Performance, 312; Fan Regulation, 318; *Feedwater Treatment:* General, 323; Zeolite Treatment, 326; Lime-Soda
Treatment, 327; Internal Treatment, 329; Demineralization, 330

8 HEAT EXCHANGERS 339

General, 339; *Condensers:* Barometric and Low-Level Jet Con-
densers, 339; Surface Condensers, 341; Condenser Calculations, 350;
Economic Selection of Condensers, 357; Circulating Pumps, 359;
Air Ejectors, 362; *Feedwater Heaters:* Surface Type of Feedwater
Heater, 366; Heat-Balance Calculations for Surface Heaters, 370;
Surface Calculations, 372; Desuperheating Zones, 375; Contact
Heaters and Deaerators, 377; Heat-Balance Calculations for De-
aerators and Contact Heaters, 382; *Evaporators and Miscellaneous
Heat Exchangers:* Evaporators, 383; Evaporator Arrangement and
Energy Calculations, 387; Surface Calculations, 390; Other Heat
Exchangers, 391

9 STEAM TURBINES 402

General, 402; Classification of Steam Turbines, 404; Theory of
Nozzles, 408; Nozzle Efficiency, 413; Supersaturation, 417; Nozzle
Designs, 417; Theory of Impulse Blades, 419; Velocity and Pressure
Compounding, 426; Impulse-Blade Efficiencies, 429; Theory of Reaction Blades, 434; Dummy Pistons, 438; Performance, 440; Governing, 447; Curves of Steam Rate and Willans Line, 454; Extraction Factor, 457; Performance of Controlled-Extraction Turbines,
458; Constructional Details, 461

CHAPTER   PAGE

10 STEAM ENGINES 484

General, 484; Steam Engine Nomenclature, 485; Slide Valves, 487;
Other Types of Steam Engines, 488; Cylinder Condensation, 493;
Governing, 494; Engine Performance, 496; Diagram Factor, 499

11 HEAT BALANCES 504

General, 504; Effect of Pressure, Temperature, and Vacuum on
Cycle Efficiency, 504; Basic Heater Arrangements, 506; Heat Rates,
510; Heat Rate Estimates, 512; Estimation of Turbine-Throttle
Steam Flow, 518; Heat-Balance Calculations, 518; Partial-Load
Heat Balances, 524; Selection of Extraction Pressures, 525; Pre-
ferred Standard Turbine Generators, 526; Actual Station-Heat
Balances, 528; Binary Cycles, 533; Location of Pumps in the
Cycle, 540; Condensate Pump Control, 541; Selection of Equip-
ment Capacities, 544

12 GAS TURBINES 556

General, 556; Gas-Turbine Cycles, 557; Performance, 564; Description of Gas Turbines, 564; Gas Turbine Prime Mover Applications,
568

13 DIESEL PLANTS 573

General, 573; Engine Descriptions, 573; Engine Performance, 580;
Auxiliary Equipment, 583; Building, 590

14 NUCLEAR PLANTS 595

General, 595; The Atom, 596; Radioactivity, 598; Fusion and Fission, 599; Nuclear Reactor Materials, 602; Classification of Reactors, 605; Nuclear-Power Reactor Steam Plants, 605; Economics
of Nuclear Electric Power, 617

15 ECONOMICS OF POWER PLANTS 623

Definitions, 623; Actual Load Curves, 627; Rates, General, 628;
Fixed Costs, 630; Operating Costs, 632; Rates, 634; Off-Peak Rates,
639; Riders, 640

16 HYDROELECTRIC POWER PLANTS 645

General, 645; Hydraulic Equations, 647; Types of Hydraulic Turbines, 649; Turbine Speed, 658; Cavitation, 660; Types of Hydro
Plants, 662; Water-Flow Curves, 663; Hydro Plant Economics
667

APPENDIX

PLATE   PAGE

1. Physical Properties of Pipe 674

2. Some Thermal Properties of Saturated Water 678

3. Mollier Diagram for Mercury . . . . *facing* 678

4. Enthalpy of Liquid Mercury 679

5. Enthalpy of Vaporization of Mercury 680

6. Saturation Temperature of Mercury 681

7. Surface Heater Dimensions 682

8. Temperature-Enthalpy-Entropy Diagram for Air at Low Pressures . . . . *facing* 682

9. Deaerator Dimensions 683

10. Fan Performance Curves 684

11. Boiler Feed Pump Dimensions 687

12. Dimensions of Direct-Acting Pumps 688

13. Boiler Dimensions 689

14. Condenser Dimensions 690

15. Turbine Dimensions 691

16. Uniflow Engine Dimensions 692

17. Dimensions of Counterflow Steam Engines 693

18. Turbine Generator Engine Efficiencies 694

19. Diesel Engine Generators 695

20. International Atomic Symbols and Weights 698

DESIGN AND LAYOUT PROBLEMS 699

INDEX 703

POWER PLANT
THEORY AND DESIGN

CHAPTER 1

INTRODUCTION

1–1. Magnitude of the Power Industry. During the year of 1957, prime movers in "kilowatt-hour factories" in the United States generated 633×10^9 kwhr (about 40% of the world's total), with an installed capacity of slightly more than 128×10^6 kw. The energy delivered to customers, after deducting for the amount consumed by producers and for transmission and distribution losses, was 561×10^9 kwhr. These figures do not include power generation for mining, manufacturing, and railway classifications. Note that these figures show an average generation of approximately 5000 kwhr per kw of installed capacity.

Prime movers are sources of power such as steam turbines, hydraulic turbines, internal-combustion engines, gas turbines, and steam engines. More prime-mover capacity is in the form of internal-combustion engines, primarily in automobiles and airplanes, than in any of the other forms. The output of such units is not included in the above figures; nevertheless, internal-combustion engine-driven generators accounted for 2.0% of the installed generating capacity in 1957, hydro units accounted for 21.2%, and steam units accounted for 76.8%.

It is interesting to note that the energy generated in 1957 is more than 1136% of that generated in 1923 and 384% of that generated in 1941. During the past five years, the annual increase in electrical energy production has been over 10% per year (compounded annually).

Another interesting point is that government-owned plants accounted for about 3% of the energy generated in 1923; the remainder was generated by privately owned plants. The figures for 1957 are radically different. In that year, cooperatives (primarily companies whose financial support is obtained from the Rural Electrification Administration, abbreviated REA) generated 0.5%; municipal, state, and public power district systems generated 6.1%; federal agency plants generated 17.3%; and privately owned utilities generated 76.1% of the electrical output. This illustrates the tremendous growth of various forms of governmental inroads into the electric power field.

The national average cost of electrical energy to residential customers has decreased from 3.84¢ per kwhr in 1940 to 2.56¢ per kwhr in 1957, while the average use has increased from 952 to 3164 kwhr during the same period.

3

Of each dollar of revenue received by private utility companies in 1957 for the sale of electrical energy, 17.0¢ was spent for fuel, 17.1¢ was spent for salaries and wages, and 22.3¢ was paid to various governments for taxes. In contrast, REA systems paid only 2.86¢ of each dollar of revenue for taxes. Probably, generation by federal, state, municipal, and public power districts was similarly devoid of taxation. The loss in tax revenues due to governmental plants must be compensated for by other tax revenues.

These advances in the magnitude of the power industry have been accompanied by advances in the size, reliability, and efficiency of stations. In 1920, 3 lb of coal were required to generate 1 kwhr. Now, the same amount of energy can be produced in large modern plants by less than 0.6 lb of coal with thermal efficiencies of 42%. The national average for coal-fueled plants in 1957 was 0.93 lb per kwhr for the 349×10^9 kwhr generated by this fuel. In 1940, this figure was 1.34 lb per kwhr of coal.

1–2. An Industrial Plant Cycle. A study of some of these modern plants will show their equipment and facilities. Fig. 1–1 represents an industrial plant that generates steam primarily for process purposes. However, by generating the steam at a higher pressure than necessary for process, some electrical energy can be obtained from the turbine, which in this case acts mainly as a means of reducing the steam pressure. The cost of a turbine generator and increased boiler pressure is a small price to pay for the electrical energy that can be developed; so, if the industrial plant needs steam for process in quantities corresponding to the need for electrical energy, it is usually cheaper to generate the electrical energy than to purchase it. However, the demand for both process steam and electric power must occur at approximately the same time. Obviously, a plant that needed steam in the afternoon and electric power in the morning would not present coincident needs.

Industrial uses for process steam are almost infinite. To mention just a few, there are: space heating; cooking, canning, and other food processes; laundries; hotels; can driers for making paper and potato flour; oil refining and other chemical processes.

In the type of plant shown in Fig. 1–1, steam is generated in the *boiler* or *steam generator 1* and leaves to enter the high-pressure (HP) *header 2*, which is merely a main steam line from which branch lines are taken. Most of the steam enters the turbine *3* through a *stop valve* or *throttle valve* not shown in the picture. This valve is used to shut off steam from the turbine and is normally in either the wide-open or tight-shut position. "Throttle valve" is a name surviving from the days when these valves were used to control the quantity of steam entering the turbine. The name "stop valve" is preferred today.

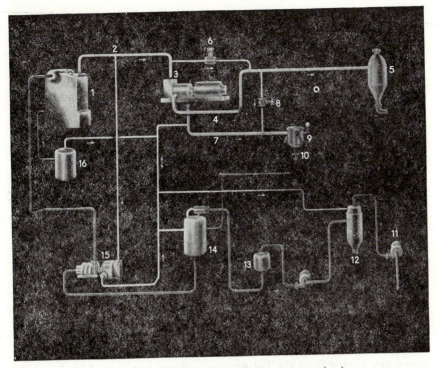

Fig. 1-1. An industrial plant. (*Power* magazine.)

Some of the steam is extracted after passing only part way through the turbine and is supplied to the intermediate-pressure header *4* and thence to a process indicated by *5*.

A line connects the HP header and the intermediate-pressure header through a pressure-reducing valve (PRV) *6*. This line has a twofold purpose: If a fault puts the turbine or generator out of operation for any reason, steam can be supplied to process *5* without interruption; and additional intermediate-pressure process steam can be supplied to process *5* when the electrical load on the turbine does not require enough steam to satisfy the process requirements.

Steam that is not extracted from the turbine continues on through the turbine, leaves from the exhaust nozzle, and goes through the low-pressure (LP) header *7*. Flanged connections to relieve or admit a fluid to any major piece of equipment are called *nozzles*, but they have no relationship to the expansion type of nozzle discussed in thermodynamics.

There is a by-pass, pressure-reducing valve between the intermediate-pressure and LP headers *8* for the same reasons as given for valve *6*. A piece of process equipment *9* is indicated for the LP header.

An important difference exists between the process equipment *5* and *9*. The former equipment does not have a pipe leading from it to any other part of the system for the return of the condensed steam, called *drains*. This indicates that either the steam was consumed in the process or the drains were so contaminated with oil or chemicals that it would not be safe to return them to the boiler. Process *9* has a small drain pipe leading from the equipment to return the drains back to the system. Immediately below *9* is a *trap* or *drainer 10* in the drain line. A trap permits intermittent flow of drains out of the equipment but does not allow steam to pass into the drain line. A drainer is slightly different in construction from a trap and permits continuous flow of drains but no steam. Usually a drainer has more capacity than a trap, but the words are often used indiscriminately.

Since water was lost from the system when the steam used by process *5* was wasted, additional water known as *make-up* is introduced by the *raw-water pump 11*. In all but exceptional cases, raw water should be treated either before or after entering the boiler (or both) to prevent scale formations inside the boiler that eventually will cause failure of the tubes. In the plant of Fig. 1–1, a *hot-process softener 12* both purifies the water and raises its temperature with steam from a lead which comes off the LP header. The softener exchanges less harmful chemicals for the harmful impurities that are in the raw water. *Effluent* is pumped to a *filter 13* where any sludge from the hot-process softener is removed by layers of sand. A filter such as this must be arranged so that the sand bed can be cleaned periodically by washing in the reverse direction of flow. Raw water may be used for this purpose.

Next, the make-up goes to the *deaerator 14* where its temperature is raised still further and objectionable gases are removed. The most objectionable gas that may be dissolved in the water is oxygen. It can cause severe corrosion in the boiler and piping.

Since drains from process *9* were treated before leaving the boiler, they are injected directly into the deaerator without being treated again in the hot-process softener. However, all drains should be deaerated before being returned to the boiler because there always will be entrained gases regardless of the pressure at which the steam is condensed.

Steam for the deaerator is taken from the LP header. Since water and steam are mixed in the deaerator, the feedwater leaving it will be at the steam pressure. Thus, another pump, *15*, is needed to force the feedwater into the boiler. This pump is known as the *boiler feed pump* (BFP) and in this case is driven by a small steam turbine. Steam for the turbine comes from the HP header and is exhausted into the LP header. Discharge from the BFP is introduced into the main drum of the boiler.

As water is evaporated in the boiler, most of the impurities from the feedwater remain in the boiler. Therefore, the boiler-water concentration will increase after several hours of operation. The concentration can be reduced by discharging some of the boiler water to waste, a process that is known as *blowing down* the boiler. Blowdown may be intermittent or continuous, depending on the plant and water conditions. When a large amount of make-up is needed, as in this case, continuous blowdown is usually expedient. However, the boiler water has received energy from the combustion of the fuel, and its waste represents a loss of energy from the system. In the plant of Fig. 1-1, this loss is reduced by taking the blowdown to a *flash tank 16.* When water at a high temperature is reduced in pressure, part of the water will flash into steam. Under proper conditions, i.e., with a flash tank designed to give off nearly dry steam, the flash steam will be nearly free of contamination and suitable for use in the system. In some cases the blowdown is passed through a tubular heat exchanger. Feedwater flows inside the tubes of the heat exchanger, and blowdown flows over the tubes. Thus, most of the energy in the blowdown is saved for the cycle, but the blowdown water is lost and must be replaced by make-up.

The steam turbine will require steam in accordance with the load on the generator. If the process steam demand is more than the steam flow through the turbine, additional steam can be supplied to process through the PRV. However, if the turbine steam demand should exceed the process demand, there must be some provision for either condensing the excess steam from the turbine or discharging it to atmosphere. An atmospheric relief valve on the exhaust piping, set to open at some predetermined exhaust pressure, will care for this contingency.

1-3. Central Station. A central station containing a *hydrogen-cooled,* 30,000-kw turbine generator that receives steam at 650 psig and 825 F is shown in Fig. 1-2. Coal from the railroad cars is transported to the coal control tower where it can be distributed to the coal reserve in the station yard or to the coal bunker. Spontaneous combustion is a hazard in the coal reserve pile. Piling the coal in thin layers and then packing with a steam roller will reduce this fire hazard. Coal from the bunker is fed to the *stoker hopper* at the front of the boiler. This station uses No. 4 buckwheat anthracite coal on a *traveling-grate stoker.* The long, low rear arch over the stoker is characteristic of anthracite installations. A long arch promotes mixing of the gases over the hot zone and reduces the loss of small carbon particles when coal fines are being used.

Ash from the grate falls into the ash hopper at the back of the boiler from which it can be removed periodically. A *forced-draft fan* supplies

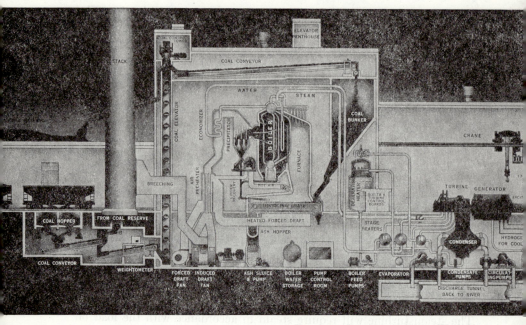

Fig. 1–2. Jennison Station. (New York State

air underneath the stoker by way of the *air preheater*. This latter piece of equipment is a tubular heat exchanger with air on one side of the tubes and hot flue gases on the other. By using the air preheater, some of the energy from the flue gases is saved for the system, and better combustion is obtained in the boiler. Some of the combustion air is supplied in jets through the front of the rear arch. This is known as *overfire air* and assists in the combustion process by creating turbulence and preventing stratification of the gases. Overfire air pressure is often supplied at about 5 psig by the auxiliary blower. The amount of air furnished by overfire jets in many cases may approximate some 15% of the total air for combustion.

Furnace gases pass over some of the boiler tubes, superheater tubes, more boiler tubes, and then out of the boiler to the precipitator. Centrifugal action in the precipitator cones, caused by forcing the gases into vortexes, deposits the fine particles that have been carried in suspension. There is a considerable amount of carbon in this refuse due to burning a small size of coal; so, it is returned from the precipitator and the hoppers at the bottom of the boiler passes at the rear of the furnace.

From the precipitator, the gases travel to the *economizer,* where some of their energy is transferred to the feedwater before it enters the boiler.

Electric & Gas Corp. and Gilbert Associates, Inc.)

Next, the gases flow through the air preheater, the *induced-draft fan,* and the *breeching* to the chimney or stack.

After the feedwater leaves the economizer it is introduced into the boiler drum. Circulation through the boiler tubes is caused by the difference between the density of the water flowing downward through the tubes in the cooler part of the boiler and the lower density of mixtures of steam and water rising through tubes in the higher temperature sections of the boiler. The high-temperature boiler tubes (*risers*) that have upward flow from the lower to the upper drums are the ones nearest to the furnace. Tubes in the back of the boiler are the *downcomers.* The front, rear, and roof of the furnace are lined with plain tubes, while the sidewalls are lined with finned tubes to give waterwalls throughout the furnace. The waterwalls are also risers.

The steam generator of Fig. 1–2 has *natural circulation.* In a few *forced-circulation* boilers, the water is forced through the tubes by special pumps.

Actually, there are two boilers in the plant of Fig. 1–2 to serve the one turbine generator. Each boiler has a capacity of 200,000 lb per hr of steam. The boilers are *two-drum, bent-tube* units.

Steam leaves the upper drum and flows through the superheater tubes into the steam header. A stop valve is shown in the steam line to the turbine. Steam, to be used for feedwater heating, is extracted from the turbine at four points.

Exhaust steam flows to the *surface-type* twin *condenser*. Circulating water is taken from the inlet tunnel by *circulating pumps* and after passing through the condenser is discharged back to the river through the discharge tunnel.

Condensate leaving the condenser is first heated by a *closed* or *surface type* of *feedwater heater* by passing the water through tubes that are surrounded by steam from the lowest pressure extraction point of the turbine. This steam is condensed by the condensate, and the temperature of the feedwater is raised. Drains from the heater are returned to the steam space of the condenser through a drainer.

Next in the water cycle is the deaerator. Boiler feed pumps take the feedwater from the deaerator and send it through two more surface heaters to the economizer, which may be considered as a form of feedwater heater using flue gases instead of steam as the hot fluid. Except for some details of construction, the condenser also is similar to the feedwater heaters. Drains from the two high-pressure heaters are cascaded to the deaerator.

An important item in the turbine hall is the crane used for dismantling the turbine and generator.

A feature of the modern plant is the automatic controls. These include water-level control in the boiler drum, automatic control of air and coal supply, automatic starting of spare pumps when an operating pump fails, and motor-operated valves. Important valves may be operated by electric motors that can be started or stopped from a control board, locally at the valve, from a control room, or from a combination of these locations.

Typical flow diagrams for large plants are shown in Fig. 11–10 and Fig. 11–16.

1–4. Package Plants and Semi-outdoor Plants. A form of plant that was developed during World War II is the package plant. These are small plants assembled at the factory and requiring little more than a foundation at the site. Since the engineering and development have already been done on these plants, their future costs may be low enough to make the units worthy of more definite consideration for industrial installations.

Semi-outdoor plants represent investment savings of from $5.00 to $10.00 per kilowatt of installed capacity. They have been constructed mostly in moderate climates, although satisfactory operation has been obtained in climates having subzero temperatures and heavy snow. Reduction of operating labor due to the compactness of the plant is another claimed advantage.

For a semi-outdoor plant, the roof of the boiler, the back of the boiler, and one side of the boiler may be used to form a part of the plant building. The firing aisle and the heads of the boiler drum that have water-level control connections and gage glasses should be inside the building. Fans and deaerators may be outside. If the steam header is well insulated and jacketed, preferably with a metal casing, it also may be located outdoors. Special care should be taken in the design of boiler casings exposed to the weather, particularly when freezing temperatures may be encountered. All casings should be watertight. Pumps, condensers, and the surface type of feedwater heater are best located inside. In extremely mild climates, a light, easily removed cover is put over the turbine and sometimes over the generator so that these units may be located outdoors.

The semi-outdoor plant is particularly suited to oil or gas firing. Coal bunkers and conveyors must be enclosed, and this may radically reduce the expected savings. In some cases experience has not been good with coal-fired semi-outdoor stations.

1-5. Small Versus Large Plants. The two plants shown in Figs. 1-1 and 1-2 were selected because they are representative of industrial and central station designs. The small plant has one stage of feedwater heating, whereas the larger one has four stages. However, it is difficult to make a distinction between large and small plants. The fact that some of the steam is used for processes is not a criterion. Likewise, size is not a criterion. Although most industrial plants are small, others are much larger than some central stations. Some large central stations supply steam for industrial purposes and for district steam-heating systems. The distinction between industrial plants and central stations lies not in the equipment or design of the station but in the ownership.

The number of stages of feedwater heating are determined by an economic balance of increased first cost against the return on the investment by fuel savings and by operating reliability. Current practice in typical stations provides for the use of one stage of feedwater heating for units of 1500 kw and less, two stages for 2000 to 3000 kw, three stages for 4000 to 7500 kw, three or four stages from 10,000 to 22,000 kw, five stages up to 100,000 kw, and six to eight stages for larger units. Units above that capacity are not so standardized, but as many as eight stages have been used for the very large turbines. It should be realized that these figures are only typical. Special conditions for a specific installation may warrant a different number of feedwater heaters. The effect on station economy of increasing the number of heaters forms a part of a later chapter of this text (Chap. 11).

By learning the function and operation of the equipment in large central stations, a student will be able to understand the small stations as well, for the small stations are merely simplifications of the larger ones.

Observe that a steam-power plant is a problem in flow of fluids and heat transfer. Condensate leaves the condenser at a low pressure. Its pressure is then raised and the water is heated and evaporated into steam. As work is done in the turbine, the pressure is reduced back to the condenser pressure. Velocities are such that this cycle is completed in a fraction of a second.

1-6. Other Types of Plants. Other developments have been made in the power field that will probably become important in the future.

Many improvements have been made in the gas turbine during and since World War II, and marine and railway installations have been put into operation. The advances in these units have been in the metallurgy of the blades and in the combustors. Use of coal in the gas turbine, if the problems of abrasion can be overcome, would be an important factor in the advancement of that prime mover.

Gas turbines have the advantages of compactness (which reduces building costs), quick starting, very low water requirements compared with the condenser of a steam plant, and the need for very little attendance during operation because they are completely automatic after they have been synchronized. At the moment their disadvantages are low thermal efficiency (maximum of about 28% with regeneration), their size limitations (maximum of about 22,000 kw), the difficulty in burning coal and oils containing vanadium and other corrosive compounds, and the necessity of an external electrical source for the starting motor.

One proposed application of the gas turbine to large central stations

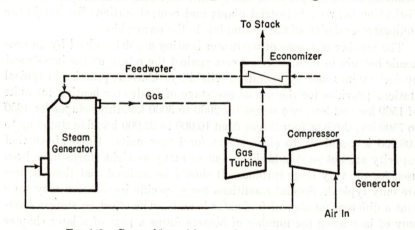

Fig. 1-3. Gas turbine with supercharged steam generator.

shows some promise. Fig. 1-3 shows a schematic arrangement of a gas turbine with a supercharged steam generator, i.e., a steam generator in which the gas pressure is above atmospheric. Air flows through the gas turbine compressor and then into the steam generator, where it combines with the fuel. Hot flue gases leaving the steam generator flow through the gas turbine to drive the compressor and an electric generator. Then the gases flow through an economizer and out the stack. This cycle provides about a 6% increase in station performance (6% decrease in net station heat rate; see Art. 1–7) over the same steam cycle without the gas turbine.

There have been several binary-cycle installations using mercury and steam. So far, mercury seems to be the only satisfactory fluid to use in conjunction with water. The efficiencies obtained with these plants were superior to anything obtained with steam cycles of the same periods. These cycles are not so attractive now as they were a few years ago, due to the increased cost of mercury, the improvements of the last few years in the steam cycle by the use of high temperatures and pressures, and the use of the reheat cycle; also, because the mercury plant is expensive to construct.

The effect of atomic energy upon the power industry is still to be determined. Some estimates indicate that with the present increase in use of electrical energy, the fossil fuels such as coal, oil, and gas will become uneconomical shortly after the year A.D. 2000. Under such conditions, power will have to be generated by other energy sources such as fissionable materials.

It has been suggested that the development of nuclear-power plants will take place in three stages. The first group of plants, those now being designed, are for the purpose of determining the most satisfactory reactors and methods of operation. These plants are not intended to compete economically with existing steam plants. The next group of plants will demonstrate the advantages and improvements that were learned from the lessons of the first group. Thereafter, however, the nuclear plants must compete with the best steam plants of their day; not with the steam plants of today nor the average plants operating in the future, but the most modern at the time.

Estimates indicate that by the year 1980, the installed nuclear generating capacity will amount to 120×10^6 kw. However, there should be an installed capacity of some 350×10^6 kw of thermal plants in the same year. Since the present installed capacity of thermal plants is about 115×10^6 kw, these figures indicate that fossil-fuel-plant capacity will more than triple in the next 25 yr.

Present indications are that the only basic difference between the atomic plant and the present thermal plant will be that today's fossil-fuel steam generator will be replaced by the atomic reactor.

1-7. Plant Performance. The term *thermal efficiency* is of practically no importance to either the designer or the operator as a means of expressing steam-plant performance. The energy input to the plant for each kilowatt-hour of output is the accepted criterion of plant efficiency. It is interesting to note that the addition of some equipment to the plant may be justified economically with an improvement in thermal efficiency of less than 0.02%.

Plant performance is usually expressed as either *gross station heat rate* or *net station heat rate*. In order to illustrate these terms, refer to the plant portrayed in Fig. 1-2. The performance of this plant could be determined by measuring the kilowatt-hours generated at the generator terminals during a suitable period, and by measuring the coal input to the boiler during the same period. Laboratory tests of the coal burned would indicate the Btu of energy released by the combustion of each pound of the coal, called the *heating value* of the coal (see Chap. 5). The performance could then be expressed as

$$\text{Gross station heat rate} = \frac{W_f \text{HV}}{\text{kwhr generated}} = \frac{\text{Btu}}{\text{kwhr}} \qquad (1\text{-}1)$$

where W_f = weight of fuel burned during the test, lb
        HV = heating value of the fuel, Btu per lb

However, some of the electrical energy generated must be used within the plant to drive the auxiliary equipment. This includes power required by circulating water pumps, condensate pumps, boiler feed pumps, fans, pulverizers or stokers, etc. At rated load on the generator, these auxiliaries require from 4.5 to 6% of the generator output for a large modern plant and somewhat more for a small plant. At lower loads, the percentage is greater.

Net station heat rate is based on the output of the station after deducting for the power consumed by the auxiliaries.

$$\text{Net station heat rate} = \frac{W_f \text{HV}}{\substack{\text{kwhr generated less} \\ \text{kwhr used in plant}}} = \frac{\text{Btu}}{\text{net kwhr}} \qquad (1\text{-}2)$$

Observe that the net station heat rate is always numerically higher than the gross station heat rate and that

$$\text{Thermal efficiency} = \frac{3413}{\text{heat rate}} \qquad (1\text{-}3)$$

Most plants maintain a continual record of their heat rate as a measure of their efficiency of operation.

Net station heat rates have decreased steadily during the past years. In the early 1920's, a net station heat rate of about 45,000 Btu per kwhr was considered satisfactory; modern units have heat rates of about 9000 Btu per kwhr.

EXAMPLE 1-1. During a test a unit was operated at its rated load of 100,000 kw for 12 hr. The coal consumption was 429 tons, with a heating value of 12,670 Btu per lb. Auxiliary power amounted to 62,700 kwhr. Find (a) the gross station heat rate and (b) the net station heat rate.

SOLUTION. (a) From Eq. (1–1),

$$\text{Gross station heat rate} = \frac{429 \times 2000 \times 12{,}670}{12 \times 100{,}000} = 9059 \text{ Btu per kwhr}$$

(b) From Eq. (1–2),

$$\text{Net station heat rate} = \frac{429 \times 2000 \times 12{,}670}{12 \times 100{,}000 - 62{,}700} = 9558 \text{ Btu per net kwhr}$$

Conventional expressions for costs of fuels, such as dollars per ton of coal, cents per gallon of oil, and cents per thousand cubic feet of gas, are more comparative when converted to cents per million Btu. Then the unit fuel cost of electrical energy is:

$$\text{Unit fuel cost} = \frac{\text{net station heat rate} \times \text{cost of } 10^6 \text{ Btu of fuel}}{10^6} \quad (1\text{–}4)$$

EXAMPLE 1-2. The cost of coal for the plant of the preceding example is $6.30 per ton delivered to the plant. What is the unit fuel cost?

SOLUTION.

$$\frac{\$6.30 \times 10^6}{2000 \times 12{,}670} = 24.9\cancel{c} \text{ per } 10^6 \text{ Btu}$$

From Eq. (1–4),

$$\text{Unit fuel cost} = \frac{9558 \times 24.9}{10^6} = 0.238\cancel{c} \text{ per kwhr, or 2.38 mills per kwhr}$$

1–8. Economic Selection of Equipment. Building a steam-power plant is a business enterprise, as is nearly all engineering work. It is anticipated that the power plant will return the original investment and in addition will show a profit. A plant that conforms to these requirements will not necessarily be the most efficient plant that can be constructed. It has been previously mentioned that turbines in the range of 2000 to 3000 kw normally have two stages of feedwater heating. These units could be constructed with more stages; the plant would be more efficient and fuel costs would be lower. However, in the typical installation, the increased cost of the additional stages of feedwater heating and their associated equipment could not be justified economically even with the increased efficiency.

A major portion of the design of any power plant is the economic justification of equipment and operating conditions. A few examples would be: steam pressures and temperatures, size of turbine generators, size of condensers and feedwater heaters, thickness of insulation, sizes of piping and duct work, and even the location of the plant.

In order to develop the concepts of economic analysis used in power-plant design, assume that the erection of a small power plant in a village of 1000 population is being considered and that the plant will cost $300,000. Regardless of the source of the capital for this plant, that is, whether the money is available or must be borrowed, the income must at least pay interest on the investment, pay the operating expenses, pay the taxes, pay the insurance on the plant, and return the initial investment during the life of the plant. If this venture should ultimately prove to be exceptionally profitable, the income could provide a very high interest on the original investment. If it should not be profitable, the income might not even pay the expenses and then the venture would fail.

Certain of the factors listed above are customarily grouped together and called *investment charges*. This group, sometimes called *fixed* charges, includes the interest on the investment, taxes, insurance, and the retirement of the original investment. This latter item, repayment of the debt, is known as *amortization* or depreciation, and the amount is a function of the life of the loan or the expected life of the equipment. For income tax purposes, minimum amortization periods have been established by the U.S. Treasury Department and range from 20 to 30 yr for most major power-plant equipment.

Nearly all electric utilities depreciate their equipment at an equal rate each year, known as *straight-line depreciation*. If a 33-yr amortization period is taken as the length of time in which to repay a $300,000 loan for the plant, the yearly depreciation will then be $300,000/33 = $9090, or 3.03%. The total yearly investment charges for the plant will include this 3.03% plus average interest which may range from 5 to 8%, insurance of about ⅓%, and taxes that may amount to about 1% for local taxes and about 5% for federal taxes. The total of these percentages will amount to about 15%, or $45,000 per year.

Observe that there can be considerable variance in each of the components of the investment charges. Industrial owners of power plants justify new equipment on a much shorter depreciation period (in the order of 10 yr) because of business uncertainties and obsolescence of equipment and processes. Government-sponsored REA plants are based on long depreciation periods, very low interest rates, and extremely low taxes.

Common values of investment charges that are encountered in power-plant work are in the order of the following: 20 to 25% for industrial plants, 12 to 15% for privately owned utilities, 10% for municipal utilities, and 5% for REA plants.

Note the effect of the value of investment charges on the minimum required yearly earnings of a power plant after deducting operating and maintenance costs (fuel, supplies, wages, etc.). At 15%, the investment charges amount to the $45,000 previously calculated, at 20% they are $60,000, and at 5% they are only $15,000 per yr.

These calculations, expressed in the form of an equation, would be

$$P \times C = R \qquad (1\text{--}5)$$

where P = original investment or present worth, dollars
  C = annual investment charges
  R = annual income after deduction of maintenance and operating costs, dollars

Many problems arise in design work where the unknown quantity in Eq. (1–5) is either C or P, rather than the quantity R, as in the preceding discussion. In considering alternate pieces of equipment, it is frequently possible to estimate the yearly savings of one piece over the other. Then the question is asked in the form of, "What investment, P, will this saving justify?" Again referring to the proposed plant, suppose that, by increasing the pressure and temperature at the boiler outlet, there is an annual saving in fuel cost of $900. With 15% investment charge, the higher steam pressure and temperature would be justified economically if the increased investment were not more than $900/15% = $6000. The increased investment could be $18,000 if 5% investment charges were used. This illustrates one of the reasons why REA plants can economically justify investments that could not be justified by a private utility or an industry.

In engineering economy, the $6000 is the *present worth* of a yearly return of $900 when the investment charges are 15%. Some engineers use the expression *capitalized cost* in place of present worth.

If both the annual return and the present worth of a piece of equipment are known or can be estimated, the investment charges or return on the investment, C, can be calculated.

When comparing two or more alternatives to determine the most economical method or piece of equipment, it is necessary to consider only those cost items that are different for the alternatives. Example 1–3 will illustrate both the methods of calculation and the effect of low and high investment charges.

EXAMPLE 1–3. A manufacturer offers a low-speed fan and a high-speed fan for a certain boiler. Both fans have equal capacity, but the low-speed fan is more efficient than the high-speed fan. Determine the most economical fan for (a) 20% investment charges on the annual cost basis, (b) 5% investment charges on the annual cost basis, (c) 20% investment charges on the present worth basis. The data are as follows:

Speed	808 rpm	1175 rpm
Power	432 bhp	484 bhp
Cost of fan	$12,900	$8700
Cost of motor	$ 7500	$6800

All other costs are assumed to be the same. Motor efficiency is 92% for either motor, power costs 5 mills (0.5¢) per kwhr, no salvage value, and 4000 hr of operation per yr.

SOLUTIONS.

(a)   Annual cost for low-speed fan:

Investment cost, 0.20($12,900 + $7500)   = $ 4080

Energy cost, $4000 \text{ hr/yr} \times \dfrac{432 \text{ bhp}}{0.92 \text{ eff}} \times 0.746 \dfrac{\text{kw}}{\text{hp}} \times \dfrac{\$0.005}{\text{kwhr}} =$ 7000

Total = $11,080

Annual cost for high-speed fan:

Investment cost, 0.20($8700 + $6800)   = $ 3100

Energy cost, $4000 \times \dfrac{484}{0.92} \times 0.746 \times \$0.005$ =   7850

Total = $10,950

There is a slight preference for the high-speed fan. In this case, items such as reduced wear and maintenance on the low-speed fan should be considered.

(b)   Annual cost for low-speed fan:

Investment cost, 0.05($12,900 + $7500)   = $ 1020

Energy cost, $4000 \times \dfrac{432}{0.92} \times 0.746 \times \$0.005$ =   7000

Total = $ 8020

Annual cost for high-speed fan:

Investment cost, 0.05($8700 + $6800)   = $ 775

Energy cost, $4000 \times \dfrac{484}{0.92} \times 0.746 \times \$0.005$ =   785ʋ

Total = $ 8625

There is now a preference for the low-speed fan.

(c)   Present worth of low-speed fan at 20% investment charges:

Investment, $12,900 + $7500   = $20,400

Energy cost, $\dfrac{\$7000}{0.20}$ = 35,000

Total = $55,400

Present worth of high-speed fan:

Investment, $8700 + $6800   = $15,500
Energy cost, $7850   = 39,250
$$\overline{0.20}$$

Total = $54,750

Note that the difference of $650 in present worths can also be obtained from part (a) by dividing the difference in annual costs of $130 by 20% investment charges.

Observe that a reduction in operating costs of a piece of equipment is always accompanied by an increase in the original cost of the equipment. This is shown graphically in Fig. 1–4 except that investment cost is used

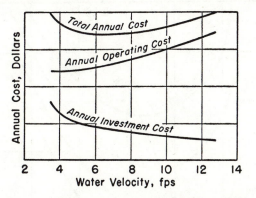

FIG. 1–4. Annual cost curves for a piping system.

in place of original cost. These curves were developed for water flow in a certain piping system. Annual investment cost is shown to decrease as velocity increases because of the consequent reduction in pipe size and therefore in original cost. However, an increase in operating cost accompanies the increase in velocity, since more pump power is required at high velocities than at low velocities.

Total annual cost is the summation of annual operating cost and annual investment cost. The velocity at which the total annual cost is at a minimum is the most economical velocity for the particular conditions assumed for the calculations.

Investment charges of 20% were used for the data of Fig. 1–4 and were applied to the original cost of the pipe, fittings, pump and motor, insulation, and labor of installation. The most economical velocity would have been lower if the investment charges or any of the equipment costs had been lower.

Annual operating costs for Fig. 1–4 included such items as number of hours of operation of the system per year, cost of electrical energy, and motor efficiency. Changes in any items that would increase operating costs would thereby decrease the most economical velocity.

Curves, such as those of Fig. 1–4, are not always smooth. A change in commercial sizes of motors, for example, can cause an irregularity in the curves.

Effects of investment charges on total annual costs can be seen in Fig. 1–5, which was calculated for the same piping system as that used for Fig. 1–4. Note that as investment charges decrease, the operating costs have a more pronounced effect on the total annual cost. Thus, low investment charges justify high initial investments with correspondingly low operating costs, or in this case, low velocities. Also, for a given

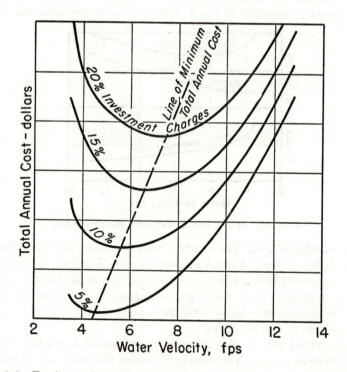

Fig. 1–5. Total annual cost curves for a piping system, showing effect of investment charges.

saving in operating costs, low investment charges permit large original costs.

The most economical velocities and the shapes of the curves would have been the same if the present worth method of analysis had been used to calculate the costs.

Another useful device for economic studies is called the *break-even* point, that is, the point at which two alternatives are exactly the same. When determining the break-even point, one of the economic factors is allowed to vary while the others remain constant. The one variable may be investment charges, hours of operation, the load on a plant or unit, the life of the equipment, efficiency, cost of fuel, or any of the other variables encountered in economic studies.

In the Example 1–3, the high-speed fan was the most economical at 20% investment charges, but at 5% investment charges the low-speed fan was the most economical. The break-even point for investment charges can be found for these fans by drawings curves of the annual cost versus investment charges for each fan. The break-even point is the point where the curves cross in Fig. 1–6. By taking proportional triangles $[(x - 5)/605 = (20 - x)/130]$ the break-even point, x, is 17.3%. Other variables that could have been used in place of investment charges are hours of operation or cost of electrical energy.

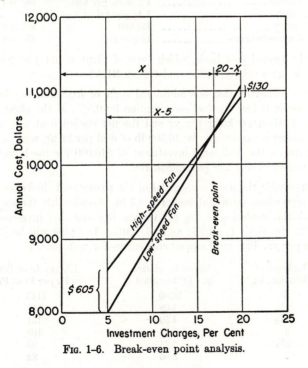

FIG. 1–6. Break-even point analysis.

PROBLEMS

1-1. A pump is to be purchased to transfer fuel oil from the storage tanks in the yard of a power plant to the boiler house. It is expected that the pump will operate for 3500 hr per yr with energy costing 5 mills per kwhr and that it will have a life of 18 yr. The ABC Manufacturing Co. offers a pump for this service at a cost of $250 that will require an input of 0.4 hp. A pump manufactured by the Pumpit Co. will cost $190 and will require an input of 0.55 hp. The ABC unit will have a freight cost of $6.00, but the Pumpit unit will have a freight cost of $10.00. A motor for either pump will have an efficiency of 79%. Which unit should be purchased by (a) an industrial plant and (b) by an REA plant? Use both annual cost and present-worth methods of calculation.

1-2. As a consulting engineer, you are to recommend to your client the type of power plant that he should install. Because of local conditions, either a steam plant using coal or a diesel plant using oil would be the most suitable. The plant capacity will be 4000 kw, and it is expected that the average generation per year throughout the life of the plant will be 17,000,000 kwhr. Data on the two types of plant are as follows:

	Diesel	Steam
Cost of plant per kw	$210	$290
Fuel cost	9.5¢ per gal	$5.10 per ton
Fuel consumption at average load	11 kwhr per gal	1.7 lb per kwhr
Plant maintenance	1.2 mills per kwhr	98¢ per ton of coal
Lubricating oil	1.4¢ per gal of fuel oil	
Labor cost per yr	$51,000	$87,000
Supplies and miscellaneous	1¢ per gal	45¢ per ton

Using the annual cost basis, which type of plant would you recommend if your client is (a) a private utility and (b) an REA?

1-3. It is estimated that one additional stage of feedwater heating in a certain plant would reduce the fuel consumption by 0.8%. If the plant is expected to generate 120,000,000 kwhr per yr and the net station heat rate without the additional heater is expected to be 16,500 lb of coal per kwhr, what would be the rate of return on the additional investment of $16,000 for the heater? Assume coal at this plant to cost 29¢ per million Btu.

1-4. Determine the most economical of the commercial thicknesses of insulation for a pipe whose outside diameter is 4.5 in. Assume that the cost of energy lost through the insulation is 30¢ per million Btu and that investment charges are (a) 5%, (b) 10%, (c) 15%, and (d) 20%. The pipe will be in operation for 8000 hr per yr. Plot total annual cost curves for each part.

Thickness of Insulation, In.	Cost of Insulation per Ft, Installed	Energy Loss, Btu per Hr per Ft of Pipe
0	$0.00	1145
1⅛	1.08	152
1½	1.59	131
2	2.43	108
2½	3.24	93
3	3.96	82

1-5. Use the same data for Prob. 1–4, except use 60¢ per million Btu as the the energy cost and 6000 hr per yr operating time.

1-6. A certain pipeline in a power plant is to handle 2000 gpm of water. If the length of the line is 450 ft, determine the most economical size of pipe for (a) an REA plant, (b) a municipal plant, (c) a privately owned utility, and (d) an industrial plant. Plot curves of the annual cost as the ordinate and velocity as the abscissa. Motor efficiency may be taken as 90% in all cases. Use 8500 hr per yr and 5 mills per kwhr.

Pipe Size, In.	Velocity, Fps	Pumping Power, Hp	Cost of Pipe Installed, Dollars per 100 Ft
6	22.2	191	$ 372
8	12.8	99	574
10	8.14	77	831
12	5.68	71	1102
14	4.17	69	1344
16	3.51	68	1773

1-7. Repeat Prob. 1–5, but use 7500 hr per yr and 9 mills per kwhr.

1-8. An industrial plant has grown to the extent that it must arrange for additional electrical energy. Two likely alternatives are: (1) buy power from the local utility company, which will cost 1.6¢ per kwhr plus an investment in sub-station facilities of $110,000; or (2) enlarge the existing power plant by installing a new 5000-kw steam turbine-generator, which will cost $268 per kw with an average net station heat rate of 18,700 Btu per kwhr. The operating costs for the steam plant will be: 32¢ per million Btu for coal, 1.4 lb per kwhr of coal, $2.99 per ton for other operating costs. Auxiliary power will use 7% of the kwhr generated. It is expected that 17,700,000 kwhr per yr will be consumed by the industrial plant. If energy is purchased, the industrial plant must pay for the transformer losses of 1½%. As consulting engineer for this industrial concern, should you recommend generating or purchasing the additional energy?

1-9. For the data of Prob. 1–8, what original cost per kilowatt (present worth) of the enlarged steam plant would make the annual cost the same as the annual cost of purchased power?

1-10. A power plant located on a river may install a wharf that will cost $80,000 and have an expected life of 20 yr. For an additional cost of $25,000, the expected life of the wharf can be increased. What must be the minimum expected life of the more costly wharf to justify the additional expenditure if interest is 5%, taxes and insurance are a total of 6%, and depreciation is on the straight-line basis? Other costs for the two alternates are the same.

1-11. Find the break-even point of investment charges for motors A and B:

Motor A: $1500 cost, 90% efficiency at average load
Motor B: $1800 cost, 91% efficiency at average load
Cost of power: 7 mills per kwhr.
Average load: 80 hp for 6000 hr per yr
Expected life of either motor: 20 yr
Maintenance for either motor: $18 per yr
Oil for either motor: $3 per yr
Salvage value for either motor: none

BIBLIOGRAPHY

CHRISTIE, A. G. "Power in 1946," *Combustion* (February, 1946).

———. "Progress in Transformation of Energy," *Combustion* (November, 1946).

Electrical World, Annual Statistical Number (January 27, 1958).

GOURDON, P. E., FRIEND, W. F., and ELLIOTT, LOUIS. "Semi-Outdoor Steam Plants," *Combustion* (April, 1946).

GRANT, E. L. *Principles of Engineering Economy.* New York: The Ronald Press Co., 1950.

SCHWERKART, H. C. "Features of the Jennison Station," *Combustion* (December, 1945).

CHAPTER 2

FLOW OF FLUIDS

2–1. General. The piping system of a power plant is an important item in the total cost of the station. Published figures indicate that the piping cost can reach 13% of the station cost, although an average figure would be about 6 or 7%. Based on a total station cost of $200 per kw of installed capacity, piping cost would run from $12.00 to $14.00 and might go as high as $26.00. Thus, proper design and selection of the piping system are of utmost importance and may well mean the difference between an economical station and one that is uneconomical.

The miles of piping in a large central station will include pipes varying in size from small ¼-in. pipes used for instrument connections to large steam pipes 14 to 30 in. in diam and circulating water pipes 7 to 8 ft in diam.

Factors that should be considered when selecting these various types of pipes are: (1) correct pipe diameter for the service conditions, (2) flexibility of the system to absorb thermal expansion, (3) materials required for satisfactory performance under the service temperature, and (4) wall thickness needed to withstand the stresses that will be imposed on the pipe.

Obviously the flow through a ¼-in. pipe will be much less than the flow through a 14-in. pipe under similar conditions. The pipe must provide sufficient cross-sectional area to permit the desired quantity of fluid to flow at a reasonable velocity. A reasonable velocity under specified conditions will usually depend on the available pressure drop, but there are conditions when diameter may be determined by other considerations. For example, if a connection is to be welded to a large heat-treated tank, the connection must be welded onto the tank before shipment to the power plant because the heat treatment cannot be done in the field. If the connection were for a pressure gage, the pipe diameter could be small (say, ¼ in.) because the fluid in a line to a pressure gage is stagnant. However, a ¼-in. pipe connection onto a large tank would be so small that it would probably be damaged during shipment. It would therefore be good engineering practice to make the connection of 1-in. pipe even though that would be much larger than required on a basis of flow only.

Piping systems are normally installed with the atmospheric temperature varying from 0 F to 100 F. However, when the pipe is in operation, its temperature is that of the fluid. In the case of steam, the fluid temperature may be as high as 1050 F. The pipe would then expand in accordance with its coefficient of expansion and the temperature change. For a temperature change of 900 F, the expansion of steel pipe would be about 8½ in. per 100 ft of length. Expansion must be permitted in the system without causing undue stress on the equipment to which the pipe connects. This may be accomplished by designing the system with sufficient flexibility, i.e., with bends and loops. This flexibility will reduce the forces due to thermal expansion, but it will not entirely eliminate them.

There must be ample metal area in the pipe cross-section to keep the stresses within allowable limits. In addition, allowance must be made for stresses caused by the bursting pressure of the fluid, the weight of the pipe, the insulation, and the fluid that is supported between the pipe hangers.

The materials used in the manufacture of the pipe must be in accordance with the applicable ASME and ASA codes for the temperatures involved. For example, cast iron may not be used when the service temperature is in excess of 450 F. Some latitude is allowed the designer in the selection of materials, provided he does not exceed certain allowable stresses.

Of the four factors previously mentioned that must be considered in the design of a piping system, this chapter will deal primarily with the first: determination of pipe diameter. Brief mention will be made of materials and proper thicknesses.

2-2. Steady-Flow General Energy Equation. The steady-flow general energy equation (abbreviated GE equation) is derived from the Law of Conservation of Energy without any limitations as to the kind of fluid or the processes through which the fluid passes.*

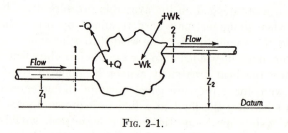

Fig. 2-1.

As the name implies, the GE equation is based on a steady flow of fluid, as shown between sections *1* and *2* in Fig. 2-1. None of the fluid may be stored in the process, nor may more be rejected by the process during any given period than the quantity which entered during that

* See any standard text on thermodynamics.

period. Therefore, the flow entering and leaving must be equal and may be represented by the same symbol. Equating all the energy entering the process to that leaving,

$$CE_1 + EE_1 + WZ_1 + \frac{WV_1{}^2}{2g} + WP_1v_1 + 778Wu_1 + 778Q$$

$$= CE_2 + EE_2 + WZ_2 + \frac{WV_2{}^2}{2g} + WP_2v_2 + 778Wu_2 + Wk \quad (2\text{-}1)$$

where the symbols used are

Z = elevation above the arbitrary datum, ft
g = acceleration of gravity, 32.2 fps^2
V = velocity, fps
P = fluid pressure, lb per sq ft
W = weight of fluid flowing, lb per unit time
v = specific volume, cu ft per lb
u = internal energy, Btu per lb
Q = heat added, Btu per unit time
Wk = work done by the fluid, ft-lb per unit time
CE = chemical energy, ft-lb per unit time
EE = electrical energy, ft-lb per unit time

and subscripts 1 and 2 indicate the sections.

Note that the unit of time has not been specified for the weight of fluid flowing, W, work done, Wk, and heat added, Q. It is important only that the same unit for time be used for these three items; it is not important whether the unit be seconds, minutes, hours, etc. The work term has been defined as plus *when work is done by the fluid*. Work done by steam in a turbine or steam engine would be plus in the GE equation, but work done on water by a pump would be minus because work is then entering the system. The opposite is true of the heat-energy term. Since heat was grouped with other terms representing energy entering the system, *heat added to the fluid* is plus and that rejected by the fluid is minus. For example, heat energy added to water in a boiler would be plus, whereas heat rejected by steam in flowing through a pipe would be minus.

Each of the items in the GE equation has units of foot-pounds per unit of time. Certain of the items have been given names that are commonly used by engineers. The elevation of the centerline of the inlet and outlet pipes above the arbitrary datum is potential energy and is often called the *static head* or gravity head. The Pv product is referred to as the *flow work* in thermodynamics, but in hydraulics this same term is called *pressure head*. Because density, ρ, is the reciprocal of specific volume, v, flow work or pressure head may be written as WP/ρ. Kinetic energy, $V^2/2g$, is universally known as the velocity head.

Chemical and electrical energy are unimportant when dealing with the flow of fluids. With these terms omitted, Eq. (2–1) reduces to

$$WZ_1 + \frac{WV_1{}^2}{2g} + \frac{WP_1}{\rho_1} + 778Wu_1 + 778Q$$
$$= WZ_2 + \frac{WV_2{}^2}{2g} + \frac{WP_2}{\rho_2} + 778Wu_2 + Wk \quad (2\text{–}2)$$

where $\rho =$ density, lb per cu ft, and other symbols are as in Eq. (2–1).

Although Eq. (2–2) applies to the flow of a fluid through a conduit, it may be arranged into a much more convenient form for determining the pressure loss in the conduit. The result of this rearrangement is the familiar equation presented by Bernoulli in his *Hydrodynamica* (1738) and also developed generally by Euler (1750).

If Eq. (2–2) is rewritten with $W = 1$ lb per unit time and for a system in which there is no mechanical energy added to or extracted from the system, i.e., with no pump or turbine in the system, then

$$778 \, dq = dZ + d\left(\frac{V^2}{2g}\right) + 778 \, du + d(Pv) \qquad (a)$$

where q is the heat added to the system per lb of the fluid, Btu per lb, and other symbols are as in Eq. (2–1).

Thus $\qquad 778 \, dq = dZ + d\left(\frac{V^2}{2g}\right) + 778 \, du + P \, dv + v \, dP \qquad (b)$

Pressure loss of a fluid flowing through a conduit is the result of the viscous * or adhesive action of the molecules causing turbulence and friction. In the GE equation this turbulence and friction convert energy from one form to another and are the reason why there is no item in Eq. (2–2) for friction loss. The energy required to overcome friction is obtained at the expense of pressure head, static head, or velocity head. But friction is always converted into internal energy or into heat as, for example, in the temperature rise of the bearings of any machine. Energy dissipated by friction may leave the system or may be returned to the fluid, thereby increasing the internal energy, u, of the fluid and the flow work, Pv. Such energy added to the fluid would not necessarily cause a change in any of the other terms of the GE equation because they are all dependent on the physical dimensions of the system. Then

$$778 \, dq + dh_f = 778 \, du + P \, dv \qquad (c)$$

where h_f is the head lost due to friction, feet of the fluid, and other symbols are as in Eq. (b).

* See Art. 2–7 for a more complete discussion of viscosity.

Substituting Eq. (c) into Eq. (b),

$$dh_f + dZ + d\left(\frac{V^2}{2g}\right) + v\,dP = 0 \qquad\qquad (d)$$

Integration yields

$$h_f = (Z_1 - Z_2) + \frac{V_1{}^2 - V_2{}^2}{2g} - \int_1^2 \frac{dP}{\rho} \qquad\qquad (2\text{–}3)$$

Integration of the last term of Eq. (2–3) will depend on the variation of the density with pressure.

Consider, for example, water flowing through the tubes of a boiler. Water enters the downcomers from the drum as a saturated liquid. The water flows through what is essentially a U tube with nearly all the heat being added in the riser. The fluid returns to the drum through the riser as a mixture of saturated liquid and steam. From Eq. (2–3), it will be found that the difference between the density of the water in the downcomer and the density of the water and steam mixture in the riser must be sufficient to overcome the friction loss and the change in velocity head. Notice, however, that not only does the density of the water and steam mixture in the riser vary throughout the height of the riser, but that it is also a function of the rate of heat transfer from the hot products of combustion to the fluid. Thus, the integration of dP/ρ may have an important influence on Eq. (2–3).

If we apply Eq. (2–3) to the flow of an incompressible liquid, density will be independent of pressure, and we obtain

$$h_f = (Z_1 - Z_2) + \frac{V_1{}^2 - V_2{}^2}{2g} + \frac{P_1 - P_2}{\rho} \qquad\qquad (2\text{–}4)$$

where the symbols are the same as in Eqs. (2–1) and (2–2). This is Bernoulli's equation with the addition of a term for friction loss.

2–3. Application of the GE and Bernoulli Equations. Throughout this text the GE equation will be applied to many of the processes involved in a steam-power plant. However, in this chapter we are interested in applications that deal with the flow of fluids through pipes or conduits. In many applications of these equations it will be found that some of the terms of the GE equation can be omitted because (1) they are negligible, (2) the difference between like terms at section *1* and section *2*, Fig. 2–1, is negligible, or (3) the term is zero for the process. There is often much confusion in the student's mind as to why certain terms are included in some applications of the GE equation and are omitted in other applications. Much of this confusion will be eliminated if the student will pay close attention to the definitions of the processes and the magnitudes of the terms.

If a liquid is flowing through pipes connecting pieces of equipment, there will be no work done on or by the fluid. For this special case, work may be omitted from the GE equation, or Bernoulli's equation may be applied.

Velocities commonly used for liquids vary from about 100 to 600 fpm (see Table 2-6). Whether this will produce a kinetic energy term of sufficient magnitude to be included in the GE equation will depend on the process and the magnitude of the other terms. Similarly, the difference in potential energy between sections 1 and 2 would be zero for a horizontal pipe. Potential energy is of considerable importance when *gravity feed* is used. Consider flow from an overhead tank of water. Even though the elevation of the tank is not high, the potential energy will be the only energy available to overcome friction and kinetic energy.

When the flow is adiabatic, the density of water for sections 1 and 2 may be assumed to be the same. Water is incompressible for most practical problems.

EXAMPLE 2-1. A water storage tank that is open to atmosphere is to be located above the floor so that there will be sufficient static head to deliver 400 gpm of 80 F water to a piece of equipment that operates at 2 psig and is located on the floor. The flow will be through a 4-in. pipe (4.00 in. ID) which will have a friction loss of 17.6 ft of water. Find the elevation of the bottom of the tank.

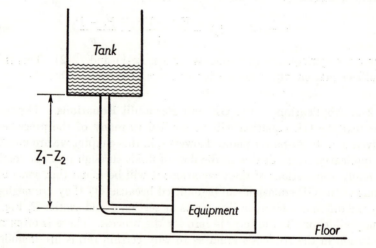

SOLUTION. From Plate 2 in the Appendix,

$\rho = 62.2$ lb per cu ft

$$W = \frac{(\text{gpm})(\text{min/hr})(\text{density})}{\text{gal/cu ft}} = \frac{(400)(60)(62.2)}{7.48} = 200{,}000 \text{ lb per hr}$$

$$V = \frac{\text{quantity}}{\text{area}} = \frac{(400)(144)}{(7.48)(60)(4)^2(\pi/4)} = 10.21 \text{ fps}$$

From Eq. (2–4),

$$Z_1 - Z_2 = \frac{V_2{}^2 - V_1{}^2}{2g} + h_f + \frac{P_2 - P_1}{\rho}$$

Since it was stated that this tank was for storage purposes, all the water in the tank must be available to the equipment at not less than 2 psig even when the tank is nearly empty. Thus, for design purposes, section *1* should be taken at the minimum water level in the tank; i.e., the bottom of the tank. Section *2* would be at the entrance to the piece of equipment that is located on the floor. Expressing all items in units of feet of the fluid, and remembering that the velocity of the water on the surface is zero,

$$Z_1 - Z_2 = \frac{(10.21)^2 - (0)^2}{64.4} + (17.6) + \frac{144(16.7 - 14.7)}{62.2} = 23.8 \text{ ft}$$

2–4. Pipe. Selection of the proper pipe materials, wall thickness, and method of fabrication is of major importance in the design of a piping system. Chemical and physical properties of pipe materials are established in the ASTM (American Society for Testing Materials) specifications, and the materials are designated by their ASTM specification number. However, the maximum allowable stress at various operating temperatures for acceptable materials are covered in the ASME *Boiler Construction Code* and the ASA (American Standards Association) *Code for Pressure Piping.* Steam piping between the boiler and the first stop valve, water piping between the regulator valve and the boiler, and boiler drain and blowdown piping are all under the jurisdiction of the Boiler Code. All other steam, water, air, gas, and oil pressure piping must follow the Pressure Piping Code. All pipe in a plant should conform to state laws in effect for the plant locality. However, most state laws are patterned after the codes mentioned above.

Copper alloys are used in power plants for small control lines, condenser tubes, and for tubular feedwater heaters. Reference to them will be made in other chapters.

Cast-iron pipe was formerly used extensively in power plants but now is used primarily for underground water lines and low-pressure and low-temperature drains and sewers. The brittle characteristics of cast iron make it unsuitable for most power-plant uses where fluid temperatures cause expansion of the piping. Most codes limit the use of cast iron, for pipe and equipment, to operating temperatures of not more than 450 F, but many designers do not recommend its use for piping when the temperatures exceed 150 F or 200 F. Cast-iron pipe has long been listed in three weight classifications: 25-psi, 150-psi, and 250-psi standards.

Where resistance to corrosion is important and temperatures and pressures are low, wrought-iron pipe may be used. This material is more resistant than steel to attack by oxygen but has a lower strength and is more expensive.

Wrought steel and alloy steel (as distinguished from cast steel) may be obtained in many varieties. Values for allowable stress, S, for the most popular types of steel pipe are listed in Table 2–1. These stresses are obtained by considering both the ultimate tensile strength (4 to 5 factor of safety) and the creep strength (1% in 100,000 hr) of the steel.

The material covered by ASTM specification A120 corresponds to the type of pipe readily obtainable in plumbing shops. It is not acceptable for work covered by the Boiler Code and should not be used for temperatures in excess of 450 F. This pipe may be obtained in either black or galvanized finish and is usually butt-welded in sizes less than 4 in. and lap-welded for 4 in. and over. Seamless pipe is available in all sizes.

ASTM specification A53 is for a common type of carbon-steel pipe obtainable as butt-welded, lap-welded, and seamless. This specification is more rigid than A120 and is recognized by the codes for power-plant use. Seamless is preferred for most uses.

The best seamless carbon-steel pipe is manufactured in accordance with ASTM specification A106 and is used for operating temperatures up to 775 F. Grade B has a higher allowable stress than Grade A, but both are used extensively for steam and water pipe throughout the plants and may be obtained in black or galvanized finish. Observe from Table 2–1 that the allowable stress for A106 decreases rapidly as temperatures exceed 750 F. To overcome this deficiency, steel containing 0.10 to 0.20% carbon and 0.45 to 0.65% molybdenum was used extensively in the high-temperature ranges during the 1930's. Although the strength of this steel is no higher than A106 at low temperatures, the carbon molybdenum steel has higher creep resistance at elevated temperatures and therefore has a higher allowable stress at elevated temperatures. However, it was found that both are subject to failure due to graphitization around welds at high operating temperatures. Therefore, A120 is not recommended for temperatures above 775 F, and carbon molybdenum, A335 Grade P1, is not recommended for temperatures above 875 F.

Slow decomposition of the iron carbide (Fe_3C) at the grain boundaries when carbon or carbon molybdenum steel is exposed to elevated temperatures produces a segregation of graphite particles, known as *graphitization*, that results in loss of tensile strength, ductility, and impact resistance. Graphitization occurs in the welded zones of the pipe and has resulted in pipe failures. There is some evidence to indicate that the carbide structure may be restored to normal by heating the pipe to 1700 F for a period of 2 hr per in. of thickness. This would be difficult to do in the field and, furthermore, could cause warpage of valves and other equipment.

TABLE 2-1

ALLOWABLE STRESSES FOR SOME SEAMLESS PIPE MATERIALS

Material	ASTM Specification	Grade	Allowable Values of Stress, psi, for Temperatures in Deg F: Not To Exceed								
			−20 to 650	700	800	850	900	950	1000	1100	1200
Carbon steel..........	A53, A106	A	12,000	11,650	9,000	7,100	5,000				
		B	15,000	14,350	10,800	7,800	5,000				
Carbon molybdenum......	A335	P1	13,750	13,750	13,450	13,150	12,500				
Chrome molybdenum											
½% Cr–½% Mo.....	A335	P2	13,750	13,750	13,450	13,150	12,500	10,000	6,250	2,800	
1% Cr–½% Mo.....	A335	P12	15,000	15,000	14,750	14,200	13,100	11,000	7,500	4,000	
1¼% Cr–½% Mo.....	A335	P11	15,000	15,000	15,000	14,400	13,100	11,000	7,800	4,000	
2¼% Cr–1% Mo......	A335	P22	15,000	15,000	15,000	14,400	13,100	11,000	7,800	4,200	
3% Cr–1% Mo.......	A335	P21	15,000	14,800	13,900	13,200	12,000	9,000	7,000	4,000	
5% Cr–½% Mo......	A335	P5	....	13,400	12,800	12,400	11,500	10,000	7,300	3,300	1,500

NOTE: Values of stress for intermediate temperatures may be obtained by interpolation.

Addition of chromium to the steel will eliminate graphitization by stabilizing the carbide. Specification A335 Grade P2 steel is a chrome molybdenum containing ½% chromium and ½% molybdenum. The added chromium does not appreciably improve the strength, corrosion resistance, or oxidation resistance, but it reduces or eliminates graphitization when the steel is used at temperatures up to 975 F. A steel containing 1% chrome is preferable for the 950 F to 975 F temperature range but may be used up to 1000 F, while 1¼% chrome is permissible up to 1050 F and is recommended for the 975 F to 1000 F range. About 2¼% chrome is recommended for temperatures up to 1050 F but may be used to 1100 F; the 3% and 5% chrome are for temperatures up to 1100 F. There is a significant improvement in corrosion resistance when the steel contains 2% or more chrome.

Two systems have been used for classifying wrought-steel pipe-wall thickness. The oldest of these lists pipe as *standard weight, extra strong,* and *double extra strong* in the order of increasing wall thickness. The newer system is more rational and uses the term *schedule number.* Plate 1 in the Appendix lists the nominal wall thicknesses and other physical data for pipe of various schedule numbers. Pipe mills require a tolerance (usually about −12½%) on the nominal wall thicknesses listed.

Note that the nominal diameter of steel pipe is approximately the same as the inside diameter (ID) of schedule 40 pipe for sizes up to and including 12-in. pipe. For sizes of 14-in. and larger, the nominal diameter is exactly the outside diameter (OD) of the pipe. This applies to all weights of pipe.

Small pipe is not available in all schedules, since it is not economical to roll small pipe with a large variety of wall thicknesses that would differ only slightly.

Certain sizes of schedule 20, 30, 40, 60, and 80 pipe correspond to the old standard weight and extra strong pipe. These are indicated by italics and asterisks, respectively, in Plate 1, Appendix.

All pipe of a given nominal diameter has the same OD regardless of schedule number or wall thickness. For example, the OD of 6-in. pipe is 6.625 in. regardless of the schedule, but the ID is 6.065 in. for schedule 40 and is 5.189 in. for schedule 160. This constant OD for any one nominal diameter is a distinct advantage during fabrication, since it permits the use of the same thread dies and taps, insulation, hangers, and holes for welding fittings regardless of wall thickness.

Pipe with a wall thickness less than schedule 40 should not be threaded.

Minimum wall thicknesses for pipe under the jurisdiction of the Power Piping Code should be determined by the following formulas:

$$t_m = \frac{p\,D}{2S + 2yp} + C \tag{2-5}$$

or
$$p = \frac{2S(t_m - C)}{D - 2y(t_m - C)} \qquad (2\text{-}6)$$

where t_m = minimum wall thickness, in.

p = maximum internal service pressure, psig

D = pipe OD, in.

S = allowable stress due to internal pressure at operating temperature, from Table 2-1, psi

C = a constant as follows:

Threaded steel, ½ in. and larger, depth of thread

Plain end steel, 1 in. and smaller, 0.05 in.

Plain end steel, larger than 1 in., 0.065 in.

Plain end nonferrous, 0.000 in.

and y = a coefficient (that may be interpolated) that is as follows for ferritic steels: 900 F and below, 0.4; 950 F, 0.5; 1000 F and above, 0.7.

Values of minimum wall thickness obtained from these equations should be increased to allow for the mill-rolling tolerance previously mentioned. This may be accomplished by dividing the calculated wall thickness by 0.875.

The formulas given above do not provide for fluid pulsations or expansion stresses.

Example 2-2. Calculate the wall thickness and select the proper schedule number for an 8-in. steam pipe that will operate at 400 psig and 500 F.

Solution. Use A106 Grade A, whose allowable stress is 12,000 psi. From Eq. (2-5),

$$t_m = \frac{400\,(8.625)}{2\,(12,000) + 2\,(0.4)\,(400)} + 0.065 = 0.207 \text{ in.}$$

$$\frac{t_m}{0.875} = 0.237 \text{ in.}$$

From Plate 1 of the Appendix, select schedule 20 pipe if the ends are not to be threaded. Since this is the minimum schedule for 8-in. pipe, there would be no point in using the more expensive Grade B material.

2-5. Joints and Fittings. (a) *Flanged Joints.* Concurrent with the increase in temperatures and pressures in use in power plants, the methods of attaching pipe to valves and equipment have improved. Flanges have long been a popular method of making a tight joint. Even with advanced welding procedures known today, flanges are universally used as a joint between pipe and equipment except for very high pressures or small low-pressure joints. Flanges have the advantage over welding of permitting the joint to be disassembled without torch-cutting of the line near valves and equipment.

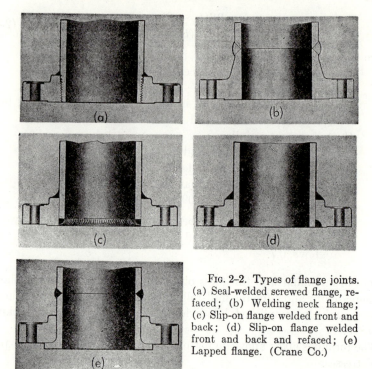

Fig. 2–2. Types of flange joints. (a) Seal-welded screwed flange, refaced; (b) Welding neck flange; (c) Slip-on flange welded front and back; (d) Slip-on flange welded front and back and refaced; (e) Lapped flange. (Crane Co.)

Steel flanges have nonshock, primary service ratings of 150, 300, 400, 600, 900, 1500, and 2500 psig. As in the case of pipe, flanges are satisfactory for increased pressures at low temperatures, but the allowable pressure decreases at high temperature. Cast-iron flanges are rated at 25, 125, and 250 psig.

Many types of flange are in use, and many other types that have been used are now of historical interest only. This discussion will be confined primarily to those types that are approved by the ASA. Three important items to consider for any pipe flange are: (1) the connection to the pipe, (2) the shape of the face, and (3) the finish on the face.

Threaded flanges, Fig. 2–2(a), have several inherent weaknesses. Threads are not only points of stress concentration but also weaken the pipe and fitting walls. Tests have shown that a threaded joint has less than 60% of the strength of the pipe. Because of the difficulty of obtaining purchase with wrenches or tongs on large pipe, it is recommended that field-fabricated, threaded joints be limited to 6-in. pipe and low pressures.

When threaded pipe and flange connections are fabricated in the shop, the flange is screwed so far onto the pipe that some of the pipe protrudes through the flange. After assembly, the face of the pipe and flange is machined, making the pipe flush with the flange face and eliminating any warpage that may have occurred in the flange face during the assembly

This also eliminates any pocket that might exist between the end of the pipe and the face of the flange. Such pockets collect moisture and are a source of corrosion. They are common in field-made joints when the flange is not screwed far enough onto the pipe. Shop-fabricated, threaded joints have been satisfactory for pressures as high as 400 to 600 psig and for large pipe sizes.

Screwed-flange joints are susceptible to leakage and corrosion. For this reason they are frequently seal-welded at the back and in the front pocket between the end of the pipe and the flange face. However, the welding is likely to cause warpage of the flange and consequent leakage between flange faces unless the flange is refaced after welding. Screwed flanges are unsuited for locations where there are high bending stresses because of possible leakage.

Welding-neck flanges, Fig. 2–2(b), have a smooth transition from the pipe wall to the flange. This is beneficial under conditions of repeated bending stress. The endurance strength of this superior joint is said to be equal to that of a butt-welded pipe, which is the same as that of unwelded pipe. The welding area is remote from the flange proper due to the tapered neck. This prevents undue distortion of the flange during welding. As in the case of the threaded flange, extreme care must be exercised to prevent misalignment of the bolt holes.

Slip-on flanges, Fig. 2–2(c) and (d), may be rotated on the pipe before welding to permit easy alignment of the bolt holes. Because they have all the disadvantages of structural weakness and welding distortion of the threaded flange, slip-on flanges are limited by the code to pressures of 300 psig. In addition, their fatigue life is much less than that of a welding-neck flange (about one-third).

The stub end of Van Stone or lapped flanges, Fig. 2–2(e), is formed by an upset process, so that after machining the lip thickness is not less than the pipe-wall thickness. However, there is a concentration of stress in the hub of lapped flanges resulting in a lower fatigue life than for the slip-on flange (about one-tenth that of a welding-neck flange). The advantage of the lapped flange lies primarily in the ease of aligning bolt holes and is therefore advantageous when the joint must be dismantled frequently.

Cast-iron flanges may be an integral part of the equipment or may be screwed on. When the casings of equipment are of cast steel, the flange may be cast integral with the casing or may be welded to the casing. For power-plant use, nearly all other flanges are of forged steel or forged alloy steel.

ASA-approved types of flange facing are shown in Fig. 2–3. These flanges are shown screwed to the pipe for pictorial purposes only and may be obtained as welding-neck, slip-on, or lapped flanges, or as castings when integral with equipment.

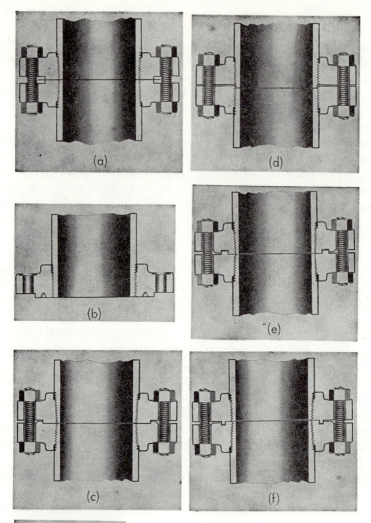

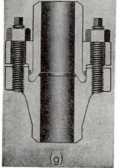

FIG. 2–3. Flange facings. (a) Raised face or double male; (b) Ring joint; (c) Large male and female; (d) Small male and female; (e) Large tongue and groove; (f) Small tongue and groove; (g) Pressure seal. (Crane Co.)

A recent survey of piping fabricators indicated that about 85% of the power plants currently being installed employ raised faces on the flanges. This form of facing, Fig. 2–3(a), shows the least tendency of leakage due to flexing or vibration of the pipe. Once assembled, raised-face flanges may be disassembled easily without springing the pipe and orifices, and nozzles may be installed between the faces. The faces of these flanges are raised $\frac{1}{16}$ in. for the 150- and 300-psig ratings and $\frac{1}{4}$ in. for higher, primary service ratings.

Cast-iron flanges are flat-faced for the 25-psi and the 125-psi series but have $\frac{1}{16}$-in. raised faces on the 250-psi series. Steel flanges that are to mate with cast-iron flanges always should have flat faces to prevent breakage of the cast iron when the bolts are tightened.

The ring-joint flange shown in Fig. 2–3(b) is next to the raised-face flange in popularity. Mechanically, it is an excellent joint, but practically, it has proved to be deficient. The two halves of the flange are identical, and a metal ring gasket of either soft steel or 5% chrome and $\frac{1}{2}$% molybdenum steel is used in the groove. The cross-section of the ring gasket may be either oval or octagonal. Very high gasket pressures can be obtained with moderate bolt pressures, but the joint has not proved to be satisfactory for steam conditions above 600 psig and 850 F. At higher pressures and temperatures the ring gasket deforms due to creep of the metal and also due to localized yielding. This is particularly important at locations in the pipeline where there is flexure and bending of the pipe. After the flange has been brought up to temperature and then cooled, it will leak and will continue to leak even after the joint is returned to operating temperature.

The large male and female flanges, Fig. 2–3(c), and the small male and female flanges, Fig. 2–3(d), are attempts to get high gasket pressures and confine the outside of the gasket. These flanges will not withstand flexing of the pipe. The joint must be pried open when being dismantled. These flanges are adaptations of the raised-face flange. Either the male and female or the tongue-and-groove flanges, Fig. 2–3(e) and (f), must be carefully handled and assembled to ensure a good joint. Tongue-and-groove flanges produce a high gasket pressure; however, the tongue is structurally weak and may be stressed beyond the yield point, thus causing failure. The joint must be pried open and will not remain tight when the line vibrates or flexes.

Pressure-seal joints, Fig. 2–3(g), are a recent development for high pressures and temperatures that are not yet recognized by the code, although they have been used successfully for the highest steam temperatures and have the advantage of greatly reduced weight and size. Observe that the fluid pressure inside the joint increases the gasket pressure and tightens the joint. Pressure-seal joints are available only with butt-welded ends.

The ASA code recognizes smooth tool finish on the flange face or on the phonographic (spiral) and the serrated (concentric) grooved faces. The grooves are usually $\frac{1}{64}$-in. deep and spaced 32 grooves to the inch.

Blind flanges have the bolt holes drilled and are spot faced and the flanges are faced, but there are no center holes for the passage of fluid. Blank flanges have neither the bolt holes nor the center hole for passage of fluid.

Spirally wound, stainless steel and asbestos gaskets are the most popular and the most satisfactory for high-pressure and high-temperature water and steam. Solid metal gaskets of ingot iron or other metals may be used, but they are not satisfactory if the joint is subjected to bending stresses. A metal-jacketed or metal-reinforced asbestos gasket gives good service. Asbestos gaskets, for steam pressures up to about 300 psig, are common. Other low-pressure gaskets are made of organic materials, vegetable fiber, rubber, neoprene, cork, and plastics.

Costs of the several types of flange facing and methods of attachment to the pipe vary with the schedule and nominal size of the pipe and with the primary service rating of the flange. Raised-face flanges are used as the base for pricing purposes and all other facings are more expensive. Tongue-and-groove or female facing costs about $1.50 extra per flange (not per pair) for 6-in. size, while the same size of ring joint facing costs about $3.00 extra, not including the ring gasket. Phonograph or concentric serrated finishes cost about 60¢ more than a smooth tool finish for a 6-in. flange.

Approximate comparative costs of 300-psi, forged, carbon steel raised-face flanges, shop-fabricated on schedule 40 carbon steel pipe (including cost of one flange, material, and labor, but excluding cost of pipe, gasket, bolts or nuts) are as follows:

	6-in Nominal Size	12-in. Nominal Size
Welding-neck flange	$45	$138
Screwed flange, seal welded front and back, and refaced	49	108
Screwed flange, refaced	35	74
Slip-on flange, welded back and front and refaced	49	105
Lapped flange	58	133

Although all these prices will vary with price trends and quantities ordered, the relative costs will remain nearly constant.

(b) *Welding.* Welded joints between sections of pipe, and between sections of pipe and fittings, valves, or equipment, reduce the weight of the pipe system as well as provide a joint that is less hazardous. Welded joints are economical and reduce insulation costs. It is frequently possible to fabricate complicated-weld header systems with close clearances that would be impossible with any other type of joint. Thus, with the exception of attachment of pipelines to equipment, welded piping systems are universal in power plants.

Piping of 2-in. size and smaller is joined by socket welding. The socket is a recessed portion in a valve or fitting into which the pipe will fit. The weld is made around the pipe at the outer edge of the socket. Butt welding of lines 2½ in. and larger is done by either arc or acetylene welding. Both techniques are used in field work, but arc welding is used in shop fabrication. Ends of adjoining portions of the pipe and fittings are beveled, as shown in Fig. 2–4. V-type bevels are recommended for wall thicknesses of ³⁄₁₆ in. and larger for acetylene welding and for arc welding when the wall thickness is from ³⁄₁₆ to ¾ in. A U type of bevel is recommended for arc welding of wall thicknesses larger than ¾ in.

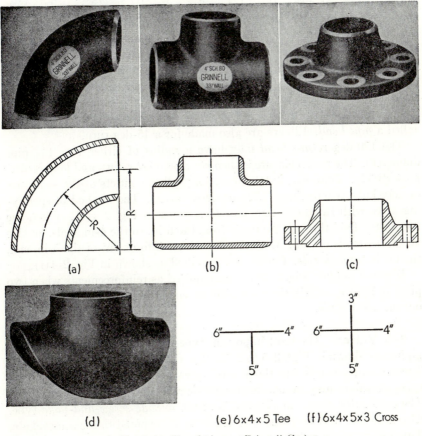

(a)   (b)   (c)

(d)   (e) 6x4x5 Tee   (f) 6x4x5x3 Cross

Fig. 2–4. Pipe fittings. (Grinnell Co.)

A welding ring or backing ring is advisable for butt welding. These rings fit inside the pipe at the weld and not only reduce beads or icicles inside the weld but also are of assistance in aligning the pipe and producing a sound weld.

Alloy steel pipe of either chrome molybdenum or carbon molybdenum should be stress-relieved after welding. Carbon steel less than ¾ in. thick need not be stress-relieved. Gamma-ray inspection of welds in important high-pressure and high-temperature lines is justifiable. Ultrasonic testing of pipes also has been used.

Exact procedures and tests for power-plant system welding are given in both the *Boiler Construction Code* and the *Pressure Piping Code*.

(c) *Fittings*. The fittings illustrated in Fig. 2–4 are shown with the ends beveled for butt welding. Cast-steel fittings have threaded ends or flanged ends using any of the types of flange facing previously described. Forged fittings usually have threaded or socket-welding ends, while cast-iron fittings have either threaded or flanged ends. Seamless-welding fittings are light in weight, with flexible elbows which are of great assistance in absorbing bending stresses due to thermal expansion. The radii, R, of the *elbow* of Fig. 2–4(a) may be equal to the nominal diameter of the pipe or may be equal to 1½ times the nominal diameter. If the radius is 1 diam, the elbow is a *short-radius, standard*, or *close-turn* elbow. With the 1½ diam radius, the elbow is known as a *long-radius* or *long-turn* elbow. At times, the radius may be 5 pipe diam and the fitting is then called a *pipe bend*. Elbows are also made for a 45-deg turn.

The 180-deg *return bend* may have a radius of either 1 or 1½ pipe diameters. The nomenclature is then *close* or *short* return and *long* return, respectively. *Tees*, Fig. 2–4(b), may have all the outlets of the same size or may have two or three different sized outlets. The method of specifying a tee with three different outlets is given in Fig. 2–4(e). When the flow of the fluid through the tee is from the 6-in. to the 4-in. ends, or vice versa, the tee is *on run*. With the flow either into or out of the 5-in. end, the tee is *through side*. *Crosses* are specified as shown in Fig. 2–4(f).

Welding saddles, Fig. 2–4(d), are required as reinforcements when two pipes of approximately the same size are connected by welding and without the use of a tee.

2–6. Valves. The most important types of valve used in power-plant piping are shown in Figs. 2–5 to 2–12. The details of design vary with different manufacturers, but the general types are common to all. All these types of valves may be obtained with screwed, flanged, or welding ends. The *gate valve* shown in Fig. 2–5 is of the *outside-screw-and-yoke* type. This indicates that the threads on the stem are exposed to the atmosphere. This construction reduces the tendency for the threads to corrode and bind from contact with the fluid so that the valve will work more easily on intermittent service. Another advantage is that the design is such that the stem will rise as the valve is opened, and operators can tell at a glance

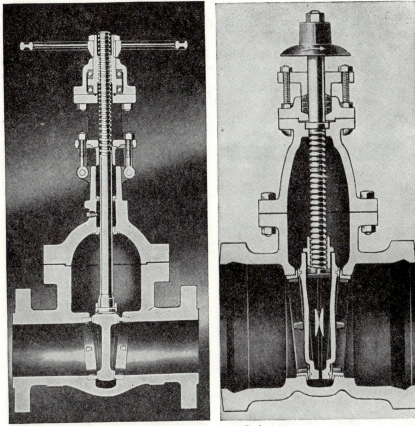

(Reading, Pratt & Cady.)

Fig. 2–5. Gate valve with outside-screw-and-yoke, rising stem, and solid wedge.

Fig. 2–6. Gate valve with inside screw, nonrising stem, and split wedge.

whether the valve is open or closed. The wedge is solid and separate seat and disc rings are screwed on. When the valve is fully open, the shoulder at the bottom of the stem will fit into the bushing at the top of the bonnet, thereby reducing steam leakage into the condensing chamber. A bleed connection (shown plugged) is located in the chamber so that leakage may be carried away if necessary. Packing and packing glands are located above the condensing chamber to guide the stem and to prevent excess leakage along the stem.

Fig. 2–6 shows a gate valve with *split-wedge* design. This type is used primarily on low-pressure installations. The stem is of the nonrising type.

Gate valves should be used for shut-off purposes only, as the wedge fits into a groove around the body of the valve to form a tight seat. The sharp edges of the wedge will soon score and wear if the valve is used for throttling or control purposes, and it will not seat tightly. In large sizes, or for high pressures, or when remote control is required for important lines, gate valves may be equipped with motor operators.

The type of valve normally used for control purposes is a *globe valve*, Fig. 2–7, or some adaptation of it. The wedge of the gate valve has been replaced by a disc with the disc ring welded on. It can be seen that the pressure drop through this valve will be larger than that for a gate valve. This disc can be replaced if necessary, as well as either the disc rings or

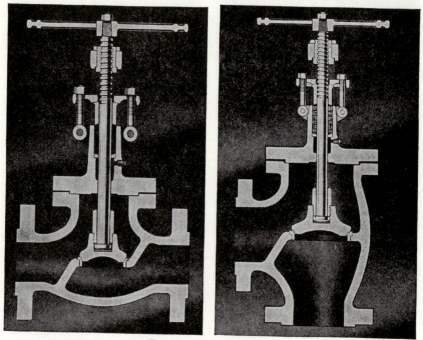

(Reading, Pratt & Cady.)

Fig. 2–7. Globe valve.   Fig. 2–8. Angle valve.

seat rings. A variation of this valve, called the *angle valve,* is shown in Fig. 2–8. Sometimes this type of valve fits into the piping system more readily than a globe valve.

In order to prevent back flow of any fluid in a pipe, *check valves* are installed, Fig. 2–9. Swing-check and horizontal and vertical piston-check valves all operate on the principle that the force of the fluid flowing in the proper direction will keep the valves open and a reversal of flow will close them. These valves are not tight-seating and can stop only a major portion of the flow in the reverse direction. The *swing-check* valve, Fig. 2–9, is not adaptable to installation in vertical pipe. The *piston type* is used horizontally and vertically. It can be seen that the pressure drop

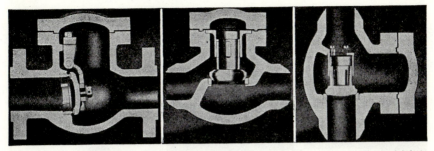

FIG. 2-9. Swing-check valve (left), horizontal (center), and vertical (right) piston-check valves. (Reading, Pratt & Cady.)

through any type of check valve is high and will change radically with design.

A section of a new form of valve bonnet design, known as a *pressure-seal* bonnet, is shown in Fig. 2-10. The principle of this design is the same as that of the pressure-seal joint previously mentioned. This bonnet may be applied to a gate, globe, angle or stop-check valve. Developed for use at pressures of 600 psig and higher, the fluid pressure acting against the seal ring compresses the packing to give a tight joint. By eliminating the bonnet flange, the weight and stem length of the valve are reduced. Large high-pressure valves of this design are up to 40% lighter than conventional valves of the same size.

Another modern development is the venturi-shaped gate valve. The inlet and outlet sections of these valves are shaped like the expanding

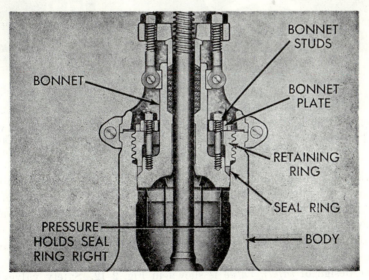

FIG. 2-10. Pressure-seal valve bonnet. (Crane Co.)

portion of a venturi. The wedge section of the valve is one size smaller than the ends; in an 8-in. line, the wedge would correspond to a 6-in. size. This design reduces the weight of the valve, but the pressure drop across the valve is about the same as the pressure drop through a valve of standard design.

The *stop-check* valve shown in Fig. 2–11 is a variation of the check valve. This type of valve is required by law on the steam outlets of

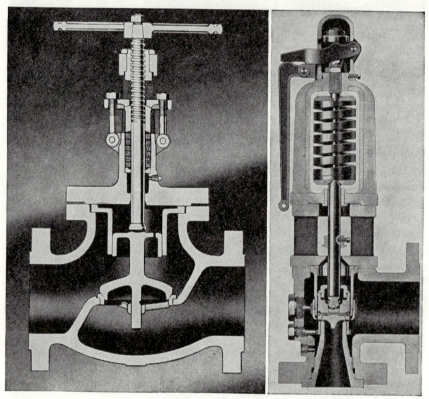

FIG. 2–11. Stop-check valve. (Reading, Pratt & Cady.)

FIG. 2–12. Safety valve with exposed spring. (Crosby Steam Gage & Valve Co.)

boilers where two or more are connected together. Its purpose is to prevent backflow into a boiler that is not in operation. When the stem of the valve is at the top of its travel, the valve acts like any other check valve. However, the valve may be forced shut, regardless of the direction of flow, by moving the stem down. Very large stop-check valves for high pressure are operated by a toggle mechanism.

One design of *safety valve* is shown in Fig. 2–12. Safety valves are required on all boilers or pressure vessels in which the pressure may rise above the design pressure. Sufficient pressure acting upward on the disc will overcome the force of the spring and open the valve. When the valve is open, there is a larger disc area for the steam to act upon, and the valve will not close again until the pressure has dropped below the opening pressure. However, the valve may be opened and closed manually by the lever. The difference between the opening and closing pressures is called *blowdown* and is expressed in percentage of the opening pressure. Safety valves for boiler service have 3% blowdown. The spring is open to atmosphere so that it will not become too hot, for a rise in temperature would change its characteristics.

Any of these types of valve may be obtained in brass, cast iron, or various kinds of steel. The pressure and temperature ratings of the metals, as established by the codes, determine the maximum service conditions. Cast-iron valves should be used with caution in steam lines for working pressures above 30 psig because cast iron cannot withstand stresses due to thermal expansion of the line. For severe wearing and cutting service, the disc and seat rings should be made of a chrome steel or should be stellited.

2–7. Viscosity. Viscosity, which is a property of all fluids, is caused by the cohesion of the molecules and the interchange of momentum between adjacent layers due to molecular agitation normal to the direction of motion of the fluid. For example, a gas will change its shape almost instantaneously if its constraining vessel changes shape. A liquid will require an appreciable amount of time under the same conditions. A heavy oil will require more time to change its shape than water. The

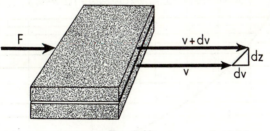

Fig. 2–13.

numerical values of absolute viscosity for steam, water, and oil increase in that order.

Consider that Fig. 2–13 represents sections of two laminae of a fluid. A force, F, must be applied to the upper lamina if it is to move at a velocity of $v + dv$ at the same time that the lower lamina is moving at a velocity of v. The force F then causes a shearing action between the two laminae over the area of contact, A. The shear stress, F/A, has been found to be proportional to the velocity gradient of the fluid, dv/dz, or

$$\frac{F}{A} \propto \frac{dv}{dz}$$

Absolute viscosity of the fluid, μ, is the constant of proportionality needed to make this an equality. Then

$$\frac{F}{A} = \mu \frac{dv}{dz}$$

$$\mu = \frac{(F)(dz)}{(A)(dv)} \tag{2–7}$$

Absolute viscosity is defined from Eq. (2–7) as that unit force required to move one layer of a fluid at unit relative velocity to another layer of the fluid which is at unit distance from the first.

The pressure existing on the fluid did not enter into the derivation of Eq. (2–7). This would indicate that absolute viscosity is independent of pressure. The viscosity of many fluids may be considered as a function of temperature only—for example, that of water, oil, and air. However, the viscosity of steam and most vapors is seriously affected by both pressure and temperature.

Absolute viscosity of saturated water may be determined from Fig. 2–14. It will be seen that the absolute viscosity decreases as the temperature increases. This has been attributed to the large cohesive forces exist-

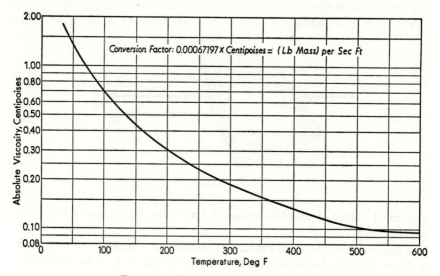

FIG. 2–14. Viscosity of saturated water.

ing between the closely spaced molecules. These forces decrease rapidly as temperature increases. Momentum exchanges caused by agitation of the molecules of the liquid are small in comparison with the cohesive forces.

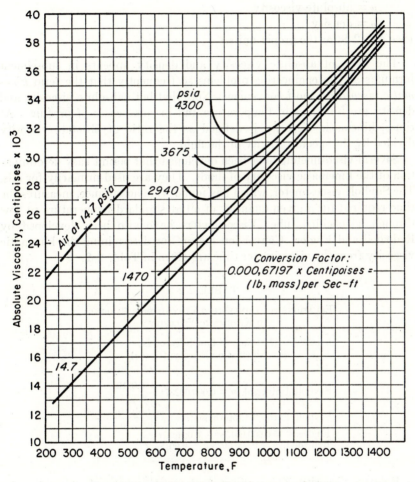

FIG. 2–15. Absolute viscosity of steam and air.

In a gas, molecular cohesion is negligible compared with the exchanges of momentum, since the molecules are already widely separated. Since molecular activity also increases with temperature, the absolute viscosity of gases reacts in just the opposite manner from liquids and increases as the temperature increases. This reasoning is confirmed by the experimental results in Fig. 2–15 of absolute viscosity for steam and air.

Kinematic viscosity is defined as the ratio of absolute viscosity divided by density:

$$y = \frac{\mu}{\rho} \tag{2-8}$$

where y = kinematic viscosity
μ = absolute viscosity
ρ = density
(See Art. 2-8 for the proper units.)

The ratio of absolute viscosity and density appears in subsequent formulas and in engineering problems so frequently that the term *kinematic viscosity* is most convenient. Note that both absolute viscosity and density are functions of temperature and in some cases of pressure. At times, kinematic viscosity is defined as the ratio of absolute viscosity divided by the specific gravity of the fluid.

Several viscosimeters have been developed that permit easy determination of the kinematic viscosity of fluids. They are used primarily for oils. The Saybolt viscosimeter, which has found wide acceptance, measures the time required for a given quantity of oil at standard temperature to flow through a specified tube. Viscosimeters have also been developed by Redwood and Engler. Conversion factors for determining the kinematic viscosity from the viscosimeter time are given in Table 2-2.

TABLE 2-2

EQUATIONS FOR CONVERTING VISCOSIMETER READINGS
TO KINEMATIC VISCOSITY

(t = time of flow in seconds;
y = kinematic viscosity, ft²/sec)

Saybolt Universal, $y = 0.000,00237t - \dfrac{0.00194}{t}$

Saybolt Furol = (approximately) $\dfrac{\text{Saybolt Universal}}{10}$

Engler, $y = 0.000,00158t - 0.00403/t$
Redwood, $y = 0.000,00280t - 0.00185/t$

Relative viscosity, Z, is a dimensionless ratio of the absolute viscosity of any fluid to the absolute viscosity of water.

2-8. **Units of Viscosity.** By referring to Eq. (2-7), it can be seen that viscosity involves the units of force, F, length, L, and time, T. A dimensional expression for viscosity would be

$$\mu = \frac{(F)(dz)}{(A)(dv)} = \frac{(F)(L)}{(L^2)(LT^{-1})} = \frac{FT}{L^2} \tag{a}$$

From Newton's equation, force is proportional to mass times acceleration. For dimensional equality, $F = MLT^{-2}$, and viscosity is

$$\mu = \frac{MLT}{L^2 T^2} = \frac{M}{LT} \tag{b}$$

Any consistent set of British or metric units may be used for these dimensional symbols. If the "absolute" metric system of units is used, the unit
of viscosity is named the *poise*, which is 100 centipoises. From Eq. (a),
the poise is

$$\mu = \text{poise} = \frac{FT}{L^2} = \frac{\text{dyne-sec}}{\text{cm}^2}$$

For the "absolute" British system using the poundal and pound-mass,

$$\mu = \frac{FT}{L^2} = \frac{(\text{poundal}) (\text{sec})}{\text{ft}^2} = 0.0672 \times \text{poises} \tag{2-9a}$$

$$= \frac{M}{LT} = \frac{\text{lb-mass}}{(\text{ft}) (\text{sec})} = 0.0672 \times \text{poises} \tag{2-9b}$$

The "gravitational" British system, or the "engineer's" system, employs
pounds-force and slugs;

$$\text{Slug} = \frac{W}{g} = \frac{(\text{lb-force}) (\text{sec}^2)}{\text{ft}}$$

Thus, $$\mu = \frac{FT}{L^2} = \frac{(\text{lb-force}) (\text{sec})}{\text{ft}^2} = 0.00209 \times \text{poises} \tag{2-9c}$$

$$= \frac{M}{LT} = \frac{\text{slugs}}{(\text{ft}) (\text{sec})} = 0.00209 \times \text{poises} \tag{2-9d}$$

Note that 0.00209 equals 0.0672 ÷ 32.2. For kinematic viscosity, take

$$\rho = ML^{-3}$$

for either system, and

$$y = \frac{\mu}{\rho} = \frac{ML^{-1}T^{-1}}{ML^{-3}} = \frac{L^2}{T} \tag{2-9e}$$

In the case of the gravitational system, mass will be in slugs.

EXAMPLE 2-3. Find the numerical values of absolute viscosity and kinematic viscosity in terms of all the units discussed above for steam at 850 psig
and 900 F.

SOLUTION. The viscosity of steam at 850 psig and 900 F is, from Fig. 2–15,
0.0267 centipoise (there are 100 centipoises in a poise).

Then absolute viscosity is

$\mu = 0.0267$ centipoise $= 0.000,267$ poise

$= 0.000,267$ dyne-sec per cm^2

$= 0.0267 \times 0.000,672 = 0.000,01795 \dfrac{\text{poundals-sec}}{\text{ft}^2}$

$= 0.000,01795$ (lb-mass) per ft-sec

$= 0.0267 \times 0.000,0209 = 0.000,000,558 \dfrac{\text{slugs}}{\text{ft-sec}}$

$= 0.000,000,558$ (lb-force) sec per ft^2

The specific volume of steam for these conditions may be found in the steam tables to be $v = 0.8873$ cu ft per lb. Then kinematic viscosity is

$$y = \frac{\mu}{\rho} = v\mu = 0.000,01795 \times 0.8873 = 0.000,01592 \text{ ft}^2 \text{ per sec}$$

2–9. Laminar and Turbulent Flow. Professor Osborne Reynolds [*] was the first to show that there are two very different conditions under which a fluid can flow through a conduit, namely, *laminar flow* and *turbulent flow*. Reynolds' apparatus is depicted schematically in Fig. 2–16,

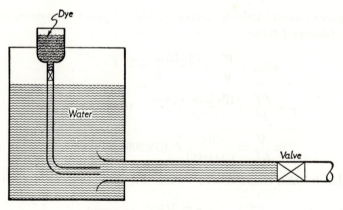

FIG. 2–16.   Reynolds' apparatus.

which shows a reservoir containing water and a bell-mouthed pipe of glass with a valve at the outlet end. A colored dye stored in a smaller reservoir may be injected into the bell mouth of the glass pipe. The velocity of the water flowing in the glass pipe is controlled by the outlet valve. Reynolds noted that, when the water velocity in the glass pipe was low, the thin stream of dye injected into the bell mouth did not diffuse with the

[*] O. Reynolds, "An Experimental Investigation of the Circumstances Which Determine Whether the Motion of Water Shall Be Direct or Sinuous and of the Law of Resistance in Parallel Channels," *Philosophical Trans. of the Royal Society,* Vol. 174 (1883), Part 3, p. 935.

water but flowed down the center of the glass pipe as a separate filament, completely intact. When the water velocity was increased, the filament of dye wavered and then broke, diffusing with the water. He also noted that once the filament of dye had diffused, it could be restored only by reducing the water velocity.

Reynolds' explanation of the phenomenon exhibited during his experiments was that at low velocities there was no intermingling of the molecules and the fluid particles moved in parallel layers or laminae; thus the name "laminar flow." The dye filament diffused with the water at higher velocities because of the intermingling of the particles or because of turbulence.

Critical velocity is that velocity at which there is a transition from laminar to turbulent flow. Critical velocity varies with the fluid and other factors, but even for a specific fluid, critical velocity is a range rather than a definite value.

Reynolds further deduced that a dimensionless term, now called Reynolds' number, N_R, could be used to define the critical range.

$$N_R = \frac{VD\rho}{\mu} = \frac{(LT^{-1})\,(L)\,(ML^{-3})}{ML^{-1}T^{-1}} \qquad (2\text{-}10)$$

where V = average fluid velocity, fps
  D = internal pipe diameter, ft
  ρ = fluid density or reciprocal of specific volume $(1/v)$, lb-mass per cu ft
  μ = fluid absolute viscosity, lb-mass per ft-sec

Note that the units for the components of Reynolds' number have been based on the absolute system of units. If the gravitational system were used, then density must be expressed in slugs per cubic foot, and viscosity would be in units of (lb-force) (sec) per ft². The various forces that may act on a fluid as it flows through a confining channel are the forces of pressure, gravity, inertia, viscosity, elasticity, and surface tension. Reynolds' number is the ratio of the force of inertia divided by the force of viscosity.

EXAMPLE 2–4. Calculate Reynolds' number for 850-psig, 900 F steam flowing at a velocity of 12,000 fpm in a 6-in. schedule 80 pipe.

SOLUTION. From Example 2–3, $\mu = 0.000{,}01795$ (lb-mass) per ft-sec, and the specific volume $= 0.8873$ cu ft per lb-mass.
From Eq. (2–10),

$$N_R = \frac{12{,}000 \times 5.761}{60 \times 12 \times 0.000{,}01795 \times 0.8873}$$
$$= 6{,}030{,}000$$

Most experiments indicate that the lower value of Reynolds' number for the critical range is 2100. The upper limit of the critical range is governed by several factors that are not a part of the Reynolds' number expression. Reynolds found that the upper value for the critical range is from 12,000 to 14,000. Ekman, in 1910, used Reynolds' original apparatus for further experiments. By keeping the water in the reservoir quiet for several days before the tests, he was able to obtain laminar flow for a value of $R = 50,000$. In addition to the initial quietness of the fluid, the shape of the pipe entrance and the roughness of the pipe may also affect the upper critical limit.

For practical applications, engineers consider that the critical range for R varies from 2100 to 4000. Between these two limits there is a region of uncertainty. Conservative estimates should be used when designing pipe systems that will operate within the critical range. However, nearly all power-plant design lies well above the critical range.

Many hybrid systems of units for Reynolds' number are in use and some are not dimensionless. The only justification for these hybrid systems is that they use more conventional dimensions. Some of the more common variations use: viscosity in centipoises with other items in the foot-pound-second system; pipe diameter in inches rather than feet; velocity in feet per minute or per hour rather than in feet per second. Obviously, the critical range will not be from 2100 to 4000 when these hybrid systems are used unless the proper conversion factors are also included in the calculation.

Care should be taken when using a strange set of curves involving Reynolds' number to determine the exact units involved.

2–10. Turbulent Flow. The economical design of pipes in the steam-power plant requires that the velocities be well above the critical velocity of the fluid. Therefore, most fluids will pass through the pipes in a highly turbulent condition.

The general formula for flow under turbulent conditions must be derived with the assumption that the fluid is compressible. Flow of an incompressible fluid will then become a special case of the general conditions with a simplified formula. As a compressible fluid flows through a confining conduit, the decrease in pressure causes a decrease in the density with a corresponding acceleration of the fluid. The work required to move a section of the fluid along a horizontal pipe will be expended to perform two functions: (1) that of overcoming frictional resistance, and (2) that of accelerating the fluid.

Consider an elementary section of the fluid in a horizontal, circular pipe to have a length dL, diameter D, density ρ, and a pressure drop along the length dL of $-dP$, as shown in Fig. 2–17. Then the work of moving

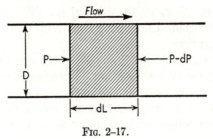

Fig. 2–17.

the section will be

$$\text{Total work done on fluid} = -(dP)\,\frac{\pi}{4}\,D^2\,dL \qquad (a)$$

The work of overcoming friction will, from experiment, vary directly as the density ρ, velocity head $(V^2/2g)$, surface of the pipe wall $\pi D\,dl$, and a friction factor c, or

$$\text{Work of overcoming friction} = \left(c\rho\,\frac{V^2}{2g}\,\pi D\,dL \right) dL \qquad (b)$$

The work of accelerating the fluid, if it enters the section with velocity V and leaves with velocity $V + dV$, will be the weight of the fluid times the increase in the kinetic energy per unit weight, or

$$\text{Work of accelerating the fluid} = \rho\,\frac{\pi}{4}\,D^2\,dL\left[\frac{(V+dV)^2 - V^2}{2g} \right]$$

$$= \frac{\rho\pi D^2\,dL\,(V\,dV)}{4g} \qquad (c)$$

since the term $(dV)^2\,dL$ is negligible. Equating Eq. (a) to Eq. (b) plus Eq. (c),

$$-dP\,\frac{\pi}{4}\,D^2\,dL = \frac{c\rho V^2 \pi D\,dL\,(dL)}{2g} + \frac{\rho\pi D^2\,dL\,(V\,dV)}{4g} \qquad (d)$$

Dividing by $(\pi/4)D^2\,dL$ and assigning the proper limits,

$$\int_{P_1}^{P_2} -dP = \int_0^L \frac{4c\rho V^2\,dL}{2gD} + \int_{V_1}^{V_2} \frac{\rho V\,dV}{g} \qquad (2\text{–}11)$$

Eq. (2–11) cannot be completely integrated, since it is not known how the density varies with velocity except by assuming that the fluid is a perfect gas. This assumption will be made in Art. 2–14 when dealing with the flow of air and other similar gases through pipes. However, consider the case of saturated water flowing through a pipe; for example, the drains from a closed type of feedwater heater. As the pressure decreases, some of the water flashes into steam so that the fluid consists of water and steam. Since the steam portion of the fluid has a much greater specific volume than the water, the integral $(\rho V \, dV)/g$ becomes an important part of the equation. Under such conditions it is necessary to resort to arithmetic integration.

For most practical problems it is sufficiently accurate to assume that the density and velocity are constant. Determination of the length of pipe, L, to be used in calculating the pressure loss will be inaccurate to such an extent that this approximation will not be serious. Of course, if the fluid is incompressible, no inaccuracy is incurred. Eq. (2–11) may now be reduced to

$$\int_{P_1}^{P_2} -dP = \int_0^L \frac{4c\rho V^2 \, dL}{2gD} \tag{2–12}$$

Integration produces

$$P_1 - P_2 = \frac{f\rho V^2 L}{2gD} \tag{2–13}$$

where P_1 and P_2 = initial and final pressure, lb per sq ft
ρ = density, lb per cu ft
V = velocity, fps
L = length, ft
D = pipe ID, ft
$f = 4c$ = friction factor, see Fig. 2–19 (a) and (b).

Eq. (2–13) will give sufficiently accurate results by using the initial velocity and density when the pressure differential does not exceed 10% of the initial absolute pressure. When the average velocity and density are used, the formula is satisfactory for pressure differentials up to 30% of the initial absolute pressure.

If the pipe cross-section is not circular, Eq. (2–13) must include the hydraulic radius, m, in feet, in place of the diameter. Hydraulic radius is defined as the cross-sectional area of the fluid stream, divided by the wetted perimeter of the conduit. Thus

$$P_1 - P_2 = \frac{f\rho V^2 L}{4m2g} \tag{2–14}$$

More convenient forms of Eq. (2–13) may be obtained by inserting the proper equivalents:

$$\Delta p = \frac{0.001295 f L \rho V^2}{d} \tag{2–15}$$

$$\Delta p = \frac{0.01214 f L W^2}{\rho d^5} \tag{2–16}$$

where Δp = pressure loss, psi

   d = pipe ID, in.

   W = flow, lb per min

and other symbols are as in Eq. (2–13).

EXAMPLE 2–5. Determine the pressure drop per 100 ft of 6-in. schedule 80 pipe when 850-psig, 900 F steam flows through it at 12,000 fpm. Assume a friction factor of 0.015.

SOLUTION. $\rho = \dfrac{1}{v} = 1.127$

From Eq. (2–15) and Plate 1,

$$\Delta p = \frac{0.001295 \times 0.015 \times 100 \times 1.127 \left(\dfrac{12{,}000}{60}\right)^2}{5.761} = 15.2 \text{ psi}$$

Eqs. (2–15) and (2–16) involve a large number of tedious trial-and-error calculations when they are applied to loop systems carrying steam or other gases. Loop systems are composed of complicated arrangements of series and parallel pipe connections and often many different pipe sizes. Since it is the average density term of Eqs. (2–15) and (2–16) that is the cause of the difficulty, a formula for pressure drop that would not involve density would be advantageous. Density may be eliminated by assuming that the fluid behaves like a perfect gas. Then the throttling process becomes one of constant temperature, since a constant enthalpy process is also a constant temperature process for a perfect gas. This assumption is very nearly true for steam, particularly in the superheat region, and it will cause less error than other inherent inaccuracies in the calculations.

From the characteristic gas equation, $v = RT/P$. Then

$$\rho = \frac{1}{v} = \frac{P}{RT}$$

$$V = \frac{Q}{A} = \frac{wv}{A}$$

where V = velocity, fps

   Q = quantity flowing, cu ft per sec

   A = conduit area, ft^2

   w = flow, lb per sec

   v = specific volume, cu ft per lb

   P = pressure, lb per sq ft

   R = gas constant, ft-lb per lb R

   T = absolute temperature, R

   ρ = density, lb per cu ft

Thus,
$$\rho V^2 = \frac{w^2 v}{A^2} = \frac{w^2 RT}{A^2 P} \qquad \text{(a)}$$

Substituting Eq. (a) into Eq. (2–12),

$$\int_{P_1}^{P_2} -P\,dP = \int_0^L \frac{fw^2RT\,dL}{2gA^2D} \qquad \text{(b)}$$

$$\frac{P_1{}^2 - P_2{}^2}{2} = \frac{fw^2RTL}{2gA^2D} \qquad \text{(c)}$$

$$P_2 = \left[P_1{}^2 - \frac{fw^2 2P_1 v_1 L}{2gA^2D} \right]^{1/2} \qquad \text{(d)}$$

It should be remembered that $P_1 v_1$ is a constant because the process has been assumed isothermal. Rearranging into more convenient terms,

$$p_2 = [p_1{}^2 - K\overline{W}^2 L]^{1/2} \qquad (2\text{–}17)$$

where p_1 and p_2 = initial and final absolute pressure, psia
$\overline{W}$ = flow, lb per hr
L = length, ft
d = pipe ID, in.
$K = \dfrac{0.01214f(2p_1 v_1)}{3600 d^5}$, see Table 2–4.

Values of K for various pipe sizes are tabulated in Table 2–4, using the Babcock and Wilcox friction factor and $p_1 v_1$ of 500. For any other values of $p_1 v_1$, K should be corrected accordingly. Note that Eq. (2–17) has the same limitations as Eq. (2–12).

EXAMPLE 2–6. Determine the pressure drop due to friction for the data of Example 2–5 by using Eq. (2–17).

SOLUTION. Using the data of Plate 1 for 6-in. schedule 80 pipe,

$$K = \frac{0.01214 \times 0.015 \times 2 \times 865 \times 0.8873}{3600 \times 6350} = 12.23 \times 10^{-9}$$

$$\overline{W} = \frac{AV}{v} = \frac{0.1810 \times 12,000 \times 60}{0.8873} = 146{,}870 \text{ lb per hr}$$

$$\Delta p = 865 - 849.8 = 15.2 \text{ psi}$$

$$p_2 = [(865)^2 - 12.23 \times 10^{-9}(146{,}870)^2 100]^{1/2} = 849.8 \text{ psia}$$

2–11. Friction Factors. Many investigators have endeavored to determine suitable values for the friction factor for both laminar and turbulent flow. Experiments have proved the validity and accuracy of the following theoretical equation for friction factor for laminar flow:

$$f = \frac{64}{N_R} \qquad (2\text{–}18)$$

where f = friction factor
N_R = Reynolds' number

This equation shows that the friction factor is independent of roughness of the pipe wall. However, the friction factor for turbulent flow does not lend itself to such easy analysis because experiments show that for this condition the roughness of the pipe wall has a major effect on the pressure drop. The condition of the pipe wall is determined by pipe material and by the manufacturing process.

It has been shown experimentally that, during turbulent flow, there is no motion of fluid particles relative to the conduit wall immediately adjacent to the wall. This has led to the theory that there is a boundary layer adjacent to the wall in which the flow is laminar, even though the majority of the fluid may be flowing under turbulent conditions. The thickness of this layer changes with the flow conditions, but it is very thin under any circumstances. At times, the thickness may be only that of a few molecules. In all probability the boundary layer is not a sharply defined region and does not have a definite line of demarcation. However, the existence of a boundary layer does help to explain the phenomenon of turbulent flow over surfaces.

A pipe surface is said to be *smooth* when its projections or protuberances do not extend through the laminar boundary layer. Conversely, a rough pipe is one whose protuberances do extend beyond the boundary layer. Since the thickness of the layer is dependent on the flow conditions, the same pipe may be either smooth or rough, depending on the flow. When the surface projections are extremely large compared with the boundary layer, they may serve to increase turbulence.

Nikuradse's classical experiments demonstrate the effect of surface roughness on friction for both the laminar and turbulent regions (Fig. 2-18). He used a roughness factor ϵ/D, which was the ratio of the height of artificial protuberances to the diameter of the pipe. The degree of roughness was obtained by coating the inside pipe surface with carefully graded sand.

Note that the friction factor in the laminar region conforms to the theoretically developed formula, Eq. (2–18), irrespective of the surface roughness. From a Reynolds' number of 2100 to 4000 there is a region of transition in which it is very difficult to determine a friction factor.

For low values of Reynolds' number in the turbulent region, there is one curve that indicates the friction factor for smooth pipe. Curves for rough pipe surfaces originate at this smooth pipe curve, but they diverge from it as the Reynolds' number increases. This may be explained by the effect of the laminar boundary layer that was previously discussed. Pipes that are smooth for the lower values of Reynolds' number become rough pipes at larger Reynolds' numbers. At very high Reynolds' numbers the friction factor becomes constant for a given pipe roughness or ϵ/D ratio.

Other investigators have used different methods of obtaining surface

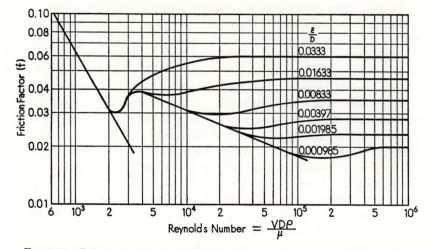

Fig. 2–18. Relation of Reynolds' number, friction factor, and relative roughness for similar pipes. (J. Nikuradse, "Strommung Gesetze in rauhen Rohren," *VDI Forschungsheft,* 1933, p. 361.)

roughness, but their results have been similar to those of Nikuradse. However, it is difficult to correlate these roughness coefficients with actual pipe surfaces. Furthermore, the pipe surface does not remain constant but changes unpredictably due to dirt, scale, corrosion, and welding beads or chill rings.

Friction factors developed by Moody from a series of tests on pipes of many materials are given in Figs. 2–19(a) and 2–19(b).* Dimensionless values of relative pipe roughness (ϵ/D) for materials ranging from drawn tubing to riveted steel are given in Fig. 2–19(a). Since the height of the protuberances on any pipe surface primarily will be a function of the manufacturing process for the particular material, values of ϵ are shown on the curves. The relative roughness decreases as the pipe diameter increases. Values of relative roughness read on the left-hand ordinate scale are used to select the proper curve on Fig. 2–19(b).

Notice that the curves of Fig. 2–19(b) conform to the general shape of the Nikuradse curves and that there are four sections; i.e., the region of laminar flow, the critical zone, the transition zone, and the region of complete turbulence. The boundary between the last two, the transition zone and the region of complete turbulence, is indicated by a diagonal, dashed line. Further, the region of complete turbulence begins at a point on each

* The curves in Fig. 2–19(b) may be expressed by the approximate formula

$$f \cong 0.0055 \left[1 + \left(20{,}000 \frac{\epsilon}{D} + \frac{10^6}{R} \right)^{\frac{1}{3}} \right]$$

within the range of values of R from 4000 to 10^7. (Moody, *Mechanical Engineering,* Vol. 69, No. 12.)

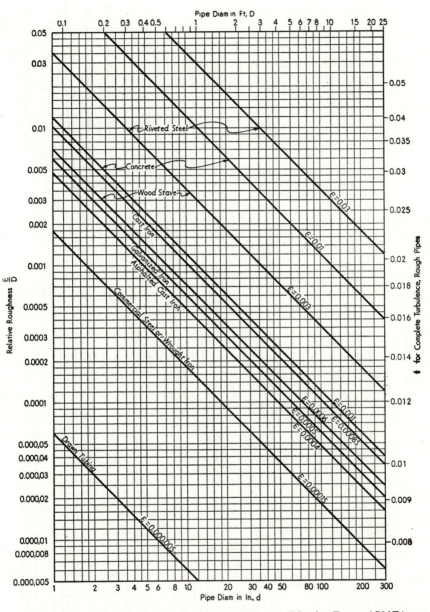

FIG 2–19(a). Friction factors and relative roughness. (Moody, *Trans. ASME.*)

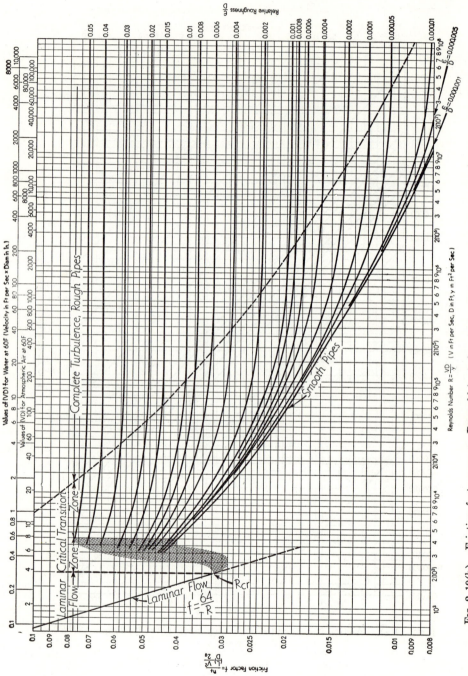

FIG. 2-19(b). Friction factors versus Reynolds' number with roughness as parameter. (Moody, *Trans. ASME.*)

relative roughness curve where the friction factor is practically independent of Reynolds' number. Values of friction factors for flow in the complete turbulence region are also given on the right-hand ordinate scale of Fig. 2–19(a).

Results of investigations by a great many others have been used by designers. Table 2–3 lists some of the most noteworthy of the friction

TABLE 2–3

FRICTION FACTORS FOR STEAM *

Investigator	G	f
Babcock	$0.0001321[1 + (3.6/d)]$	$0.01090[1 + (3.6/d)]$
Hawksley	0.000337	0.0278
Gutermuth	0.0003557	0.0293
Babcock & Wilcox (B. & W.)	$0.0001310[1 + (3.6/d)]$	$0.01080[1 + (3.6/d)]$
Carpenter-Unwin	$0.0001306[1 + (3.6/d)]$	$0.01076[1 + (3.6/d)]$

* d = ID of pipe, in.; f = friction factor in Eq. (2–16); G = a factor in equation $\Delta p = \dfrac{GLW^2}{\rho d^5}$; symbols as for Eq. (2–16).

factors that have been proposed for steam. Note that the value G includes both the friction factor and the constant of Eq. (2–16). None of the values listed in Table 2–3 has the rational background of the Moody curves since the values do not indicate any variation of friction factor with either viscosity or velocity. Values by Babcock, Babcock and Wilcox, and Carpenter and Unwin, show the effect of pipe roughness, in terms of pipe diameter, on the friction factor.

The Babcock and Wilcox friction factor has been used in computing K in Eq. (2–17), as listed in Table 2–4. Note that the table is based on $p_1 v_1 = 500$. For other values of $p_1 v_1$, multiply K by $p_1 v_1 / 500$.

Williams and Hazen's formula, Eq. (2–19), gives very satisfactory results for water and other fluids having similar viscosities. Thus,

$$h_f = \left(\frac{V}{1.318 C R^{0.63}}\right)^{1.852} = \frac{0.002068 (\text{gpm})^{1.852}}{(d)^{4.871}} \qquad (2\text{--}19)$$

where h_f = friction loss per ft of pipe, ft of the liquid
  V = velocity, fps
  C = a constant depending on the condition of the pipe surface (see Table 2–5)
  R = hydraulic radius, ft
  gpm = flow, gal per min
  d = ID of pipe, in.

TABLE 2–4

VALUES OF K FOR EQ. (2–17) USING B. & W. FORMULA BASED ON $p_1 v_1 = 500$

Nominal Pipe Size (In.)	Schedule 10 K	Schedule 20 K	Schedule 30 K	Schedule 40 K	Schedule 60 K
½	....	....	....	2.653×10^{-3}	....
¾	....	....	....	514.3×10^{-6}	....
1	....	....	....	127.0×10^{-6}	....
1¼	....	....	....	26.24×10^{-6}	....
1½	....	....	....	10.89×10^{-6}	....
2	....	....	....	2.644×10^{-6}	....
2½	....	....	....	974.9×10^{-9}	....
3	....	....	....	291.0×10^{-9}	....
3½	....	....	....	130.4×10^{-9}	....
4	....	....	....	65.17×10^{-9}	....
5	....	....	....	19.04×10^{-9}	....
6	....	....	....	7.066×10^{-9}	....
8	....	1.483×10^{-9}	1.536×10^{-9}	1.631×10^{-9}	1.826×10^{-9}
10	....	434.6×10^{-12}	460.9×10^{-12}	489.7×10^{-12}	565.5×10^{-12}
12	....	170.7×10^{-12}	182.8×10^{-12}	195.3×10^{-12}	224.4×10^{-12}
14 OD	102.8×10^{-12}	107.9×10^{-12}	113.3×10^{-12}	119.0×10^{-12}	134.9×10^{-12}
16 OD	50.12×10^{-12}	52.25×10^{-12}	54.53×10^{-12}	59.42×10^{-12}	66.28×10^{-12}
18 OD	26.73×10^{-12}	27.73×10^{-12}	29.89×10^{-12}	32.26×10^{-12}	35.53×10^{-12}
20 OD	15.29×10^{-12}	16.34×10^{-12}	17.48×10^{-12}	18.39×10^{-12}	20.77×10^{-12}
24 OD	5.855×10^{-12}	6.185×10^{-12}	6.723×10^{-12}	7.113×10^{-12}	7.979×10^{-12}
30 OD	1.867×10^{-12}	1.994×10^{-12}	2.085×10^{-12}		

Nominal Pipe Size (In.)	Schedule 80 K	Schedule 100 K	Schedule 120 K	Schedule 140 K	Schedule 160 K
½	5.694×10^{-3}	....	....	....	14.45×10^{-3}
¾	946.8×10^{-6}	....	....	....	2.862×10^{-3}
1	215.9×10^{-6}	....	....	....	548.2×10^{-6}
1¼	40.74×10^{-6}	....	....	....	71.10×10^{-6}
1½	16.29×10^{-6}	....	....	....	31.32×10^{-6}
2	3.793×10^{-6}	....	....	....	8.290×10^{-6}
2½	1.372×10^{-6}	....	....	....	2.263×10^{-6}
3	397.7×10^{-9}	....	....	....	690.9×10^{-9}
3½	174.9×10^{-9}	....	....	....	
4	86.15×10^{-9}	....	115.7×10^{-9}	....	155.1×10^{-9}
5	24.63×10^{-9}	....	32.91×10^{-9}	....	44.74×10^{-9}
6	9.318×10^{-9}	....	11.95×10^{-9}	....	16.38×10^{-9}
8	2.078×10^{-9}	2.370×10^{-9}	2.844×10^{-9}	3.276×10^{-9}	3.789×10^{-9}
10	625.9×10^{-12}	719.8×10^{-12}	831.1×10^{-12}	1.001×10^{-9}	1.167×10^{-9}
12	251.4×10^{-12}	290.9×10^{-12}	338.3×10^{-12}	382.9×10^{-12}	463.3×10^{-12}
14 OD	153.6×10^{-12}	180.0×10^{-12}	200.7×10^{-12}	237.6×10^{-12}	274.4×10^{-12}
16 OD	75.79×10^{-12}	87.05×10^{-12}	100.3×10^{-12}	119.0×10^{-12}	131.6×10^{-12}
18 OD	40.82×10^{-12}	47.08×10^{-12}	53.36×10^{-12}	62.04×10^{-12}	70.87×10^{-12}
20 OD	23.53×10^{-12}	26.73×10^{-12}	31.06×10^{-12}	36.25×10^{-12}	40.82×10^{-12}
24 OD	9.107×10^{-12}	10.44×10^{-12}	11.82×10^{-12}	13.86×10^{-12}	15.80×10^{-12}

TABLE 2-4 (*Continued*)

Values of K for Eq. (2-17) Using B. & W. Formula Based on $p_1v_1 = 500$

$$p_F = \sqrt{p_1{}^2 - \overline{W}^2 LK}$$

where p_F = final pressure, psia
p_1 = initial pressure, psia
$\overline{W}$ = flow, lb per hr
L = length, ft
K = friction factor
d = ID, in.
v_1 = initial specific volume, cu ft per lb

$$K = \frac{0.000{,}03639(d + 3.6)}{d^6}$$

$$= \frac{0.000131 \left(1 + \dfrac{3.6}{d}\right) 2p_1v_1}{3600d^5}$$

An advantage of the Williams and Hazen formula is the constant which is determined by the age or material of the pipe. Water piping in steam-power plants is frequently designed on the basis of old pipe ($C = 100$) in order to allow for possible deterioration of the pipe surface after several years of operation. This factor also provides an allowance to cover the many uncertainties that inherently exist in calculating pressure loss. However, some designers prefer to consider all pipe that carries treated water, such as condensate and boiler feedwater, as having the same loss as new pipe with $C = 130$. This value is based on the contention that the boiler feedwater and condensate for modern high-pressure plants are practically oxygen-free; of course these waters carry no entrained solids and are relatively noncorrosive.

TABLE 2-5

Commonly Used Values of C and Conversion Factors for the Williams and Hazen Formula

Type of Pipe	C	Correction Factor
Fiber, bitumastic-enamel-lined steel	140	0.54
Copper, copper alloy, glass pipe, or tubing	130	0.62
Welded and seamless steel, cast-iron, wrought-iron, new	130	0.62
Concrete	120	0.71
Vitrified	110	0.84
Steel, cast-iron, wrought-iron, old	100	1.00
Corrugated steel	60	2.57

Note: Correction factor to be multiplied by h_f at $C = 100$ to obtain h_f for other values of C.

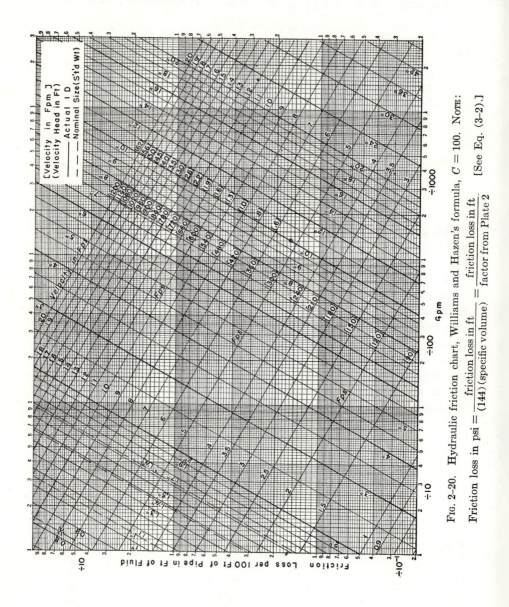

Fig. 2-20. Hydraulic friction chart, Williams and Hazen's formula, $C = 100$. Note:

$$\text{Friction loss in psi} = \frac{\text{friction loss in ft}}{(144)\,(\text{specific volume})} = \frac{\text{friction loss in ft}}{\text{factor from Plate 2}} \quad [\text{See Eq. (3-2).}]$$

Fig. 2-20 is a plot of Eq. (2-19) using $C = 100$. It will be seen that the exponent in the empirical Williams and Hazen formula corresponds closely to the exponents in the theoretical equation, Eq. (2-16). Correction factors to be used with Fig. 2-20 are given in Table 2-5.

Although Fig. 2-20 should be used only for water, it illustrates an important principle that applies to all fluids. Consider any one line of constant velocity, such as the line representing 6 fps. At this velocity, water in a 1-in. ID pipe will have a friction loss of 30 ft per 100 ft of pipe; in a 4-in. ID pipe there will be a loss of 6 ft per 100 ft, while in a 20-in. ID pipe there will be a loss of 0.92 ft per 100 ft. Thus, the pressure drop for a unit length of pipe decreases radically with an increase in pipe diameter at a given velocity. Fluid velocities are a guide to be used in selecting the proper size, but suitable velocities will vary with the diameter to prevent excessive pressure loss.

EXAMPLE 2-7. Calculate the pressure drop for the flow of 146,870 lb per hr of 850-psig, 900 F steam in a 100-ft length of 6-in. schedule 80 pipe. Use (a) the B. & W. friction factor in Eq. (2-16), (b) Eq. (2-17) and Table 2-4, and (c) the Moody friction factors given in Fig. 2-19.

SOLUTION. Use data from previous examples.

(a) From Table 2-3, $f = 0.0108 \left(1 + \dfrac{3.6}{5.761}\right) = 0.01755$

From Eq. (2-16),

$$\Delta p = \frac{0.01214 \times 0.01755 \times 100(146,870)^2}{1.127 \times 6350 \times (60)^2}$$

$$= 17.85 \text{ psi}$$

(b) $p_1 v_1 = 865 \times 0.8873 = 768$

From Table 2-4,

$$K = 9.318 \times 10^{-9}$$

Corrected for $p_1 v_1$,

$$K = 9.318 \times 10^{-9} \times \frac{768}{500} = 14.303 \times 10^{-9}$$

$$p_2 = [(865)^2 - 14.303 \times 10^{-9}(146,870)^2 \times 100]^{1/2} = 847 \text{ psia}$$

$$\Delta p = 865 - 847 = 18 \text{ psi}$$

(c) From Example 2-4, $R = 6,030,000$

From Fig. 2-19(a), $\epsilon/D = 0.00032$

Note from Fig. 2-19(b), that this condition is in the range of complete turbulence. Then, from either Fig. 2-19(a) or (b),

$$f = 0.015$$

From Eq. (2-16),

$$\Delta p = \frac{0.01214 \times 0.015 \times 100(146,870)^2}{1.127 \times 6350(60)^2}$$

$$= 15.2 \text{ psi}$$

EXAMPLE 2-8. A 6-in. schedule 40 pipe passes 500,000 lb per hr of 250 F water. From Fig. 2-20, determine (a) friction loss for 100 ft of pipe, both in feet of the fluid and in psi, (b) velocity, and (c) velocity head.

SOLUTION. For 60 F water, gpm × 60 × 8.33 = 500 gpm = lb per hr. From Plate 2, in the Appendix, specific gravity at 250 F = 0.9425.

$$\text{gpm} = \frac{\text{lb per hr}}{(500)(\text{specific gravity})} = \frac{(500{,}000)}{(500)(0.9425)} = 1060$$

From Plate 1, ID = 6.065 in.

(a) Entering Fig. 2-20 at 1060 gpm on the abscissa, proceed vertically until intersecting a diagonal line (interpolated) for ID = 6.065 in. Read h_f = 12.7 ft of 250 F water on the ordinate.

$$\frac{\text{Head in ft}}{(144)(\text{specific volume})} = \text{psi}$$

From Plate 2 in the Appendix for 250 F, (144)(specific volume) = 2.447 ft head per psi

$$\Delta p = \frac{12.7}{2.447} = 5.2 \text{ psi}$$

(b) At the intersection of the lines for 1060 gpm and 6.065 in. ID, read 710 fpm velocity. (Values of velocity are in brackets in Fig. 2-20.)

(c) For the intersection found in (b), read 2.1-ft velocity head. (Values of velocity head are in parentheses in Fig. 2-20.)

2-12. Equivalent Lengths. The term *length of pipe* used in all the friction loss formulas was intended to mean the length of straight pipe. However, in almost any pipe installation in a power plant there is a considerable number of fittings such as elbows, tees, and valves, which have a higher friction drop than the same length of straight pipe. Additional loss in fittings is caused by the irregularities of the confining walls and the changes in the direction of flow of the fluid.

Equivalent lengths have been determined by experiment for most of the common pipe fittings in order to simplify calculations. Equivalent lengths are given on a so-called *no-length* basis, i.e., the equivalent length of a fitting is that length of straight pipe which would have the same friction loss as the fitting over and above the friction loss in a straight piece of pipe physically the same length as the fitting. For example, consider a standard elbow (one diameter radius) 4-in. nominal diam. The actual length of the centerline of this elbow is about 6¼ in. and the equivalent length is 11 ft. The friction loss for this elbow would be the same as that of a straight pipe of the same diameter but 11 ft 6¼ in. long. Thus, to determine the equivalent length of a pipe system, the actual linear length of the pipe and fittings should be calculated and the equivalent length of the fittings should be added to this to get the equivalent length of the system.

It is difficult to determine accurately the values of equivalent lengths of fittings for all conditions of flow and at the same time to obtain an expression that will be easy to use. Actually, the equivalent lengths for a given type of fitting vary considerably with velocity, pipe size, and radius for elbows and bends. In many cases, slight inaccuracies in the development of a pressure-drop formula or inaccuracies in determining a friction factor are insignificant compared with the inaccuracies of the equivalent lengths.

The values for equivalent lengths determined by Dean Foster [*] have found wide use in practice. Of more recent date are the values developed by the Crane Co.[†] and others, Fig. 2–21, from a series of extensive tests using water and steam. Although the fittings shown in this diagram are portrayed with screwed ends, some of the tests indicated that there was no difference in the loss through screwed or flanged fittings. Therefore, these values might also be used for welded fittings. In any event, if an error is incurred, it is on the safe side. These tests also indicated that there was no appreciable difference when using either steam or water.

Fig. 2–22 shows in curve form the equivalent lengths of standard radius elbows. Also included thereon is a tabulation of the equivalent resistance for some other fittings in terms of elbows, all plotted primarily from the data in Fig. 2–21.

EXAMPLE 2–9. Find the equivalent length of 133 ft of 8-in. schedule 160 pipe, with two gate valves, one globe valve, two tees on-run, eight standard elbows, two 45-deg elbows.

SOLUTION. ID of 8-in. schedule 160 pipe is 6.813 in.

8 elbows @ 1 elbow each	8	elbows
2 gate valves @ ¼ elbow each	½	"
1 globe valve	10½	"
2 tees on run @ ⅝ elbow each	1¼	"
2 45-deg elbows @ ½ elbow each	1	"
Total	21¼	elbows

The equivalent length of one elbow, from Fig. 2–22, equals 18.7 ft. Equivalent length of pipe equals $133 + 18.7 \times 21\frac{1}{4} = 530$ ft.

The value of L in any of the applicable pressure-drop formulas would then be 530 ft.

2–13. **Determination of Pipe Size.** Pipe is sized on the basis of both velocity and allowable pressure loss as determined by economic analysis. In certain instances, velocity is a very real criterion in the

[*] Dean E. Foster, "Effects of Fittings on the Flow of Fluids through Pipe Lines," *Trans. ASME*, 1922, p. 1219.

[†] Crane Co., *Engineering Data on Flow of Fluids in Pipes and Heat Transmission*, 1935.

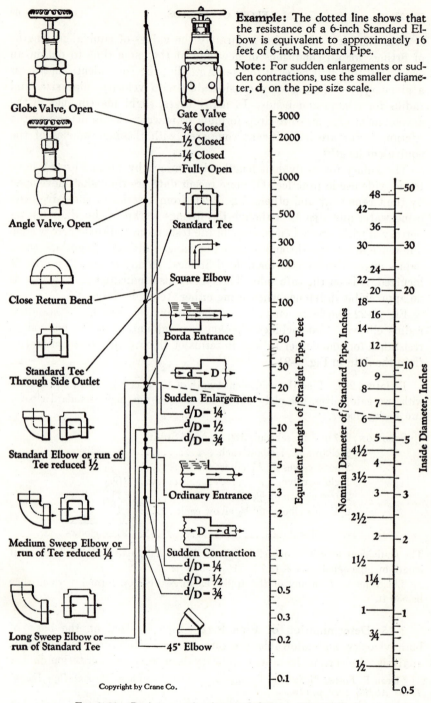

Example: The dotted line shows that the resistance of a 6-inch Standard Elbow is equivalent to approximately 16 feet of 6-inch Standard Pipe.

Note: For sudden enlargements or sudden contractions, use the smaller diameter, **d**, on the pipe size scale.

Globe Valve, Open

Angle Valve, Open

Close Return Bend

Standard Tee Through Side Outlet

Standard Elbow or run of Tee reduced ½

Medium Sweep Elbow or run of Tee reduced ¼

Long Sweep Elbow or run of Standard Tee

Gate Valve
¾ Closed
½ Closed
¼ Closed
Fully Open

Standard Tee

Square Elbow

Borda Entrance

Sudden Enlargement
d/D – ¼
d/D – ½
d/D – ¾

Ordinary Entrance

Sudden Contraction
d/D – ¼
d/D – ½
d/D – ¾

45° Elbow

Equivalent Length of Straight Pipe, Feet

3000
2000
1000
500
300
200
100
50
30
20
10
5
3
2
1
0.5
0.3
0.2
0.1

Nominal Diameter of Standard Pipe, Inches

48
42
36
30
24
22
20
18
16
14
12
10
9
8
7
6
5
4½
4
3½
3
2½
2
1½
1¼
1
¾
½

Inside Diameter, Inches

50
30
20
10
5
3
2
1
0.5

Copyright by Crane Co.

Fig. 2–21. Resistance of valves and fittings. (Crane Co.)

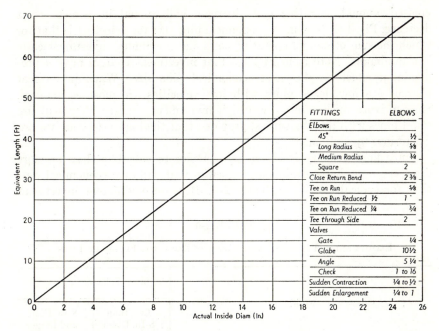

Fig. 2–22. Equivalent lengths of standard elbows. (Adapted from data by Crane Co. and others.)

selection of a pipe size, since very high velocities can cause serious damage to equipment. Water at high velocity has been known to cut through the thickest of pipes. Steam, particularly moist steam, can cause erosion and vibration of pipes and tubes. Steam entering the surface type of feedwater heater should be limited to velocities of about 6000 fpm to reduce the possibility of tube erosion even when the tubes are protected by baffles.

In connection with Fig. 2–20, it was shown that velocity cannot be considered indicative of pressure drop. At a given velocity, the pressure drop of a fluid through a small pipe may be several times the pressure drop of the same fluid flowing through a large pipe at the same velocity. Loss in pressure due to friction is an economic factor in pipe design. Consider a pipe that conveys steam from a boiler to a turbine. A small pipe will require a low initial investment in pipe, but because of a high pressure drop, the small pipe will increase the boiler cost and the pumping power required to supply water to the boiler. A large steam pipe will have an opposite effect on all these costs. An economic analysis is required to determine the proper size.

Pipe for a water system was used in Chap. 1 to develop curves typical of an economic analysis. These curves indicated that as velocity increased, annual operating costs increase and annual investment costs decrease, Fig. 1–4. The most economical pipe size (or velocity) is that for which the sum of annual operating costs and the annual investment costs is a minimum. Note from Fig. 1–5 that investment charges have a predominant effect on the most economical pipe size. When investment charges are low, large pipes with low velocities are most economical. Since private utilities must use high investment charges, they will use small pipes with high velocities, provided that the high velocities are not deleterious to equipment.

Since fluid velocity is easier and quicker to calculate than pressure drop, engineers have established velocities that will give economical pipe sizes for typical conditions, Table 2–6. Because of different investment

TABLE 2–6

Recommended Velocities and Pressure Drops for Fluids in Pipes

Fluid	Velocity, fpm	
	Less than 8-In. Diam	8-In. Diam and Larger
Steam, moist, subatmospheric–10 psig	2000–4000	3000–4500
moist, 10–100 psig	2000–5000	4000–8000
superheated, up to 100 psig	2000–6000	5000–12,000
superheated, 100–600 psig	3000–8000	5000–14,000
superheated, over 600 psig	4000–10,000	5000–15,000
Water, general service, pump suction *	100–300	100–300
general service, pump discharge	300–500	400–600
condensate and boiler feed, pump suction *	80–250	80–250
condensate and boiler feed, pump discharge	350–500	400–600
circulating water	...	250–500
control and desuperheating	600–10,000	600–10,000
flashing (based on water only)	80–200	80–200
Compressed air, up to 25 psig	2000–5000	4000–6000
26 to 250 psig	2500–8000	4000–10,000
Fuel oil, light	200–300	200–400
heavy, hot	200–300	200–400
	Pressure Drop (%)	
Steam, subatmospheric and low pressure	5–10	
medium and high pressure	5	

* These velocities are governed by net positive suction head (see Chap. 3).

charges and because of the effect of velocity of pressure drop with various pipe sizes, the values are listed as ranges rather than specific values.

When selecting a pipe for a system, it is important to consider the normal and maximum flow for the system. Turbine generators are designed to carry 10 to 25% over the rated load. Since pressure drop in a pipe varies as the square of the flow [see Eq. (2–16)], a pipe designed for a reasonable pressure drop at rated load would have a 56% higher pressure drop at 25% turbine overload. Such a large pressure drop at overload conditions could limit the capacity of the turbine and prevent it from carrying 25% overload.

EXAMPLE 2–10. Select a schedule 40 pipe for a flow of 100,000 lb per hr of steam at 200 psig and 500 F.

SOLUTION. Assume that the pipe will be 8 in. or over. From Table 2–6, try a velocity of 10,000 fpm; $v = 2.525$ cu ft per lb.

$$A = \frac{\overline{W}v}{V} = \frac{100,000 \times 2.525 \times 144}{60 \times 10,000} = 60.6 \text{ in.}^2$$

From Plate 1, this would indicate a pipe somewhere between 8 and 10 in.

Try 8 in., $\quad V = \dfrac{\overline{W}v}{A} = \dfrac{100,000 \times 2.525 \times 144}{60 \times 50} = 12,120$ fpm

Try 10 in., $\quad V = \dfrac{100,000 \times 2.525 \times 144}{60 \times 78.9} = 7685$ fpm

Try 12 in., $\quad V = \dfrac{100,000 \times 2.525 \times 144}{60 \times 111.9} = 5414$ fpm

Since the pipe will be near 8 in., and the range given in Table 2–6 for 8-in. diam and larger is 5000 to 14,000 fpm, the pipe should have a velocity in the low part of this range. Therefore select a 10-in. diam pipe unless knowledge of the system should indicate that a very low-pressure drop is required; then a 12-in. diam pipe would be selected.

2–14. Flow of Air Under High Pressure.
Another application of turbulent flow of a compressible fluid (see Art. 2–10) that deserves special attention is that of the flow of high-pressure gases. In power-plant design these gases may be compressed air for power or control or they may be gaseous fuel for the boiler. The flow of compressed air is distinguished from the flow of air under low pressures (for example, the low-pressure air supplied to the boilers for combustion purposes) by the pressure drop in percentage of the initial pressure. The relatively large-pressure drops encountered with compressed air cause an appreciable increase in the specific volume.

Again, it is advantageous to consider the flow process as one of constant temperature. This is more nearly true in the case of compressed air than in the case of steam because the air discharged from the compressor is stored in an uninsulated tank. In this way it reaches room temperature before leaving the tank and remains at approximately room temperature during its journey through the piping. Eq. (2–17) was expressed in the differential form as

$$-\int_{P_1}^{P_2} P \, dP = \int_0^L \frac{fw^2RT \, dL}{2gA^2D}$$

and, when integrated, this becomes

$$\frac{P_1^2 - P_2^2}{2} = \frac{fw^2RTL}{2gA^2D} \qquad (2\text{–}20)$$

Then

$$w = \frac{Q}{v60} = \frac{QP_0}{60RT_0}$$

where Q = cfm measured at P_0 and T_0
P_0 = reference pressure, psfa
T_0 = reference temperature, R
R = gas constant

and for air

$$Q = 3.920 \frac{T_0}{p_0} \sqrt{\frac{(p_1^2 - p_2^2) \, d^5}{fTL}} \qquad (2\text{–}21)$$

where p_1 = initial pressure, psia
p_2 = final pressure, psia
d = ID of pipe, in.,
L = length of pipe, ft
T = temperature of air, R
f = friction factor
T_0 = reference temperature, R
p_0 = reference pressure, psia
Q = cfm measured at T_0 and p_0

It is often convenient to express the volume flow of a gas in terms of a standard pressure (p_0) and temperature (T_0) called *free air*. There is some difference of opinion among engineers as to the exact definition of free air, but we shall define it as air at 14.7 psia and 60 F. Using these values for p_0 and T_0, Eq. (2–21) becomes

$$Q = 138.67 \sqrt{\frac{(p_1^2 - p_2^2) \, d^5}{fTL}} \qquad (2\text{–}22)$$

If it is preferable to express the flow in terms of weight (w = lb per sec),

$$w = 0.1762 \sqrt{\frac{(p_1^2 - p_2^2) \, d^5}{fTL}} \qquad (2\text{–}23)$$

Friction factors for Eqs. (2–21), (2–22), and (2–23) may be determined from the Reynolds' number and Fig. 2–19 or from the investigations listed in Table 2–3. If an approximate solution is satisfactory, f may be assumed to range from 0.02 to 0.03. It will be noted that this range covers the normal range of the Reynolds' number shown on Fig. 2–19. By correcting the equations for the proper value of R, Eqs. (2–21), (2–22), and (2–23) may be used for any gas that may be assumed to be a perfect gas.

Weymouth * developed a similar equation that has found wide application in design of high-pressure natural and artificial gas mains:

$$Q = 21.874 \frac{T_0}{P_0} \sqrt{\frac{(p_1{}^2 - p_2{}^2) d^{5.33}}{GTL}} \qquad (2\text{–}24)$$

where G is the specific gravity of the gas and other symbols are as for Eq. (2–21). From tests that he conducted, Weymouth arrived at a friction factor of

$$f = \frac{0.032}{\sqrt[3]{d}}$$

which has been included in Eq. (2–24).

All these equations are approximations because the term

$$\int_{V_1}^{V_2} \frac{\rho V \, dV}{g}$$

of Eq. (2–11) has been assumed negligible. This assumption cannot be made when the pressure drop is extremely large. Although this function could not be integrated with any ease for steam, it may be integrated for gases approximating a perfect gas if the temperature is assumed to be constant. For the same units as those of Eqs. (2–11) and (2–13), the proper mathematical manipulation will result in

$$P_1{}^2 - P_2{}^2 = \frac{w^2 R T}{g A^2} \left(2 \ln \frac{V_2}{V_1} + \frac{fL}{D} \right) \qquad (2\text{–}25)$$

2–15. Flow of Air Under Low Pressure. Cases of air flow that fall under this category are the flow of air in ducts used for ventilating, heating, air conditioning, and induced- and forced-draft ducts for boilers. This class of flow problem has two distinguishing features: (1) the change in specific volume is negligible because the pressure loss is small, and (2) the velocity pressure or head, $V^2/2g$, becomes important because of the small pressure loss.

Velocity pressure, when expressed in inches WG (to indicate *water gage*) p_v, is

$$p_v = \frac{V^2 \rho}{334} \qquad (2\text{–}26)$$

where p_v = velocity pressure, in. of water

ρ = air density, lb per cu ft

V = air velocity, fps

* T. R. Weymouth, "Problems in Natural Gas Engineering," *Trans. ASME,* Vol. 34 (1912), p. 185

If the pressure loss due to friction in Eq. (2–13) is divided by the density of the water used as the gage fluid (assumed 62.4 lb per cu ft), then

$$p_f = \frac{f\rho V^2 L}{334D} = \frac{fp_v L}{D}$$ (2–27)

where p_f = loss of head due to friction, in. of water
  D = diam of the confining conduit, ft
  L = length of conduit, ft

Eq. (2–27) is based on the assumption that the duct is circular in cross-section. A similar formula may be developed from Eq. (2–14) for non-circular ducts. Friction factors may vary greatly for various shapes of noncircular ducts, but shallow rectangular ducts offer much more resistance to flow than do square ducts. If the ratio of length to width of a rectangular duct is not too great, the friction factor may be determined from Reynolds' number, using D equal to four times the hydraulic radius.

The amount of soot collected on the walls of the duct, corrosion, and other similar factors will affect the friction factor. Reasonably clean ducts may be assumed to have 0.0005 as the value of ϵ, corresponding to galvanized iron in the data of Fig. 2–19(a). However, the run of straight ducts and breeching in most power plants is short, and most of the friction loss is encountered in bends and other fittings. Therefore, satisfactory accuracy may be obtained by assuming friction factors of about 0.013 for large smooth ducts, up to 0.02 for smaller rough ducts.

If the friction losses per foot of duct for circular and rectangular ducts are made equal at the same flow, the use of consistent friction factors in the basic flow equations will produce the relationship between rectangular ducts and equivalent circular ducts. Thus, a rectangular duct may be converted into an equivalent circular duct that will have the same friction loss:

$$d = 1.265 \left[\frac{(ab)^3}{(a+b)} \right]^{1/8}$$ (2–28a)

where a, b = the two sides of the duct
  d = diam of the equivalent circular duct

If allowance is made for the frictional variations, Eq. (2–28a) becomes

$$d = 1.297 \frac{(ab)^{5/8}}{(a+b)^{1/4}}$$ (2–28b)

Any units may be used for the duct dimensions in Eqs. (2–28a) and (2–28b), provided all dimensions are expressed in the same units.

EXAMPLE 2–11. Find the velocity pressure and friction loss when 75,000 cfm of 320 F flue gas flow through a horizontal duct that is 7 ft high, 4 ft wide, and 30 ft long.

SOLUTION. Since flue gas and air are both predominantly nitrogen, their apparent molecular weights are nearly the same and R may be taken as 53.3.

$$\rho = \frac{P}{RT} = \frac{144 \times 14.7}{53.3(460 + 320)} = 0.05092 \text{ lb per cu ft}$$

$$V = \frac{\text{cfm}}{60A} = \frac{75,000}{60 \times 4 \times 7} = 44.6 \text{ fps}$$

Use Eq. (2-26):

$$p_v = \frac{(44.6)^2 \times 0.05092}{334} = 0.3033 \text{ in. of water}$$

Use Eq. (2-28a):

$$d = 1.265 \left[\frac{(4 \times 7)^3}{4 + 7}\right]^{1/5} = 5.782 \text{ ft}$$

$$\text{Velocity for equivalent duct} = \frac{75,000}{60 \times \frac{\pi}{4}(5.782)^2} = 47.6 \text{ fps}$$

$$p_v \text{ for equivalent duct} = \frac{(47.6)^2 \times 0.05092}{334} = 0.346 \text{ in. of water}$$

Assume $f = 0.015$, and use Eq. (2-27):

$$p_f = \frac{0.015 \times 0.346 \times 30}{5.782} = 0.0269 \text{ in. of water}$$

Then the velocity pressure in the duct will be 0.3033 in. of water and the friction will be 0.0269 in. of water.

As in any other flow problem, the proper sizing of ducts and breeching is a problem in economic selection. Since noise is not a factor to be considered in power plants, except for ventilation air in offices, laboratories, and control rooms, economic velocities are usually high and are determined by the cost of power required to overcome the pressure loss in the duct. For large plants equipped with fans and having large ducts and breeching, economical velocities for cold air are in the order of 2100 to 2400 fpm. Eq. (2-27) shows that at constant duct size and velocity, friction loss is directly proportional to density or inversely proportional to the absolute temperature of the air or gas. Therefore, economical velocities for hot air and flue gases range from 3000 to 3600 fpm. These velocities should be reduced by about 50% if the plant is not equipped with fans but depends solely on stacks or chimneys to overcome the duct and breeching losses.

The pressure drop across a duct elbow is considered to be a function of the velocity pressure and the size, shape, angle of bend, and the number and style of internal division plates. The arrangement of duct leading to and from the elbow also will have an important effect on the loss through the elbow; however, this effect cannot be calculated and must depend on engineering judgment.

The first configuration factor to be considered will be that of curvature because this has the greatest effect upon the elbow loss. The curva-

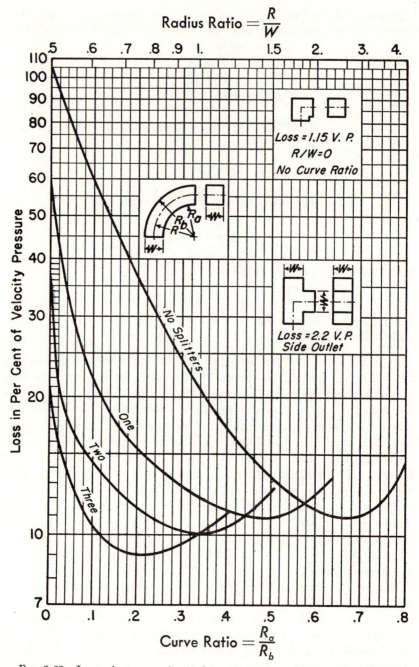

FIG. 2–23. Loss of pressure in 90-deg rectangular elbows with and without splitters. (Reproduced from *Fan Engineering* by permission of the publishers, Buffalo Forge Co.)

ture is expressed in terms of either *curve ratio* (CR) or *radius ratio* (RR). As indicated on Fig. 2–23, these ratios are defined as

$$CR = \frac{R_a}{R_b} = \frac{\text{inside radius}}{\text{outside radius}} \qquad (2\text{--}29)$$

$$RR = \frac{R}{W} = \frac{\text{centerline radius}}{\text{width}} \qquad (2\text{--}30)$$

Note that there is a direct relationship between these two ratios. Because CR is usually the easiest to calculate from duct drawings, it will be used throughout this text.

Values given on the ordinate scale of Fig. 2–23 are the losses in elbows having various curve ratios. The loss is expressed as a percentage of velocity pressure and includes both the loss due to curvature and the loss due to surface friction through the elbow. Curves are shown in this figure for plain elbows without splitters and for the use of one, two, and three splitters. Irregularly spaced division plates, extending from the inlet to the outlet of an elbow, are known as splitters, Fig. 2–24. Extreme caution should be used in providing splitters for an elbow because they may increase the loss through the elbow, rather than decrease it, if the flow in the duct leading to the elbow is not evenly distributed across the duct. This is also true for the turning vanes shown in Fig. 2–24. Therefore, splitters and turning vanes will not be considered further.

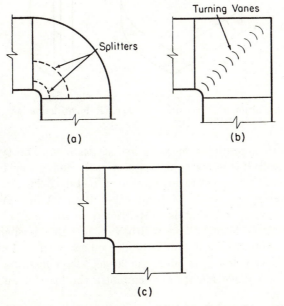

FIG. 2–24.   Elbows for rectangular ducts with (a) splitters, (b) turning vanes, and (c) rounded inside and square outside.

Elbows of the style shown in Fig. 2–24(c) are economical to construct for rectangular ducts. These elbows, for the sizes of ducts used in power plants, have no more loss than those with rounded outside corners.

Returning to Fig. 2–23, observe that the loss through an elbow diminishes as the curve ratio increases, until the curve ratio reaches a value of 0.67; then the loss increases. However, curved ductwork is expensive, and the reduction in pressure loss for curve ratios greater than about 0.3 is proportionately small. Thus economical curve ratios for power-plant ducts vary from 0.25 to 0.4.

Losses for miter elbows and miter side outlets are also given in Fig. 2–23.

Aspect ratio (AR) is the other configuration factor. It is defined, as shown in Fig. 2–25, as the ratio of the depth of the duct to the width.

$$AR = \frac{D}{W} \tag{2-31}$$

When duct elbows have a numerically low aspect ratio, like the elbow in Fig. 2–25(a), centrifugal force causes the gas to crowd to the outside of

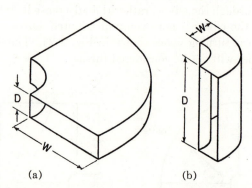

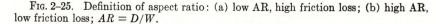

(a)   (b)

FIG. 2–25. Definition of aspect ratio: (a) low AR, high friction loss; (b) high AR, low friction loss; $AR = D/W$.

the elbow with a resultant increase in friction loss. The type of bend shown in Fig. 2–25(b) has a high aspect ratio and a low friction loss. Factors for varying aspect ratios are shown in Fig. 2–26.

Elbows having an angle of bend other than 90 deg are generally assumed to have a loss that is proportional to the angle. This is not exactly correct. A 45-deg elbow will have a loss that is slightly greater than one-half the loss for a similar 90-deg elbow, while a 180-deg elbow will have somewhat less than twice the 90-deg elbow loss.

When all the above factors are combined, the loss for an elbow may be expressed as

$$\Delta p = p_v \times F_{CR} \times F_{AR} \times F_{AB} \tag{2-32}$$

where Δp = pressure loss for an elbow, in. of water

   p_v = velocity pressure, Eq. (2–26), in. of water

   F_{CR} = curve ratio factor, Fig. 2–23

   F_{AR} = aspect ratio factor, Fig. 2–26

   F_{AB} = angle of bend factor

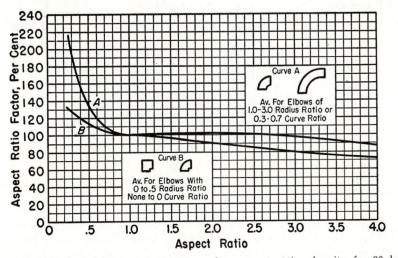

Fig. 2–26. Pressure loss in percentage of an aspect ratio of unity for 90-deg elbows followed by ducts. (Reproduced from *Fan Engineering* by permission of the publishers, Buffalo Forge Co.)

Although the data given above for the losses to be expected in elbows were from experiments conducted at 1800-fpm velocity, they may be used with only minor error for the velocities encountered in power-plant service.

Changes in duct cross-sectional area are necessary in many systems. Whether these changes are abrupt or gradual, there is always a conversion from velocity head to static head, or vice versa, with an accompanying loss in total head. An abrupt contraction with curves illustrating the variation in total and static pressures is shown in Fig. 2–27(a). Due to the reduction in area, the velocity pressure p_{v2} is much larger than p_{v1}. The increase in velocity energy from section 1 to section 2 has been obtained by a conversion from pressure energy to velocity energy. Therefore, the static pressure has diminished by the amount of the increase in velocity pressure plus the loss. The pressure loss in an abrupt contraction will be

$$\Delta p = C p_{v2} \qquad (2\text{–}33)$$

where Δp = loss of total pressure in an abrupt contraction

   p_{v2} = velocity pressure in the *smaller* duct, in. of water

   C = coefficient defined, in terms of the area ratio, as

Area ratio	0.1	0.2	0.3	0.4	0.5	0.6	0.7
C	0.48	0.46	0.42	0.37	0.32	0.26	0.20

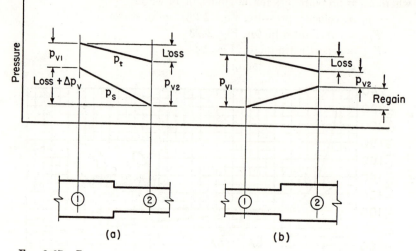

Fig. 2-27. Pressure variations in (a) abrupt contraction, and (b) abrupt enlargement.

An abrupt enlargement, Fig. 2-27(b), causes a reduction in velocity. However, there is a loss in the conversion from velocity energy to pressure energy so that regain in static pressure is less than the reduction in velocity pressure. In this case the loss in total pressure is

$$\Delta p = \left(1 - \frac{A_1}{A_2}\right)^2 p_{v1} \tag{2-34}$$

where Δp = loss of total pressure in an abrupt enlargement
A_1 = area of smaller duct
A_2 = area of larger duct
p_{v1} = velocity pressure in smaller duct, in. of water

Gradual changes in duct area provide a more orderly conversion between velocity and pressure or static energy than do abrupt changes. In a gradual contraction, the loss in total pressure should not exceed 4 or 5% of velocity pressure in the smaller duct if the included angle is less than 45 deg.

A gradual enlargement is frequently used in ductwork when space permits. When used on the outlet of a fan, a gradual enlargement is known as an evasé; this will be discussed later. There is more recovery through a gradual enlargement than through an abrupt one, and consequently there is less loss. However, in order to obtain this greater recovery, there must be no discontinuity in the slope of the sides. Losses to be expected from gradual enlargements or evasés are given in Fig. 2-28. From this data, the loss in total pressure will be given by Eq. (2-35).

$$\Delta p = C(p_{v1} - p_{v2}) \qquad (2\text{-}35)$$

where Δp = loss in total pressure in a gradual enlargement

C = coefficient from Fig. 2–28

p_{v1} = velocity pressure in smaller duct, in. of water

p_{v2} = velocity pressure in larger duct, in. of water

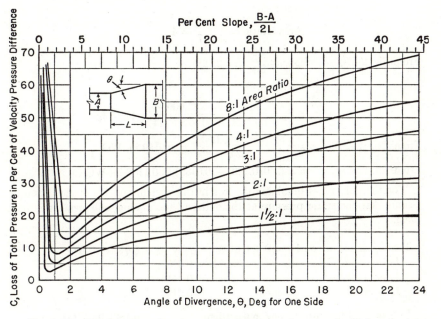

FIG. 2–28. Total pressure loss in gradual enlargement or evasé. (Primarily from data in *Fan Engineering*, 5th ed., Buffalo Forge Co.)

Because of possible errors in accounting for losses and for conversions between static and velocity pressures when calculating duct systems, all losses have been expressed in terms of total pressure.

EXAMPLE 2–12. Calculate the loss in a breeching that is 7 ft high by 4 ft wide when there is a flow of 75,000 cfm of 320 F flue gas. The breeching contains 30 ft of straight duct, one 45-deg elbow (high AR) with an inside radius of 18 in., and discharges into the bottom of a chimney with a gradual enlargement that is 8 ft wide by 14 ft high at the outlet and 9 ft long. All the breeching is in the horizontal plane.

SOLUTION. Use the data obtained in the preceding example.
For the 45-deg elbow:

From Fig. 2–25(b), $\text{AR} = \dfrac{7}{4} = 1.75$

From Fig. 2–26, curve A, $F_{\text{AR}} = 1.01$

From Fig. 2–23,

$$CR = \frac{1.5}{4 + 1.5} = 0.273$$

$$F_{CR} = 0.27$$

Use Eq. (2–32),

$$p = 0.3033 \times 0.27 \times 1.01 \times \frac{45}{90}$$

$$= 0.0414 \text{ in. WG}$$

For the gradual enlargement:

$$\text{Area Ratio} = \frac{8 \times 14}{4 \times 7} = 4$$

From Fig. 2–28,

$$\text{Slope} = \frac{8 - 4}{2 \times 9} = 22.2\%$$

$$C = 0.40$$

$$V_2 = \frac{\text{cfm}}{60 \times A} = \frac{75,000}{60 \times 8 \times 14} = 11.15 \text{ fps}$$

From Eq. (2–26),

$$p_{v2} = \frac{(11.15)^2 \times 0.05092}{334} = 0.0189 \text{ in. WG}$$

From Eq. (2–35), $p = 0.40(0.3033 - 0.0189) = 0.1138$ in. WG

For the purpose of obtaining the loss in this system and the static pressures, assume that the static pressure at the outlet of the gradual enlargement is zero. Then,

Total pressure at outlet $= 0 + 0.0189$	0.0189 in. WG
Loss in duct, from preceding example $= 0.0269$	
Loss in elbow $\qquad\qquad = 0.0414$	
Loss in gradual enlargement $\qquad = 0.1138$	
Loss in total pressure throughout system	0.1821 in. WG
Total pressure at inlet to system	0.2010 in. WG
Velocity pressure at inlet to system, from preceding example	0.3033 in. WG
Static pressure at inlet to system, based on zero static pressure at outlet	−0.1023 in. WG

The regain from the gradual enlargement was sufficient to overcome much of the system losses.

PROBLEMS

2–1. Determine the elevation of the bottom of a tank that is to supply 100 gpm of 60 F water through a pipe whose ID is 2 in. The pressure in the tank is atmospheric, and the water must be discharged from the pipe at a pressure of 4 psig and at an elevation of +27 ft. The friction loss through the pipe will be 38 ft.

2–2. Can a pipe be designed to supply 750 gpm from tank A sitting on the floor at elevation +389 ft to tank B at elevation +401 ft under the following conditions: pressure in tank A is atmospheric, pressure in tank B is 1½-in. Hg abs, water temperature is 80 F, friction drop through the pipe will be 9 ft?

2-3. What is the allowable friction drop through an 8-in. ID pipe that transfers 1200 gpm of 60 F water from a reservoir at elevation +243 ft to a tank at elevation +209 ft? There is atmospheric pressure in the tank. The water entering the tank leaves the pipe through a spray nozzle that has a pressure drop of 1½ psi.

2-4. Select the proper schedule of pipe for each of the following conditions: (a) feedwater at 1400 psig and 410 F, 12-in. diam; (b) condensate at 265 psig and 220 F, 6-in. diam; (c) feedwater at 2100 psig and 480 F, 8-in. diam; (d) steam at 275 psig and 550 F, 18-in. diam; (e) steam at 900 psig and 900 F, 10-in. diam; (f) steam at 1850 psig and 1000 F, 12-in. diam.

2-5. Calculate the absolute viscosity for water at 275 F in units of (a) (poundals)(sec) per sq ft; (b) lb-mass per ft-sec; (c) slugs per ft-sec; and (d) (lb-force)(sec) per sq ft, and the kinematic viscosity in sq ft per sec.

2-6. Same as Prob. 2-5, except for air at 120 F.

2-7. Same as Prob. 2-5, except for steam at 175 psig saturated.

2-8. Same as Prob. 2-5, except for steam at 850 psig and 900 F.

2-9. Calculate Reynolds' number for the data of Prob. 2-1.

2-10. Calculate N_R for oil flowing through a 4-in. schedule 40 pipe at a velocity of 110 fpm and a viscosity of 300 Saybolt-sec universal.

2-11. Determine Reynolds' number for 230-psig, 550 F steam flowing through a 6-in. schedule 80 pipe at a velocity of 10,000 fpm. What is the flow in pounds per hour?

2-12. What is N_R for 100,000 lb per hr of steam at 135-psig, 400 F, flowing through an 8-in. schedule 40 pipe?

2-13. Determine the flow of water in pounds per hour and in gallons per minute for a 4-in. schedule 80 pipe. Temperature of the water is 400 F and the velocity is 475 fpm. Also find R.

2-14. If the maximum value of N_R for laminar flow is assumed to be 2100, find (a) the flow in pounds per hour of 100 F water in a 6-in. ID pipe with laminar flow, (b) same for 50-psia saturated steam, (c) the pressure loss per 100 ft of pipe for part (a), and (d) the pressure drop per 100 ft of pipe for part (b).

2-15. Use Eq. (2-13) and check the numerical constants in Eqs. (2-15) and (2-16).

2-16. Find the pressure drop through 300 ft of 12-in. schedule 40 pipe for steam at 350 psia, 600 F, with a velocity of 15,000 fpm by using the friction factors established by (a) Babcock, (b) Hawksley, (c) Gutermuth, and (d) Moody.

2-17. Use Eq. (2-16) to find the pressure drop caused by 275,000 lb per hr of 850-psig, 900 F steam flowing through 585 ft of 12-in. schedule 120 pipe.

2-18. Use Eq. (2-17) to work Prob. 2-17.

2-19. What is the capacity (pounds per hour) of a 10-in. schedule 80 pipe 1145 ft long when steam enters at 425 psig, 750 F, and leaves at 400 psig?

2–20. Calculate the friction drop through 20 ft of ¾-in. copper tubing that has a thickness of 0.049 in. when handling 3500 lb per hr of 60 F water.

2–21. Find the friction loss through 780 ft of 3-in. schedule 40 pipe when handling 60,000 lb per hr of 310 F water.

2–22. Find the equivalent length of a 6-in. schedule 40 pipe that is 130 ft long and contains the following fittings: two gate valves, one swing check valve, one tee through-side, three tees on-run, nine 90-deg long-radius elbows.

2–23. Select a pipe and calculate the friction loss for the following conditions: general service water, 800 gpm at 275 psig and 80 F, 345 ft, thirteen 90-deg long-radius ells, three gate valves, one swing check valve, four tees through-side outlet, five tees on-run.

2–24. Check the value given in Fig. 2–22, by using the data of Fig. 2–21, for (a) tee on-run and (b) gate valve.

2–25. Select a pipe to carry 850-psig, 925 F steam from a boiler to a turbine. The maximum flow will be 330,000 lb per hr and the normal flow will be 260,000 lb per hr.

2–26. Select a pipe to carry 945,000 lb per hr of 410 F water with a velocity of approximately 500 fpm. The water pressure is 650 psi. Determine the head loss per 1000 ft of pipe.

2–27. Select pipes to carry steam from two boilers to one turbine. The steam pressure is 1250 psig and the temperature is 925 F. Each boiler is rated at 550,000 lb per hr maximum. Determine the pressure loss per 100 ft of pipe.

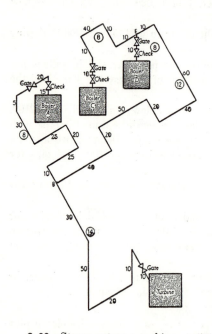

2–28. A pipeline is made up of 700 ft of 6-in. schedule 40 pipe and 300 ft of 6-in. schedule 80 pipe (allowances for fittings included) in series. What is the pressure drop when 30,000 lb per hr of 250-psig, 550 F steam is flowing?

2–29. An existing 12-in. schedule 80 pipeline 200 ft long (containing 8 elbows, 2 gate valves, 2 tees on-run, 1 tee through-side, and 1 globe valve) is to be enlarged by adding a parallel line of 8-in. schedule 80 pipe. The 12-in. and 8-in. pipes will then connect into a section of 10-in. schedule 80 pipe. The 8-in. pipe will be 275 ft long and will have 2 gate valves, 1 globe valve, and 14 ells. The 10-in. line is 80 ft long and has 11 ells, 1 gate valve, and 1 tee through-side. The steam entering the system will be at 90 psig and 75 F superheat. What will be the flow if the pressure drop is limited to 8 psi? Use Eq. (2–17).

2–30. Steam enters a turbine at 475 psig, 600 F, at the rate of 450,000 lb per hr supplied equally by boilers *A*, *C*, and *D*. Determine the highest pressure at which any one of the boilers must operate for the piping arrangement shown. Circled figures on the isometric sketch indicate nominal diameters; others indicate measured length; *check* indicates stop-check valve, *gate* indicates gate valve. All elbows are long radius.

2-31. Calculate the pressure at the turbine for the steam line shown in the sketch by using friction factors determined by Babcock and Wilcox and by Moody. Steam leaves the boilers at 176 psig and 530 F when the flow is 100,000 lb per hr per boiler. Select pipe sizes that will give velocities of not more than 9000 fpm. Figures on the sketch are linear distances in feet.

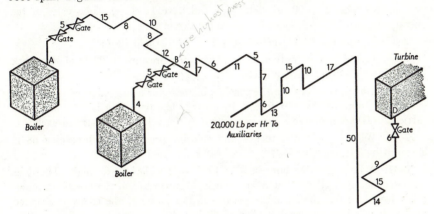

2-32. Would you recommend designing all steam pipes in a power plant for the same velocity? How about all the water pipes? Why?

2-33. Water flows from two tanks (deaerators) to three pumps. The flow from each tank is 600,000 lb per hr, and any two of the three pumps may handle the total water flow. The tanks contain saturated water at 5-psig pressure. Figures indicate nominal diameters of schedule 40 pipe and the equivalent lengths. What is the maximum pressure drop from the tanks to any pump?

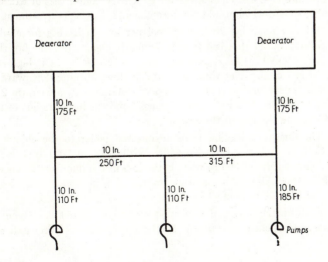

2–34. Same as Prob. 2–33 except that only one pump and one tank are in operation.

2–35. Determine the size of pipe required to supply air to a reciprocating water pump delivering 50 gpm of water against a 250-ft head. It is estimated that 120 cfm of free air are required at a pressure of 68 psig at the pump. The minimum pressure in the compressed air storage tank will be 90 psig and the maximum 110 psig. The pipe will have an equivalent length of 530 ft.

2–36. Air is to be used for boiler soot blowing at the rate of 1500 cfm of free air. The air in the storage tank will be at 300 psig and the drop through the pipe will be limited to 20 psi. The pipe will have 11 ells, 1 globe valve, 1 tee on-run, and will be 145 ft long. Estimate the size of pipe.

2–37. Derive Eq. (2–31). *Hint:* Use Eqs. (2–13) and (2–14).

2–38. Calculate the friction loss and velocity pressure in 20 ft of ductwork that is 5 ft by 8 ft and carries 72,000 cfm of 80 F air.

2–39. How much does the friction loss increase for a breeching when the size is changed from 8 ft by 8 ft to 4 ft by 16 ft? The breeching is 50 ft long and carries 224,000 cfm of 375 F flue gas. What is the velocity pressure in each?

2–40. What is the pressure loss and velocity pressure in a breeching 38 ft long, 4 ft by 5 ft, carrying 58,000 cfm of 345 F flue gas?

2–41. A duct carrying hot air at 500 F is 6 ft wide by 8 ft high. The duct contains one miter elbow (low AR) turning downward from the horizontal duct. What would be the annual dollar savings in fan power if the elbow is changed so that it will have an inside radius of 22 in. when electrical energy costs 0.3¢ per kwhr, the flow is 163,000 cfm, 5000 hr per yr operation, and 1.4 kw change in motor input for each 0.1 in. of water change in the pressure loss?

2–42. A designer has the option of using an abrupt enlargement with an area ratio of 2.1 or a gradual enlargement 8 ft long and with the same area ratio. The small duct has a flow of 90,000 cfm of 100 F air and is 7 ft wide by 6.5 ft high. For 3000 hr per year of operation, 1.3 kw of motor input for each 0.1 in. of water pressure loss, 0.82¢ per kwhr for electricity, and a cost of $1220 for the gradual enlargement, which would you recommend? Use 20% investment charges.

2–43. A boiler requiring 20 tons of coal per hr will have a forced-draft fan with a horizontal outlet that is 4 ft wide by 5.5 ft high. After the fan there is a gradual enlargement 12 ft long to a duct that is 8 ft wide by 11 ft high and 20 ft long. At the end of the duct there is a 90-deg elbow turning downward so that the air enters an air preheater with a sudden enlargement, area ratio 2:1. The boiler requires 13.7 lb of 75 F air per lb of coal, and the inside radius of the bend is 14 in. Calculate the loss in the system.

2–44. The ductwork leading from an oil-fired boiler to the chimney has a length of 45 ft and is 3 ft high by 4 ft wide. With a flow of 1200 lb of flue gas per min and a temperature of 365 F, what is the loss in the ductwork? There is one 45-deg elbow (low AR) with an inside radius of 8 in. and two 90-deg elbows (high AR) with inside radii of 11 in.

2–45. Early in the design of a steam-power plant, a 7500-kw steam turbine was purchased to operate at 600 psig, 825 F; the boiler was purchased normally to deliver steam at 635 psig. This allowed 35-psi pressure loss in the pipe. Several months later, the design had progressed to the point where the steam piping between the boiler and turbine was being designed, and it was determined that the line would have 135 ft of schedule 80 pipe, seven long-radius ells, one tee on-run, one check valve, and one gate valve. A large-sized pipeline would permit operation of the turbine at slightly higher than design pressure with a reduction of 20 Btu of fuel input to the boiler per kilowatt hour for each 10-psi increase in turbine pressure. If the installed cost of 100 ft of piping with fittings, valves, and insulation is $1000 + 400(d)^{1.5}$ where d is the nominal diameter of the pipe, what would be the most economical pipe diameter? The plant is expected to operate 4500 hr per yr at 7500 kw with fuel costing 33¢ per million Btu and investment charges at 10%. Steam flow at that load is 73,000 lb per hr.

2–46. Same as Prob. 2–45 except use 15% investment charges and 40¢ per million Btu fuel cost.

2–47. When purchasing a boiler to supply steam to a 33,000-kw, preferred standard, turbine generator, an allowance of 45 psi was included for piping friction at a load of 30,000 kw and a steam flow 265,000 lb per hr. Rated conditions at the entrance to the turbine are 850 psig and 900 F. When the drawing was prepared for the schedule 80 steam line, it was found that the line would contain 155 ft of pipe, 11 long-radius elbows, 1 check valve, 1 gate valve, and 2 on-run tees. The installed cost of 100 ft of piping including valves, fittings, and insulation was estimated at 700(d)^{1.5}$, where d is the nominal pipe diameter. If fuel costs 37¢ per million Btu when the plant operates at an average equivalent to 30,000 kw for 4700 hr per yr, what is the most economical pipe size when the fuel input to the boiler is reduced 5.8 Btu per kwhr for each 10-psi increase in turbine inlet pressure due to a reduction in pressure drop?

BIBLIOGRAPHY

BUFFALO FORGE Co. *Fan Engineering.* Buffalo, 1949.

NATIONAL BUREAU OF STANDARDS. *Tables of Thermal Properties of Gases.* Circular 564.

BINDER, R. C. *Fluid Mechanics.* Englewood Cliffs, N. J.: Prentice-Hall, Inc., 1955.

CRANE Co., *Flow of Fluids Through Valves, Fittings, and Pipe.* Technical Paper No. 409. Chicago, 1942.

GOODENOUGH, G. A. *Principles of Thermodynamics.* New York: Henry Holt & Co., Inc., 1920.

KENT, R. T. *Mechanical Engineers' Handbook.* New York: John Wiley & Sons, Inc., 1950.

MARKS, L. S. *Mechanical Engineers' Handbook.* New York: McGraw-Hill Book Co., Inc., 1951.

MOODY, LEWIS F. "Friction Factors for Pipe Flow," *Trans. ASME,* Vol. 66, No. 8, 1944.

VENNARD, JOHN K. *Elementary Fluid Mechanics.* New York: John Wiley & Sons, Inc., 1954.

ASME Codes for:
   Pipe and Pipe Fittings
   Pressure Piping
   Unfired Pressure Vessels
   Boiler Construction

CHAPTER 3

PUMPS

3–1. General. The purpose of a pump is to transfer a fluid from a region of low pressure to another region at the same or higher pressure. Because of the many requirements in the power plant for transferring fluids, the pump is probably the most important auxiliary of the steam power plant.

There are several ways of classifying pumps, but the most useful method is based on the principle of operation of the pump; that is,

(1) Reciprocating   (2) Rotary
   (a) Direct-acting   (3) Jet
   (b) Indirect-acting   (4) Centrifugal

Reciprocating and rotary pumps may also be grouped together under the general title of *positive displacement pumps*.

3–2. Hydraulic Definitions. *Velocity head,* $h_v = V^2/2g$, *static head,* $h_s = Z$, and *pressure head,* $h_p = Pv$ or P/ρ, have been discussed in connection with the GE and Bernoulli equations, but we should note again that they must be expressed in units of *feet of the fluid being pumped.*

Total dynamic head or *dynamic head* is the sum of the pressure and velocity heads at a given section stated in units of *feet of the fluid flowing.* Obviously, the dynamic head of a fluid may change from point to point in a conduit. Therefore, the values of pressure and velocity heads used to determine the dynamic head must both be measured at the same section.

Total dynamic suction lift is applied to pumps handling cold water and is the reading of a manometer or vacuum gage (converted to feet of the fluid flowing). The manometer is connected to the suction pipe at the pump inlet; its reading is corrected to the pump centerline, and the velocity head at the point of attachment of the manometer is added.

Net positive suction head (NPSH) is the difference between the absolute dynamic pressure of the liquid measured at the centerline of a pump and the saturation pressure corresponding to the temperature of the liquid at the same point, all expressed in terms of *feet head of the fluid flowing.* For practical purposes, the NPSH can never be a negative value, for the liquid would then flash and vapor-bind the pump. NPSH may also be defined as the pressure at the pump suction flange, corrected to the pump centerline, that prevents vaporization of the water.

Developed head (DH) is the difference between the sum of the absolute pressure head and velocity head (or absolute dynamic head) at the outlet of the pump and the sum of the absolute pressure head and velocity head (or absolute dynamic head) at the inlet, both corrected to the centerline of the pump and expressed in *feet head of the fluid*. This definition is nothing more than a statement of the GE equation. Refer to the schematic diagram, Fig. 3–1, and apply the GE equation as stated in

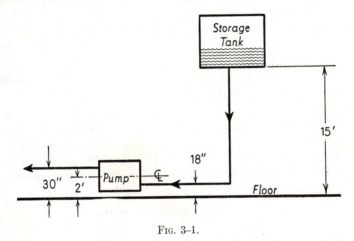

F$_{IG}$. 3–1.

Eq. (2–2). We can assume that there is no heat added to or rejected by the water flowing through the pump and that there is no change in the internal energy. Then

$$Wk = w\left(\frac{P_1 - P_2}{\rho} + \frac{V_1{}^2 - V_2{}^2}{2g} + Z_1 - Z_2\right) \qquad (a)$$

Inspection of this formula will reveal that work will be a negative value in all applications to a pumping process. However, this is in accordance with our original definition of work in the GE equation, i.e., work done on a fluid is a negative value. Work input to a pump is considered as a positive value with the understanding that this is a change in the original definition purely for the sake of convenience. Also it is customary to correct suction and discharge pressures to the pump centerline in order to eliminate any differences in static elevation $(Z_1 - Z_2)$. Then,

$$Wk = w\left(\frac{P_2 - P_1}{\rho} + \frac{V_2{}^2 - V_1{}^2}{2g}\right) = w(H_2 - H_1) = w(\mathrm{DH}_f) \qquad (3\text{–}1)$$

where Wk = work per min or power, ft-lb per min

w = weight of fluid flowing, lb per min

P = pressure corrected to the pump centerline, psfa

V = velocity, fps

ρ = fluid density, lb per cu ft

H = dynamic head corrected to pump centerline, ft

DH_f = developed head, ft

and subscripts *1* and *2* indicate suction and discharge, respectively.

The definition of developed head is obtained directly from Eq. (3-1).

The expression *feet head of the fluid* has been used extensively throughout the above definitions and will bear further discussion. There is a distinct difference between pressure expressed in terms of pounds per square inch and pressure expressed in terms of feet of the fluid. Note that the GE and Bernoulli equations do not use pressure to represent head but use the term Pv for head. For example, assume that a fluid is flowing through a conduit under a pressure of 100 psia. If this fluid is water at 60 F, the pressure head will be $h_p = Pv = 100 \times 144 \times 0.01604 = 231$ ft of 60 F water. If the fluid is water at 280 F, then the pressure head will be $h_p = 248.5$ ft of 280 F water. If the fluid is an oil with a specific gravity of 0.70, then the pressure head will be $h_p = (100 \times 144 \times 0.01604)/0.70 = 330$ ft of oil. A convenient formula for conversion from pounds per square inch to pressure head, deduced from the above, is

$$h_p = \frac{p\,144 \times 0.01604}{\text{sp gr}} = \frac{2.31p}{\text{sp gr}} \qquad (3\text{-}2)$$

where p = pressure, psi

sp gr = specific gravity of the fluid referred to water at 60 F

Values of specific gravity and pressure head for water at various temperatures may be obtained from the Appendix, Plate 2.

3-3. Pump Horsepower. Since power is the rate at which work is done, Eq. (3-1) may be used to determine the *hydraulic horsepower* of any pump. Hydraulic horsepower is that power required to deliver a given quantity of fluid against a given developed head, with no losses in the pump.

$$\text{Hydraulic hp} = \frac{w}{33,000}\,DH_f \qquad (3\text{-}3)$$

$$= \frac{\text{gpm} \times DH_f \times \text{sp gr} \times 8.33}{33,000} = \frac{\text{gpm} \times DH_f \times \text{sp gr}}{3960} \qquad (3\text{-}4)$$

$$= \frac{\text{gpm} \times DH_p \times 2.31 \times 8.33}{33,000} = \frac{\text{gpm} \times DH_p}{1714} \qquad (3\text{-}4a)$$

$$= \frac{W \times \mathrm{DH}_p \times 2.31}{60 \times 33{,}000 \times \mathrm{sp\ gr}} = \frac{W \times \mathrm{DH}_p}{857{,}140 \times \mathrm{sp\ gr}} \qquad (3\text{-}4b)$$

where w = weight flowing, lb per min
W = weight flowing, lb per hr
DH_f = developed head, ft
DH_p = developed head, psi
8.33 = weight of 60 F water, lb per gal
sp gr = specific gravity of the fluid referred to water at 60 F
gpm = gal per min

The conversion from pounds per hour to gallons per minute is obtained from

$$\mathrm{gpm} = \frac{W}{60 \times 8.33 \times \mathrm{sp\ gr}} = \frac{W}{500 \times \mathrm{sp\ gr}} \qquad (3\text{-}5)$$

$$\mathrm{Shaft\ hp\ or\ bhp} = \frac{\mathrm{hydraulic\ hp}}{\mathrm{pump\ efficiency}} \qquad (3\text{-}6)$$

EXAMPLE 3-1. For the pump of Fig. 3-1, determine: (a) velocity, pressure, and dynamic heads in suction and discharge pipes, (b) developed head, (c) NPSH, (d) hydraulic hp, (e) bhp. Data: 200 gpm, 4-in. schedule 40 suction connection, 3-in. schedule 40 discharge pipe, 110-psig discharge pressure, saturated water at 10 psia in the storage tank, 4-ft. friction loss in suction pipe, 70% pump efficiency.

SOLUTION. The most severe operating condition for this pump will occur when the water level is at the bottom of the tank. From Plate 1 (Appendix) and Fig. 2-20, or by calculation,

(a) Velocity in suction pipe = 300 fpm, velocity head in suction pipe = 0.4 ft, velocity in discharge pipe = 520 fpm, velocity head in discharge pipe = 1.2 ft. Pressure head at pump discharge corrected to pump centerline equals

$$(110 + 14.7)2.39 + (2.5 - 2.0) = 298.5 \text{ ft abs}$$

Pressure head or static pressure at pump suction corrected to pump centerline equals (absolute pressure at water level) + (pressure due to elevation of the water level above pump centerline) − (friction loss) − (velocity head), or

$$10 \times 2.39 + (15 - 2.0) - 4 - 0.4 = 32.5 \text{ ft abs}$$

Dynamic head at pump discharge = H_2 = 298.5 + 1.2 = 299.7 ft abs; dynamic head at pump suction = H_1 = 32.5 + 0.4 = 32.9 ft abs.

(b) Developed head = DH_f = 299.7 − 32.9 = 266.8 ft

(c) NPSH = absolute dynamic suction pressure at pump centerline minus the vapor pressure of the water = 32.9 − 10 × 2.39 = 9 ft.

(d) Hydraulic hp, Eq. (3-4) = $\dfrac{200 \times 266.8 \times 0.966}{3960}$ = 13.0

(e) bhp, Eq. (3-6) = $\dfrac{13.0}{0.70}$ = 18.6

3–4. Descriptions of Reciprocating Pumps. (a) *Types.* For many years reciprocating pumps were the only type of pump used in power stations. Today they are used only where the flow is too small for the satisfactory design of centrifugal pumps or where there is moderate capacity with high developed heads. Approximate ranges of applications of reciprocating and centrifugal pumps are indicated in Fig. 3–2. Specific locations of reciprocating pumps in modern power plants include condensate and boiler feed pumps in small installations, evaporator-feed pumps, fuel-oil pumps, chemical pumps for feedwater treatment, etc.

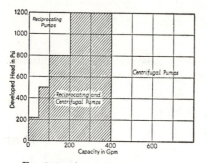

Fig. 3–2. Approximate comparative service ranges for centrifugal and reciprocating pumps.

Among the advantages of reciprocating pumps are their flexibility of operation, their nearly constant efficiency for wide changes in capacity and head, and their ability to handle small volumes at high heads. Disadvantages include oil contamination of the steam, packing and valve troubles, pulsating flow and head, higher cost, greater floor space, and complexity of moving parts.

Reciprocating pumps have been classified previously as (a) direct-acting and (b) indirect-acting. This is a distinction in the method of driving the water pistons or plungers. The direct-acting pump of Fig. 3–3 is motivated by the steam force on the steam piston. The steam piston is directly connected to the water piston by a rod, so that any force exerted on the steam piston is transmitted to the water piston. Compare this with the indirect-acting or power pump shown in Fig. 3–4, where the power for driving the water piston is obtained from an electric motor through gears. Internal-combustion engines and steam engines may also be used to drive indirect-acting pumps. It should be noted that power-driven pumps have high efficiency, and because of the type of drive, are of constant speed. Thus they have the ability to deliver a constant quantity of fluid against variable heads. This is a feature neither direct-acting nor centrifugal pumps possess. The discharge pressure of the power-driven pump will become high enough to break the pump if the discharge valve is closed during operation. For this reason, power pumps should be equipped with relief valves on the discharge.

Water and steam pistons of direct-acting pumps are typically double-acting, i.e., work is performed on both faces of the pistons, and the fluids are in contact with both faces of their respective pistons. Therefore, water is pumped during each stroke of the pump. Indirect-acting pumps may

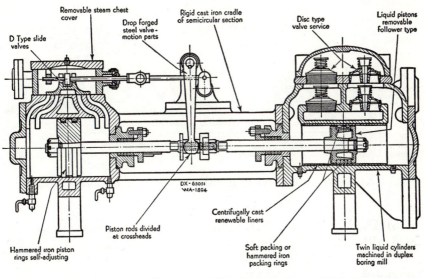

FIG. 3–3. Direct-acting duplex pump. (Worthington Pump & Machinery Corp.)

be either single-acting or double-acting. The pump shown in Fig. 3–4 is double-acting.

Another classification of reciprocating pumps indicates the number of water cylinders contained in the pump. Those with only one water cylinder are known as *simplex pumps* (Fig. 3-5); those with two water cylinders are *duplex pumps* (Fig. 3–6); those with three water cylinders, *triplex pumps*; etc. When the pump is direct-acting, there must be a steam cylinder for each water cylinder.

Cylinder and stroke dimensions for reciprocating pumps are always designated in inches with the diameter of the steam cylinders first, the liquid cylinders next, and the stroke last. A $6 \times 4 \times 10$ pump would have 6-in. diam steam cylinders, 4-in. diam water cylinders, and a 10-in. stroke of both.

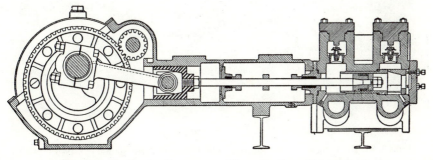

FIG. 3–4. Indirect-acting pump. (Dean Brothers Pumps, Inc.)

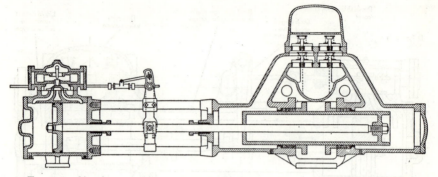

FIG. 3–5. Simplex, outside-center-packed, plunger pump. (Dean Brothers Pumps, Inc.)

(b) *The Liquid End.* Water cylinders of reciprocating pumps may be fitted with either *pistons* or *plungers*, depending on the location of the packing. In Fig. 3–3 the packing for the water piston is fastened to the piston and is carried along with it throughout the stroke. Packing may be made of iron, leather or rawhide, flax, etc.; the piston moves in a cylinder liner. When the packing is stationary, as in Figs. 3–5 and 3–7, the pump is of the plunger type. The advantage of the plunger pump is that any undue leakage of fluid past the packing may be readily noticed, since the packing is on the outside of the pump.

Plunger pumps can be obtained in two designs, *outside-center-packed* (Fig. 3–5) and *outside-end-packed* (Fig. 3–7). Outside-end-packed pumps have only two points at which packing is required, whereas outside-center-packed pumps are packed at three points and are slightly more difficult to pack. The two plungers of an outside-end-packed pump are connected by side rods (Fig. 3–6). When the casing of an outside-end-packed pump is made of heavy forged steel, the pump may be used for

FIG. 3–6. Plan view of a duplex, outside-end-packed plunger pump. (Dean Brothers Pumps, Inc.)

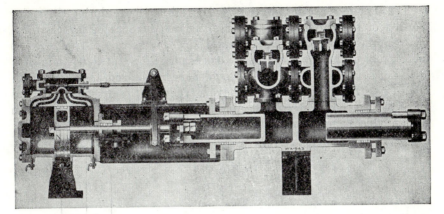

FIG. 3–7. Duplex, outside-end-packed plunger pump with side pot valves. (Worthington Pump & Machinery Corp.)

pressures up to 15,000 psi. Piston pumps and outside-center-packed pumps are seldom used for pressures of more than a few hundred pounds.

(c) *Steam End of a Direct-acting Duplex Pump.* Design features of the steam ends of duplex pumps are similar, regardless of the type of liquid ends employed. Very often the steam valves are of the **D** slide-valve type or the **B** slide-valve type. These names are derived from the shape of the valve. Steam valves for the duplex pumps of Figs. 3–3 and 3–7 are the **D** type.

Steam enters the steam chest through the flanged connection at the upper left side, Fig. 3–3, and surrounds the **D** slide valve located in the steam chest. When the valve uncovers either of the ports or passages at the extreme ends of the cylinders, steam will be admitted to the cylinder through the ports. The two middle ports are for exhaust, which leaves the pump through the large triangular section in the center of the steam end. Each steam cylinder has its own set of passageways and its own slide valve. However, the slide valve for the right-hand cylinder (right-hand cylinder is so designated when viewing the pump from the steam end) is operated by the rod and linkage for the left-hand cylinder, Fig. 3–6. When the slide valve shown in Fig. 3–3 (assumed to be the left-hand cylinder) is moved to the right by the piston rod for the right-hand cylinder, steam is admitted to the head end of the cylinder shown. The exhaust port for the inboard end of the steam cylinder shown is connected to the exhaust chamber. The left-hand piston shown in the figure will move toward the right. When the position of the slide valve is reversed by the piston rod for the right-hand cylinder, steam is admitted to the inboard end of the cylinder and the outboard end is connected to the exhaust. The steam piston will therefore move to the left.

The **D** slide valve for either cylinder will not move during the major portion of the stroke of the other piston rod because of the clearance between the nut on the valve rod and the lugs on the valve. Thus, the two piston rods of the duplex pump are out of phase, i.e., the left-hand piston does not start its stroke until the right-hand piston rod has nearly completed its stroke. Duplex pumps have a tendency toward uneven lengths of stroke, called short stroking, but the alternate motion of the piston rods produces a nearly constant delivery from the pump.

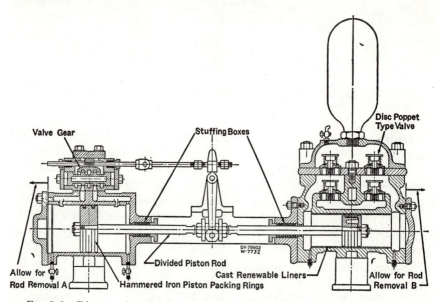

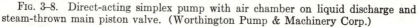

Fig. 3–8. Direct-acting simplex pump with air chamber on liquid discharge and steam-thrown main piston valve. (Worthington Pump & Machinery Corp.)

D slide valves are limited to steam temperatures of 450 F. At higher temperatures the **D** slide valves have a tendency to warp with resulting steam leakage.

3–5. Calculations of Cylinder Sizes of Direct-acting Pumps. Steam pressure acting on the face of the piston of a direct-acting pump produces the force which causes the piston to move. Since there is no flywheel attached to the mechanism, the steam pressure must be sufficient to move the piston throughout the entire stroke. Such an engine, in which steam is admitted with 100% cutoff, is known as a nonexpansion engine. Indicator cards from the steam cylinder of a direct-acting pump during one cycle are theoretically rectangular in shape, as are the indicator cards from the water cylinder. Actual cards are close approximations to the theoretical Because of the inability of the engine to take advantage of the internal energy of the steam through an expansion process, the work done by the steam is the PV product. Thus, the nonexpansion engine will give maximum power for a given bore and stroke but will have extremely poor thermal efficiency.

The force exerted on the piston rod is equal to the piston area times the difference in steam pressure acting on the two faces of the piston. Thus,

$$F = (p_i - p_e)(d_s)^2 \frac{\pi}{4} \qquad (a)$$

in which F = force, lb
$\qquad p_i$ = inlet steam pressure, psia
$\qquad p_e$ = exhaust steam pressure, psia
$\qquad p_i - p_e$ = mean effective pressure
$\qquad d_s$ = steam piston diam, in.

Only a part of this force is transmitted to the liquid piston because of the frictional resistance of the packing around the shaft and possibly because of some leakage of steam past the piston rings. The force transmitted to

TABLE 3–1

MECHANICAL EFFICIENCIES OF DIRECT-ACTING PUMPS

Stroke, In.	Piston Type	Plunger Type	Stroke, In.	Piston Type	Plunger Type
3	50	47	8	70	66
4	55	52	10	75	71
5	60	57	12	77	73
6	65	61	18	82	78

the liquid piston will be $E_m F$, where E_m is the mechanical efficiency, Table 3–1. This force must be sufficient to increase the liquid pressure from suction to discharge pressures, or

$$E_m F = (p_d - p_s)(d_L)^2 \frac{\pi}{4} \qquad (b)$$

where p_d = discharge pressure, psia
$\qquad p_s$ = suction pressure, psia
$\qquad d_L$ = diam of the liquid piston, in.,

and other symbols are as in Eq. (a).

Then $\qquad E_m F = E_m \frac{\pi}{4}(d_s)^2 (p_i - p_e) = \frac{\pi}{4}(d_L)^2 (p_d - p_s) \qquad$ (c)

$$d_s = d_L \sqrt{\frac{p_d - p_s}{E_m(p_i - p_e)}} \qquad (3\text{–}7)$$

Eq. (3–7) indicates that the diameter of the steam piston is a direct function of the diameter of the liquid cylinder, but the liquid cylinder diameter, obviously, must be determined from the quantity of liquid, length of stroke, and number of strokes per minute. A stroke may be defined as the travel in one direction of the connecting rod from one extreme position to the other extreme position. A piston will require two strokes, or one cycle, to return to its original starting point.

Maximum piston speed, or strokes per minute, at which a pump may operate is determined by the strains imposed on the mechanism and the wear that takes place. Boiler feed pumps must have reliability of service and long wearing qualities, and they should, therefore, operate at relatively slow speeds. Pumps used for intermittent service, such as fire pumps or pumps operated during an emergency, may operate at double the speed of boiler feed pumps. Recommended maximum continuous operating speeds for feedwater service are given in the Appendix, Plate 12.

Piston displacement for a multicylinder pump is

$$PD = \frac{\frac{\pi}{4}(d_L)^2 LSn}{231} = \frac{LSn(d_L)^2}{294} \tag{3–8}$$

where PD = piston displacement, gpm
$\quad\quad d_L$ = liquid piston diam, in.
$\quad\quad L$ = length of stroke, in.
$\quad\quad S$ = strokes per min
$\quad\quad n$ = number of cylinders

Even when a reciprocating pump is in good condition, it will not discharge a quantity of liquid equal to the piston displacement because of leakage past the packing and valves. Also, the pistons do not always travel the same distance during each stroke of a duplex pump. Using the customary definition of *volumetric efficiency* (E_v) as the ratio of the actual volume discharged to the piston displacement, *slip is one* minus *the volumetric efficiency.*

$$\text{Slip} = 1 - E_v \tag{3–9}$$

Modifying Eq. (3–8), the actual discharge from the pump is

$$\text{gpm} = PD\,(1 - \text{slip}) = \frac{LSn(d_L)^2(1 - \text{slip})}{294} \tag{3–10}$$

Pertinent data for the selection of reciprocating pumps are listed in the Appendix, Plate 12. Actual discharge for the pumps listed may be determined from Eq. (3–10). The magnitude of the slip will depend on the type of valve and the condition of the pump. Slip will vary from 2% for pumps in good condition to as much as 40% or more for pumps in poor repair. Taking into account the types of valve employed in pumps for power-plant service and the average state of repair of power-plant pumps, slip may be estimated at 10 to 20% when selecting the pump. It is always wise to choose a reciprocating pump that is slightly too large, rather than one that will just meet the capacity conditions.

3–6. Rotary Pumps. One design of rotary pump using double helical gear teeth is shown in Fig. 3–9. This pump is built for counterclockwise

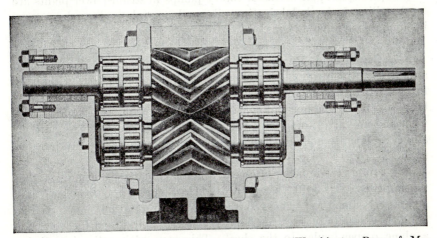

Fig. 3–9. Rotary pump with double helical gears. (Worthington Pump & Machinery Corp.)

rotation when facing the end of the pump having the shaft extension. In that position, the suction would be on the right side of the pump and the discharge on the left side.

Fluid entering the pump is trapped in the voids between gear teeth and is carried around the pump next to the casing to the discharge side. There it is compressed by the discharge pressure and expelled.

Double helical gear teeth have two advantages: (1) there is no end thrust, and (2) none of the fluid is compressed between the top land of one gear and the bottom land of the other when the gears mesh. Any fluid trapped between the two lands has its pressure relieved by the channel effect of the helical teeth leading back to the discharge space. Leakage from the discharge to the suction can be reduced by cutting the teeth with little or no root clearance.

Gear pumps formed from spur teeth tend to be noisy and have low efficiency, particularly at high speeds, because fluid is trapped between the gear lands when the gears mesh.

Other rotary pumps have impellers formed as lobes, eccentric pistons, eccentric guide vanes, or single and double screws.

Rotary pumps have some of the characteristics of reciprocating pumps in that they are positive displacement and that they will develop dangerously high pressures if operated against a closed discharge. However, they are particularly suitable for small capacities. Since the rotary pumps have a constant, nonpulsating delivery and operate at higher speeds than the reciprocating pumps (up to 1800 rpm), they have some of the characteristics of centrifugal pumps.

The principal applications of rotary pumps in steam-power plants are for handling fuel oil, lubricating oils, and control fluids such as noninflammable synthetic resins.

3–7. Centrifugal Pumps. Centrifugal pumps have been adopted almost exclusively for all power-plant pumping applications within the limits of their satisfactory design (see Fig. 3–2). Their popularity has been due to the simplicity of construction (there is only one rotating element and no valves to adjust), their small space requirements, and their ability to operate at high speeds when directly connected to turbines or motors.

Fluid enters the impeller of a centrifugal pump at the *eye*, or center, of the impeller, Fig. 3–10. There it is picked up by the impeller vanes and travels along the vanes. With the impeller rotation shown on Fig. 3–10, the absolute path of the fluid resembles a spiral, and the flow is in a plane perpendicular to the shaft. This type of impeller is called a *radial-flow* impeller. The fluid pressure and velocity are increased during travel through the impeller. After leaving the impeller, the fluid enters the volute where a major part of the velocity energy is transformed to pressure energy. The fluid leaves the pump through the outlet nozzle with some velocity energy and with more pressure energy than it had upon entering the pump.

The increase in pressure energy acquired as the fluid flows through the impeller is due to the centrifugal force exerted on the fluid and is a function of the velocity squared. Since the kinetic energy developed by the impeller is also a function of the velocity, the total pressure developed is proportional to the velocity squared.

Water entering the eye of the impeller axially has its direction changed by the shape of the impeller, so that it enters the vane section with approximately radial velocity, vector V_1 of Fig. 3–10. Vector U_1 is the velocity of the eye of the impeller; thus, the relative velocity of the water with respect to the vane is C_1. The angle β_1 is determined by the relative velocity vector and must be the angle of the vane at the point of contact with the water to prevent a loss from shock.

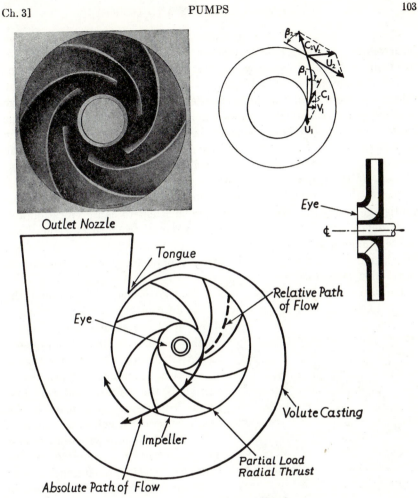

FIG. 3–10.   A volute-type centrifugal pump.

At the periphery of the impeller the water leaves in the direction indicated by the vane producing angle β_2 and vector C_2 relative to the rotor. By combining the velocity of the impeller periphery U_2 with the relative liquid velocity, we obtain the absolute velocity V_2. As might be expected, the direction and magnitude of the vector C_2 of an actual impeller are somewhat different from that for the theoretical vector, due to imperfect hydraulic conditions and to turbulence.

The velocity energy of the stream at the impeller periphery must be substantially converted into pressure energy if suitable efficiency is to be obtained. The two methods of accomplishing this conversion, volute casings and diffusor casings, provide one means of classifying centrifugal pumps.

The pump shown in Fig. 3–10 is a volute pump wherein the conversion from velocity to pressure energy takes place only in the outlet nozzle of the volute. At rated capacity, approximately the point of highest efficiency, the average velocity throughout the volute is constant. Then the velocity is reduced at the outlet nozzle by a divergence in the passage ranging from 8 to 13 deg. At flows other than rated capacity, the average velocity in the volute is not constant because, with the reduced flow, the area of the volute is too large and energy conversion takes place as the velocity decreases. This results in pulsating pressures and a radial thrust whose resultant is toward the volute tongue, as shown in Fig. 3–10.

Radial thrusts created in small pumps are unimportant, but in large multistage pumps the radial thrusts are sufficient to cause vibration and shaft deflections that create rubbing of the wearing rings. High-head volute pumps used in power plants employ twin volutes for each stage. By placing the two volutes diametrically opposed, the radial thrust at partial flow is balance.

Diffusors, Fig. 3–11, are fastened to the casing and surround the impellers. The expanding passages of the diffusor have an included angle of about 11 degrees and convert much of the velocity energy into pressure energy with negligible radial thrust. A collecting chamber, of volute shape, surrounds the diffusor and the final energy conversion takes place in the nozzle outlet of the collecting chamber.

FIG. 3–11. A diffusor for a centrifugal pump. (Worthington Pump & Machinery Corp.)

Multistage pumps with diffusor casings have certain manufacturing advantages. Optimum efficiencies may be maintained for a wide variation in capacity, without changing casing patterns, by modifying the diffusors. In addition, the casing may be smaller for large impeller diffusor designs than for efficient volute designs. Also it is easier to machine diffusors than the volutes of multistage pumps. For single-stage pumps, the volute design is smaller and cheaper to manufacture. Commercially, volute and diffusor multistage power-plant pumps have about equal efficiencies at design capacity, but the efficiency curve for the volute pump is flatter, giving slightly better efficiency at partial capacities.

Further consideration of the vector diagram for the entrance of the water into the impeller will show why some centrifugal pumps must operate at low speeds. Horizontal condensate pumps located immediately below a condenser, as in Fig. 1–2, are one example of low-speed pumps. The NPSH of such a pump is small because excavation expense prevents a large static head from the pump centerline to the minimum condenser water level.

Kinetic energy to establish the inlet velocity V_1, Figs. 3–10 and 3–12, must not exceed the NPSH, since the water cannot be allowed to flash into

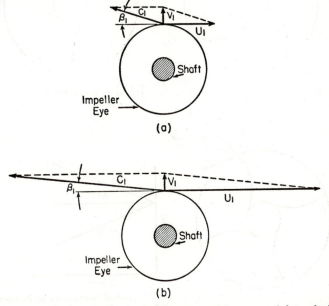

FIG. 3–12. Vector diagrams showing the effect of low pump-inlet velocity on the inlet vane angle β_1 for the same impeller at different speeds. (a) Low-speed impeller; (b) high-speed impeller.

steam. Thus with low NPSH there must be a low entering velocity V_1. Consider the impeller eye operating at low speed as in Fig. 3–12(a), and note that a reasonable value of vane angle β_1 is obtained. If the speed of the same impeller is increased as in Fig. 3–12(b), angle β_1 becomes impractically small for the given low velocity V_1 and the high peripheral velocity U_1. Thus, pumps operated with low values of NPSH (1 to 3 ft for horizontal condensate pumps) must operate at low speeds.

3–8. Characteristics of Centrifugal Pumps. Curve shapes for the characteristics of centrifugal pumps are a function of the inclination of the impeller vane employed. Most impellers have backward-curved vanes, but the vanes may also be straight, radial, or forward-curved, Fig. 3–13. Fans are frequently manufactured with forward-curved vanes.

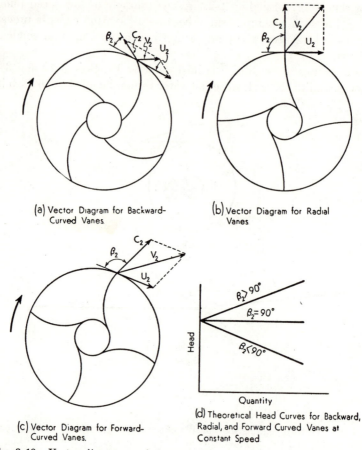

(a) Vector Diagram for Backward-Curved Vanes

(b) Vector Diagram for Radial Vanes

(c) Vector Diagram for Forward-Curved Vanes.

(d) Theoretical Head Curves for Backward, Radial, and Forward Curved Vanes at Constant Speed

Fig. 3–13. Vector diagrams and theoretical head curves for radial and backward- and forward-curved impeller vanes.

The inlet of the vane must correspond with the resultant of the vector diagram as previously discussed. But the outlet angle β_2 will determine the magnitude of the vectors in the outlet triangle and the shape of the characteristic curve. Backward-curved vanes are those with β_2 less than 90 deg. The angle for radial vanes is 90 deg; for forward-curved vanes it is greater than 90 deg. It can be shown that the head developed by a centrifugal pump is proportional to the tangential component of the resultant vector in the outlet diagram.* Since the relative vector (C_2) is proportional to the flow of fluid out of the impeller, the relative velocity will be zero at zero flow. The point on the head-capacity curve where there is no flow is referred to as the shut-off head, since this is the con-

* A. H. Church, *Centrifugal Pumps and Blowers* (New York: John Wiley & Sons, Inc., 1944), p. 38.

dition resulting from a closed discharge valve. As the discharge valve is opened and flow increases, C_2 becomes larger and the tangential component of the absolute leaving velocity (V_2) becomes smaller for backward-curved vanes. Thus, the theoretical head curve for backward-curved vanes decreases as the capacity increases, Fig. 3–13(d).

Following the same reasoning, the head-capacity curve for radial vanes is constant for varying capacity because the tangential component of the absolute leaving velocity is constant regardless of capacity. Similarly, the head increases with capacity for forward-curved vanes.

The curves of Fig. 3–13(d) have been referred to as theoretical, since it is obvious that there are losses in the impeller and volute or diffusor due to turbulence and friction. It has also been found that there is circulatory flow between the vanes which further reduces the head obtained from the impeller. An actual head-capacity curve for an impeller with vanes inclined backwards is shown in Fig. 3–14, together with the effects of the losses that occur in the impeller.

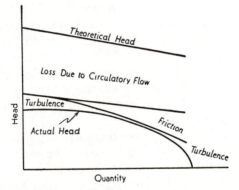

Fig. 3–14. Theoretical and actual head-capacity pump curves for constant speed.

Pumps for boiler feed service, or any service where pumps may be required to operate in parallel throughout the head-capacity curve, should be required to have a constantly drooping head curve from shut-off to design flow, Fig. 3–15. Curve A represents a stable characteristic, while curve B is for a pump that is unstable when operated at any head in excess of the shut-off head. The latter pump is unstable because in some cases it could deliver two different capacities at a given head (see parallel operation of fans, Art. 7–6). The shape of the vane at the impeller periphery may cause a backward-curved vane pump to have a characteristic such as B of Fig. 3–15.

The rise in the head curve from design point to shut-off for the types of pump used for boiler feed service varies from 10 to 25% of the head at design conditions. A smaller rise than 10% would be undesirable, for a slight change in the pressure required by the system would mean a large change in the quantity of fluid delivered by the pump. Also, a pump having 10% or less rise is probably oversized and is not rated for its maximum capacity. When a pump has more than 20% rise, the shut-off

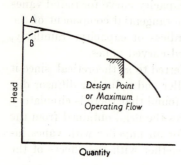

pressure may be so high that a thick-walled pipe will be required to withstand the pressure. A rise that is too high indicates that the pump may be over-rated and may operate near its cavitation point. Therefore a rise of about 15 to 20% indicates good design.

Fig. 3–15. Stable A and unstable B head curves at constant speed.

Another important matter to discuss in connection with characteristic curves is the speed at which the pump operates. Curves are generally plotted for constant speed, but pumps seldom, if ever, operate at constant speed. If the pump is driven by an electric motor, the speed of the motor will decrease 1 to 1.5% of synchronous speed from no load to full load, depending on size, etc. Thus, if a pump curve indicated the shut-off head at 3550 rpm, the actual shut-off head with a motor drive would occur at a speed of approximately 3600 rpm, and the head would be about 2.8% higher than at 3550 rpm. So-called *constant-speed turbines* have similar or even greater speed variations from full load to no load.

All the characteristic head curves that have been shown had the units of *head in feet* and *capacity in volume* for coordinates. This is in accordance with general custom, since a centrifugal pump *displaces volume of fluid against feet head of the fluid pumped.* That statement cannot be emphasized too strongly, since it is the basis for the calculation of many pump problems. As noted before, there is a vast difference between volume flow and weight flow and between head in feet and head in pounds per square inch. A given volume of flow will represent fewer pounds per hour at a high temperature than at a low temperature. Likewise, a given head in feet will represent fewer pounds per square inch at a high temperature than at a low temperature. But regardless of the temperature of the fluid, a centrifugal pump will displace a definite volume against a certain feet of head at a particular point on its characteristic curve.

From Eq. (3–4), water horsepower or brake horsepower varies as the specific gravity when flow is expressed in gallons per minute and head in feet. Therefore, the *horsepower is a function of temperature.*

EXAMPLE 3–2. Determine the rated capacity in pounds per hour and head in pounds per square inch for the pump in Fig. 3–16 when the water is at 350 F.

SOLUTION. From Eq. (3–5),

$$\text{Flow} = \text{gpm} \times 500 \times \text{sp gr}$$
$$= 450 \times 500 \times 0.890$$
$$= 200{,}250 \text{ lb per hr}$$

$$\text{Head} = \frac{\text{ft}}{\text{ft/psi}} = \frac{1000}{2.59} = 386 \text{ psi}$$

From Eqs. (3–4) and (3–6),

$$\text{bhp} = \frac{\text{gpm} \times \text{ft head} \times \text{sp gr}}{3960 \times \text{efficiency}}$$
$$= \frac{450 \times 1000 \times 0.890}{3960 \times 0.71} = 142.4$$

By calculating several points, the horsepower curve could be replotted for water at this temperature.

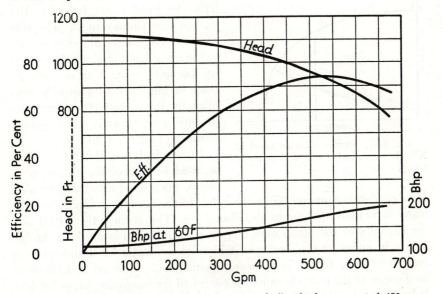

FIG. 3–16. Performance curves for a four-stage boiler feed pump rated 450 gpm, 1000 feet head, 71% efficiency, and 3550 rpm.

NPSH is the maximum head available for conversion into kinetic energy at the inlet of the first stage of the pump, or the head available to produce the absolute velocity V_1. Of course, the manufacturer cannot use all this head for conversion into velocity energy, for any slight inaccuracies in manufacture would cause the liquid to flash. However, the NPSH is an indication of the energy available for introduction of the fluid into the impeller.

The phenomenon of flashing in the pump is not unusual, but it limits the capacity and life of the pump. When the liquid flashes, small cavities of vapor form in the liquid stream. As the cavities reach a region of higher pressure in their travel through the impeller, they collapse or implode with accompanying shock on the confining walls. These implosions cause a noise sounding like so many pebbles being forced through the pump and cause pitting or erosion of the impeller. Flashing in the pump is called *cavitation*; it reduces the flow because of the large volume occupied by the vapor compared with that occupied by an equal weight of water.

Refer to Fig. 3–17 and note the head curve *AB* which may be obtained by operating the pump with the required NPSH as indicated by curve *CD*. Reduction of the NPSH from *D* to *G* will cause the maximum capacity of the pump to be reduced in accordance with the dashed curve beginning at *E*, called the *break-off* curve. Still further reduction in the NPSH will cause cavitation at a lower capacity, with the head curve following the dashed line similar to *F*. The efficiency of the pump will reduce to zero at the same flow for which the capacity becomes zero. Curves such as those of Fig. 3–17 are normally supplied by the manufacturer for condensate pumps and other pumps operating under low NPSH.

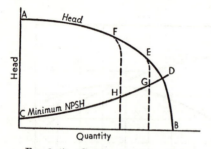

FIG. 3–17. Curves showing effect of NPSH on capacity of a centrifugal pump.

In one instance, the NPSH may be used to control the output of a pump. Such self-control may be used for a condensate pump. When the inflow to the condenser diminishes, the water level above the pump also diminishes. This reduces the NPSH and the capacity of the pump until a state of equilibrium is reached. An increase in inflow to the condenser will raise the water level and increase pump capacity. In this instance, cavitation has been found to do little or no damage to the pump because the energy of the imploding vapor bubbles is low at the low absolute pressures encountered in condensate pumps. No special metals are used for these pumps, but the impeller vanes extend well into the eye to pick up the water quickly and implode the vapor as soon as possible. These pumps are rated at a head about 10% greater than system head for design flow and design NPSH.

3–9. Impeller Contours. There are four types of impellers: *radial* or *conventional, Francis, mixed flow,* and *propeller* or *axial flow,* shown in that order in Fig. 3–18. The pictures of Fig. 3–19 will help to visualize a high-speed, Francis screw-vane impeller (a), a mixed-flow impeller (b), and a propeller type of impeller (c).

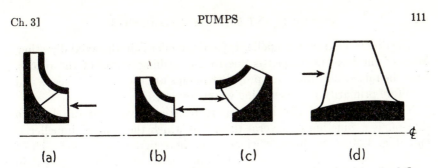

FIG. 3–18. Sections of pump impellers: (a) radial, (b) Francis, (c) mixed flow, and (d) axial flow.

Radial impellers should not be confused with radial vanes. The fluid enters each of these types of impeller in an axial direction. However, the direction of the fluid is changed in the conventional impeller from axial to radial. A large part of the impeller is radial so that the ratio of the outside diameter to the diameter of the impeller eye may be as much as 3½ to 1. Some of the head developed by the wheel is due to centrifugal force. Radial impellers are used extensively in power-plant pumps, for they will develop heads in excess of 300 psi per stage at 3600 rpm with a flat characteristic curve.

Francis impellers are similar in shape to radial impellers except for the straight radial section. With the lower discharge to eye diameter ratio for the Francis type, the developed head is also lower but, consequently, the speed may be higher. Also, the characteristic curve rises more sharply from full load to shut-off.

Mixed-flow impellers discharge the fluid in a partly radial and partly axial direction. Some of the head developed by this type of runner is due to the pushing action of the vanes and some is due to the centrifugal force. The mean discharge diameter of the impeller may be about equal to the eye diameter.

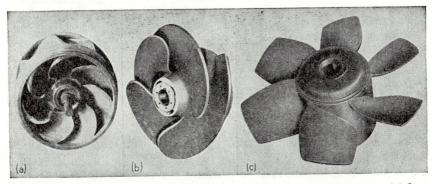

FIG. 3–19. Centrifugal pump impellers: (a) Francis, (b) mixed flow, (c) axial flow. (Worthington Pump & Machinery Corp.)

The discharge from a propeller type of impeller is in the axial direction. Most of the head developed is due to the pushing action of the blades. Like the mixed-flow impeller, propeller pumps are used for low heads and therefore primarily for circulating water service.

The percentage of rise in the characteristic curve from full load to shut-off is a function of several things, of which the type of impeller is one. Fig. 3–20 shows head, efficiency, and horsepower curves for each of the four types of impeller, plotted in percentage of full load conditions. These curves may be considered typical for each class of pump but may be varied somewhat by changes in design details. For example, the head curve may be made steeper by (1) decreasing the discharge to eye diameter ratio, (2) reducing the number of vanes in the impeller, (3) reducing the outlet angle, (4) using a narrower runner, or (5) reducing the casing area. However, changes in these factors will also change the stage efficiency. The pump designer must select a happy compromise for all these factors and produce a pump that will most nearly satisfy the customer's desires.

It will be seen from Fig. 3–20 that propeller pumps (and some mixed-flow pumps) have the interesting phenomenon of requiring maximum horsepower input at shut-off conditions. If the driver is selected for shut-off power, there is no possibility of overloading it at any other flow and constant speed. Such is not the case with radial or Francis runners. Axial-flow pumps may be arranged for starting with open discharge or at reduced speed to lessen the starting horsepower, thereby permitting a smaller motor at constant speed. However, a large motor would operate at one-half to one-third load after starting and would therefore be costly. In order to reduce the starting load on the motor, the piping system may be partially evacuated before starting the pump or, if the system static head is not too high, the pump may be arranged to start only when the discharge valve is open. Another method

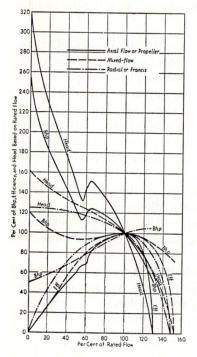

Fig. 3–20. Approximate characteristic curves for various types of centrifugal pump impeller based on percentage of the conditions at rated flow.

that can be used to reduce the starting load is to reduce the starting speed.

Francis and radial impellers may be made in a double-suction design, Fig. 3–21, to increase capacity and to improve hydraulic balance. Double-suction impellers are the equivalent of two impellers placed back to back. They have twice the capacity of single-suction wheels.

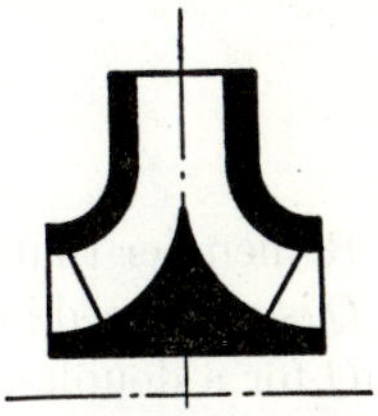

3–10. Specific Speed. The concept of specific speed has been developed for design purposes to show the relationship of head, capacity, and speed. *Specific speed of an impeller may be defined as the speed in revolutions per minute at which a geometrically similar impeller would operate to develop 1 ft of head when displacing 1 gpm.* It is customary to give the

Fig. 3–21. Section view of a double-suction impeller.

specific speed of an impeller for design conditions. Since specific speed is a function of wheel proportions, it is a constant for any group of impellers having similar angles and dimensions. All wheels with the same specific speed would have the same efficiency except for variations due to viscosity of the fluids.

There are two steps in the derivation of the specific speed formula. First, the speed of the impeller must be changed at constant diameter to reduce the head to 1 ft; second, the dimensions must be changed at constant head or velocity to get a capacity of 1 gpm.

For a constant impeller diameter, flow, Q, will vary as the speed, N, and head, h, will vary as the square of the speed.

$$\frac{N_2}{N_1} = \frac{Q_2}{Q_1} = \sqrt{\frac{h_2}{h_1}} \tag{a}$$

Taking the subscript 1 to indicate the original impeller and subscript 2 the conditions after securing a head of 1 ft,

$$N_2 = N_1 \sqrt{\frac{h_2}{h_1}} \quad \text{and} \quad Q_2 = Q_1 \sqrt{\frac{h_2}{h_1}} \tag{b}$$

For the second step, we must change the physical size of the impeller to reduce Q_2 to a flow of 1 gpm, Q_3, but at the same time maintain the 1-ft head. That is the same as maintaining the same peripheral velocity and the same velocity at the impeller eye. The volume of flow will vary as area and as the square of the diameter. However, if the diameters are reduced, the speed must be increased to maintain an impeller velocity that will give the desired 1 ft of head. Therefore, at constant head,

$$\frac{Q_2}{Q_3} = \left(\frac{d_2}{d_3}\right)^2 = \left(\frac{N_3}{N_2}\right)^2 \tag{c}$$

where the subscript 2 has the same significance as in Eq. (b) and the subscript 3 indicates the properties after conversion to 1 gpm. Substituting Eq. (b) into Eq. (c),

$$N_3 = N_2 \sqrt{\frac{Q_2}{Q_3}}$$

$$N_3 = N_1 \sqrt{\frac{h_2}{h_1}} \sqrt{Q_1 \sqrt{\frac{h_2}{h_1}}}$$

Remember that $Q_3 = 1$ gpm and $h_2 = 1$-ft head. N_3 is the specific speed, Q is measured in gallons per minute, and h is the head in feet for *one stage*. Q for a double-suction impeller is taken as one-half the total flow for the impeller. Therefore,

$$N_s = \frac{N\sqrt{Q}}{(h)^{3/4}} \tag{3-11}$$

The specific speed for any one impeller, calculated throughout its head curve, would range from zero to infinity.

Each of the types of impeller previously discussed has been found to have a range of specific speeds in which it will give the best performance. Radial impellers have specific speeds up to about 3000 rpm, while Francis wheels go up to 4500 rpm. Mixed-flow impellers range from the specific speed of the Francis wheel to about 10,000; for propeller types the range is from 10,000 to 14,000 rpm.

3–11. Descriptions of Centrifugal Pumps. An exceedingly popular type of centrifugal pump for capacities up to at least 1000 gpm and heads up to 500 ft is the close-coupled pump with either one or two stages, Fig. 3–22. This particular pump has a radial wheel and backward-curved

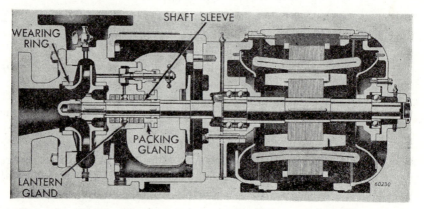

Fig. 3–22. Details of a close-coupled, single-stage, radial-impeller centrifugal pump. (Ingersoll-Rand Co.)

vanes. Several important features of construction that are applicable to both small and large pumps may be learned from an examination of this figure. Starting with the shaft, notice that there is a *sleeve* around it to prevent wearing, scoring, or corrosion. The shaft sleeve may be replaced more easily than the shaft. The packing is held in place, compressed into the *stuffing box*, by the *gland*. If the packing is compressed too tightly around the shaft or sleeve, the packing itself will wear quickly. It will also cause undue wear of the sleeve and will increase the friction loss. There should be a slight leakage of fluid along the shaft. The stuffing box in this pump has been designed particularly for use with a suction pressure below atmospheric, as can be seen by the inclusion of a *lantern gland* between the third and fourth rings of packing. A lantern gland has an H-shaped cross-section and fits around the shaft sleeve. Water from the pump discharge is introduced into the lantern gland and will flow to both the right and left. The quantity of water admitted to the lantern gland is controlled by a small needle valve. As long as water leaks to atmosphere and to the impeller, there can be no air leakage into the pump along the shaft to reduce the suction pressure. In some cases, lantern glands have been found unsatisfactory when the packing seals from a high pressure down to atmospheric, for they tend to reduce the compression on the innermost rings of packing.

The eye of the impeller is rounded (resembling a nozzle) to give smooth flow into the impeller.

This pump has two renewable *wearing rings* between the impeller and the pump casing. Clearances at these points must be kept to a minimum to reduce leakage from the discharge volute back to the suction. Larger pumps also have wearing rings on the impeller eye to permit easy replacement after the pump has been in use for some time and the clearances have increased from wear.

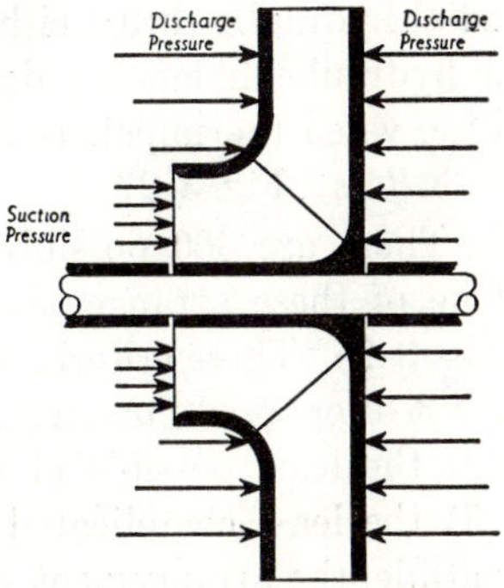

Fig. 3–23. Pressure distribution on a single-suction impeller.

One additional item on this pump is worth close attention, namely, the means provided for hydraulic balance. Balancing holes have been drilled through the impeller from the eye to the similar space in the back of the wheel to equalize the pressures in the front and back. This design will theoretically result in perfect hydraulic balance insofar as forces due to pressure are concerned. In practice, the pressures on the two sides of the impeller may not be exactly equal because of differences in front and back wearing-ring leakage. However, this inequality in balance may be accommodated by the thrust bearing. Consider the impeller in Fig. 3–23. Discharge pressure acts on all outside surfaces of this impeller except the eye, which is under suction pressure. The pressure on the two shrouds is not uniform because of centrifugal force and because of the liquid entrapped between the shrouds and the casing. The resultant force to the left is the algebraic summation of the force caused by the difference between suction and discharge pressures acting on the area of the eye and the force of the change in direction of the fluid from axial to radial. This latter force is small and may be neglected for practical purposes.

Double-suction impellers have almost perfect hydraulic balance, but the width of the impeller and the inlet passages required on both sides of each stage tend to produce a long bearing span for multistage pumps, Fig. 3–24 (numbers indicate numbers of stages). Other disadvantages of multistage double-suction pumps are the very complicated casing and the high pressure against the packing on the discharge end of the pump. Since it is impractical to arrange the impeller in any way other than in the order of ascending pressure, the packing on the discharge end must withstand discharge pressure less the pressure developed by the last stage. Nonetheless, pumps like those in Fig. 3–24 have given good service for plants operating in the lower part of the high-pressure range.

Pumps having several single-suction impellers may be balanced for axial hydraulic thrust either by using opposed impellers or by employing a hydraulic balancing device, such as a balancing drum or balancing disc, when the impellers all face in one direction (thus, called *sequence impellers*), Fig. 3–27.

There are 360 possible arrangements for a pump having six stages. One of these arrangements for an opposed impeller pump is shown in Fig. 3–25. Three criteria are used to select the most satisfactory arrangement of opposed impellers: (1) the least possible pressure on the packing, (2) the least possible pressure differential on the interstage shaft seals, (3) the least complicated casing passages. As these criteria are incompatible, the arrangement selected must represent a compromise.

Observe the arrangement of stages for the pump of Fig. 3–25, in view of these three criteria. The numbers shown on the picture indicate the numbers of the stages. With the first and second stages on opposite ends of the pump, the shaft packing must withstand the least possible pressure differential. In a similar manner, the least possible pressure differential is exerted on the interstage seal between stages five and six.

Although the terms *balancing drum* and *balancing disc* are used somewhat interchangeably, the cylindrical portion extending from A to E and without the flanged section from D to E of Fig. 3–26 would be a balancing drum. Likewise, a balancing disc would be the portion from D to E and without the portion A to B. Thus the balancing drum shown in Fig. 3–26 is actually a combination balancing drum and disc. The action of this

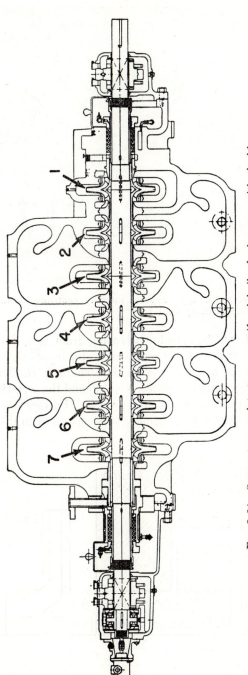
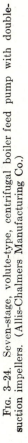

Fig. 3-24. Seven-stage, volute-type, centrifugal boiler feed pump with double-suction impellers. (Allis-Chalmers Manufacturing Co.)

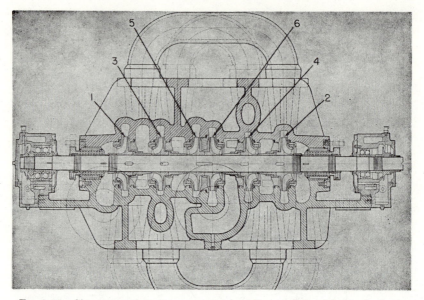

FIG. 3–25. Six-stage, volute-type, opposed impeller, centrifugal boiler feed pump. (Worthington Pump & Machinery Corp.)

combination, which will be called a balancing drum, is as follows: pump discharge pressure acts on area A and some water passes through the serrated *labyrinth* to the intermediate pressure chamber B. The clearance between the serrations and the casing is small, and the serrations act as a series of annular orifices to throttle the pressure. Further pressure reduction takes place from B to C. Chamber C is connected to the pump suc-

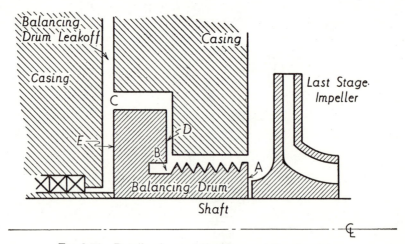

FIG. 3–26. Details of a centrifugal-pump balancing drum.

tion so that the pressure in C will be suction pressure plus a few pounds of friction drop in the pipe. The force at D is the area of D times the average pressure between B and C, while the force at E is essentially suction pressure times the area of E. Obviously, the net force for all surfaces of the balancing drum is to the left, so that it counteracts the unbalance of the impellers.

The particular feature of this design is that the rotor may float slightly to either the right or the left, thereby increasing or decreasing the area of the flow passage from B to C. If the impeller unbalance is greater than the algebraic sum of the balancing drum forces, the rotor will move to the right and reduce the area of the passage from B to C. Therefore the pressure on D will be increased. Thus the total force for the balancing drum will increase and balance the force from the impellers.

If the balancing-drum force is greater than the impeller unbalance, the opposite action will occur.

A boiler feed pump employing the balancing drum principle is shown in Fig. 3–27. Other features of this pump bear discussion. In order to diminish packing wear, a water jacket surrounds the packing gland. Cold water at low pressure from the plant service water system is used for the gland cooling as well as for the *smoothering gland* at the outside end of the packing. The smoothering gland will quench any vapor formed from the leakage of hot boiler feedwater leaking through the packing.

The inboard end of the pump (the end next to the driver) is pinned to the base. However, the outboard end is free to move in the slots provided as the pump expands due to the fluid temperature. In this pump the impellers are keyed to the shaft but have a slight clearance to permit easy removal. As the pump warms up, there will be differential expansion between the Monel metal shaft and the chrome steel impellers. This will tighten the impellers to the shaft during operation and prevent leakage and cutting along the shaft due to the pressure difference between stages. In other makes of pumps, the impellers are keyed to the shaft and are either press-fitted or shrink-fitted to the shaft.

Observe the diffusor vanes at the impeller outlet and the guide vanes to prevent vortexes as the water flows downward into the next stage.

One method of classifying boiler feed pumps is by the design of the casing. The pump in Fig. 3–27 is known as a barrel type of pump, while the pump in Figs. 3–24 and 3–25 is a horizontally split pump. The casing of a horizontally split pump is made in two parts, an upper half and a lower half, which are bolted together at the pump centerline. In order to assure perfect matching of the two halves, they must be machined blind, i.e., the internal machining is done while the two halves are bolted together. Experience has shown that the flanged joint of this type of pump is subject to leakage when there is a high pressure inside the pump

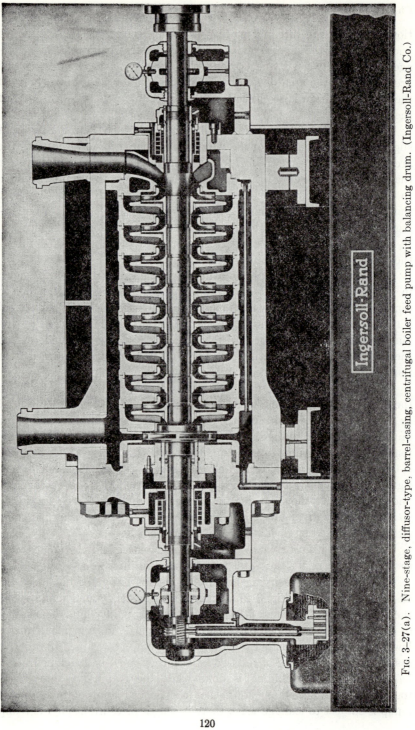

Fig. 3-27(a). Nine-stage, diffusor-type, barrel-casing, centrifugal boiler feed pump with balancing drum. (Ingersoll-Rand Co.)

because the pressure will "spring" the flanges. For discharge pressures above 1200 psig, it is recommended that the barrel type of construction, shown in Fig. 3–27, be used. Note that in the barrel type there is a water space, at discharge pressure, immediately inside the barrel casing and surrounding the inner portion of the pump. This pressure, acting on the *pancake* sections of each stage, holds the sections together. In another

type of barrel construction, the inner casing is made of two horizontally split sections, and the water at discharge pressure between the barrel and the inner casing prevents springing of the flanged joint. The stay rods, shown in Fig. 3–27, are used only for assembly purposes; the pancake sections are held together by water pressure and the force exerted by the vertical head at the outboard end of the pump.

Casings of boiler feed pumps are made of cast iron for low discharge pressures (not over 1200 psig) or 4 to 6% or 11 to 13% chrome steel. Cast steel has been found to be subject to corrosion and erosion attack from the high-velocity feed-water.

The discharge pipe of a large boiler feed pump should include a gate valve, a check valve, a small by-pass line around the check valve to permit warming the pump if the fluid is hot, and a recirculating con-

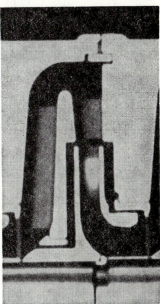

Fig. 3–27(b). Detail of barrel-type boiler feed pump.

nection. At low capacity, the efficiency of a centrifugal pump is very poor. This means that a large amount of work is required to compress each pound of water, and therefore there is considerable rise in temperature of the water flowing through the pump. In order to prevent damage to a high-pressure pump, a minimum flow through the pump is assured by automatically recirculating some of the water from the pump discharge to a low-pressure section of the piping. The amount of recirculating, or minimum, flow will usually be about 10 to 15% of the pump rating. This will limit the temperature rise in the pump to from 10 F to 20 F. The temperature rise of the water flowing through a pump can be determined at any given flow by multiplying the difference between the brake horsepower and the water horsepower by 2545 and dividing by the flow in pounds per hour. This calculation assumes that the specific heat of the water is unity and that there are no radiation losses for the pump, both of these assumptions being sufficiently accurate for practical purposes.

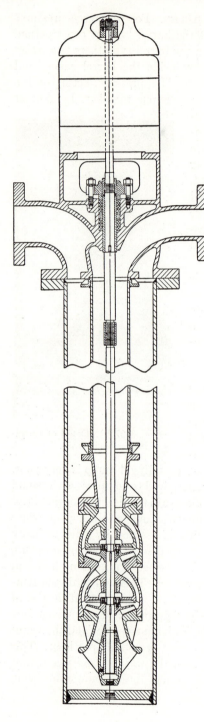

There is a trend to high-speed boiler feed pumps for the very high-pressure stationary plants. Speeds of 5000 to 9000 rpm are being considered, the exact speed being dependent on the type of drive and determined by an economic study. When motor-driven, speed-increasing gears are used with a 3600-rpm motor and a hydraulic coupling between the motor and gears if variable speed is required. When driven by steam turbines, high-speed boiler feed pumps are direct-connected to the turbine so that the advantages of high speed are then also realized in the turbine design.

High-speed pumps permit smaller pump-impeller diameters, smaller shaft-bearing spans, fewer stages, and smaller casings. They also have a manufacturing advantage in that a pump with a given number of stages and a given impeller diameter may be used for a variety of capacities and heads by varying the design speed. Speeds of 6000 rpm and higher have been used in marine installations for many years as a means of reducing the weight of the pump.

Vertical condensate and circulating water pumps have found wide application in recent plants. A multistage vertical condensate pump is shown in Fig. 3–28. The advantages of this style of pump are: the motor can be located well above flood level; there cannot be any air leakage into the pump,

Fig. 3–28. Vertical, multistage condensate pump. (Ingersoll-Rand Co.)

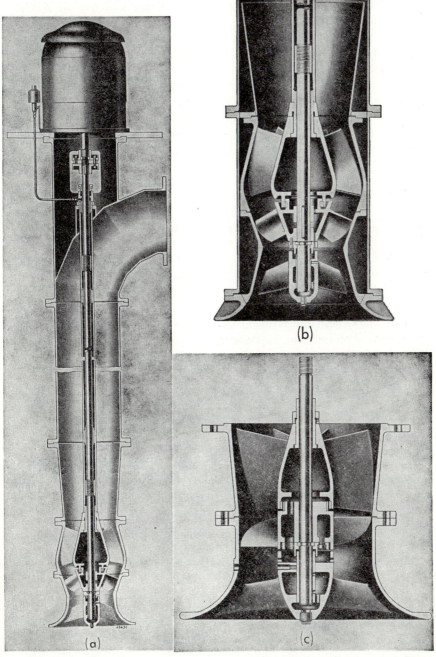

FIG. 3–29. Vertical circulating-water pumps: (a) nonpull-through mixed flow; (b) pull-through mixed flow; (c) nonpull-through axial flow. (Ingersoll-Rand Co.)

since the packing is subject to discharge pressure; less head room and excavation underneath the condenser is required; and a high NPSH, ranging from 10 to 20 ft can be obtained easily. For these reasons, over 50% of the condensate pumps for new plants are of the vertical type. Note that the impellers shown in Fig. 3–28 are not exactly radial and do not have a shroud on the inlet side.

Vertical circulating-water pumps, Fig. 3–29, need no priming because the pump impeller is submerged in the water. They, too, can be operated during floods, since the motor is placed above flood level. Both a mixed flow and a propeller type of construction are shown. In most instances, bronze impellers are used for circulating pumps.

3–12. System Head Curves. Further inspection of the GE equation shows that the head required to force a fluid through any system of conduit may be divided into two categories: (1) heads that are constant regardless of flow, and (2) heads that vary as the square of the flow. In the first class are static elevation from the reservoir to the receiver and pressure difference between the reservoir and the receiver; this latter item may vary as a function of flow but is usually constant. The second category includes velocity head and friction drop, which both vary as the square of the flow.

If we know these four heads at one flow, it is easy to find them for any other flow. For example, suppose that a pump is to transfer water at 190.8 F and 9.5 psia from one tank at constant pressure to another having a constant pressure of 60 psia, Fig. 3–30. We must calculate the system head for the pump from the minimum possible water level in the reservoir, since the pump should be able to operate when the level is at the bottom of the tank. As further assumptions, take the maximum velocity at any point in the system as 8 fps and the friction loss as 65 ft at a flow of 500 gpm. The minimum velocity head would, of course, be zero as the water is taken from a storage tank. The constant losses through this system would be the static head and the pressure differential, or 175.5 ft. Variable losses of friction and velocity head together are 66 ft at the given flow. These values are plotted in the figure. The head required at one-half flow would be the sum of the constant head of 175.5 ft and ¼ of 66 ft, or 192.0 ft.

The required head may be determined in a like manner for all other flows and may be plotted as in Fig. 3–30.

Any pump for this system, regardless of its type, must be so controlled that it will produce the head given by this curve for any flow. If the pump supplies more head, then the flow must increase, and conversely, if the pump cannot develop sufficient head, the flow will decrease until the pump head and the system head match.

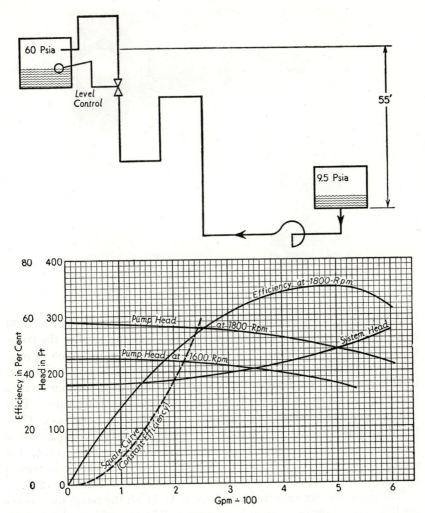

FIG. 3–30. Development of a system head curve.

Typical system head curves have the shape of the one that has been developed, but it is possible to have a drooping system head curve. This might occur if one of the tank pressures were to change with flow.

A head curve for a centrifugal pump that might be used for this system has also been included in Fig. 3–30. Note that the pump curve will coincide with the system head curve at only one point. Therefore, the pump would always operate at this point unless some means of control were used. There are two methods of controlling a centrifugal pump: (1) a throttling valve, and (2) variable-speed operation of the pump.

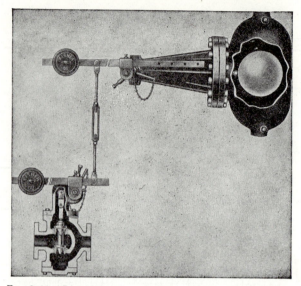

FIG. 3–31. Liquid level controller. (Fisher Governor Co.)

A float type of level control has been shown schematically on Fig. 3–30, and an actual controller is shown in Fig. 3–31. As the level in the 60-psi tank drops, the float follows and opens the valve. The reduction in pressure permits more water to flow through the system. Conversely, if the level in the 60-psi tank rises, then the pump is supplying too much water. The float will then follow the level and close the throttling valve. That will put more resistance in the system, and the flow will decrease. Thus, the throttling valve is actuated by the float to reduce the difference in pressure between the system head curve and the pump head curve. For example, at a flow of 300 gpm the throttling valve would be in such a position that it would have a pressure drop of 272 ft − 200 ft = 72 ft.

The procedure for calculating the speed at which the pump must operate if it is to follow the system head curve will be the subject of Art. 3–13.

If the pump used for this system were of the direct-acting type, it would be necessary to throttle the steam pressure rather than the water pressure. Therefore, the valve would be located in the steam line to the pump.

3–13. Variable-Speed Calculations for Centrifugal Pumps. The fundamental formulas for calculation of variable-speed performance were developed in the article on specific speed [Art. 3–10, Eq. (a)]; that is,

$$\frac{\text{rpm}_2}{\text{rpm}_1} = \frac{\text{quantity}_2}{\text{quantity}_1} = \sqrt{\frac{h_2}{h_1}} = \sqrt[3]{\frac{\text{hp}_2}{\text{hp}_1}} \qquad (3\text{–}12)$$

for constant fluid temperature and constant pump efficiency. Any convenient units, such as gallons per minute or pounds per hour, may be used for the quantity, and feet or pounds per square inch may be used for head, provided that the ratios are dimensionless. Horsepower has been added to the formula previously given. The fact that the horsepower varies as the cube of the speed can be readily understood when it is remembered that horsepower is the product of quantity and head. Quantity varies as the speed, and head as the square of the speed; so the product of the two, horsepower, should vary as the cube of the speed.

Note that there is a variation in *both head and quantity* for a variation in speed. For the typical coordinates of a centrifugal pump head curve (quantity and head), the curve of the relation of these two would be a family of square curves all passing through zero-zero. Any of the points on one of these square curves theoretically would have the same efficiency.

Fig. 3–32 shows a head curve for a centrifugal pump and also one square curve.

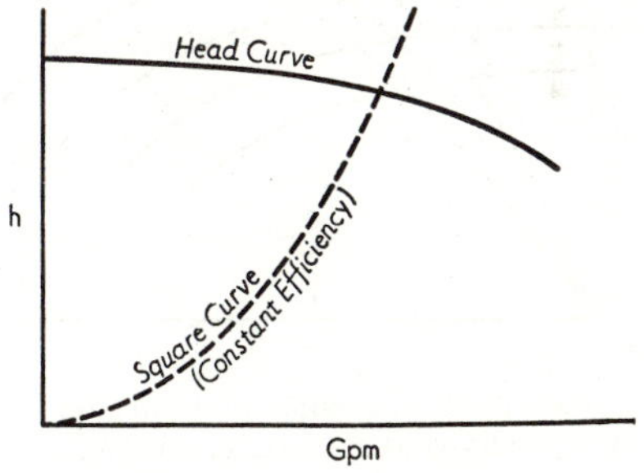

To demonstrate the procedure to be followed in making variable-speed calculations, the head curve for the pump of Fig. 3–30 will be calculated for 1600 rpm.

Substitute into Eq. (3–12) for $rpm_1 = 1800$, $rpm_2 = 1600$, $gpm_1 = 500$, and $h_1 = 241.5$. Gpm_2 will then be 445 and h_2 will be 191. The A and B scales of the slide rule can be used

Fig. 3–32. Head and square curves for variable-speed calculations.

in conjunction with the C and D scales for speed and quantity to solve Eq. (3–12) in only one setting. By a similar procedure, other points on the 1600-rpm head curve can be calculated and plotted.

Since the head curve for 1600 rpm intersects the system head curve at 355 gpm, the pump would have to operate at 1600 rpm to satisfy the system when the flow was 355 gpm.

Instead of calculating all the possible head curves at all speeds, a procedure similar to the above may be used to determine the operating conditions for one point. As an example, suppose it is desired to find the speed at which this pump should operate when there is a flow of 200 gpm. There is no known speed ratio to substitute into Eq. (3–12) as there was in the preceding case. Therefore, it is necessary to use a trial-and-error procedure. Set up Eq. (3–12) with the known values and estimate the point at which a square curve passing through zero-zero and the 200-gpm point on the system curve would intersect the 1800-rpm head curve. Substitute these estimated values into the equation and cut-and-try until the equation balances at $gpm_1 = 245$ and $h_1 = 278$ ft. From the ratios now established, rpm_2 is found to be 1470.

Efficiency can be found for any of the partial speed points that have been calculated, by remembering when solving Eq. (3–12) that eff_1 equals eff_2. For example, the efficiency of the pump will be the same, 70%, at 1600 rpm and 445 gpm as at 500 gpm and 1800 rpm. In this manner, the efficiency curve for 1600 rpm could be plotted. Likewise, the efficiency at 200 gpm and 1470 rpm would be the same, 65.5%, as for 245 gpm and 1800 rpm.

Horesepower at partial speed may be determined from Eq. (3–12) if it is known for the original speed or from the horsepower formulas involving head, quantity, and efficiency.

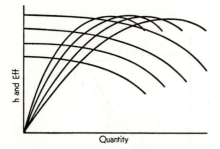

A family of constant rpm curves for a centrifugal pump would be as indicated in Fig. 3–33. The head curves recede toward zero-zero, and the efficiency curves move horizontally toward the ordinate as speed decreases.

The procedure just described is approximate but satisfactory for most power-plant calculations. The deviation from actual conditions may be seen by inspection of Fig. 3–34. Part (a) of this figure gives head capacity at 1750 rpm for an 8×6 single-stage pump (8-in. suction and 6-in. discharge) for impeller diameters varying from 16 to 11 in. Lines of constant efficiency are also shown. Part (b) of Fig. 3–34 shows the performance of the 13-in. impeller at three speeds. Note that the constant efficiency lines are not exact square curves as shown in Fig. 3–32 and as assumed for Eq. (3–12). However, the deviation from theoretical is not large for average speed changes. Head-capacity curves for small changes in the diameter of a given impeller may be predicted at *constant speed* and *constant temperature* from the formula

FIG. 3–33. Families of head and efficiency curves at four different speeds.

$$\frac{d_1}{d_2} = \frac{gpm_1}{gpm_2} = \sqrt{\frac{h_1}{h_2}} = \sqrt[3]{\frac{hp_1}{hp_2}} \qquad (3\text{–}13)$$

which is derived from the specific speed relations.

3–14. Variable-Speed Drives.

Several drives for variable-speed operation of pumps and fans are available. They include d-c motors, turbines, variable-speed a-c motors (such as the Rossman drive), and slip-ring motors, magnetic couplings, and hydraulic couplings. Most large plants

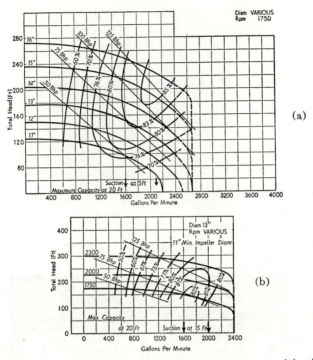

Fig. 3–34. Characteristic curves for an 8 × 6 single-stage pump (a) with varying impeller diameters and (b) with a 13-in. impeller at three speeds. (Ingersoll-Rand Co.)

do not have direct current available in sufficient quantities to permit the use of that type of motor, and special motor-generator sets cannot be justified economically. Turbines are ideal as variable-speed drives, but the cost of piping and particularly the inability of the cycle to absorb the exhaust steam during low loads limit their use.

The Rossman drive consists of an a-c motor whose stator is mounted on bearings so that it can rotate. Variation of the stator speed by a belt drive from a d-c motor causes variation of the absolute armature speed. This type of drive is limited to low speeds because of the large bearings required by the rotating stator.

Slip-ring motors are inefficient and have poor speed regulation.

Magnetic couplings have not been used extensively in stationary plants, but hydraulic couplings have been economical and have found wide use, Fig. 3–35. In a hydraulic coupling, a runner attached to the driven shaft receives a vortex of oil from the impeller that is fastened to the driving shaft. There is no mechanical connection between the runner and the impeller, and the two are almost identical in shape. Kinetic energy is imparted to the oil by the impeller, and the oil flows radially

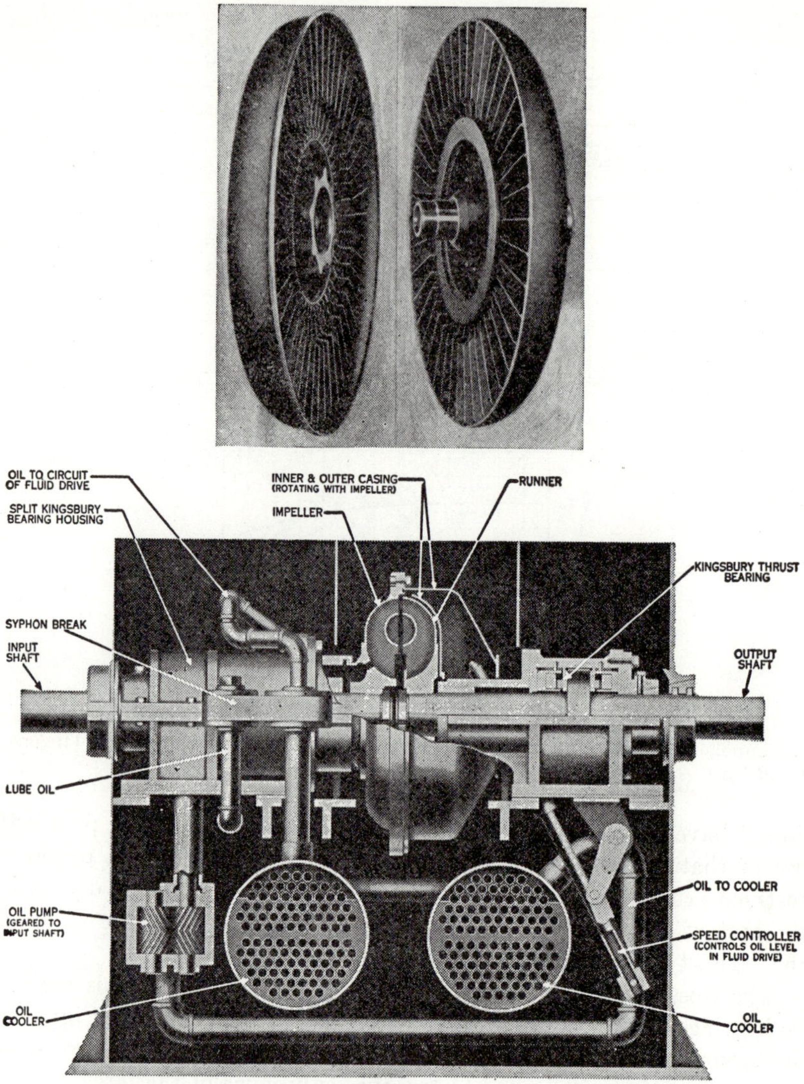

FIG. 3-35. Hydraulic coupling. (American Blower Corp.)

outward and into the vanes of the runner. The oil then flows from the outer periphery of the runner toward the center and transfers its energy to the runner in much the same way as steam imparts energy to a turbine blade.

Since the amount of energy received by the runner at any given impeller speed is a function of the oil flow, a scoop tube is provided to regulate oil level. Oil leaves the working circuit of the coupling through the scoop tube. The position of the scoop tube determines the quantity of oil in the coupling and therefore the speed of the runner and output shaft.

After leaving the working circuit, the oil goes to the reservoir. A gear pump, direct-driven from the input shaft, takes the oil from the reservoir and returns it to the working circuit via the oil coolers. The energy from the coupling inefficiency raises the oil temperature.

Efficiencies of hydraulic couplings range from 92 to 98% at maximum output speed. The efficiency may be approximated by

$$\text{Hydraulic coupling efficiency *} = \frac{\text{output speed}}{\text{input speed}} \qquad (3\text{-}14)$$

3-15. Economics of Variable-Speed Drives. If the savings in pump power are equal to or greater than the cost of a variable-speed drive, then it is economical to use a variable-speed drive. Usually, there is no increase in the cost of a pump or motor for variable-speed hydraulic coupling application, although there may be a small increase in the cost of the longer pump bed plate needed to accommodate the coupling. Although the coupling will occupy considerable space, it is seldom that the plant building must be enlarged to accommodate the coupling. Therefore, the factors that must be included in an economic analysis of variable-speed operation of a pump are: cost of the coupling, maintenance on the coupling, value of the power savings, the hours of operation at various loads, and the investment charges. Possible lower maintenance costs for the pump (or fan) because of the lower speed are almost impossible to determine accurately.

It will generally be found that variable-speed drives cannot be justified when the plant is expected to operate near its rated load (high-capacity factor, Chap. 15) most of its life. Of course low-investment charges assist in justifying the increased equipment cost necessary to obtain the reduced operating cost provided by variable-speed operation of pumps and fans.

Frequently three half-sized constant-speed pumps (one spare) will be as economical as two full-sized variable-speed pumps (one spare).

In order to account for the load fluctuations on a unit, an economic evaluation of variable-speed vs constant-speed operation should be made for not less than three loads The following example illustrates a comparison of three constant-speed pumps with two variable-speed pumps.

EXAMPLE 3-3. Calculate the annual operating cost for three small boiler feed pumps operating at a constant speed of 3570 rpm (one pump will be a spare and one or both of the remaining pumps will supply the boiler, depending on water flow) and for two large pumps (one spare). Assume 7500 hr per yr of

---

* This neglects the losses due to bearing friction and oil pump power. These range up to 50 hp for large couplings.

operation at a 650,000 lb per hr, 1200-psi developed head, 250 F, a heat rate of 12,000 Btu per kwhr, and a fuel cost of 28¢ per 10^6 Btu (only one load point will be used to simplify the example). Pump performance may be taken from Fig. 3–36; 13% investment charges should be used. Use the annual cost method.

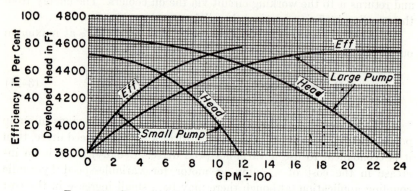

Fig. 3–36. Characteristic curves for two boiler feed pumps.

Other data are as follows:

	Constant-Speed Pump	Variable-Speed Pump
Pump cost, each	$35,000	$50,000
Motor cost, each	30,000	54,000
Motor circuit cost, each	4,500	4,500
Hydraulic coupling, each	...	23,000 (installed)
Additional piping cost, total	3,000	...
Motor efficiency	90%	83%
Full-load motor speed, rpm	3,570	3,570
Full-load pump speed, rpm	3,570	3,480

SOLUTION. From Eq. (3–5),

$$\text{Flow} = \frac{650,000}{500 \times 0.942} = 1380 \text{ gpm}$$

From Eq. (3–2)   System head $= 1200 \times 2.45 = 2940$ ft

From Fig. 3–36, at 1380 gpm the two constant-speed pumps will operate at 690 gpm each, 4310-ft developed head, and 63.5% efficiency. Other calculations for the constant-speed pump are [from Eq. (3–4)]:

$$\text{Shaft hp} = \frac{690 \times 4310 \times 0.942}{3960 \times 0.635} = 1114 \text{ each}$$

$$\text{Motor input} = \frac{1114 \times 0.746}{0.90} = 923 \text{ kw}$$

The annual power cost for the two pumps will be equal to the product of (heat rate)(power)(hours per yr)(fuel cost), or

$$\frac{2 \times 12,000 \times 923 \times 7500 \times 0.28}{10^6} = \$46,500$$

For the variable-speed pump at the head and flow calculated above, use Eq. (3–12) to obtain

$$\frac{\text{rpm}_1}{\text{rpm}_2} = \frac{3480}{2890} ; \quad \frac{Q_1}{Q_2} = \frac{1660}{1380} ; \quad \frac{h_1}{h_2} = \frac{4255}{2940} ; \quad 74.5\% \text{ efficiency}$$

From Eqs. (3–4) and (3–6),

$$\text{Shaft hp} = \frac{1380 \times 2940 \times 0.942}{0.745 \times 3960} = 1295$$

The coupling efficiency, from Eq. (3–14), is

$$\frac{2890}{3570} = 81.0\%$$

$$\text{Motor input} = \frac{1295 \times 0.746}{0.810 \times 0.83} = 1437 \text{ kw}$$

The annual power cost for the pump will be

$$\frac{12,000 \times 1437 \times 7500 \times 0.28}{10^6} = \$36,200$$

Total investment for the three half-size pumps is

$$3 \times \$35,000 + 3 \times \$30,000 + 3 \times \$4500 + \$3000 = \$211,500$$

The annual investment charges are

$$0.13 \times \$211,500 = \$27,500$$

and the total annual cost is

$$\$27,500 + \$46,500 = \$74,000$$

The total investment for the two full-sized pumps is

$$2 \times \$50,000 + 2 \times \$54,000 + 2 \times \$4500 + 2 \times \$23,000 = \$263,000$$

The annual investment charges are

$$0.13 \times \$263,000 = \$34,200$$

and the total annual cost is

$$\$34,200 + \$36,200 = \$70,400$$

Thus there is an annual saving of $\$74,000 - \$70,400 = \$3600$ by using the two full-sized variable-speed pumps.

PROBLEMS

3–1. Convert 75 psi to feet of water at 80 F, 250 F, and 400 F. Use the steam tables and not the constants given in the Appendix.

3–2. What is the equivalent pressure in pounds per square inch of a head of 130 ft of oil, sp gr 0.82?

3–3. A pump receives water from an open tank whose minimum water level is 5 ft below the centerline of the pump. The water is at 60 F and the friction loss in the pipe is 3.5 ft. For a water velocity of 200 fpm, what is the NPSH?

3–4. Water is supplied to a pump from an overhead tank whose pressure is 5 psig. The water level in the tank is 17 ft above the floor, and the water is saturated. The pump centerline is 30 in. above the floor. If the suction pipe carries 120 gpm and is standard 4-in. pipe, calculate the NPSH when the friction loss is 2 psi.

3–5. Use the same data as in Prob. 3–4 and calculate the NPSH when the water temperature is 5 F less than the saturation temperature.

3–6. Calculate the NPSH for a pump that receives 300 F water at the rate of 1250 gpm through a pipe whose ID is 10 in. A suction gage located 20 ft below the pump centerline indicates a pressure of 77 psig, and the barometer indicates a pressure of 28.2 in. Hg.

3–7. A pump receives water from a tank. If the pump needs 79 ft of NPSH, how high must the water level be above the pump centerline if the water is saturated at 250 psig? The friction loss in the pipe averages 1 ft for each 20 ft of height and the flow is 350 gpm in a 6 in. ID pipe.

3–8. If a pump requires a minimum of 9 ft of NPSH, how high can the centerline of the pump be located above the top of an open tank that supplies 80 F water to the pump? The barometer varies from 27.0 to 28.8 in. Hg throughout the year. The tank is 6 ft high, the pipe friction is 2.3 ft, and the velocity is 240 fpm.

3–9. Plot the percentage of change in horsepower for a pump against water temperatures of 60 F to 500 F, using 60 F as unity. Assume constant flow in pounds per hour and constant developed head in pounds per square inch.

3–10. A centrifugal pump displaces 875 gpm of 350 F water with a developed head of 534 psi. If the efficiency is 68.4%, what is the water horsepower and the brake horsepower?

3–11. A quantity of 650,000 lb per hr of 212 F water is pumped from a tank at atmospheric pressure that is located 19 ft above the centerline of the pump. The pump discharge gage reads 421 psig, and the velocity in the discharge pipe for this flow is 525 fpm. The equivalent length of the 12-in. schedule 40 suction pipe is 187 ft. Specify the size of motor required for the pump if the pump efficiency is 71.0%.

3–12. A centrifugal pump receives 75,000 lb per hr of 90 F water when the pump centerline is 9 ft above the water level in a vessel that has a pressure of 4.5 psia. Calculate the water horsepower, neglecting friction and velocity heads, if the discharge pressure gage reads 125 ft.

3–13. A fire pump delivers 1050 gpm on test. The suction gage reads 6 in. Hg vacuum and is 4 ft below the centerline of the pump. The discharge gage reads 86 psi and is 3 ft above the pump centerline. What is the pump efficiency if the water is at 77 F and the input is 97 hp?

3–14. The following data were taken during the rest of a pump: 85,000 lb per hr, inlet pipe 4-in. schedule 40, 3-in. schedule 40 discharge pipe, gage in suction line located 34 in. below the pump centerline reads 3.5 in. Hg vacuum, discharge gage located 18 in. below the pump centerline reads 175 psi, water temperature 120 F, input to pump 24 hp. Find the developed head and the pump efficiency.

3–15. What is the developed head and water horsepower for the following pump: 145,000 gpm, 60 F water, 96-in. ID suction and discharge pipes, suction gage located 8 ft above the pump centerline reads 7.5 in. Hg vacuum, discharge gage located 8 ft above the pump centerline reads 21 ft of water?

3–16. Determine the discharge in pounds per hour and gallons per minute of a 10 × 6 × 12 duplex pump operating at 35 strokes per min if the slip is 7% and the water is at 150 F.

3–17. What is the actual displacement of a 6 × 4 × 10 direct-acting simplex pump in gallons per minute when the inlet pressure is 15 psig and the outlet pressure is 200 psia? Water enters at 230 F, slip is 12%, and the speed is 40 strokes per min.

3–18. Select an outside-end-packed plunger pump (direct-acting duplex) for boiler feed service. The pump is to handle 15 gpm of 210 F water from 3 to 150 psig with steam entering at 125 psig and exhausting at 5 psig. Use 75% volumetric efficiency. Determine the necessary steam pressure and the strokes per minute.

3–19. Select a duplex pump for boiler feed service. Suction pressure is 12 psia, water temperature 180 F, discharge pressure 165 psig. Assume 70% volumetric efficiency, 150 gpm, 140 psig to 8 psig steam.

3–20. Determine the minimum steam pressure at which the pump selected in Prob. 3–16 will need to develop the required head.

3–21. Estimate the size of duplex, direct-acting, plunger-type boiler feed pump equipment to be used on a ship as an emergency feed pump and fire pump. When being used as a feed pump, its capacity should be not less than 80,000 lb per hr of cold water with 400-psi developed head. As a fire pump, its capacity should be about 200,000 lb per hr but with 200-psi developed head. In either case, the effective steam pressure is 340 psi. What will be the speed in strokes per minute for each condition of operation?

3–22. Explain how a direct-acting pump operates at partial load.

3–23. Select a duplex, direct-acting, boiler feed pump for the following conditions: 40 gpm of 175 F water from 0 to 225 psig, steam from 175 to 2 psig. Using 15% slip, find the pump speed and minimum steam pressure.

3–24. Use the data of Prob. 3–18, and select a simplex pump.

3–25. Explain why the pipe used for the suction of a centrifugal pump is very often larger than the pipe used for the discharge.

3–26. Sketch curves similar to those in Fig. 3–14 for a pump with radial vanes.

3–27. Sketch curves similar to those in Fig. 3–14 for a pump with back-curved vanes.

3–28. Determine the specific speed at rated capacity for the pump in Fig. 3–30.

3–29. Calculate the specific speed for the pump for Fig. 3–16.

3-30. Should pumps with either radial or Francis impellers be started with the discharge valve open or closed? Would your answer be different if the pump had either a mixed-flow or axial wheel?

3-31. Plot the pump characteristics listed below and calculate and plot the horsepower input to the pump for 160 F water. Use gallons per minute as abscissa.

Performance of One Pump at 1750 rpm

Gpm	Ft head	Efficiency (%)
850	2005	74.8
800	2100	76
750	2180	76.8
700	2260	77
650	2325	76.3
600	2385	75
550	2425	72.8
500	2460	70
400	2510	62
300	2545	52
200	2565	39
100	2568	21
0	2570	0

3-32. Calculate and plot the efficiency, head, and horsepower for the pump data in Prob. 3-31 if the speed is changed to 1800 rpm.

3-33. Calculate and plot the efficiency, head, and horsepower for the data of Prob. 3-31 for a speed of 1700 rpm.

3-34. Calculate and plot the efficiency, head in pounds per square inch, quantity in pounds per hour, and horsepower for the data of Prob. 3-31 with a water temperature of 250 F.

3-35. Plot head, efficiency, and horsepower curves at 1700 rpm for the pump in Fig. 3-30. Use gallons per minute and foot of head as the coordinates.

3-36. Plot head, efficiency, and horsepower curves at 3000 rpm for the pump of Fig. 3-16.

3-37. Plot head, efficiency, and brake horsepower curves for the pump in Fig. 3-30, using pounds per square inch and pounds per hour for water at 60 F.

3-38. Calculate the speed at which the pump of Fig. 3-30 should operate when the system flow is 100 gpm.

3-39. At what speed should the pump of Fig. 3-30 operate for a system flow of 420 gpm?

3-40. What would be the speed of the pump whose characteristics are shown in Fig. 3-16 if it were handling 200 F water, displacing 300 gpm, and developing a head of 900 ft? Also calculate the brake horsepower.

3-41. Plot a system head curve for a piping arrangement schematically the same as that of Fig. 3-30 with the following data: saturated water at 235 F;

static elevation difference, 28 ft; pressure in final receiver, 725 psig; 400-fpm maximum velocity in the system; 65-psi pressure loss; pressure in the low-pressure tank decreases linearly with flow to zero pressure at zero flow. All the above data are for a maximum flow of 300 gpm.

3-42. If two pumps, each with the characteristics given in Prob. 3-31, are required to feed a boiler, plot the combined head, efficiency, and horsepower curves for the two pumps operating in parallel. On the same curve, plot the system head curve for the following data: heater pressure constant at 4.7 psia; static elevation difference, 27 ft; boiler pressure, 650 psig; friction and velocity heads, 245 ft at 1600 gpm. Also plot on the same sheet the speed at which the pump should operate at all flows from zero to maximum flow and the horsepower at variable speed.

3-43. From which company would you recommend purchase of a pump to operate 4000 hr per yr at 200 gpm, 600 ft head, 60 F water? Company A charges $935 for a pump of 69% efficiency, Company B charges $885 for a pump of 67% efficiency, and Company C charges $1100 for one of 71.5% efficiency. Electricity costs 1.25¢ per kwhr; the motor efficiency for all pumps is 94%, and the charges are 15% annually.

3-44. For the example of Art. 3-15, determine the return on the increased investment.

3-45. For the example of Art. 3-15, determine the maximum permissible investment for two large pumps with hydraulic couplings for (a) 15%, (b) 10%, and (c) 5% investment charges.

3-46. Rework the example of Art. 3-15, using two of the large-sized pumps at constant speed and two of the large-sized pumps at variable speed. Assume maintenance on each coupling to be $300 per yr and the developed head for the variable-speed pump to be 1250 psi. All other data to be taken as given in the example.

3-47. Use the data for the small pump of Fig. 3-36 at a constant speed of 3570 rpm, and determine the annual monetary savings for two of these pumps (one spare) at constant speed versus two at variable speed. Use the same characteristic curve for variable-speed operation, but assume the speed for the curve to be the maximum hydraulic-coupling output speed of 3450 rpm. Other data are: 300 F at all flows; 19¢ per 10^6 Btu fuel cost; 3570-rpm motor speed; 89% average motor efficiency at all loads; 2000 hr per yr operation at 320 gpm, 1150-psi developed head, 14,350 Btu per kwhr heat rate; 4000 hr per yr operation at 640 gpm, 1240-psi developed head, 13,100 Btu per kwhr heat rate; 1800 hr per yr operation at 960 gpm, 1360-psi developed head, 12,700 Btu per kwhr heat rate.

3-48. A 3550-rpm characteristic curve of a pump shows that the pump develops 2500-psi head at shut-off when the water is at 350 F. When the pump is in service, shut-off conditions will be 150 F water, 3900 rpm, 275-psi suction pressure. What will be the pressure to which the discharge piping will be subjected?

3-49. Plot the pump characteristics listed below and calculate and plot the shaft horsepower for 60 F water. Use gallons per minute as the abscissa. The data for one pump at 3550 rpm are:

Gpm	Developed Head, Ft	Efficiency, %
0	7440	0
200	7420	26
400	7330	44.5
600	7200	57
800	7010	66
1000	6730	73
1250 (design point)	6200	76
1400	5800	76

3–50. Plot the efficiency, head, and shaft horsepower of the pump of Prob. 3–46 for a speed of 3000 rpm (use feet and gallons per minute).

3–51. Plot the efficiency, head, and shaft horsepower of the pump of Prob. 3–46 for a water temperature of 300 F. Use pounds per hour and pounds per square inch.

3–52. What would be the speed of the pump of Prob. 3–46 to displace 300,000 lb per hr of 240 F water at a developed head of 2000 psi?

3–53. If investment charges are 12%, how much can be paid for two hydraulic couplings to operate two boiler feed pumps (one spare) whose characteristics are given in Prob. 3–46? Use the same curve for a constant speed of 3550 rpm and for 3450 rpm. Motor speed will be 3550 rpm. Other data are: 90% motor efficiency; 40¢ per 10^6 Btu fuel cost; 275 F water; 4000 hr per yr operation at 400 gpm, 1900-psi developed head, 10,900 Btu per kwhr heat rate; 3500 hr per yr operation at 1000 gpm, 2250-psi developed head, 9300 Btu per kwhr heat rate.

3–54. Calculate the minimum flow for the pump of Prob. 3–31 for a temperature rise of 20 F.

3–55. Calculate the minimum flow for the pump of Prob. 3–49 for a temperature rise of 15 F.

BIBLIOGRAPHY

CHURCH, AUSTIN H. *Centrifugal Pumps and Blowers.* New York: John Wiley & Sons, Inc., 1944.

HYDRAULIC INSTITUTE. *Standards of the Hydraulic Institute.* New York, 1954.

KARASSIK, IGOR J. "So You Are Going to Buy a Boiler Feed Pump?" *Southern Power and Industry* (April to July, 1943).

KENT, WILLIAM. *Mechanical Engineers' Handbook.* New York: John Wiley & Sons, Inc., 1950.

MARKS, LIONEL S. *Mechanical Engineers' Handbook.* New York: McGraw-Hill Book Co., Inc., 1951.

SPILLMANN, MAX. *Axial and Radial Thrust in Multi-stage Centrifugal Pumps.* Bulletin W-318-E3A. Harrison, N. J.: Worthington Pump & Machinery Corp.

CHAPTER 4

THEORY OF HEAT TRANSFER

4–1. Introduction. Heat is defined as energy in transition due to a temperature difference. There are three methods by which this transition of energy may take place: (1) conduction, (2) convection, and (3) radiation. When the transition takes place because of contact of the particles of one or more bodies, conduction occurs. When the energy is transferred because of the motion or mixing of the particles of a fluid, convection takes place. All matter may receive or reject energy to some degree as a wave motion; such rays may even pass through a vacuum or the atmosphere. Energy received by this means is known as radiant energy.

Theoretically, heat transfer takes place by all three methods. However, in many power-plant heat exchangers, the heat flow takes place predominantly by one or two of these methods. Feedwater heaters perform their function in the plant both by conduction through the metal of the tubes and by convection of the fluids confined by the wall. On the other hand, boiler tubes which can "see" the flame receive a large portion of their energy by radiation.

Throughout this discussion we shall be concerned only with *steady* transfer of heat. That is, the temperature at any point in any of the bodies is constant with time.

The purpose of this chapter is to discuss briefly the theories of these three forms of heat transfer and their practical application to insulation of piping. Since heat transfer takes place in all the apparatus in the steam-power plant—although it is not of major importance in all the equipment—other practical applications of the subject will be discussed in detail when the design of that equipment is considered.

4–2. Conduction. Fourier's equation shows the fundamental relationship for heat transfer by conduction:

$$q = -kA \frac{dt}{dx} \tag{4-1}$$

where q = heat transferred, Btu per hr

A = area of the material perpendicular to the path of heat flow, sq ft and $t =$ temperature, F

    $x =$ thickness of the material along the path of heat flow, ft

    $k =$ constant of proportionality, called *thermal conductivity*, at the arithmetic mean temperature, Btu (ft) per (ft)^2(hr) (F).

The minus sign in this equation merely signifies that heat is flowing in the direction of decreasing temperature.

Before we can integrate this equation, it is necessary to make certain assumptions. We have already stated that we shall consider only steady-flow conditions, i.e., there is no storage of energy, or the temperature of any point in the conducting material is constant with time. Obviously, we must assume that the material is homogeneous and that the flow is unidirectional.

The constant of proportionality, thermal conductivity k, will bear further thought. It has been stated that flow of heat energy by conduction is due to contact of the particles in the body. Conductivity of gases serves to illustrate an important point. At very low absolute pressures the particles of gases are much farther apart than they are at higher pressures. Although the conductivity of gases at high vacua is nearly zero, at partial or atmospheric pressure the conductivity is appreciable. Fortunately, the conductivity of gases in the range of atmospheric pressure can be considered independent of pressure for practical purposes. Also, as might be expected, the conductivity of gases increases roughly as the molecular weight increases.

The temperature of a substance generally has an important effect upon conductivity, with conductivity increasing with temperature. However, water is a notable exception. Experiments have shown that k for water increases proportionally to temperature from 32 F to 200 F but then decreases sharply when the temperature increases above 250 F.

Conductivity for many insulating materials represents a much more complicated picture than that for metals, liquids, and gases. The insulating effect, or low conductivity of some materials, is due primarily to the porous nature of the material. There are many voids—small dead-air pockets—in insulating materials; and because these spaces are small, there can be little, if any, movement of air particles, i.e., there can be little convection. However, if the spaces are enlarged, there may be convection currents set up that would reduce the insulation value of the material. Similarly, an insulation that becomes wet has a higher value of thermal conductivity than it has when dry, because the voids have become filled with water, the conductivity of which is higher than that for dead air.

Values of thermal conductivity for some substances are given graphically in Fig. 4–1.

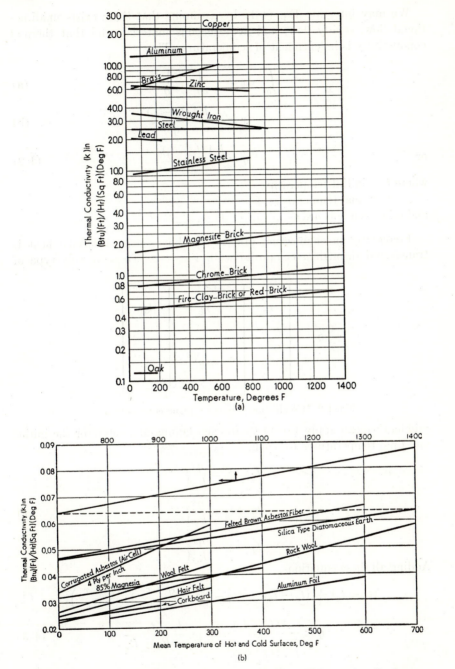

FIG. 4-1. Thermal conductivities of some materials. (Taken primarily from an article by W. J. King, *Mechanical Engineering*, 1932, Vol. 54.)

We may integrate Eq. (4–1) by assuming that there exists unidirectional flow, steady flow, and homogeneous material, and that thermal conductivity is independent of temperature. Thus

$$\int_0^x dx = -\int_{t_i}^{t_f} \frac{kA\,dt}{q} \tag{a}$$

$$x = -\frac{kA\,(t_f - t_i)}{q} \tag{b}$$

or

$$q = \frac{kA\,(t_i - t_f)}{x} \tag{4–2}$$

where t_i = initial temperature, F
$\quad t_f$ = final temperature, F
and other symbols are as in Eq. (4–1).

Frequently in engineering practice, problems arise in which heat is transferred through a composite wall, Fig. 4–2. To solve this type of

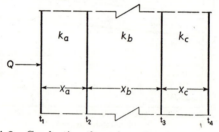

FIG. 4–2. Conduction through a plane composite wall.

problem we can apply Eq. (4–2) to each lamina and solve for the initial and final temperatures for each.

$$t_1 - t_2 = \frac{qx_a}{k_a A}$$

$$t_2 - t_3 = \frac{qx_b}{k_b A} \tag{c}$$

$$t_3 - t_4 = \frac{qx_c}{k_c A}$$

Adding all these equations,

$$(t_1 - t_2) + (t_2 - t_3) + (t_3 - t_4) = \frac{q}{A}\left(\frac{x_a}{k_a} + \frac{x_b}{k_b} + \frac{x_c}{k_c}\right) \tag{d}$$

$$q = \frac{A\,(t_1 - t_4)}{\dfrac{x_a}{k_a} + \dfrac{x_b}{k_b} + \dfrac{x_c}{k_c}} = \frac{t_1 - t_4}{R_a + R_b + R_c} \tag{4–3}$$

where $R = \dfrac{x}{kA}$ for any lamina = resistance, $\dfrac{\text{(hr) (F)}}{\text{Btu}}$.

Resistance, R, receives its name from its similarity to electrical resistance in Ohm's law.

The same law and assumptions may be applied when the area through which heat flows is not a plane area, as was used previously, but a cylindrical wall. Also, the length of the path of flow will be a radial distance. Referring to Fig. 4–3,

$$q = -k2\pi rL \frac{dt}{dr} \tag{e}$$

$$\int_{t_i}^{t_f} dt = -\int_{r_i}^{r_f} \frac{q\,dr}{k2\pi rL} \tag{f}$$

$$q = \frac{2\pi kL(t_i - t_f)}{\ln \dfrac{r_f}{r_i}} \tag{4-4}$$

where r_f = final or outside radius, ft or in.

r_i = initial or inside radius, ft or in.

(The units of feet or inches may be used interchangeably in the ratio if *both* radii are expressed in the same units.)

L = length of the cylinder, ft

and other symbols are as in Eq. (4–3).

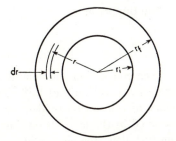

FIG. 4–3. Conduction through a cylindrical wall.

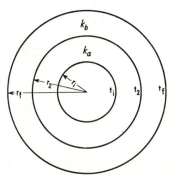

FIG. 4–4. Conduction through a composite cylindrical wall.

Formulas for conduction through composite cylindrical walls may be derived in the same manner as Eq. (4–3) for plane composite walls. Remembering that the area is cylindrical rather than plane, and using Fig. 4–4,

$$q = \frac{2\pi L(t_i - t_f)}{\dfrac{\ln \dfrac{r_2}{r_i}}{k_a} + \dfrac{\ln \dfrac{r_f}{r_2}}{k_b}} \tag{4-5}$$

where the symbols have the same units as in Eqs. (4–3) and (4–4) and Fig. 4–4.

EXAMPLE 4–1. Determine the heat transferred through a composite plane-wall area of 1 sq ft of carbon steel 1 in. thick and magnesite brick 8 in. thick. Take the temperatures at 700 F and 300 F.

SOLUTION. From Fig. 4–1(a), select thermal conductivities of 25 for steel and 2.0 for the magnesite brick at a mean temperature of 500 F. Using Eq. (4–3),

$$q = \frac{1(700 - 300)}{\dfrac{1}{12 \times 25} + \dfrac{8}{12 \times 2.0}} = 1188 \text{ Btu per hr}$$

Now find the temperature of the hot surface of the magnesite brick by Eq. (4–2),

$$t = 300 + \frac{1188 \times 8}{12 \times 2.0 \times 1} = 696 \text{ F}$$

The mean temperature of the magnesite brick is 498 F and the assumed value of thermal conductivity is correct. Note that an almost identical answer could be obtained by neglecting the metal wall.

EXAMPLE 4–2. Use the data in the previous example to determine the heat transferred through a steel pipe 1 ft long and 1 in. thick with an ID of 1.12 ft and covered with 8 in. of magnesite brick.

SOLUTION. From Eq. (4–5),

$$q = \frac{2\pi 1(700 - 300)}{\dfrac{\ln \dfrac{0.644}{0.56}}{25} + \dfrac{\ln \dfrac{1.310}{0.644}}{2.0}} = 6940 \text{ Btu per hr}$$

4–3. Convection. The distinction between convection and conduction is not exact because there are some features of convection that fall within the category of conduction. The film theory of convection is identical to the boundary-layer concept that was discussed in the chapter on flow of fluids (Chap. 2). It was stated that during turbulent flow of a fluid over a confining wall, there is a thin boundary layer of the fluid adjacent to the wall which has no relative velocity with respect to the wall. The thickness of this lamina is dependent on many factors, including the roughness of the wall and the turbulence of the main body of the fluid. The fact that there is heat transfer through the wall does not alter the existence of this boundary layer. In heat transfer, this layer is called the *stagnant film*. Regardless of the name, if this layer has little or no motion, then heat transfer through it comes under the province of conduction.

Much of the resistance to the flow of heat encountered during convection is due to the very thin stagnant film. But because of its minute and variable thickness, it is convenient to combine the effects of both conduc-

TABLE 4–1

SIMPLIFIED FORMULAS FOR FILM COEFFICIENTS *

(During turbulent flow and forced convection except for Eqs. I and J)

Fluid	Equation	Conditions	Formula for h Btu/(sq ft)(hr)(F)
Air or flue gas	A †	Flow inside a cylindrical pipe. For rectangular ducts use $D = \dfrac{2ab}{a+b}$ where a and b are the sides	$h = \dfrac{0.32 V_0^{0.8}}{D^{0.25}}$
	B	Vertical or horizontal rolled metal walls	$h = 0.3 \left(\dfrac{530V}{T}\right)^{0.78}$
	C	Flow of gases perpendicular to a pipe	$h = 0.7 \dfrac{V_0^{0.56}}{D^{0.44}}$
	D	Flow over 4 or more rows of evenly arranged pipes	$h = 0.65 \dfrac{V_0^{0.654}}{D^{0.346}}$
	E	Flow over 4 staggered rows of pipes	$h = 0.86 \dfrac{V_0^{0.69}}{D^{0.31}}$
	F	Flow over 5 or more staggered rows of pipes	$h = 0.91 \dfrac{V_0^{0.69}}{D^{0.31}}$
Water	G	Flow inside a pipe	$h = 163 V^{0.85}(1 + 0.0104t)$
Oil	H	Flow inside a pipe	$h = 0.087 V(t - 32)$
Steam	I	Condensing on or inside horizontal pipes	$h = \dfrac{1225 + 3.15(t_s + t_w)}{\sqrt[4]{D(t_s + t_w)}}$
Water	J	Evaporating from a pipe	$h = \dfrac{(p)^{1/8}(\Delta t)^{4.08}}{252}$
Steam	K	Superheated steam in pipes	$h = \left(0.432 + 0.151\dfrac{t}{1000}\right)\dfrac{V_0^{0.79}}{D^{0.16}}$

(handwritten margin note: 10–20 Gas)

(handwritten margin note: 2000–3000 liquid)

Symbols: V_0 = velocity of gas based on 14.7 psia and 32 F, fps

$\quad = \dfrac{G}{\rho_0}$ where G = mass flow, lb per (sq ft)(sec)
$\qquad\qquad\quad \rho_0$ = density at 32 F and 14.7 psia, lb per cu ft
$\qquad\qquad\qquad$ = 0.0502 for superheated steam *(handwritten: .0802 air)*

$\quad V$ = actual fluid velocity, fps
$\quad D$ = diam, ft
$\quad T$ = absolute temperature of fluid, R
$\quad t$ = mean fluid temperature, F
$\quad t_s$ = steam temperature, F
$\quad t_w$ = wall temperature, F
$\quad p$ = pressure of fluid being evaporated, psia
$\quad \Delta t$ = temperature difference between wall and liquid, F

* All equations, except Eq. J, approximately as given in *Industrial Heat Transfer* by Schack, Goldschmidt, and Partridge; published by, and reproduced by permission of, John Wiley & Sons, Inc.

† If flow is parallel to tubes and on the outside, use $D = \dfrac{4S}{P}$, where S is the cross-sectional flow area in sq ft and P is the part of the circumference through which heat is transferred, in ft.

tion through the boundary layer and convection due to mixing, for the remainder of the fluid into one heat-transfer coefficient (h). Then

$$q = hA(t_f - t_i) \tag{4-6}$$

where h = factor of proportionality or film coefficient of heat transfer,

$$\frac{\text{Btu}}{(\text{hr})\,(\text{sq ft})\,(\text{F})}$$

and other symbols are as in Eq. (4–2).

Convection has been divided into two classes: (1) natural and (2) forced. Motion during natural convection is caused by differences in fluid density, as in a natural-convection, hot-water heating system for a home. Forced convection requires some form of fan or pump to compel fluid motion.

Film coefficients vary to a large degree and are dependent on several factors. For turbulent flow, the *film coefficient* is a function of the temperature and pressure of the fluid. If the fluid does not change state, then the coefficient is also a function of velocity and a function of the direction of flow if the fluid is on the outside of a tube. Film coefficients are best correlated by means of certain dimensionless relationships that may be developed by dimensional analysis. Further explanations of this method may be found in the Bibliography listed at the end of this chapter. Values of film coefficients for use in problems may be obtained from the formulas of Table 4–1.

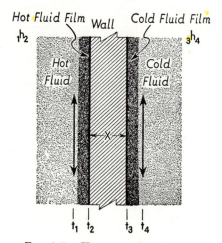

FIG. 4–5. Heat transfer through a plane wall by convection and conduction.

4–4. Combined Convection and Conduction.

Few cases exist in practice where either convection or conduction prevails singly or where one effect is negligible compared with the other. All heat-exchange equipment in a power plant combines convection and conduction, and in some cases combines all three types of heat transfer.

The typical conditions of heat transfer in a power plant include a hot fluid flowing over one side of a plane wall and a cold fluid passing over the other side, Fig. 4–5. Thermal energy is transferred from the hot, turbulent fluid to the hot fluid stagnant film by convection. The transfer through the hot, stagnant fluid is a matter of conduction but is included with convection, as explained above. The thermal energy is then transferred through the wall in pure conduction and thence to the cold fluid through its stagnant film by convection.

Proceeding in the manner previously used for conduction through walls in series,

$$q = {_1}h_2 A (t_1 - t_2)$$

$$q = \frac{{_2}k_3 A (t_2 - t_3)}{x} \qquad \text{(a)}$$

$$q = {_3}h_4 A (t_3 - t_4)$$

Solving each of the equations for the temperature difference, and adding the resulting equations,

$$t_1 - t_4 = \frac{q}{A} \left[\frac{1}{{_1}h_2} + \frac{x}{{_2}k_3} + \frac{1}{{_3}h_4} \right] \qquad \text{(b)}$$

$$q = \frac{A (t_1 - t_4)}{\left[\dfrac{1}{{_1}h_2} + \dfrac{x}{{_2}k_3} + \dfrac{1}{{_3}h_4} \right]} \qquad \text{(4-7)}$$

$$q = UA (t_1 - t_4) \qquad \text{(4-8)}$$

where U = over-all heat-transfer coefficient, Btu per (sq ft) (hr) (F), and other symbols are as in Eqs. (4–1) and (4–6) and Fig. 4–5. Then

$$U = \frac{1}{\dfrac{1}{{_1}h_2} + \dfrac{x}{{_2}k_3} + \dfrac{1}{{_3}h_4}} \qquad \text{(4-9)}$$

This new term, the over-all heat-transfer coefficient, U, has the same units as the film coefficient and is the reciprocal of the sum of all resistances of convection films and conducting materials in the path of thermal energy flow. Eq. (4–7) has been derived for the specific case of the flow of thermal energy through two films and one conducting wall. Additional conducting walls would merely add more expressions for resistance in the denominator of Eq. (4–7).

The over-all coefficient of heat transfer for cylindrical walls will differ from Eq. (4–7) in the same way that the formulas for conduction through cylindrical walls differed from those for plane walls. To obtain the necessary expression, refer to Fig. 4–6 and represent the temperature drops as follows:

$$t_1 - t_2 = \frac{q}{h_o \pi L (\text{OD})}$$

$$t_2 - t_3 = \frac{q}{\dfrac{k 2 \pi L}{\ln (\text{OD/ID})}} \qquad \text{(c)}$$

$$t_3 - t_4 = \frac{q}{h_i \pi L (\text{ID})}$$

Thus $\qquad t_1 - t_4 = q \left[\dfrac{1}{h_o \pi L (OD)} + \dfrac{1}{\dfrac{k 2 \pi L}{\ln (OD/ID)}} + \dfrac{1}{h_i \pi L (ID)} \right]$ $\qquad$ (d)

and $\qquad q = \dfrac{t_1 - t_4}{\dfrac{1}{h_o \pi L (OD)} + \dfrac{\ln (OD/ID)}{2 \pi L k} + \dfrac{1}{h_i \pi L (ID)}}$ $\qquad$ (e)

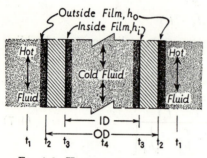

FIG. 4-6. Heat transfer through a cylindrical wall by convection and conduction.

If the over-all coefficient of heat transfer, U_o, is based on the outside area of the pipe, A_o, in accordance with common practice for power-plant heat exchangers, then

$$q = U_o A_o (t_1 - t_4) \quad (4\text{--}10)$$

$$A_o = \pi L (OD) \quad (4\text{--}11)$$

Equating Eqs. (e) and (4–10) and substituting Eq. (4–11) into the resultant,

$$U_o = \dfrac{1}{\dfrac{1}{h_o} + \dfrac{(OD) \ln (OD/ID)}{2k} + \dfrac{(OD)}{(ID) h_i}} \quad (4\text{--}12)$$

where symbols are as in Eqs. (4–1) and (4–6) and Fig. 4–6, with the diameters in feet. Subscript o indicates that the equation refers to the outside surface of the pipe.

EXAMPLE 4–3. Calculate the over-all coefficient of heat transfer, U, and the quantity of heat transferred per hour through a lead tube 1 ft long with an OD of 1.000 in. and an ID of 0.870 in. Assume the hot fluid outside the tube to be water at 150 F with a film coefficient of 400 Btu per (sq ft)(hr)(F), and the cold fluid inside the tube also water but at 130 F with a velocity of 5 fps.

SOLUTION. From Table 4–1,

$$h_i = 163 \times 5^{0.85}(1 + 0.0104 \times 130) = 1505 \frac{\text{Btu}}{(\text{sq ft})(\text{hr})(\text{F})}$$

From Eq. (4–12),

$$U_o = \dfrac{1}{\dfrac{1}{400} + \dfrac{\dfrac{1.000}{12} \ln \dfrac{1.000}{0.870}}{2 \times 20} + \dfrac{1.000}{0.870(1505)}} = 282 \frac{\text{Btu}}{(\text{sq ft})(\text{hr})(\text{F})}$$

From Eq. (4–11),

$$A_o = \frac{\pi (1)(1)}{12} = 0.2615 \text{ sq ft}$$

From Eq. (4–10),

$$q = 282 \times 0.2615(150 - 130) = 1480 \text{ Btu per hr}$$

4–5. Mean Temperature Difference. Thus far the temperatures of the two fluids involved in heat-transfer processes have been considered constant throughout the lengths of the surfaces. All fluids receiving or rejecting heat by convection and conduction will experience a temperature change unless there is a change of state during the process. Therefore, the equations which have been developed are applicable only to heat transfer from a condensing fluid to a fluid which is evaporating. Fig. 4–7(a) illustrates this process by showing the temperature history of both

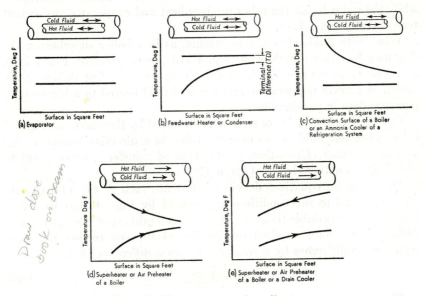

FIG. 4–7. Temperature-surface diagrams.

the hot and cold fluids as they flow along the transfer surface from 0 to 100% of the total surface in the heat exchanger. These are the types of curve that will be found to exist in evaporators where water is boiled on the outside of tubes while steam is condensed at a higher temperature on the inside of the tubes. The temperature difference between the two fluids is the thermal potential that is causing the flow of heat and is called *heat head* when the two fluids are changing state. The confining conduits are shown schematically in this figure.

Fig. 4–7(a) also represents approximate conditions existing for heat loss from an insulated pipe. Usually, the loss through insulated pipe is so small that there is little appreciable change in the temperature of the fluid inside the pipe. Furthermore, the air passing over the pipe is of such large mass that its temperature may be considered constant and the same as atmospheric temperature.

In all, there are five combinations of curves that may be drawn for heat exchangers involving one or both of the following: (1) constant-temperature heat addition or rejection, and (2) variable-temperature heat addition or rejection. All five of these possible combinations may be found in the steam-power plant. The hot fluid condensing and the cold fluid increasing in temperature, Fig. 4–7(b), is representative of the conditions existing in surface condensers or feedwater heaters. The hot fluid is, of course, steam, and the cold fluid is water. However, the hot gases passing over the convection surface of a boiler change temperature, while the water inside the tubes receives heat and evaporates at constant temperature, Fig. 4–7(c).

When both fluids change temperature, as in a boiler superheater, the flow may be either parallel flow, Fig. 4–7(d), or counterflow, Fig. 4–7(e). In many cases counterflow is preferable because the hot fluid may be cooled to a lower temperature and the cold fluid heated to a higher temperature than in parallel flow.

One final case that is not strictly represented by the curves of Fig. 4–7 exists when one fluid passes over the tubes in a direction normal to the axis of the tubes. Cross-flow, as this type of transfer is known, will be discussed in another chapter (Art. 8–17).

Heat-transfer equations developed thus far, such as Eq. (4–10), cannot be applied to the conditions represented by Figs. 4–7(b) to 4–7(e) because of the variable temperature difference existing at each element of surface. Therefore, we shall rewrite Eq. (4–10) and replace the exact temperature difference by a *mean temperature difference* (θ_m).

$$q = UA\theta_m \qquad (4\text{–}13)$$

There are two methods of calculating mean temperature difference: arithmetic and log. Arithmetic mean temperature difference (abbreviated AMTD) assumes that all the curves of Fig. 4–7 are straight lines, Fig. 4–8. Since this assumption may obviously lead to erroneous results, it is recommended that the student never use AMTD except when log mean temperature difference (abbreviated LMTD) gives an indeterminate answer, as in the case of an evaporator. Thus

$$\text{AMTD} = \theta_m = \frac{(t_1 + t_2) - (t_3 + t_4)}{2} = \frac{\theta_1 + \theta_2}{2} \qquad (4\text{–}14)$$

where symbols are as defined in Fig. 4–8.

Eq. (4–14) will give the AMTD for any of the types of temperature-surface diagram in Fig. 4–7 if θ_1 is always the temperature difference at one end of the diagram and θ_2 is taken as the temperature difference at the other end of the diagram.

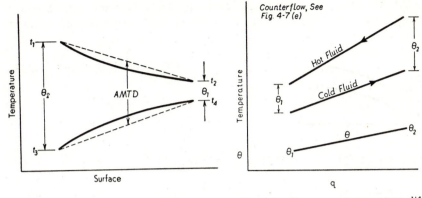

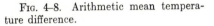

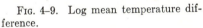

FIG. 4–8. Arithmetic mean tempera-   FIG. 4–9. Log mean temperature dif-
ture difference.   ference.

To derive an equation for LMTD, plot curves, using temperature as the ordinate and heat as the abscissa, for each of the two fluids by assuming that specific heat is a constant. This has been done in Fig. 4–9 for counterflow. Next, calculate and plot the temperature difference (θ) between the two curves. Then the slope of the θ curve is

$$\frac{d\theta}{dq} = \frac{\theta_2 - \theta_1}{q} \tag{a}$$

but

$$dq = U\theta \, dA \tag{b}$$

Combining Eqs. (a) and (b),

$$\int_{\theta_1}^{\theta_2} \frac{d\theta}{\theta} = \frac{U(\theta_2 - \theta_1)}{q} \int_0^A dA \tag{c}$$

$$q = \frac{UA(\theta_2 - \theta_1)}{\ln \frac{\theta_2}{\theta_1}} = UA\theta_m \tag{d}$$

$$\text{LMTD} = \theta_m = \frac{\theta_2 - \theta_1}{\ln \frac{\theta_2}{\theta_1}} \tag{4-15}$$

Eq. (4–15) for LMTD applies to any of the temperature-surface diagrams shown in Fig. 4–7(b) to (e). However, it should be emphasized that θ_1 is the temperature difference at one end of the diagram and θ_2 is the temperature difference at the other end of the diagram. Either θ_1 or θ_2 may be taken as the larger value.

In subsequent chapters, the smaller of the temperature differences θ_1 and θ_2 will be referred to as the *terminal difference* (TD).

EXAMPLE 4–4. Calculate the LMTD and the AMTD when the hot fluid enters a heat exchanger at 150 F and leaves at 90 F. The cold fluid enters at 80 F and leaves at 95 F.

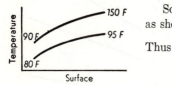

SOLUTION. Draw the temperature-surface diagram, as shown in the illustration here.

Thus $\theta_1 = 150 - 95 = 55$ F

               $\theta_2 = \ 90 - 80 = 10$ F

From Eq. (4–15), $\text{LMTD} = \dfrac{55 - 10}{\ln \frac{55}{10}} = 26.4$ F

From Eq. (4–14), $\text{AMTD} = \dfrac{55 + 10}{2} = 32.5$ F

4–6. Scale. Fluids containing certain impurities often deposit these impurities in the form of scales on the tubes. One of the hardest scales produced in power-plant equipment is the glasslike scale that forms on evaporator tubes that have been operating in water containing silicon. Some waters form other scales that are not so hard and which can be removed more easily.

Slag and soot deposits on the outside of boiler tubes, as well as scale deposits on the inside, act in much the same fashion as insulation to reduce the heat transfer.

Although allowances should be made in designing apparatus which may become scaled, it is difficult to predict the scale thickness. Some boiler tubes have been found to be almost completely clogged by inside scale formations. As time progresses, such deposits will grow in thickness unless cleaned at periodic intervals.

Since it is customary to use a film coefficient for scales,

$$q = h_s A (t_1 - t_2) \qquad (4\text{--}16)$$

where h_s = scale coefficient, Btu per (sq ft) (hr) (F).

We may now write a typical formula for the over-all coefficient of heat transfer (U_o), including scale, film, and conduction coefficients for a circular pipe:

$$U_o = \cfrac{1}{\dfrac{1}{h_o} + \dfrac{1}{h_{so}} + \dfrac{(\text{OD}) \ln (\text{OD/ID})}{2k} + \dfrac{(\text{OD})}{(\text{ID}) h_{si}} + \dfrac{(\text{OD})}{(\text{ID}) h_i}} \qquad (4\text{--}17)$$

where subscripts i and o refer to inside and outside, respectively, and other symbols are as previously defined.

Recommended minimum scale coefficients for some waters are listed in Table 4–2.

TABLE 4–2

SCALE COEFFICIENTS FOR SOME WATERS *

Temperature of heating medium	Up to 240 F	240–400 F †
Temperature of water	125 F or Less	Over 125 F
Types of water	$\dfrac{1}{h_s}$	$\dfrac{1}{h_s}$
Sea water	0.0005	0.001
Brackish water	0.001	0.002
Cooling tower and artificial spray pond:		
Treated make-up	0.001	0.002
Untreated	0.003	0.005
City and well water (such as Great Lakes)	0.001	0.002
River water:		
Minimum	0.001	0.002
Mississippi	0.002	0.003
Delaware, Schuylkill	0.002	0.003
East River and N. Y. Bay	0.002	0.003
Muddy or silty	0.002	0.003
Hard (over 15 grains per gal)	0.003	0.005
Distilled	0.0005	0.0005
Treated boiler feedwater	0.0005	0.001
Boiler blowdown	0.001	0.001

  * Standards of Heat Exchange Institute, Tubular Exchanger Section.
  † These ratings are based on a temperature of the heating medium of 240–400 F. If the heating medium is over 400 F, and the cooling medium is known to scale, these ratings should be modified accordingly.

4–7. Condensing and Boiling. Condensation of a vapor on a metal surface may take place in dropwise or filmwise form. Dropwise condensation occurs when the vapor condenses as drops which are eventually removed from the surface by gravity. This condensation is unstable and normally occurs in practice only when the surface is polished or contaminated by oil. Film condensation, which occurs when a vapor condenses and forms a liquid film over the cool surface, has a much lower film coefficient for steam than the former. However, the effect of condensation on the over-all coefficient is not large because of the other component factors involved in the over-all heat-transfer coefficient U.

In a similar manner, there are two types of vaporization—film boiling and nuclear boiling. The latter consists of the formation of vapor bubbles on the hot surface. The size of the bubbles and the rate at which they break away from the tube will determine the film coefficient. Bubble size is a function of the ability of the liquid to "wet" the surface. The rate of breaking away from the surface is affected by agitation.

Film vaporization is analogous to film condensation and is evident when the heat head is too high. A film of vapor surrounds the tube and acts like an insulator because of the low conductivity of a vapor. This effect has an important bearing on evaporator design, which will be discussed in another chapter (Art. 8–16).

4–8. Radiation. Radiant energy is a form of wave motion that obeys the same laws as other waves—light waves for example. Radiant energy travels in a straight line and is subject to reflection. Probably the best-known example of this type of heat transfer is the energy received by the earth from the sun. The waves pass through the almost perfect vacuum between the sun and the earth and also through the layer of atmosphere surrounding the earth. Yet they have no appreciable effect upon the atmosphere even though they pass through it.

Some of the radiant energy striking a body is reflected, some is absorbed, and the remainder is transmitted through the body without affecting it. Thus the fraction of the radiant energy reflected by a body (*reflectivity*), plus the fraction absorbed by the body (*absorptivity*), plus the fraction transmitted through the body (*transmissivity*) must equal *unity*. The fraction of radiant energy that a body will absorb, transmit, or reflect depends upon its composition and temperature and upon the wave length of the energy. Window glass transmits most of the short-wave energy (light) it receives but absorbs most of the longer wave length energy. Some materials reflect most of the radiant energy they receive.

Any body that would absorb all the radiant energy it received would be a *perfect black body*. Obviously, no such body exists. This perfect black body would have the *ability* to radiate a maximum of energy. In fact, Kirchhoff's law in effect states that the ratio of the rate of emitting and absorbing energy is a constant for any body for a given temperature and wave length. A perfect black body is *capable* of emitting radiant energy at the same rate at which it is *capable* of absorbing it. Since it can absorb more energy than any other body, it can also emit more radiant energy than any other body.

The emissivity of any body is the ratio of the rate of emission of radiant energy for that body to the rate of emission for a perfect black body under the same conditions. But since the rate of emission of energy from a body is proportional to the rate at which the body is capable of absorbing radiant energy (Kirchhoff's law), emissivity may also be defined in terms of the ratio of absorptivity of the given body and the absorptivity of a perfect black body. Observe that the concept of a perfect black body serves as a standard of comparison for other bodies and that emissivity may be thought of as the efficiency of the actual body in absorbing or emitting radiant energy.

Emissivity for many surfaces increases with temperature.

Any surface will radiate some energy, but the amount depends on the material and its temperature. For a perfect black body, the total radiation may be calculated from the Stefan-Boltzmann law (named after its discoverers):

$$q = 0.174A \left(\frac{T}{100}\right)^4 \tag{4-18}$$

where T = the absolute temperature of the body

 q = Btu per hr

and other symbols are as before.

By applying our definition of emissivity (ϵ), the energy radiated by an actual body is

$$q = 0.174A\epsilon \left(\frac{T}{100}\right)^4 \tag{4-19}$$

While a perfect black body, or any other body, is absorbing radiant energy at temperature T_1, it is also emitting energy to the other body at another absolute temperature T_2. In most cases we shall be interested in the net energy received by a particular body, and so for convenience we shall assume for the body under consideration that T_1 is a higher absolute temperature than the temperature T_2. We may now write an equation for the *net* energy absorbed by a body whose emissivity is ϵ:

$$q = 0.174\epsilon A \left[\left(\frac{T_1}{100}\right)^4 - \left(\frac{T_2}{100}\right)^4\right] \tag{4-20}$$

Note that Eq. (4-20) is valid *only* for a small body of any shape completely enclosed by a large body. The reason for this is that one body may not always intercept all the rays emitted by another body, but a large enclosure must intercept all the rays emitted by a small body completely within the enclosure.

Values of emissivity for a few surfaces that will be of interest in power-plant problems are listed in Table 4-3.

EXAMPLE 4-5. Calculate the radiant-heat transfer to a room from 12 ft of bare 1½-in. schedule 40 pipe carrying steam at 225 F. The room is at 70 F and the pipe is painted a light color ($\epsilon = 0.90$).

SOLUTION.   OD = 1.9 in.

From Eq. (4-20),

$$q = 0.174 \times 0.90 \times 12 \times \frac{1.9}{12} \pi \left[\left(\frac{685}{100}\right)^4 - \left(\frac{530}{100}\right)^4\right] = 1318 \text{ Btu per hr}$$

4-9. Insulation Materials. Piping and other equipment in steam-power plants may be insulated for any one or a combination of the follow-

TABLE 4-3

EMISSIVITY VALUES FOR SOME SURFACES *

Surfaces	Values of ϵ		
	70 F	1000 F	3000 F †
Clean Surface:			
Aluminum	0.05	0.075	....
Brass	0.05	0.06	....
Copper	0.04	0.08	0.15
Iron, cast or wrought	0.20	0.25	0.28
Lead	0.08	....	....
Monel metal	0.07	0.10	....
Nickel	0.06	0.10	....
Platinum	0.036	0.10	0.20
Silver	0.025	0.035	....
Steel	0.20	0.25	0.28
Tin	0.08	....	....
Tungsten	0.03	0.09	0.25
Zinc	0.10	....	....
Oxidized Surface:			
Aluminum	....	....	0.10–0.20
Brass	....	....	0.25–0.60
Copper	....	....	0.55–0.75
Iron and steel	....	....	0.60–0.90
Monel metal	....	....	0.40–0.50
Nickel	....	....	0.40–0.60

* *Mechanical Engineering*, Vol. 54, p. 494.
† Or molten, if the melting point is below 3000 F.

ing reasons: (1) to reduce heat loss, (2) to prevent sweating of cool surfaces, (3) to reduce the possibility of injury to personnel from contact with hot surfaces. Insulations used for any of these reasons are composed of cork, cattle hair, wool felt, glass wool, rock wool, diatomaceous earth, magnesium carbonate, asbestos, aluminum foil, or refractories. A brief summary of the salient properties of the more common of these materials follows.

Magnesium carbonate, made from limestone, is extremely popular for temperatures up to 600 F. It has a high re-use value and low conductivity, and it is resistant to shock and vibration. Usually it is mixed with asbestos in the proportions of 15% asbestos and 85% magnesium. In this form it is referred to as "85% magnesia," and it may be used as a cement, molded into blocks, or molded to fit pipe, Fig. 4–10. Magnesia pipe coverings are held in place by canvas and wire or metal strips. The magnesia blocks are used on large surfaces, Fig. 4–11, and are held in position by wire netting.

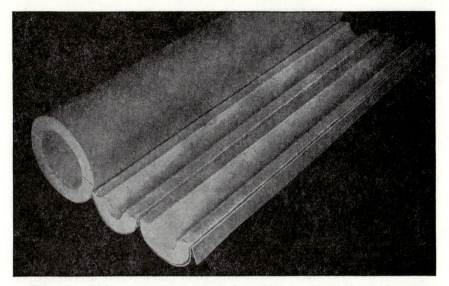

FIG. 4–10. Eighty-five per cent magnesia molded pipe insulation with canvas covering. (Ehret Magnesia Manufacturing Co.)

Asbestos is a fibrous mineral used as a mixture with magnesium, as a felt, or as a paper. In its paper form, asbestos may be wound in strips on pipe or may be used as corrugated board, as cement, or in built-up laminar sections. The air-cell type (cellular asbestos) consists of layers of corrugated asbestos paper cemented together, Fig. 4–12. Because of shrinkage and lack of mechanical strength, cellular asbestos does not have a long life. It is not recommended for temperatures exceeding 300 F.

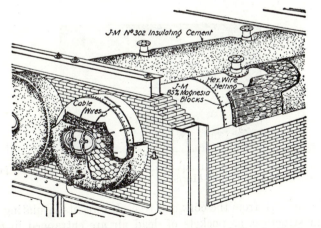

FIG. 4–11. Application of 85 per cent magnesia insulation in the form of blocks and cement to a boiler drum. (Johns-Manville Corp.)

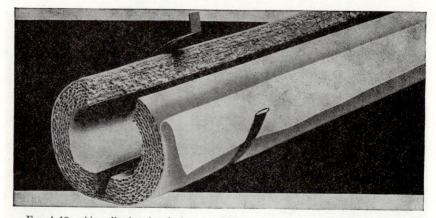

Fig. 4–12. Air-cell pipe insulation consisting of eight corrugated asbestos paper plies and ten plain asbestos paper plies per inch with canvas covering. (Keasbey and Mattison Co.)

Laminated asbestos felts are composed of thin layers of asbestos paper cemented together. The maximum temperature is about 600 F. The material should not be used in places where the insulation may become wet.

Diatomaceous earth is composed of the siliceous skeletons of minute diatoms that were deposited on the bottoms of lakes and seas many years ago. When mixed with asbestos fiber for a binder, it is an excellent insulating material for the high-temperature range up to about 2000 F. Diatomaceous earth may be used in the natural form, called Celite, by being cut into bricks. It may also be used after pulverizing and firing in kilns. Usually, a layer of 85% magnesia is used in combination with the diatomaceous earth, Fig. 4–13.

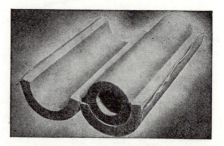

Fig. 4–13. Combination high temperature and 85 per cent magnesia pipe covering with canvas covering. (Ehret Magnesia Manufacturing Co.)

Cork is the bark from the cork oak tree. It owes its insulating value to its cellular structure, as pockets of dead air are entrapped in the cells. Cork is particularly suitable for the insulation of cold surfaces. In the power plant, it is used in the granular or molded form to prevent sweating of cold surfaces.

Glass wool and *rock wool* are formed by melting the basic materials and blowing them into fine threads. Glass wool has found some importance in the power-plant field for insulation of high temperatures. It is

Fig. 4-14. Fiberglas blanket insulation with different types of facing materials. (Owens-Corning Fiberglas Corp.)

often used in the blanket form, Fig. 4-14, but may be obtained in blocks, cement, or preformed sleeves to fit the pipe, Fig. 4-15.

Aluminum foil is worthy of mention particularly because of the method by which it reduces heat loss. The foil is crumpled and applied to the surface in sheets which entrap small quantities of air. The insulation

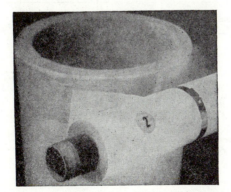

Fig. 4-15. Fiberglas-formed pipe insulation with canvas covering. (Owens-Corning Fiberglas Corp.)

value is then obtained by the low conductivity of dead-air spaces and also because of the low emissivity of the aluminum foil surface. If the foil becomes tarnished from acids, etc., its insulating value may be seriously reduced.

Hair felt, made from cattle hair, is suitable for subzero temperatures and up to 200 F. Temperatures recommended for *wool felt* range from freezing of water to 200 F.

Refractories are made from fire-clay, alumina, silica, magnesite, and chrome. They are used in furnaces, kilns, etc., where resistance to extremely high temperature is essential. Because of the effects of the high temperatures, erosion, and chemical action of slag, the life of furnace refractory linings is limited.

There are many methods of applying these various types of insulation, depending on the nature of the material and the shape of the surface. Blankets of insulation materials are especially adaptable to large surfaces or to turbine and pump casings which must be removed for inspection of the internal parts at frequent intervals. Some methods of applying insulation to pipe fittings may be visualized from the sketches of Fig. 4–16.

Insulation thicknesses to give a surface temperature of 125 to 140 F represent good practice for hot piping, boilers, and equipment. The lower

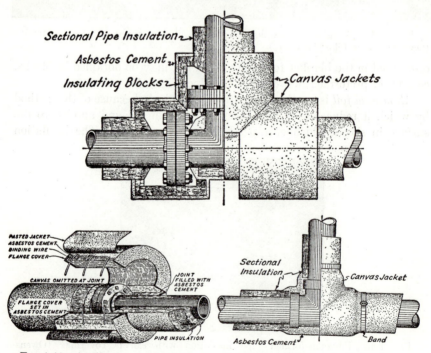

FIG. 4–16. Applications of insulation to pipe fittings. (Johns-Manville Corp.)

temperature frequently represents more than the most economical thickness but aids in maintaining comfortable building temperature and promotes safety.

4–10. Pipe Insulation. Heat transmission through a pipe wall and the insulation surrounding the pipe is a matter of conduction, convection, and radiation. For convenience, radiation and convection coefficients are combined into one coefficient:

$$q = h_{cr}A(t - t_f) \qquad (4\text{–}21)$$

where h_{cr} = combined convection and radiation coefficient, Btu per (hr) (sq ft) (F)

A = area of outside insulation surface, sq ft

t and t_f = surface and air temperatures, respectively, F

Values for the combined coefficient, which may be used either for insulation surfaces or for bare steel pipe in still air at the usual room temperatures, are given in Fig. 4–17. Pipes located in an air stream will

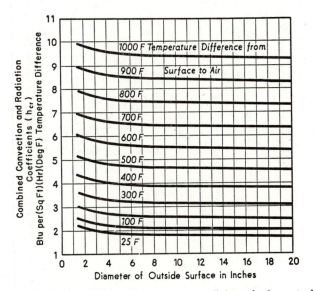

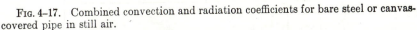

Fig. 4–17. Combined convection and radiation coefficients for bare steel or canvas-covered pipe in still air.

experience heat losses several times larger than pipes located in still air, the increase being due to convection.

Total loss of energy from an insulated pipe due to convection, conduction, and radiation may be calculated by using the procedure of Art. 4–4. Thus

$$q = U_oA_o(t_i - t_f) \qquad (4\text{–}22)$$

TABLE 4-4

RECOMMENDED INSULATION THICKNESSES

Pipe Size	Under 2 in.		2–4 in.		5–8 in.		10 in. and over		Ducts and Equipment	
Temp Range, F	M	D	M	D	M	D	M	D	M	D
150–350	1	...	1	...	1½	...	1½	...	2	...
351–450	1½	...	1½	...	2	...	2½	...	2½	...
451–550	2	2	2	1½	2½	1½	3	1½	3	1½
551–650	...	2	1½	1½	2	1½	2½	2	2½	2
651–800	...	2	2	2	2½	2½	2½	3	2½	...
801–900	...	2	2	2	2	2½	2½	3	...	...
901–1000	...	2	2½	2	2½	2½	3	3	...	...
1001–1200	...	2½	2½	2	2½	2½	3	3½	...	...
40–60	Two ½-in. layers of wool felt with inner surface of each layer lined with asphalt-impregnated asbestos paper.									
60–90	One 1-in. layer of wool felt with inner surface lined with asphalt-impregnated asbestos paper and outer surface of cotton jacket.									

Actual Pipe Insulation Thicknesses, Inches

Pipe Size, in.	$\frac{1}{2}$	1	2	3	4	6	8	10	12	14-33
Nominal Thickness, in.										
1	1	$1\frac{3}{32}$	$1\frac{1}{16}$	$1\frac{1}{32}$	$1\frac{1}{16}$	1	...	...	...	...
$1\frac{1}{2}$	$1\frac{9}{16}$	$1\frac{19}{32}$	$1\frac{19}{32}$	$1\frac{9}{16}$	$1\frac{9}{16}$	$1\frac{1}{2}$	$1\frac{17}{32}$	$1\frac{19}{32}$	$1\frac{19}{32}$	$1\frac{1}{2}$
2	$2\frac{1}{16}$	$2\frac{1}{8}$	$2\frac{1}{8}$	$2\frac{1}{16}$	$2\frac{1}{16}$	$2\frac{1}{32}$	$2\frac{1}{32}$	$2\frac{3}{32}$	$2\frac{3}{32}$	2
$2\frac{1}{2}$	$2\frac{7}{8}$	$2\frac{21}{32}$	$2\frac{5}{8}$	$2\frac{9}{16}$	$2\frac{9}{16}$	$2\frac{17}{32}$	$2\frac{11}{16}$	$2\frac{19}{32}$	$2\frac{5}{8}$	$2\frac{1}{2}$
3	$3\frac{11}{32}$	$3\frac{5}{8}$	$3\frac{1}{8}$	$3\frac{1}{16}$	$3\frac{3}{32}$	$3\frac{3}{32}$	$3\frac{1}{8}$	$3\frac{3}{32}$	$3\frac{3}{32}$	3
$3\frac{1}{2}$	$3\frac{7}{8}$	$3\frac{21}{32}$	$3\frac{5}{8}$	$3\frac{19}{32}$	$3\frac{19}{32}$	$3\frac{11}{16}$	$3\frac{5}{8}$	$3\frac{19}{32}$	$3\frac{19}{32}$	$3\frac{1}{2}$

NOTES: M = 85% magnesia or equal
D = diatomaceous silica
Insulation thicknesses given are nominal thicknesses in inches and do not include cement finish, mesh, lath, canvas, etc.

The pipe wall and the fluid film inside the pipe offer very little resistance to the heat flow, compared with the resistance offered by the insulation and the film on the outside of the pipe (compare the h_{cr} and k values for insulation with h and k values for the inside film and pipe wall). Therefore, neglect the effects of the inside film and the pipe wall and obtain

$$U_o = \frac{1}{\dfrac{(OD)\ \ln\ (D/OD)}{2k} + \dfrac{(OD)}{Dh_{cr}}} \qquad (4\text{--}23)$$

in which U_o = over-all heat-transfer coefficient based on the outside diameter of the pipe, Btu per (sq ft) (hr) (F)

A_o = outside area of pipe, sq ft

OD = outside diam of pipe, ft

D = diam of outside surface of insulation, ft

k = average thermal conductivity of the insulation, Btu per (hr) (F) (ft)

h_{cr} = combined convection and radiation coefficient, Btu per (hr) (sq ft) (F)

t_i = initial temperature or fluid temperature, F

t_f = final temperature or air temperature, F

Thermal conductivity values and combined convection and radiation coefficients must be selected for the mean temperature of the material. Therefore it is necessary to use a trial-and-error procedure. First, select values of k and h_{cr} from Figs. 4–1 and 4–17 and estimate the heat loss by Eqs. (4–22) and (4–23). Then use Eq. (4–4) or Eq. (4–21) to find the temperature of the insulation surface. It is then possible to check the assumed values of k and h_{cr}. When the insulation is composed of several layers of different materials, the procedure is the same except that there are more terms in the equations.

Insulation manufacturers express the effectiveness of their product in terms of an efficiency:

Insulation efficiency

$$= \frac{(\text{loss from bare pipe}) - (\text{loss from insulated pipe})}{(\text{loss from bare pipe})} \qquad (4\text{--}24)$$

EXAMPLE 4–6. Calculate the heat loss from, and the efficiency of, 1½ in. of canvas-covered rock wool insulation on a 4-in. pipe with 380 F working temperature and 80 F room temperature.

SOLUTION.   Pipe OD (Plate 1) = 4.500 in.

For the bare pipe, using Eq. (4–21) and $h_{cr} = 3.25$ from Fig. 4–17,

$$q = 3.25\pi\ \frac{4.50}{12} \times 1(380 - 80) = 1147\ \frac{\text{Btu}}{(\text{hr})\ (\text{linear ft})}$$

Assume the surface temperature of the canvas as 110 F. Then $h_{cr} = 1.8$ and k for the insulation (Fig. 4–1b) is 0.037.

From Eq. (4–23),

$$U_o = \cfrac{1}{\cfrac{4.50}{12} \cfrac{\ln \frac{7.5}{4.5}}{2 \times 0.037} + \cfrac{4.5}{7.5 \times 1.8}} = 0.3435$$

From Eq. (4–22),

$$q = 0.3435\pi \frac{4.5}{12} \times 1(380 - 80) = 121 \frac{\text{Btu}}{(\text{hr})(\text{linear ft})}$$

From Eq. (4–21),

$$t - t_f = \frac{121}{1.8\pi \dfrac{7.5}{12} \times 1} = 34.2$$

$$t = 80 + 34.2 = 114.2 \text{ F}$$

This is sufficiently close to the assumed surface temperature of 110 F to have little effect on either h_{cr} or k.

From Eq. (4–24),

$$\text{Efficiency} = 1 - \frac{121}{1147} = 89.5\%$$

Some recommended thicknesses of 85% magnesia and diatomaceous earth insulation for hot pipes are given in Table 4–4.

PROBLEMS

4–1. Determine the heat transferred through a brick wall 9 in. thick, 8 ft long, for wall surface temperatures of 180 F and 65 F.

4–2. The heat loss through a wall ($k = 37$) 5 in. thick is 450 Btu per hr per sq ft. What is the temperature on one side if the temperature on the other is 700 F?

4–3. Convert the value of k given in Fig. 4–1 for steel at 150 F into the units of (Btu)(in.) per (hr)(sq ft)(F).

4–4. What is the value of resistance for the data of Prob. 4–1?

4–5. Calculate the value of resistance for the data of Prob. 4–2.

4–6. A composite furnace wall is formed of 8 in. of fire brick, 6 in. of diatomaceous earth, 4 in. of 85% magnesia, and $\frac{1}{64}$-in. steel. Calculate the heat loss through the wall for 1 sq ft of surface with surface temperatures of 1400 F and 100 F.

4–7. How much heat would be conducted through a steel pipe carrying water at 110 F when the ID is 1½ in., the thickness is ½ in., and the surface temperature of the pipe is 80 F?

4–8. What is the energy conducted through 100 ft of schedule 80 pipe, nominal diam 4 in., when the outside surface temperature is 250 F and the inside surface temperature is 180 F?

4-9. Derive a formula, similar to Eq. (4–12), for the over-all coefficient of heat transfer, U, based on the inside area of the pipe.

4-10. Derive a formula for U, based on the OD of the pipe, for a pipe covered with two layers of insulation. Include h_o and h_i.

4-11. Calculate the film coefficient for air at 1 atm and 32 F flowing through a pipe (4-in. ID) at 10 fps.

4-12. Same as Prob. 4–11 except for 10 fps at an air temperature of 180 F.

4-13. Same as Prob. 4–11 except for a rectangular duct 3 by 4 in.

4-14. Determine U for a steel-tube air preheater (flue gas inside the tube and air on the outside) with staggered 3-in. tubes (more than 5 rows deep), 10 ft long and placed 6 in. on centers, and 0.180 in. thick. The tubes are located in an area (perpendicular to the air flow) 6 × 10 ft and the gas flow is 250,000 lb per hr with an average temperature of 530 F. Air flow is 230,000 lb per hr at an average temperature of 230 F. The actual gas velocity inside the tubes is 20 fps.

4-15. Same as Prob. 4–14 except neglect the conductivity of the metal wall.

4-16. An air preheater is formed by alternate passages for air and gas. The passages are divided by steel plate 0.190 in. thick. Air enters the preheater at 60 F and leaves at 390 F, while gas enters at 700 F and leaves at 415 F. The velocity of each fluid is 15 fps. Calculate U.

4-17. Water flows through condenser tubes at an average temperature of 70 F and a velocity of 7 fps. Steam is condensed at 1 in. of Hg abs. Calculate U_o for a copper tube ¾-in. OD and 0.049 in. thick.

4-18. The water velocity in a feedwater heater is 8 fps for an average temperature of 346 F. For copper tubes 1-in. OD and 0.065 in. thick with steam at 180 psig, determine U_o.

4-19. Repeat Prob. 4–18 for steel tubes 0.134 in. thick.

4-20. An economizer receives water at 210 F and gas at 635 F. Leaving temperatures are 280 F for the water and 470 F for the gas with counterflow. The 2-in. OD steel tubes are staggered. For the gas, the mass flow is 9000 lb per (sq ft)(hr) and the water velocity is 7.5 fps. The metal is 0.109 in. thick. What is the value of U_o?

4-21. Water enters an evaporator at 60 F, but leaves as vapor at 39 psia. Motivating steam enters at 88 psia and 390 F. Find the heat head.

4-22. Calculate θ_m for the data of Prob. 4–20.

4-23. Calculate θ_m for the data of Prob. 4–20 if there is parallel flow.

4-24. Repeat Prob. 4–18, using a suitable scale coefficient.

4-25. Calculate the radiant energy from an oxidized iron pipe of 2-in. nominal diam and 25-ft length at 180 F to a room at 72 F.

4-26. Considering radiant energy only, would it be better to paint a radiator in a home black ($\epsilon = 0.90$) or paint it with aluminum paint ($E = 0.30$)? For temperatures of 145 F and 70 F, estimate the percentage of change in radiant energy based on black paint.

4-27. A hot-air duct of oxidized copper passes through a room. If the metal is at the same temperature as the air (175 F) and the room is at 65 F, estimate the radiant energy received by the room.

4-28. Repeat Prob. 4-27, using oxidized steel.

4-29. What is the heat loss per linear foot from a bare 1-in. schedule 40 pipe carrying 150 F water to a room at 70 F? If the water velocity is 2 fps, what will be the drop in water temperature per 100 lin ft of pipe?

4-30. Steam enters a 100-ft section of bare 4-in. schedule 40 pipe saturated at 25 psia. Room temperature is 80 F, and the steam velocity is 6000 fpm. What is the quality of the steam leaving the pipe (neglect pressure drop)?

4-31. Calculate the heat loss from a bare 8-in. schedule 80 pipe carrying steam at 550 F to a room at 100 F.

4-32. What is the efficiency of ⅞-in. air-cell insulation on a 1½-in. schedule 40 pipe carrying a fluid at 180 F? Room temperature is 70 F.

4-33. Repeat Prob. 4-32, using 85% magnesia.

4-34. Determine the heat loss for 3 in. of 85% magnesia on a 6-in. schedule 40 pipe at a temperature of 375 F. Room temperature is 70 F.

4-35. One utility company specifies 2 in. of high-temperature insulation (diatomaceous earth) and 2 in. of 85% magnesia insulation for a working temperature of 800 F on all pipes 6 in. or larger. Estimate the efficiency of this insulation for 10-in. schedule 160 pipe. Room temperature is 70 F.

4-36. Repeat Prob. 4-35 for 14-in. schedule 120 pipe.

4-37. What is the efficiency and maximum allowable investment for 2 in. of 85% magnesia insulation on an 8-in. pipe for a fluid temperature of 375 F and a room temperature of 80 F when energy costs 36¢ per million Btu, investment charges are 15%, and the pipe is in use 8200 hr per yr?

4-38. The installed cost of 85% magnesia insulation for a 3-in. pipe is $2.67 per ft for 1-in. thickness, $3.27 per ft for 1½-in. thickness, and $3.79 per ft for 2-in. thickness. What would be the economical thickness of this insulation for a fluid temperature of 400 F when the plant uses coal having a heating value of 13,500 Btu per lb? Boiler efficiency is 70%, coal costs $7.50 per ton fired, investment charges are 15%, and there are 6500 hr per yr of operation.

4-39. An industrial plant operates a certain steam pipe (10-in. schedule 80) 10 hr per day for 300 days per yr. Steam is approximately saturated at 80 psig and is produced in a 72% efficient boiler using 13,000 Btu per lb coal costing $8.35 per ton fired. Determine the most economical thickness of 85% magnesia if investment charges are 20% and insulation installed costs $4.69 per ft for 1¼-in. thickness, $5.64 per ft for 1½-in. thickness, $6.80 per ft for 2-in. thickness, and $7.88 per ft for 2½-in. thickness.

4-40. Find the monetary savings per year and the allowable investment per foot of 6-in. schedule 40 pipe for a fluid temperature of 425 F with 2½ in. of rock wool insulation. Steam costs 43¢ per million Btu.

4-41. A 12-in. steam line from the boiler to the turbine operates at 800 F for 8200 hr per yr. If fuel costs 26¢ per million Btu and investment charges are

12.5%, which insulation should be selected, 1.5 in. of diatomaceous earth and 2 in. of 85% magnesia costing $12.15 per ft and having an efficiency of 97.3%, or 1.5 in. of diatomaceous earth and 2.5 in. of 85% magnesia costing $14.80 per ft and having an efficiency of 97.7%? Use 80 F room temperature.

4–42. The steam line from a boiler to a turbine is 8-in. and can be covered either with: (a) 1.5 in. of diatomaceous earth and 2 in. of 85% magnesia having an efficiency of 96.84% and costing $9.45 per ft, or (b) 1.5 in. of diatomaceous earth and 2.5 in. of 85% magnesia having an efficiency of 97.14% and costing $11.50 per ft. The fluid temperature is 825 F, the plant operates 7700 hr per yr, and investment charges are 5%.

BIBLIOGRAPHY

HEILMAN, R. H. "Heat Loss Through Insulating Materials," *Mechanical Engineering*, Vol. 46 (1924).
———. "Heat Insulation in Air-Conditioning," *Industrial and Engineering Chemistry*, Vol. 28 (1936), No. 7.
JAKOB, MAX, and HAWKINS, G. A. *Elements of Heat Transfer and Insulation.* New York: John Wiley & Sons, Inc., 1950.
KING, W. J. "The Basic Laws and Data of Heat Transmission," *Mechanical Engineering*, Vol. 54 (1932).
MCADAMS, WILLIAM H. *Heat Transmission.* New York: McGraw-Hill Book Co., Inc., 1954.
PHILLIPS, L. W., and ROBINSON, H. A. "Mean Temperatures in Thermal Conductivity Measurements," *Refrigerating Engineering*, Vol. 51 (1946).
STOEVER, H. J. *Applied Heat Transmission.* New York: McGraw-Hill Book Co., Inc., 1941.

CHAPTER 5

FUELS AND COMBUSTION

5-1. Introduction. The generation of power requires the combustion or reaction of certain chemicals with oxygen at a sufficient rate to release the desired amount of thermal energy. We have discussed the methods of transferring this thermal energy from one body to another. It is now logical to discuss the generation of this energy.

Steam power plants use a great variety of fuels for combustion. The two most common fuels are coal and oil. However, gaseous fuels, particularly natural gas, have become prominent in the last few years. The selection of a fuel is a problem of economics and sometimes of politics.

Fuel oil has become one of the important fuels for power generation. In the marine field it has practically replaced coal. All modern naval vessels are equipped with oil-fired boilers because of the cleanliness, ease of handling, high heating value per pound (about 19,000 Btu per lb), low weight of oil burners compared with stokers or pulverizers for coal, and flexibility of operation. An oil fire may be shut off almost instantaneously, but coal stokers (except spreader stokers) require considerable time before the fire dies out. Oil is also adaptable to rapid load fluctuations and fewer operating personnel are required.

Stationary plant designers and operators usually select the fuel that will be the most economical, which is usually the fuel that is found in the general locality of the plant. Utility customers also feel that their money is being retained in the community when the company purchases its fuel from local mines or wells.

5-2. Origin of Coal. Coal geologists believe that coal has been produced by an accumulation of vegetable matter consisting of wood, grass, shrubs, and waxy parts of swamp vegetation. These deposits have been covered with earth and heated or compressed, or both. All changes that took place after the vegetation was deposited and during the formation of the coal are known as *metamorphism*.

The structure of the coal will depend on both the nature and duration of metamorphism and also on the type of vegetation deposited at the bottom of the coal swamp. Where the vegetation consisted of logs of wood, the coal will often have a woody texture. However, grass and shrubs produce a coal of very fine structure. In some cases the layers of

coal are divided by layers of charcoal, which indicates that the vegetation
had experienced periods of drought or that it had been burned. Charcoal
is nearly indestructible during metamorphism.

The extent and duration of compression and heating have caused more
drastic differences in the coal composition than has the type of vegetation. Movements of the earth's crust may cause intense pressure and sensible heat which will drive off much of the volatile matter and leave only
carbon and minerals. Volcanic conditions will have a similar effect.

Decay of the vegetation produced peat which was then subjected to heat
and pressure. Longer and more intense periods of metamorphism produce
the higher ranks of coal. In order of increasing rank, coals are known as
lignite, subbituminous, bituminous, semibituminous, semianthracite, anthracite, and superanthracite. Since this classification is based on the
chemical composition of the coal, it will be wise to consider next the
methods of analyzing the fuel, i.e., proximate analysis and ultimate
analysis.

5–3. Proximate Analysis. A typical proximate analysis of coal determines the percentage of moisture, volatile matter, fixed carbon, and ash.
In some cases the percentage of sulfur is also obtained, but we shall consider the proximate analysis to include only the first four items.

Moisture is determined by subjecting a 1-g sample of the coal to a
temperature of 220 F to 230 F for a period of exactly 1 hr. The loss in
weight of the sample during this period, in percentage, is an indication of
the moisture content of the coal.

Volatile matter consists of hydrogen and certain hydrogen-carbon compounds that can be removed from the coal merely by heating it. Thus, a
1-g sample is placed in a covered platinum crucible and heated to 1740 F
for about 7 min to drive off the volatile matter. The loss in weight during
the heating period is due to the elimination of moisture and volatile matter. The latter may be determined since moisture has been calculated
from the previous test.

The test for percentage of ash in the fuel is performed by heating the
sample of coal used in the moisture determination to a temperature of
1290 F to 1380 F in an uncovered crucible, with good air circulation, until
the coal is completed burned. Complete combustion of the coal is determined by the repeated weighings of the sample. A constant weight indicates that there is only ash remaining in the crucible.

Fixed carbon is the difference between 100% and the sum of the percentages of moisture, ash, and volatile matter. However, that difference
does not represent all the carbon that was in the coal. Some of the carbon
may have been in the form of hydrocarbons which may have been distilled off during the determination of volatile matter. Also, it is possible

that some of the so-called *fixed* carbon may include hydrogen, nitrogen, sulfur, and oxygen.

Observe that the tests for moisture and volatile matter are somewhat arbitrary in that the duration of the test is fixed, i.e., there is no certainty that all the moisture or volatile matter has been removed by the end of the prescribed period of heating. Also, there is no assurance that some of the more volatile of the hydrocarbons were not removed during the moisture test.

5–4. Ultimate Analysis. An accurate proximate analysis can be made by anyone and the equipment is simple. However, an ultimate analysis must be made by a chemist. The ash content of the coal is independent of the type of analysis and is therefore the same for both. The ultimate analysis divides all the remaining part of the coal into the elements carbon, hydrogen, oxygen, sulfur, and nitrogen in percentages by weight. Note that moisture is proportioned by weight into hydrogen and oxygen.

In most cases the percentage of carbon obtained from the ultimate analysis is numerically larger than the percentage of fixed carbon reported by the proximate analysis. The exceptions are explained by the theory that the fixed carbon may contain elements other than pure carbon, such as hydrogen, nitrogen, sulfur, and oxygen.

Proximate and ultimate analyses of some coals are listed in Table 5–1.

5–5. Basis of Reporting Analysis. The Bureau of Mines reports coal analysis in the following manner:

    (a) *As Received* or *As Fired*
    (b) *Dry* or *Moisture Free*
    (c) *Moisture* and *Ash Free* or *Combustible*

Another basis that has some use is

    (d) *Moisture, Ash, and Sulfur Free*

The first three methods are applicable to either proximate or ultimate analysis.

The first basis, as fired, is of most use to the power-plant operator because it shows the constituents of the fuel in the same condition as it was weighed and supplied to the furnace. The remaining methods are used primarily for comparing coals from various sources.

To transfer an analysis from *as received* to *dry*, it is only necessary to deduct the weight of hydrogen and oxygen in the water from these terms and to divide each of the remaining constituents by one minus the decimal equivalent of the moisture. A similar procedure may be used for converting to any of the other methods.

TABLE 5-1

Selected Analyses of United States Coals *

State	County	Rank	Proximate				Ultimate					Heating Value
			Moisture	Volatile Matter	Fixed Carbon	Ash	Sulfur	Hydrogen	Carbon	Nitrogen	Oxygen	
Alabama	Bibb	Bit.	2.71	34.67	57.43	5.19	1.26	5.29	79.71	1.42	7.13	14,157
	Jefferson	Bit.	2.39	24.41	68.41	4.79	0.74	4.74	81.74	1.48	6.51	14,623
Alaska	Matanuska Field	Bit.	4.84	41.64	46.58	6.94	0.54	5.92	71.20	1.64	13.76	12,892
	Mount Hamilton	Semibit.	1.08	15.97	65.02	17.98	2.38	3.58	72.21	1.25	2.60	12,525
	Nenana Field	Lig.	25.66	32.28	22.80	19.26	0.21	5.93	37.86	0.53	36.21	6,478
Arkansas	Sebastian	Semibit.	2.89	19.29	67.34	10.48	1.10	4.10	77.39	1.62	5.31	13,271
	Washington	Bit.	1.84	29.69	57.85	10.62	3.71	4.95	75.45	1.31	3.96	13,561
California	Monterey	Subbit.	8.02	48.46	36.01	7.51	4.09	5.90	65.21	1.24	16.05	12,130
Colorado	Boulder	Subbit.	19.14	33.44	42.07	5.35	0.27	5.97	57.68	1.21	29.52	10,017
	Gunnison	Bit.	4.60	40.42	51.60	3.38	0.43	5.72	75.09	1.56	13.82	13,453
	Las Animas	Bit.	2.08	32.59	52.55	12.78	0.70	4.96	71.73	1.23	8.60	12,955
Idaho	Teton	Bit.	7.67	39.61	48.19	4.53	0.54	5.94	70.82	1.37	16.80	12,575
Illinois	Christian	Bit.	13.04	36.98	39.25	10.73	3.66	5.58	59.83	1.14	19.06	10,856
	Franklin	Bit.	9.77	34.15	47.45	8.63	0.75	5.34	66.23	1.51	17.54	11,725
Indiana	Sullivan	Bit.	10.84	38.42	42.78	7.96	2.63	5.76	65.36	1.44	16.85	11,882
Kansas	Cherokee	Bit.	5.09	34.47	52.17	8.27	3.34	5.23	71.81	1.20	10.15	13,082
Kentucky	Letcher	Bit.	3.00	36.62	57.25	3.13	0.65	5.39	79.81	1.53	9.49	14,200
	Webster	Bit.	5.35	34.92	50.39	9.34	1.05	5.12	70.39	1.55	12.55	12,501
Maryland	Allegany	Semibit.	2.15	17.78	69.13	10.94	1.10	4.30	76.87	1.86	4.93	13,432
Missouri	Boone	Bit.	11.16	34.95	43.32	10.57	3.71	5.23	62.92	0.95	16.62	11,327
Montana	Carbon	Subbit.	10.47	34.67	43.65	11.21	1.13	5.58	59.77	1.29	21.02	10,525
	Valley	Lig.	32.62	27.43	30.85	9.10	1.28	6.46	40.35	0.70	42.11	6,712
New Mexico	Socorro	Bit.	3.19	37.96	45.10	13.75	0.82	5.25	68.36	1.32	10.50	12,307

* Compiled from *Bulletins* of the Bureau of Mines; all data on "as received" basis.

TABLE 5-1 (*Continued*)

SELECTED ANALYSES OF UNITED STATES COALS

State	County	Rank	Proximate				Ultimate					Heating Value
			Moisture	Volatile Matter	Fixed Carbon	Ash	Sulfur	Hydrogen	Carbon	Nitrogen	Oxygen	
North Dakota	Ward	Lig.	36.93	24.92	27.72	10.43	0.22	6.39	37.36	0.61	44.99	6,010
Ohio	Columbiana	Bit.	3.60	37.80	51.22	7.38	3.21	5.35	73.68	1.85	8.53	13,390
Oklahoma	Jefferson	Bit.	2.21	39.63	47.50	10.66	4.96	5.28	71.04	1.34	6.72	13,025
	Haskell	Bit.	3.02	22.75	68.22	6.00	0.99	4.76	80.85	1.80	5.60	14,148
	Latimer	Bit.	3.68	37.41	53.85	5.06	0.96	5.22	75.19	1.65	11.92	13,687
Pennsylvania	Allegheny	Bit.	2.86	33.83	53.11	10.20	1.12	5.25	73.77	1.45	8.21	13,217
	Allegheny	Bit. (Cannel)	0.85	28.94	35.31	35.20	0.95	4.02	54.07	1.06	4.70	9,787
	Cambria	Semibit.	3.32	18.25	72.03	6.40	1.77	4.73	80.66	1.42	5.02	14,171
	Lackawanna	Anth.	3.43	6.79	78.25	11.53	0.46	2.52	78.85	0.77	5.87	12,782
	Luzerne	Anth.	1.31	5.68	85.87	7.14	0.42	2.35	86.76	0.68	2.65	13,777
South Dakota	Perkins	Lig.	42.46	23.24	25.30	9.00	1.16	7.07	35.22	0.62	46.93	5,954
Tennessee	Morgan	Bit.	2.19	32.53	58.25	7.03	2.52	5.28	78.24	1.61	5.32	14,108
Texas	Webb	Bit. (Cannel)	3.98	48.87	34.91	12.24	1.96	6.25	65.54	1.27	12.74	12,227
	Webb	Bit. (Cannel weathered)	3.64	31.61	20.96	43.79	1.35	4.36	38.98	0.65	10.87	7,234
Utah	Carbon	Bit.	7.49	39.72	47.17	5.62	0.64	6.09	69.12	1.35	17.18	12,521
Virginia	Montgomery	Semianth.	2.20	9.96	81.63	6.21	0.47	3.65	84.95	0.92	3.80	14,272
	Tazewell	Semibit.	3.00	20.25	72.24	4.51	0.47	4.65	84.00	1.17	5.20	14,605
Washington	Lewis	Subbit.	30.78	33.06	27.73	8.43	0.75	6.81	42.62	0.80	40.59	7,513
West Virginia	Fayette	Bit.	2.57	25.34	68.94	3.15	0.50	5.06	83.86	1.68	5.75	14,789
	McDowell	Semibit.	2.90	17.34	75.29	4.47	0.60	4.45	83.89	1.09	5.50	14,636
Wyoming	Lincoln	Bit.	3.58	38.43	52.51	5.48	0.97	5.28	74.45	1.30	12.52	13,313
	Sheridan	Subbit.	22.88	32.51	36.74	7.87	1.11	6.12	50.59	1.25	33.06	8,996

EXAMPLE 5-1. Convert the ultimate analysis of Cherokee County, Kansas, coal given in Table 5-1 to dry, to moisture and ash free, and to moisture, ash, and sulfur free bases.

SOLUTION. Since water is 8 parts oxygen and 1 part hydrogen by weight, deduct $5.09 \times \frac{1}{9} = 0.56\%$ from hydrogen and $5.09 \times \frac{8}{9} = 4.53\%$ from oxygen, giving the as received analysis as:

Sulfur	3.34%	Nitrogen	1.20%
Hydrogen	4.67%	Oxygen	5.62%
Carbon	71.81%	Moisture	5.09%
		Ash	8.27%

To obtain the dry ultimate analysis, divide each item by $(1.0000 - 0.0509) = 0.9491$:

Sulfur	3.52%	Nitrogen	1.26%
Hydrogen	4.92%	Oxygen	5.92%
Carbon	75.67%	Ash	8.71%

Divide each item except ash by $(1.0000 - 0.0871) = 0.9129$ to get the combustible basis:

Sulfur	3.86%	Nitrogen	1.38%
Hydrogen	5.39%	Oxygen	6.48%
Carbon	82.89%		

Similarly divide all but sulfur by $(1.0000 - 0.0386) = 0.9614$ for the moisture, ash, and sulfur free analysis:

Hydrogen	5.61%	Nitrogen	1.44%
Carbon	86.21%	Oxygen	6.74%

5–6. Heating Values of Fuels. The energy released by any fuel when it is completely burned and when the products of combustion are cooled to the original fuel temperature is known as the heating value of the fuel. The numerical values are expressed by engineers in the United States as Btu per pound for solid and liquid fuels and Btu per cubic foot for gases. Since the energy released by any substance is a function of the weight and not the volume, care must be taken to indicate the pressure and temperature of the gas when the customary units of Btu per cubic foot are used.

Combustion of any fuel having hydrogen as one of its constituents produces water vapor. If the products of combustion are at a high temperature, the water will leave the system as a vapor and will carry with it the energy represented by the energy of superheated steam. However, if the gases are cooled to a low temperature, the vapor will condense and reject this energy. Thus, it is possible to obtain two distinctly different heating values for fuels containing hydrogen—the *higher heating value* and the *lower heating value*. The higher heating value assumes that the water vapor from combustion has been condensed to a liquid while the lower heating value does not assume condensation of the vapor.

There has been considerable difference of opinion as to the merits of the two values. Advocates of the lower heating value argue that no actual engine can reduce the products of combustion to a low enough temperature to take advantage of the latent energy of the water vapor, but opponents claim that such an inability on the part of the heat engine shows an inefficiency of the engine. While the latter argument is more valid, the efficiency of any plant is used only for comparison with other plants. Therefore, the main consideration is that everyone use the same value so that all plants may be compared on an equal basis.

Power-plant engineers in the United States follow the recommendations of the ASME and use the higher heating value. The calorimeters which will be described give the higher heating value of the fuel.

Certain empirical formulas have been developed from which the higher heating value of coals may be estimated. One of the most popular of these formulas is Dulong's equation, in which the higher heating value is calculated from the ultimate analysis by multiplying the weight of each of the combustible elements in 1 lb of the coal by the heating value of that element. Thus,

$$HHV = 14,600C + 62,000 \left(H_2 - \frac{O_2}{8} \right) + 4050S \qquad (5\text{-}1)$$

where HHV = higher heating value, Btu per lb
      C = carbon in 1 lb of fuel, lb
     H_2 = hydrogen in 1 lb of fuel, lb
     O_2 = oxygen in 1 lb of fuel, lb
     S = sulfur in 1 lb of fuel, lb

Dulong's formula approximates the heating value of the fuel on the as received, moisture free, or moisture and ash free basis in accordance with the basis used for determining the weight fractions substituted into the equation. The disadvantage of this formula is that an ultimate analysis is required. The term $(H_2 - O_2/8)$ assumes that all the oxygen in the fuel has previously combined with the hydrogen in the fuel.

Similar formulas may be developed for any fuel by selecting the appropriate heating values for each of the constituents or compounds in the fuel.

Another empirical relationship for coals has been developed by Evans for use with the proximate analysis, Table 5-2. Sufficient data to estimate the ultimate analysis from the data of the proximate analysis are included in this table. This procedure should be used only when analytical methods are not available.

The only exact method of determining the heating value of a fuel is by means of a calorimeter. Constant-volume calorimeters, employing either an oxygen bomb, Fig. 5-1, or a sodium peroxide bomb, Fig. 5-2, are used

TABLE 5-2

EMPIRICAL RELATIONS FOR COALS OF THE UNITED STATES

(Based on Proximate Analyses)

Property of Coal Desired	States	Range of Volatile Matter in Combustible (%)	Equation *
Heating value of the combustible	Pa., Ohio, W. Va., Md., Va., Ky., Ga., Tenn., Ala., Ind., Iowa, Nebr., Kans., Mo., Okla., Ark., and Tex.	0–16	HHV = 14,550 + 7810V
		16–36	HHV = 16,160 − 2250V
		36 up	HHV = 18.750 − 9440V
	Ill. and Mich.	All values	HHV = 16,062 − 3830V
Total carbon in the combustible	Pa., Ohio, W. Va., Md., Va., Ky., Ga., Tenn., and Ala.	0–36	C = 0.943 − 0.242V
		36 up	C = 1.095 − 0.663V
	Ill., Ind., Mich., Iowa, Nebr., Kans., Mo., Okla., Ark., and Tex.	All values	C = 0.953 − 0.362V
	Colo., Utah, N. Mex., Ariz., Wyo., Mont., Wash., Ore., and Calif.	36–60	$C = \dfrac{(HHV + 7544)(0.0099 + 0.0208V)}{737.5V + 200}$
	All states	60 up	$C = \dfrac{(HHV + 7544)(0.0099 + 0.0045V)}{11.3V + 200}$
Hydrogen in the combustible	All states but Ark.	4–16	H = 0.013 + 0.225V
		16 up	H = 0.0457 + 0.0206V
	Ark.	All values	H = 0.0327 + 0.056V

* HHV = higher heating value of combustible, Btu per lb. Combustible = coal − ash − moisture. V = volatile matter in combustible. C = carbon in combustible. H = hydrogen in combustible. Nitrogen ranges from 0.75 to 1.75% and may be assumed as 1.25% average. Sulfur ranges from 0.5 to 2.0% in eastern coals and up to 11.0% as the extreme in midwestern coals.

for coals and oils. Constant-pressure or steady-flow calorimeters, Fig. 5–3, are particularly adaptable for use with gaseous fuels. This calorimeter also may be used for liquid fuels by employing a slightly different burner.

Approximately 1 g of coal, which has been pulverized fine enough to pass through a 100-mesh screen, is placed in a crucible inside the oxygen-

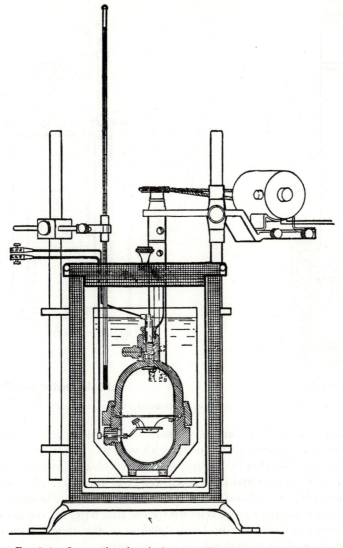

FIG. 5–1. Oxygen bomb calorimeter. (Emerson Apparatus Co.)

bomb calorimeter, Fig. 5–1. A fuse wire is connected to terminals
and is suspended immediately above the coal sample. The bomb is
assembled and compressed oxygen is admitted to a pressure of about
300 psi. After connecting the ignition circuit, the bomb is immersed in a
water bath. Thorough circulation of the water is assured by an electrical
stirring device. The fuel sample is ignited by passing an electric current
through the fuse wire. Energy liberated by the combustion raises the

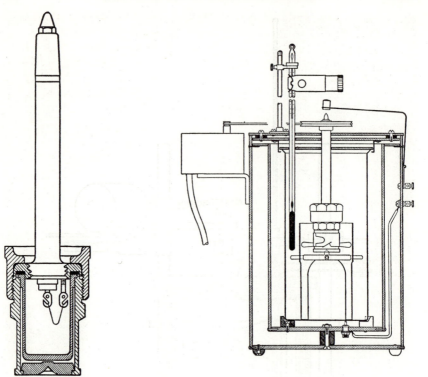

(a) Bomb   **(b) Calorimeter with Bomb**

Fig. 5–2.   Sodium peroxide calorimeter.   (Parr Instrument Co.)

temperature of the water and metal parts of the calorimeter, as indicated by an accurate mercury thermometer with very fine graduations.

The energy liberated by the fuel may be calculated from the temperature rise of the water and the known weight of the water plus the water equivalent of the metal parts. Corrections should be made to the resultant heating value for the energy released by the fuse wire, for certain chemicals encountered only in the bomb because of the almost pure oxygen atmosphere, and for radiation from the apparatus.

ASTM standards permit a difference of not more than 0.3% of the thermal value between duplicate tests performed with the oxygen-bomb calorimeter.

A sodium peroxide bomb, Fig. 5–2, differs from the oxygen bomb in that the oxygen required for combustion is supplied from the peroxide mixed with the fuel. Therefore, the peroxide type is suitable for many industrial purposes where compressed oxygen is not readily available. Furthermore, the sodium chemicals resulting from the combustion absorb

Fig. 5–3(a). Sargent type of gas calorimeter with accessories. (American Meter Co.)

the products of combustion and thereby permit the use of a thin-walled bomb.

A small-sized bomb may be used for peroxide since large volumes of oxygen are not required.

Oxygen and peroxide calorimeters burn the fuel in a container at constant volume, but the Sargent type of calorimeter operates on the steady-flow, constant-pressure principle. Fig. 5–3 shows the required apparatus and a cross-sectional view of the calorimeter. Gas flows through a meter where the volume, pressure, and temperature are measured. A small gas holder assists in maintaining constant pressure at a burner which is inserted into the combustion chamber of the calorimeter. The products of combustion travel upward and then down through a nest of tubes. A damper at the outlet permits regulation of the flow of air into the combustion chamber so that the gases leaving the calorimeter will be at the same temperature as the air entering. Thus, there is no loss of energy due to the gases leaving, and all the energy of combustion has been transferred to the water surrounding the nest of gas tubes.

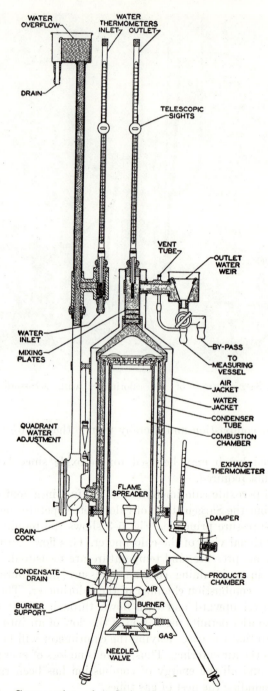

FIG. 5–3(b). Cross-section of a Sargent type of gas calorimeter. (American Meter Co.)

A constant-level tank at the water entrance provides constant water pressure and therefore constant water flow through the calorimeter, once the regulating valve has been set. Inlet and outlet water temperatures are measured by finely graduated thermometers. The water leaving the instrument is caught in buckets and weighed. The weight of water times its temperature rise and divided by the volume of gas—corrected to standard temperature and pressure—is the heating value of the fuel. Corrections may be made for the humidity of the air and the calibration of the thermometers.

Water which has condensed because of the burning of the hydrogen in the fuel is collected from the bottom of the exhaust gas chamber and measured so that the lower heating value may be determined.

5–7. Classification of Coals. As already mentioned, the composition of coals is dependent on the pressure and temperature to which the vegetable matter has been subjected and on the age of the field. There are indications that the transformation from vegetable matter to coal starts only under conditions of heavy rainfall and temperate climate. Heavy rainfall promotes luxuriant vegetation and large bodies of water on the earth's surface for decomposition of the vegetable matter. During some geological periods these ideal conditions existed into the Arctic region. The changes in the composition of coal during metamorphism may be visualized from Fig. 5–4.

Peat is characterized by high moisture content with only small percentages of volatile matter and fixed carbon. Peat is not used as a fuel for power plants in this country because of its high moisture content. However, it may be partially dried and compressed for use as a fuel in other applications.

Lignite represents the next stage in the development of coal. Its heating value is considerably higher than that for peat, and its fixed carbon and volatile matter are also larger. This coal is often suitable for power generation. After drying, lignite *slacks* or disintegrates into small flakes, making shipment difficult. Some lignite is brown in color.

The various ranks of bituminous coal form the most extensive part of the supply of coal in the United States. The lowest rank of bituminous coal is *subbituminous,* which is the next rank above lignite. Shipment and storage of this coal are also hampered by slacking, a quality that serves to distinguish it from other grades of bituminous. Subbituminous is black in color.

Although *bituminous coals* cover a wide variety, they are all characterized by low moisture content and nonslacking properties. Also, the heating value is much higher than that for any type previously discussed. One particular variety, *cannel coal,* contains a large percentage of volatile

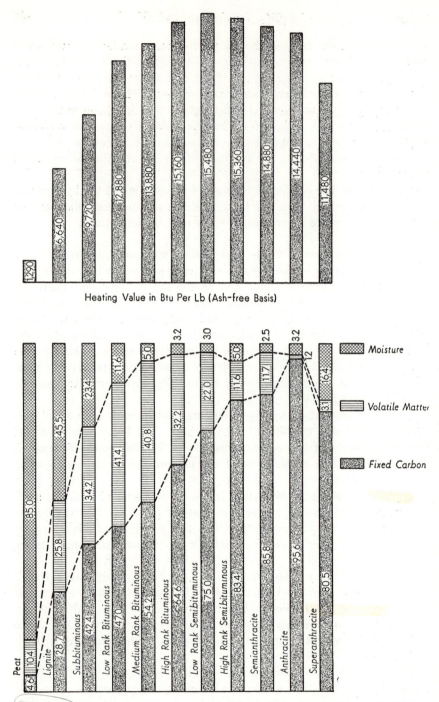

Heating Value in Btu Per Lb (Ash-free Basis)

Fig. 5–4. Classification of coal on the ash free basis by M. R. Campbell. (Taken from a paper, "Our Coal Supply: Its Quantity, Quality, and Distribution," International Conference on Bituminous Coal. Reproduced by permission of Carnegie Institute of Technology.)

matter and hydrogen. Cannel coal ignites readily and will continue to burn once it is ignited, but it is not important for power plants.

Low-rank bituminous coals are rich in gas and other coal by-products. However, they are not suited for domestic purposes or for power plants located in residential locations because they burn with a smoky flame and deposit an oily soot on surrounding buildings. Most bituminous coals are of the *caking* type while others are *noncaking* or *free-burning*. The former variety fuses into a plastic mass when heated with little or no air. Lignite, subbituminous, and anthracite coals are free-burning. Caking coals are essential to the production of good coke.

Semibituminous coals have high fixed-carbon contents and the highest heating values. They have been prized in the past as a Navy fuel because of being nearly smokeless. During metamorphism, they lost most of the volatile matter. Semibituminous coal is one of the best for power generation.

Semianthracite is a much harder coal than semibituminous and is found only in a limited section of the United States. Because of its scarcity the price of this fuel is usually too high for power generation.

Anthracite is also scarce and is found principally in Pennsylvania. Because of its high cost, it is used primarily for domestic purposes. The ash content is usually low and the coal is hard with a shiny surface. Although the fixed-carbon content of anthracite is higher than that for any other coal, it does not have the highest heating value. It burns with a short and bluish flame. Some plants use anthracite *culm*, which is the refuse from the screenings.

Superanthracite is very difficult to ignite and therefore has little significance in the power field. It borders on graphite.

Many schemes in addition to Campbell's have been proposed for the classification of coals. Frazer, of the Second Geological Survey of Pennsylvania, proposed a *fuel ratio*, which he defined as the ratio of the percentages of fixed carbon to volatile matter. According to this scheme, the maximum fuel ratio for bituminous is 2.5, and the other coals range as follows: semibituminous 2.5 to 5, semianthracite 5 to 10, and anthracite 10 and over.

Still another scheme is shown in Table 5–3, taken from the ASTM standards. If the fixed carbon from the proximate analysis is less than 69%, then the moist heating value is used as the criterion. For more than 69% fixed carbon, the fixed carbon is the criterion.

Ralston's chart, Fig. 5–5, is a plot of the percentages of carbon, hydrogen, and oxygen from the ultimate analysis. The abscissa of this chart is carbon by per cent, the ordinate is hydrogen, and the diagonal lines are constant percentages of oxygen. Ralston found that the various ranks of coal generally fall within the areas indicated. Cannel coal is usually high

TABLE 5–3

CLASSIFICATION OF COALS BY RANK

(FC = fixed carbon, VM = volatile matter, Btu = British thermal units)

Class	Group	Limits of Fixed Carbon or Btu, Mineral-Matter-Free Basis
I. Anthracitic	1. Meta-anthracite	Dry FC*, 98% or more (dry VM, 2% or less)
	2. Anthracite	Dry FC, 92% or more and less than 98% (dry VM, 8% or less and more than 2%)
	3. Semianthracite	Dry FC. 86% or more and less than 92% (dry VM, 14% or less and more than 8%)
II. Bituminous	1. Low-volatile bituminous coal	Dry FC, 78% or more and less than 86% (dry VM, 22% or less and more than 14%)
	2. Medium-volatile bituminous coal	Dry FC, 69% or more and less than 78% (dry VM, 31% or less and more than 22%)
	3. High-volatile A bituminous coal	Dry FC, less than 69% (dry VM, more than 31%); and moist Btu, 14,000 † or more
	4. High-volatile B bituminous coal	Moist Btu,* 13,000 or more and less than 14,000 †
	5. High-volatile C bituminous coal	Moist Btu, 11,000 or more and less than 13,000 †
III. Subbituminous	1. Subbituminous A coal	Moist Btu, 11,000 or more and less than 13,000 †
	2. Subbituminous B coal	Moist Btu, 9500 or more and less than 11,000 †
	3. Subbituminous C coal	Moist Btu, 8300 or more and less than 9500 †
IV. Lignitic	1. Lignite 2. Brown coal	Moist Btu, less than 8300 Moist Btu, less than 8300

* Dry mineral-matter-free fixed carbon $= \dfrac{FC - 0.15S}{1.00 - (M + 1.08A + 0.55S)}$

Moist mineral-matter-free Btu $= \dfrac{Btu - 50S}{1.00 - (1.08A + 0.55S)}$

in which FC, M, A, and S are the weights of fixed carbon, moisture, ash, and sulfur in 1 lb of coal.

† Coals having 69% or more fixed carbon on the dry, mineral-matter-free basis shall be classified according to fixed carbon, regardless of Btu.

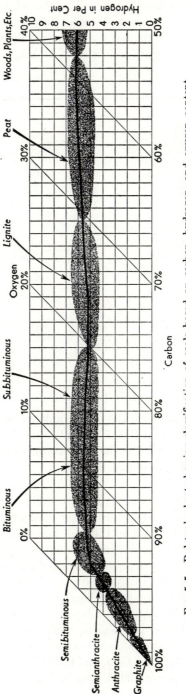

Fig. 5-5. Ralston's chart showing classification of coals based on carbon, hydrogen, and oxygen content.

in hydrogen and points plotted for it will lie above bituminous. The curve plotted on the chart is a mean of all coals plotted. An interesting feature of this method is that woods and plants have a definite position on the diagram. This chart is some verification of Dulong's formula, which was discussed in Art. 5–6.

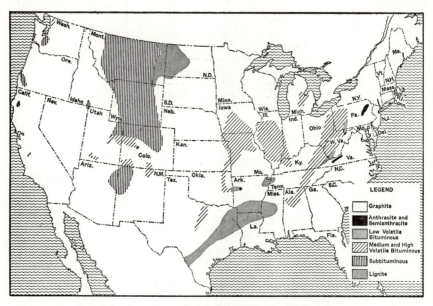

Fig. 5–6. Locations of coal deposits in the United States.

Locations of fields in which the many ranks of coals are found in the United States are shown on Fig. 5–6. Note that, in general, the higher rank or oldest coals are found in the eastern part of the country and that the rank decreases going from east to west.

5–8. Oil. Petroleums are said to be the result of marine growths and vegetation that were deposited during past ages when large areas of the United States were submerged by the sea. Like coal, these deposits were covered by sediment. Later, shifting and movements of the earth's crust compressed and heated the deposits, causing them to decay and to form hydrocarbons. Some of the hydrocarbons are light and either gaseous or very volatile, while others are heavy liquids.

Most oil deposits are located over layers of water and beneath domes of gas. Regions of oil and gas bearing strata in the United States are shown on Fig. 5–7. In addition, many deposits are known to exist under the ocean floor off the Atlantic, Gulf, and California coasts. Drillings have already been made in water 100 ft deep in Lake Maracaibo, Venezuela.

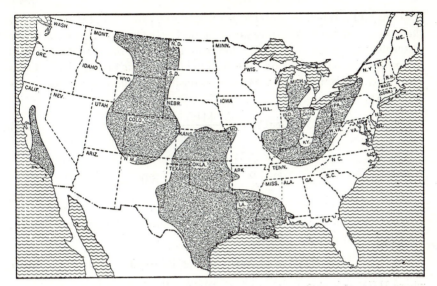

FIG. 5-7. Regions where oil and gas fields are located in the United States.

Chemical analyses of a few oils are listed in Table 5-4. Note that most petroleums have about 85% carbon so that the ultimate analysis cannot be used as the criterion in classifying and specifying this type of fuel.

Density is usually expressed in the form of specific gravity as the ratio of the weight of a cubic foot of the oil to the weight of a cubic foot of

TABLE 5-4

ULTIMATE ANALYSES OF SOME PETROLEUMS

Petroleums	Higher Heating Value, Btu/Lb	Specific Gravity at 60 F	Composition in Percentage by Weight				
			C	H₂	S	N₂	O₂
Baku (Russian).........	. . .	0.884	86.3	13.6	. . .	. . .	0.1
California crude........	18,920	0.917	84.0	12.7	0.75	1.35	1.20
Canada crude.........	20,420	0.857	84.3	13.4	. . .	. . .	2.3
Kansas crude..........	19,100	0.921	84.15	13.00	1.90	0.45	. . .
Mexican crude........	18,750	0.975	83.70	10.20	4.15	. . .	. . .
Ohio crude............	21,600	0.887	80.2	17.1	. . .	. . .	2.7
Oklahoma crude........	19,500	0.869	85.70	13.11	0.40	0.30	. . .
Pennsylvania crude.....	19,500	0.813	86.06	13.88	0.06	. . .	. . .
Texas crude............	19,450	0.875	85.05	12.30	1.75	0.70	. . .
Virginia...............	19,200	0.860	87.1	11.7	. . .	. . .	1.2
West Virginia crude.....	21,240	. . .	86.6	12.9	. . .	. . .	0.5
Wyoming crude........	19,700	0.996	82.0	14.2	. . .	. . .	3.6

water at 60 F. When the specific gravity is given for $^6\%_0$, the term means that the oil density was measured at 60 F (the upper figure) and the water density taken at 60 F (the lower figure). Two other scales are used to specify the density of oil: the API scale (American Petroleum Institute) and the Baumé scale. The relationships of these three scales are as follows:

$$\text{Deg API} = \frac{141.5}{\text{sp gr} \frac{60}{60}} - 131.5 \qquad (5\text{-}2)$$

$$\text{Deg Bé} = \frac{140}{\text{sp gr} \frac{60}{60}} - 130$$

Heating value should be determined by a bomb or Sargent calorimeter, but empirical formulas may be used:

$$\text{HHV} = 18,250 + 40\,(\text{deg Bé} - 10) \qquad (5\text{-}3)$$

where HHV = higher heating value, Btu per lb.

5–9. Gaseous Fuels. Gaseous fuels have all the advantages of petroleum except the ease of storage. Power plants located near natural gas fields or industries manufacturing gas as a by-product have found this type of fuel economical.

Natural gas is formed in a manner similar to that of oil. In fact, natural gas is usually found in a dome just above crude petroleum. Both these fuels are found in fields located in the areas shown by the shaded portions of the map in Fig. 5-7. Analyses of some natural and artificial gases are listed in Table 5-5.

5–10. Combustion. Energy is liberated within the boiler by chemical combination of the combustible elements of the fuel—carbon, hydrogen, and sulfur—with oxygen. Petroleums, gases, and the volatile portions of coal contain hydrocarbons and some carbon monoxide.

The first step in determining the quantity of air required for the combustion of a fuel is to write the equation expressing the chemical reaction that will take place. Suppose that hydrogen is to be burned with oxygen. The reaction equation is

$$2H_2 + O_2 = 2H_2O \qquad (5\text{-}4)$$

This equation states that two volumes of hydrogen combine with one volume of oxygen to form two volumes of water vapor. Each side of this equation contains the same number of atoms of hydrogen and the same number of atoms of oxygen; i.e., there are four atoms of hydrogen on each

TABLE 5-5

COMPOSITION OF SOME NATURAL AND ARTIFICIAL GASES

Gases	Heating Value Btu per Cu Ft *	Composition in Percentage by Volume						
		CH_4	C_2H_6	CO_2	H_2	CO	O_2	N_2
Natural:								
Alabama	983	97.6	...	0.3	...	...	...	2.1
Arkansas	1000	99.2	...	0.2	...	...	...	0.6
California	1062	77.5	16.0	6.5	...	...	...	...
Illinois	963	95.6	...	0.5	...	...	...	3.9
Indiana	1171	75.4	23.4	...	...	...	...	1.2
Kansas	988	98.0	...	1.2	...	...	...	0.8
Kentucky	1178	75.0	24.0	...	...	...	...	1.0
Missouri	965	84.1	6.7	0.8	...	...	...	8.4
New York	1111	84.0	15.0	...	...	...	...	1.0
Ohio	1066	83.5	12.8	...	...	...	...	3.7
Oklahoma	1064	73.5	18.4	...	...	...	...	8.1
Pennsylvania	1065	90.0	9.0	0.2	...	...	...	0.8
West Virginia	1169	76.8	22.5	...	...	...	...	0.7
Artificial:								
Producer gas								
From anthracite	138	...	...	5.5	15.5	22.7	0.3	56.0
From bituminous	151	3.7	0.1	4.8	11.6	24.4	0.6	54.8
Illuminating gas	504	23.6	10.5	7.2	11.7	13.7	0.7	32.6
Blast furnace gas	98.3	0.2	...	12.7	3.6	26.5	...	57.0

* Measured at 14.7 psia and 60 F.

side and two atoms of oxygen on each side. There must always be the same number of atoms of each constituent on both sides of the equal sign.

At any given pressure and temperature the volume of a mol of any gas will be the same as the volume of a mol of all other gases. At 14.7 psia and 32 F, the volume of a mol of any gas is 359 cu ft. The weight in pounds of a mol of any gas is the same as the numerical value of the molecular weight of the element. Thus, the approximate weight of a mol of hydrogen (H_2) may be taken as 2 lb, oxygen (O_2) as 32 lb, carbon (C) as 12 lb, sulfur (S) as 32 lb, methane (CH_4) is 16 lb, etc., based on the approximate molecular weights of the elements.

Since there is a fixed ratio between the mol and the volume of a gas, Eq. (5-4) may be written

$$2H_2 + O_2 = 2H_2O$$

2 vols of hydrogen + 1 vol of oxygen = 2 vols of water vapor

2 mols of hydrogen + 1 mol of oxygen = 2 mols of water vapor

Multiplying the mols of each constituent by its molecular weight,

4 lb of hydrogen $+$ 32 lb of oxygen $=$ 36 lb of water vapor

Therefore, every pound of hydrogen requires 8 lb of oxygen for combustion, which will form 9 lb of water vapor. Or, every pound of water requires $\frac{1}{9}$ lb of hydrogen and $\frac{8}{9}$ lb of oxygen for its formation.

The mols or volumes of constituents on each side of the equation do *not* necessarily balance; but, in accordance with the Law of Conservation of Mass, the weights of all constituents entering into the process must equal the weights of all constituents in the products of combustion.

Air is a *mechanical* mixture of oxygen, nitrogen, water vapor, argon, neon, helium, krypton, carbon dioxide, and hydrogen. For most engineering problems it is sufficient to assume that air is composed of 23.1% oxygen and 76.9% nitrogen (N_2) by weight. Application of the characteristic gas equation for conversion to the volumetric basis shows that air is then 21.0% oxygen and 79.0% nitrogen. The molecular weight of air may be taken as 28.9 based on the volume percentages of oxygen and nitrogen used above.

If we consider that methane is burned with oxygen supplied from air, our chemical equation will be

$$CH_4 + 2O_2 + 2\left(\frac{79.0}{21.0}\right) N_2 = CO_2 + 2H_2O + 7.52N_2 \qquad (5\text{--}5)$$

1 vol methane $+$ 2 vols oxygen $+$ 7.52 vols nitrogen
$\quad =$ 1 vol carbon dioxide $+$ 2 vols water vapor $+$ 7.52 vols nitrogen

1 mol methane $+$ 2 mols oxygen $+$ 7.52 mols nitrogen
$\quad =$ 1 mol carbon dioxide $+$ 2 mols water vapor $+$ 7.52 mols nitrogen

16 lb methane $+$ 64 lb oxygen $+$ 210.5 lb nitrogen
$\quad =$ 44 lb carbon dioxide $+$ 36 lb water vapor $+$ 210.5 lb nitrogen

Then $CH_4 + 9.52$ air $= CO_2 + 2H_2O + 7.52N_2$ $\qquad (5\text{--}6)$

16 lb methane $+$ 274.5 lb air $=$ 44 lb carbon dioxide
$\qquad\qquad\qquad\qquad + $ 36 lb water vapor $+$ 210.5 lb nitrogen

Thus far the chemical equations have been written on the assumption that each molecule of fuel will come in contact with the proper amount of oxygen to support combustion. Such perfect mixing of the fuel and oxygen cannot be realized in a practical combustion chamber. An excess of air must be supplied to the boiler to assure that all molecules of fuel will find the necessary oxygen for complete combustion. An appreciable quantity of carbon monoxide or soot in the flue gas indicates that insufficient oxygen is being supplied to the furnace.

The amount of excess air that is required by a boiler depends on many factors, including the type of burner, fuel, and combustion chamber. Well-designed pulverized-coal boilers may operate with complete combustion on slightly less than 15% excess air; i.e., 15% more than the theoretical air calculated from the chemical equations. Other boilers may require 50% or even 100% excess air.

Consider that methane is to be burned with 25% excess air. Our combustion equation will then be

$$CH_4 + (1.25)(2)O_2 + (1.25)(2)\left(\frac{79.0}{21.0}\right)N_2$$
$$= CO_2 + 2H_2O + (0.25)(2)O_2 + (1.25)(2)\left(\frac{79.0}{21.0}\right)N_2$$

This equation indicates that 344 lb of air are supplied to burn 16 lb of methane at 25% excess air.

The volume of flue gases at any temperature may be calculated from the combustion equations and the characteristic gas equation.

EXAMPLE 5–2. Determine (a) the theoretical air required for combustion of Cherokee County, Kansas, coal, (b) the volume of wet flue gases at 20% excess air and 540 F, and (c) the percentage of CO_2 by volume in dry flue gases from 0% excess air to 100% excess air.

SOLUTION.   (a)

Ultimate Analysis				Oxygen
C = 0.7181 lb × ³²⁄₁₂		=		1.9149 lb
H₂ = 0.0523	×	⁸⁄₁	=	0.4184
O₂ = 0.1015		=		− 0.1015
S = 0.0334	×	¹⁄₁	=	0.0334
N₂ = 0.0120				....
Ash = 0.0827				....
Total 1.0000 lb				2.2652 lb

Theoretical air = 2.2652 ÷ 0.231 = 9.8061 lb per lb of coal

(b) Wet products of combustion (including water vapor) at 0% excess air:

	Weight	$\dfrac{RT}{P}$	Volume
$CO_2 = 0.7181 + 1.9149$	$= 2.6330$ lb $\times$	$\dfrac{35 \times 1000}{144 \times 14.7}$	$= 43.54$ cu ft
$H_2O = 0.0523 + 0.4184$	$= 0.4707$	$\times \dfrac{85.7 \times 1000}{144 \times 14.7}$	$= 19.06$
$SO_2 = 0.0334 + 0.0334$	$= 0.0668$	$\times \dfrac{24.1 \times 1000}{144 \times 14.7}$	$= 0.76$
$N_2 = 0.0120 + (0.769)(9.8061) = 7.5529$		$\times \dfrac{55 \times 1000}{144 \times 14.7}$	$= 196.24$
			$\overline{259.60}$ cu ft

Each 10% of excess air supplied to the combustion chamber is an addition of

$$0.10 \times 9.8061 \times 0.231 = 0.2265 \text{ lb of oxygen}$$
$$0.10 \times 9.8061 \times 0.769 = 0.7541 \text{ lb of nitrogen}$$

At a temperature of 1000 R, the volumes of these gases are

$$\frac{0.2265 \times 48.3 \times 1000}{144 \times 14.7} = 5.17 \text{ cu ft of oxygen}$$

$$\frac{0.7541 \times 55 \times 1000}{144 \times 14.7} = 19.59 \text{ cu ft of nitrogen}$$

Therefore, the total volume of wet gas with 20% excess air will be

CO_2	$=$	43.54 cu ft
H_2O	$=$	19.06
SO_2	$=$	0.76
$N_2 = 196.24 + 2 \times 19.59$	$=$	235.42
$O_2 = 2 \times 5.17$	$=$	10.34
Total		309.12 cu ft per lb of fuel

(c) Since percentage volumetric analysis is independent of temperature, we may use the volumes that have already been calculated for 1000 R to determine the percentage of CO_2 required in part (c). Deduct the volume of the water to reduce the gases to the dry basis. For each 10% increase in excess air the volume of gases will increase $5.17 + 19.59 = 24.76$ cu ft per lb of fuel:

Excess Air, %	Total Volume of Dry Gases, Cu Ft per Lb Fuel	CO_2 in Dry Gases, %
0	240.54	18.10
20	290.06	15.01
40	339.58	12.82
60	389.10	11.19
80	438.62	9.93
100	488.14	8.92

The theoretical air required for complete combustion of a fuel containing oxygen and the combustible elements carbon, hydrogen, and sulfur may be calculated from a summation of the proper combustion equations:

$$W_{ta} = \frac{(32)\,C}{(12)\,(0.231)} + \frac{8\left(H_2 - \frac{O_2}{8}\right)}{0.231} + \frac{S}{0.231}$$

$$= 11.53C + 34.36\left(H_2 - \frac{O_2}{8}\right) + 4.32S \qquad (5\text{-}7)$$

where W_{ta} = theoretical air, lb per lb fuel
$\quad C$ = carbon, lb per fuel *
$\quad H_2$ = hydrogen, lb per lb fuel
$\quad O_2$ = oxygen, lb per lb fuel
$\quad S$ = sulfur, lb per lb fuel

Nitrogen and ash are inert and therefore do not enter into the combustion of the fuel. Oxygen in the fuel is assumed to be combined with hydrogen in the form of water, as in Dulong's formula. Eq. (5-7) may be used for any fuel containing the three combustible elements listed, provided the ultimate analysis of the fuel is known.

Calculations of theoretical air requirements for many coals and oils have shown that there is an approximate relationship between the theoretical air and the higher heating value of the fuel. This relationship,

$$W_{ta} = \frac{7.65 \text{HHV}}{10,000} \qquad (5\text{-}8)$$

may be used for rough estimates. The constant in this equation will vary for different coals and may be slightly different for oil. The value given is an average for typical coals.

5-11. Flue-Gas Analysis. Thus far we have discussed calculations for determining the air required for combustion and the products that would be formed during combustion. It is equally essential that we be able to test any boiler and to determine the amount of air that has been used for combustion and the quantity of products that have been produced.

The most satisfactory means of estimating the air used during combustion and the quantity of dry flue gases resulting from the combustion is to analyze the flue gases. The Orsat analyzer, Fig. 5-8, is a convenient portable apparatus for determining the volumetric percentage of CO_2, O_2, and CO in the dry flue gas. All other constituents in the flue gas besides these three are classified as nitrogen. Sulfur dioxide appearing in the flue gas

---

* When using Eq. (5-7) in calculating the results of a boiler test in accordance with the *ASME Test Code for Steam-Generating Units,* the carbon actually burned per pound of fuel (C_{ab}) should be used [see Eq. (5-11)].

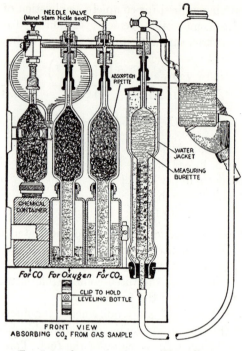

NEEDLE VALVE
(Monel stem Nickle seat)

ABSORPTION
PIPETTE

WATER
JACKET

MEASURING
BURETTE

CHEMICAL
CONTAINER

For CO   For Oxygen   For CO₂

CLIP TO HOLD
LEVELING BOTTLE

FRONT VIEW
ABSORBING CO₂ FROM GAS SAMPLE

FIG. 5–8.   Orsat apparatus.   (Hays Corp.)

will be absorbed largely by the leveling water or by the reagent in the CO_2 pipette.

Any water vapor in the gas sample will be absorbed by either the leveling water or the chemicals in the pipettes, producing a *dry* gas analysis.

It would not be amiss to mention some of the more common sources of error in the analysis of flue gas. One of the most difficult problems is that of securing an average sample of gas to analyze. The sample should be drawn from the main gas stream. For large ducts it may be advisable to sample the stream at several points over its cross-section. Leakage of air into the flue gases will cause erroneous results.

Absorption reagents should be fresh and should be given time to absorb the gases. The older the reagents, the longer will be the time required for them to remove all the CO_2, O_2, and CO from the gas.

5–12. Dry Flue Gases from Actual Combustion. The results of an Orsat analysis of flue gases may be used to calculate the weight of dry flue gases produced by the combustion of a pound of fuel. The Orsat analysis may be expressed as

$$CO_2 + O_2 + CO + N_2 = 100 \qquad (5-9)$$

where the chemical symbols represent the volumetric percentages of each constituent. Also

$$W_{dg} = \frac{\text{weight of dry gases}}{\text{lb carbon burned}} \times \frac{\text{lb carbon burned}}{\text{lb fuel}} \qquad (a)$$

in which W_{dg} is the weight of dry flue gases per pound of fuel burned. The last term of Eq. (a) will bear some discussion. When coal is put into the boiler, the carbon available for combustion per pound of fuel is the amount given by the ultimate analysis. However, some of the coal will fall through the grate and will not be completely burned. Oil, gas, and pulverized coal do not have a loss from fuel falling through the grate, but the refuse from a pulverized coal steam generator may contain some unburned combustible. There is also some unburned combustible in the fly-ash and soot, regardless of the type of fuel. This unburned combustible associated with the flue gas is normally neglected.

The 1946 *ASME Test Code for Steam Generating Units* permits the determination of the weight of dry refuse from the coal analysis when it is impractical to weigh the total refuse. Since the

$$\frac{\text{lb refuse}}{\text{lb coal}} \times \frac{\text{lb ash}}{\text{lb refuse}} = \frac{\text{lb ash}}{\text{lb coal}}$$

and since $(1 - C_r)$ is the pounds of ash per pound of refuse,

$$W_r = \frac{A}{1 - C_r} \qquad (5\text{-}10)$$

in which W_r = dry refuse per lb coal as fired, lb
$\qquad A$ = ash in coal, lb
$\qquad C_r$ = combustible in 1 lb of refuse, lb

Combustible in refuse will be determined in the same manner as the ash content of a coal.

The carbon actually burned per pound of as fired coal may be calculated either from the heating value of the refuse or by assuming that all the combustible in the refuse is carbon. The carbon actually burned will be the carbon in the coal minus the unburned carbon. Since the unburned carbon is $(W_r - A)$, then

$$C_{ab} = C - W_r + A \qquad (5\text{-}11)$$

where C_{ab} = carbon actually burned per lb of fuel, lb *
$\qquad C$ = carbon in 1 lb of fuel, lb

and other symbols are as in Eq. (5–10); or

$$C_{ab} = C - W_r \frac{\text{HV}_r}{14,600} \qquad (5\text{-}11a)$$

* Many gaseous fuels contain carbon dioxide. Since the carbon dioxide in the flue gas will be composed of the carbon dioxide from combustion and the carbon dioxide that was in the fuel, it is necessary to take C_{ab} as all the carbon in the fuel, including the carbon in the noncombustible carbon dioxide.

by the refuse heating value method, where HV_r is heating value of the dry refuse, Btu per lb.

In order to develop an expression for the weight of dry gases per pounds of carbon actually burned, the Orsat analysis must be converted into a weight analysis. Per cent volume may be considered the same as per cent mols, since all gases have the same volume (359 cu ft at 32 F and 14.7 psia) per mol. To convert from mols to weight in pounds, multiply mols of a constituent by the molecular weight:

$$CO'_2 = 44CO_2 \tag{b}$$

$$O'_2 = 32O_2 \tag{c}$$

$$CO' = 28CO \tag{d}$$

$$N'_2 = 28N_2 \tag{e}$$

in which the prime represents gravimetric percentages and other symbols are as in Eq. (5–9). Then

$$\text{Weight of dry gas} \propto CO'_2 + O'_2 + CO' + N'_2$$
$$\propto 44CO_2 + 32O_2 + 28(CO + N_2) \tag{f}$$

All the carbon burned must form CO_2, CO, or hydrocarbons. But since hydrocarbons cannot be determined by the standard Orsat apparatus, all the carbon burned must be assumed to leave the boiler in the form of CO_2 and CO, of which $12/44$ of the weight of carbon dioxide is carbon and $12/28$ of the carbon monoxide is carbon. Thus

$$\text{Pounds of carbon burned} \propto \frac{12}{44}CO'_2 + \frac{12}{28}CO'$$

$$\propto \frac{12}{44} \times 44CO_2 + \frac{12}{28} \times 28CO \tag{g}$$

Then
$$W_{dg} = \frac{44CO_2 + 32O_2 + 28(CO + N_2)}{12(CO_2 + CO)} C_{ab} \tag{h}$$

$$= \frac{11CO_2 + 8O_2 + 7(CO + N_2)}{3(CO_2 + CO)} C_{ab} \tag{5-12}$$

From Eq. (5–9), $\quad CO + N_2 = 100 - CO_2 - O_2$

$$W_{dg} = \frac{4CO_2 + O_2 + 700}{3(CO_2 + CO)} C_{ab} \tag{5-13}$$

The gas analysis used in Eqs. (5–12) and (5–13) must be volumetric. Either decimals or percentages may be used for the volumetric analysis in Eq. (5–12), but only percentages can be used in Eq. (5–13).

The boiler test code uses these formulas but corrects to account for the SO_2:

$$W_{dg} = \frac{11CO_2 + 8O_2 + 7(CO + N_2)}{3(CO_2 + CO)}\left[C_{ab} + \frac{S}{267}\right] + \frac{S}{160} \quad (5\text{-}12a)$$

The value for sulfur must be expressed as a percentage in Eq. (5-12a).

EXAMPLE 5-3. A flue-gas analysis from the combustion of Cherokee County, Kansas, coal shows 14.2% CO_2, 0.3% CO, 4.3% O_2. There was 0.115 lb of refuse collected for each pound of coal fired. Calculate the weight of dry gas per pound of coal.

SOLUTION. From Table 5-1, $C = 0.7181$, $A = 0.0827$, and $S = 0.0334$.

From Eq. (5-11), $C_{ab} = 0.7181 - 0.115 + 0.0827 = 0.6858$

From Eq. (5-12a),

$$W_{dg} = \frac{11(14.2) + 8(4.3) + 7(0.3 + 81.2)}{3(14.2 + 0.3)}\left(0.6858 + \frac{3.34}{267}\right) + \frac{3.34}{160}$$

$$= 12.04 \text{ lb dry gas per lb of fuel}$$

5-13. Incomplete Combustion. Any part of the flue gas that will burn represents incomplete combustion. More energy could have been released by mixing the partially burned substances with oxygen and continuing the combustion process. The only flue-gas constituent measured by the Orsat apparatus that would support further combustion is carbon monoxide.

A formula for the pounds of carbon supplied to the boiler that formed carbon monoxide may be derived in a manner similar to that used for the weight of dry flue gases per pound of fuel:

$$C_i = \frac{\text{weight of carbon in CO}}{\text{lb carbon burned}} \times \frac{\text{lb carbon burned}}{\text{lb fuel}} \quad \textbf{(a)}$$

$$= \frac{\frac{12}{28}(28)CO}{\frac{12}{44}(44)CO_2 + \frac{12}{28}(28)CO} \times C_{ab}$$

$$= \frac{CO}{CO_2 + CO} \times C_{ab} \quad (5\text{-}14)$$

in which C_i is the pounds of carbon in the CO per pound of fuel burned and other symbols are as in Eq. (5-9).

EXAMPLE 5-4. How much carbon burned to CO for the data in Example 5-3?
SOLUTION. From Eq. 5-14,

$$C_i = \frac{0.3}{14.2 + 0.3}(0.6858) = 0.01419 \text{ lb per lb fuel}$$

5–14. Air Actually Used During Combustion. To derive the formula given in the ASME code for the air supplied to the steam generator, apply the Law of Conservation of Mass to the combustion process. The weight of dry gases (W_{dg}) leaving the boiler includes the following: (1) all the carbon actually burned, (2) nitrogen in the fuel, (3) sulfur in the fuel, (4) excess oxygen from the air, (5) oxygen used to burn the carbon and sulfur, (6) oxygen supplied with the fuel, (7) nitrogen in the air supply. The quantity W_{dg} does not include the oxygen used to burn the hydrogen in the fuel, $8H_2$. Subtracting items (1), (2), (3), and (6) from W_{dg} and adding $8H_2$ will give the actual weight of air supplied for combustion.

$$W_{aa} = W_{dg} + 8\left(H_2 - \frac{O_2}{8}\right) - C_{ab} - S - N_2 \qquad (5\text{--}15)$$

Values of H_2, O_2, S, and N_2 are obtained from the ultimate analysis of the fuel and all values are expressed as decimals. Any moisture that entered with the fuel or air does not enter into the equation because the Orsat gives a dry analysis.

5–15. Boiler Heat Balance. The proper operating ratio of air to fuel for a boiler can be established only by computing heat losses from test data. Items computed in a boiler heat balance are in terms of Btu per pound of fuel for solid, liquid, and gaseous fuels. To make a heat balance from one boiler comparable to one from another boiler, the energy terms are converted into percentages where 100% is the heating value of the fuel. The items in the boiler heat balance may be computed per pound of coal on the *as fired* or *dry basis*, but the energy values will be different for each. However, when transferred to percentages, there will be no difference for the two bases.

Energy supplied to the boiler by 1 lb of fuel is distributed among the following items in the ASME short-form heat balance, all expressed in units of Btu per pound of fuel:

Q_1 = energy absorbed by boiler fluid
Q_2 = energy loss due to dry flue gases
Q_3 = energy loss due to moisture in fuel
Q_4 = energy loss due to evaporating and superheating moisture formed by combustion of hydrogen
Q_5 = energy loss due to incomplete combustion of carbon to CO
Q_6 = energy loss due to combustible in refuse
Q_7 = energy loss due to radiation and unaccounted for

An explanation of each of these items follows:

Energy Absorbed by Boiler Fluid. The useful output of the steam generator is the heat transferred to the fluid. Sometimes it is advantageous to divide this item into the heat transferred to the fluid by the boiler proper, the air preheater, the economizer, and the superheater. For our purposes, we shall combine all these subdivisions into

$$Q_1 = \frac{W_w(h_2 - h_1)}{W_f} \tag{5-16}$$

in which W_w = weight of fluid flowing through the boiler during the test, lb

h_1 and h_2 = fluid enthalpies entering and leaving the boiler, respectively, Btu per lb

W_f = weight of fuel burned during test

It is preferable to operate the boiler without blowdown during the test. However, if the test is of such long duration that blowdown is necessary, Eq. (5-16) should be altered appropriately.

Q_1 expressed as a percentage of the higher heating value of the fuel is the boiler efficiency.

Energy Loss Due to Dry Flue Gas. This loss is the greatest of any of the boiler losses for a properly operated unit. Thus,

$$Q_2 = 0.24W_{dg}(t_g - t_a) \tag{5-17}$$

in which 0.24 = specific heat of the flue gas at constant pressure, Btu per lb per deg F

t_g = temperature of the gas leaving the boiler, F

t_a = temperature of the air entering the boiler, F

Obviously, this loss is a function of the flue-gas temperature, but it is sometimes uneconomical to reduce the flue-gas temperature to too low a value. A low gas-outlet temperature can be obtained only by a large heat-transfer surface and a low-temperature fluid to which the energy may be transferred. Air preheaters and economizers furnish the low-temperature fluid and additional surface for reduction of the flue-gas temperature.

Since the incoming air temperature is beyond human control, the only other variable is the weight of dry gases per pound of fuel. W_{dg} should be kept as small as possible, consistent with complete combustion, by control of the excess air.

Energy Loss Due to Evaporating and Superheating Moisture in Fuel. Moisture entering the boiler with the fuel leaves as a superheated vapor in the same way as does the moisture from the combustion of hydrogen. Therefore the formula for calculating this loss may be derived in the same way as that for Q_4:

$$Q_3 = M_f(1089 + 0.46t_g - t_f), \quad \text{when } t_g < 575 \text{ F} \tag{5-18}$$

$$Q_3 = M_f(1066 + 0.5t_g - t_f), \quad \text{when } t_g > 575 \text{ F} \tag{5-18a}$$

where $M_f =$ moisture in fuel, lb per lb of fuel
   $t_f =$ temperature of fuel, F

Energy Loss Due to Evaporating and Superheating Moisture Formed by Combustion of Hydrogen. This loss is higher for gaseous fuels containing relatively large percentages of hydrogen than for the average low-hydrogen coal. Water formed by burning hydrogen leaves the boiler in the form of superheated vapor, and its energy cannot be released to the boiler fluid until the vapor can be condensed. With flue-gas temperatures of 300 F or more and the vapor at a partial pressure less than atmospheric, condensation is impossible within the boiler. Q_4 represents the loss of energy due to the inability of the boiler to condense this superheated vapor to a liquid at a temperature corresponding to the temperature of the incoming air. Thus,

$$Q_4 = 9\mathrm{H}_2(h - h_{ff}) \tag{a}$$

in which $\mathrm{H}_2 =$ weight of hydrogen in the fuel, lb per lb fuel
   $h =$ enthalpy of superheated vapor, Btu per lb
   $h_{ff} =$ enthalpy of liquid at the incoming fuel temperature

Since the partial pressure of the superheated vapor would be difficult to determine, and since this loss of energy is usually small, Eq. (a) may be simplified by assuming that the vapor pressure corresponds to a saturation temperature of 150 F. Then the enthalpy of the superheated vapor is equal to the enthalpy of the saturated vapor (1126.1 Btu per lb) plus the energy needed to superheat the vapor. The latter term is taken as $0.46(t_g - 150)$ when the gas temperature is less than 575 F. The enthalpy of the liquid (h_{ff}) is taken as ($t_f - 32$). Combining these terms we arrive at the expressions

$$Q_4 = 9\mathrm{H}_2(1089 + 0.46t_g - t_f), \quad \text{when } t_g < 575 \text{ F} \tag{5-19}$$

$$Q_4 = 9\mathrm{H}_2(1066 + 0.5t_g - t_f), \quad \text{when } t_g > 575 \text{ F} \tag{5-19a}$$

The proper value of H_2 to be used in the equation for Q_4 is the amount of hydrogen in the fuel that is available for combustion. Ultimate analyses given in Table 5-1 list all the hydrogen in the fuel, including the hydrogen present in the fuel in the form of moisture. To obtain the value of H_2 for Eqs. (5-19) and (5-19a), deduct one-ninth of the weight of moisture from the hydrogen listed in Table 5-1. The weight of moisture may be found from the proximate analysis. Note that $[\mathrm{H}_2 - (\mathrm{O}_2/8)]$ does not represent the amount of hydrogen available for combustion.

Energy Loss Due to Incomplete Combustion. Products formed by incomplete combustion may be mixed with oxygen and burned again with a further release of energy. Such products of incomplete combustion that are present in flue gas are CO, H_2, and various hydrocarbons. Carbon monoxide is the only one of these gases that can be determined conveniently in the power-plant test. Therefore, the loss due to incomplete combustion refers specifically to the incomplete combustion of carbon to carbon monoxide. A formula for the weight ratio of carbon burned to CO per pound of fuel was developed as Eq. (5-14). The difference in the energy release due to burning carbon to carbon monoxide rather than to carbon dioxide is given as 10,160 Btu per lb of carbon.

$$Q_5 = 10,160C_i$$

$$= 10,160C_{ab} \frac{CO}{CO_2 + CO} \tag{5-20}$$

Energy Loss Due to Unconsumed Carbon. All combustible in the refuse may be assumed to be carbon, since the other combustible parts of coal would probably be distilled out of the fuel before live embers would drop into the ashpit. Any unburned carbon in the flue gas (fly ash) or in the ashpit refuse is included.

$$Q_6 = 14,600(C - C_{ab}) \tag{5-21}$$

If the unburned combustible is determined from the heating value of all of the refuse, then

$$Q_6 = W_r HV_r \tag{5-22}$$

Unaccounted-for and Radiation Loss. This loss is due to radiation, incomplete combustion resulting in hydrogen and hydrocarbons in the flue gas, and unaccounted-for losses. Under the ASME code, the radiation loss may be estimated separately and not combined with the unaccounted-for loss. However, when they are combined,

$$Q_7 = HHV - Q_1 - Q_2 - Q_3 - Q_4 - Q_5 - Q_6 \tag{5-23}$$

EXAMPLE 5-5. Calculate the boiler heat balance on the *as fired* basis for the following data (this example has the same data as the other illustrative problems in this chapter):

> Fuel: Cherokee County, Kansas, coal
> Gas analysis: 14.2% CO_2, 0.3% CO, 4.3% O_2
> Coal fired: 22,260 lb per hr
> Refuse: 2,560 lb per hr
> Water: 202,030 lb per hr
> Water entering: 324.7 F
> Steam leaving: 476 psig, 743 F
> Fuel and room temperature: 82 F
> Gas temperature: 463 F

SOLUTION.
$$W_r = \frac{2,560}{22,260} = 0.1150 \text{ lb per lb of fuel}$$

$C_{ab} = 0.6858$ lb per lb fuel (Example 5–3)

$W_{dg} = 12.04$ lb per lb fuel (Example 5–3)

$C_i = 0.01419$ lb per lb fuel (Example 5–4)

$$H_2 \text{ burned to } H_2O \text{ in boiler} = 5.23 - \frac{5.09}{9} = 4.67\% \quad \text{(Example 5–1.)}$$

From Eq. (5–15),

$$W_{aa} = 12.04 + 8\left(0.0467 - \frac{0.0562}{8}\right) - 0.6858 - 0.0334 - 0.0120$$

$$= 11.63 \text{ lb per lb fuel}$$

In Example 5–2, W_{ta} was calculated to be 9.806 lb per lb of fuel. However, using C_{ab} in accordance with the Test Code [see footnote for Eq. (5–7)],

$$W_{ta} = 11.53 \times 0.6858 + 34.36\left(0.0523 - \frac{0.1015}{8}\right) + 4.32 \times 0.0334$$

$$= 9.412 \text{ lb per lb fuel}$$

and the excess air equals

$$\frac{W_{aa}}{W_{ta}} - 1 = \frac{11.63}{9.412} - 1 = 23.57\%$$

BOILER HEAT BALANCE

Item	Equation	Calculation	Energy, Btu per Lb Fuel	Percentage
Q_1	16	$\frac{202,030}{22,260}(1381.0 - 295.1)$	9856	75.34
Q_2	17	$0.24(12.04)(463 - 82)$	1097	8.38
Q_3	18	$0.0509(1089 + 0.46 \times 463 - 82)$	62	0.48
Q_4	19	$9(0.0467)(1089 + 0.46 \times 463 - 82)$	513	3.92
Q_5	20	$10,160(0.01419)$	144	1.10
Q_6	21	$14,600(0.115 - 0.0827)$	472	3.61
Q_7	23	$13,082 - 9856 - 1097 - 62 - 513 - 144 - 472$	938	7.17

PROBLEMS

5–1. Convert the ultimate analysis of Franklin County, Illinois, coal in Table 5–1 to (a) an ultimate analysis showing the amount of moisture, (b) moisture free basis, (c) moisture and ash free basis, (d) moisture, ash, and sulfur free basis. Determine the heating value in each case.

5–2. Same as Prob. 5–1 except for Valley County, Montana, lignite.

5–3. Same as Prob. 5–1 except for Montgomery County, Virginia, semianthracite.

5–4. Same as Prob. 5–1 except for Luzerne County, Pennsylvania, anthracite.

5–5. Same as Prob. 5–1 except for Boulder County, Colorado, subbituminous.

5–6. Use Dulong's formula to check the higher heating value of Nenana Field, Alaska, lignite given in Table 5–1.

5–7. Same as Prob. 5–6 except for Boone County, Missouri, bituminous.

5–8. Same as Prob. 5–6 except for McDowell County, West Virginia, semibituminous.

5–9. Same as Prob. 5–6 except for Columbiana County, Ohio, bituminous.

5–10. Estimate the higher heating value of Haskell County, Oklahoma, bituminous on the moisture free basis by Dulong's formula.

5–11. Same as Prob. 5–10 except for Christian County, Illinois, bituminous.

5–12. The following data were taken from an oxygen-bomb calorimeter test: weight of fuel pan and coal, 6.9359 g; weight of fuel pan, 6.0502 g; weight of water, 1600 g; water equivalent of apparatus, 490 g; temperature rise, 2.837 C; radiation temperature correction, 0.009 C (additive); thermometer calibration correction +0.003 C. Calculate the higher heating value of the coal in calories per gram and in Btu per pound.

5–13. A coal sample weighing 0.9858 g produced a temperature rise (corrected) of 5.793 F in an oxygen-bomb calorimeter having a water equivalent of 510 g; 1900 g of water were used. Before being tested the coal had been air-dried and the moisture reduced from 8.50% on the as-fired basis to 1.40%. Determine the heating value on the as-fired basis.

5–14. Estimate the ultimate analysis and heating value of Lackawanna County, Pennsylvania, anthracite from the proximate analysis given in Table 5–1 by using Evans' equations from Table 5–2.

5–15. Same as Prob. 5–14 except for Monterey County, California, bituminous.

5–16. Determine the rank of Garrett County, Maryland, coal having the following proximate analysis: moisture, 3.18%; volatile matter, 19.20%; fixed carbon, 67.65%; ash, 9.97%; HHV, 13,464 Btu per lb.

5–17. What is the rank of Routt County, Colorado, coal with a proximate analysis of: moisture, 3.00%; volatile matter, 3.00%; fixed carbon, 85.60%; ash, 7.50%; HHV, 13,500 Btu per lb?

5–18. Classify Thurston County, Washington, coal whose analysis is: moisture, 21.69%; volatile matter, 34.77%; fixed carbon, 33.29%; ash, 10.25%; HHV, 8696 Btu per lb.

5–19. For the analysis of Alabama natural gas, Table 5–5, determine (a) the density at 68 F and 29.92 in. Hg abs, (b) the gravimetric percentage of each constituent, (c) molecular weight of the gas, and (d) heating value in Btu per pound.

5–20. Same as Prob. 5–19 except for Illinois natural gas.

5–21. Same as Prob. 5–19 except for Oklahoma natural gas.

5–22. Same as Prob. 5–19 except for blast furnace gas.

5-23. Calculate the percentage of carbon and hydrogen available for combustion, on the gravimetric basis, for Indiana natural gas, Table 5-5.

5-24. Same as Prob. 5-23 except for Pennsylvania natural gas.

5-25. Same as Prob. 5-23 except for New York natural gas.

5-26. Same as Prob. 5-23 except for bituminous producer gas.

5-27. How many cubic feet of (a) oxygen and (b) air are theoretically required for complete combustion of 1 cu ft of Ohio natural gas, Table 5-5?

5-28. Same as Prob. 5-27 except for California natural gas.

5-29. Same as Prob. 5-27 except for West Virginia natural gas.

5-30. Same as Prob. 5-27 except for illuminating gas.

5-31. What are the theoretical air requirements in pound per pound of Sheridan County, Wyoming, coal?

5-32. Same as Prob. 5-31 except use Tazewell County, Virginia, coal.

5-33. Same as Prob. 5-31 except use Lackawanna County, Pennsylvania, coal.

5-34. What must be the capacity in cubic feet per minute at 14.7 psia and 90 F of a fan to supply 25% excess air to 20 lb per min of Bibb County, Alabama, coal?

5-35. What size of duct should be recommended to supply air at 14.7 psia and 80 F to a boiler that burns 2 tons per hr of Sullivan County, Indiana, coal at 35% excess air? Use a velocity of 2000 fpm.

5-36. Estimate the rating in horsepower of a motor to drive a fan that will supply 25% excess air to a boiler that burns 100 lb per min of Oklahoma oil, Table 5-4. The air will be at 100 F and the fan will require 1.58 bhp per 1000 cfm.

5-37. If a plant operates 3500 hr per yr at an average load corresponding to an input of 4000 lb per hr of Ohio oil (Table 5-4), what would be the present worth of fan power at 20% investment charges? Use a motor efficiency of 91%, 6.5 mills per kwhr, 1.68 bhp of fan power per 1000 cfm of 90 F air.

5-38. A boiler uses an average of 2500 cfm of Oklahoma natural gas. What would be the annual savings in forced-draft fan power alone, if the excess air were reduced from 25 to 20% when electrical energy cost 4.5 mills per kwhr, motor efficiency is 90%, and the fan requires 2.16 bhp per 1000 cfm? Use 6500 hr per yr of operation. The gas and air are at the same pressure and temperature.

5-39. Estimate the annual investment cost of a duct to supply 90 F air to a boiler using 2000 cfm of Arkansas natural gas. The boiler requires 20% excess air, 2100 fpm velocity, 60 ft of duct that has a 50% larger width than height, costs $650 per ton of steel erected, and weights 8 lb per sq ft. Use 12% investment charges. The gas and air are at the same temperature and pressure.

5-40. Tabulate the weight of wet products of combustion per pound of fuel, cubic feet of the wet products of combustion at 475 F per lb of fuel, percentage by volume of CO_2 in the dry products of combustion, and plot the percentage of CO_2 versus excess air, at intervals of 10% excess air from 0 to 100%, from Virginia crude oil.

5–41. Use 1 cu ft of Kentucky natural gas at 14.7 psia and 68 F and at temperature of 575 F for the products of combustion in Prob. 5–40.

5–42. Same as Prob. 5–40 except use Franklin County, Illinois, coal and a gas temperature of 500 F.

5–43. Same as Prob. 5–40 except for Allegheny County, Pennsylvania, coal (not cannel coal) and 450 F gas temperature.

5–44. Same as Prob. 5–40 except use Letcher County, Kentucky, coal, and 550 F as the gas temperature.

5–45. An Orsat analysis of flue gas from the combustion of Gunnison County, Colorado, coal was: CO_2, 14.2%; O_2, 4.8%; and CO, 0.3%. Determine the weight of dry gases per pound of coal and the percentage of excess air if there were 8.77% combustible in the refuse.

5–46. Combustion of Morgan County, Tennessee, coal produced flue gas containing 14.10% CO_2, 5.40% O_2, and 0.42% CO. Refuse was 7.17% of the coal fired. $HV_r = 979$ Btu per lb. What was the percentage of excess air?

5–47. Blast-furnace gas (Table 5–5) was burned with what percentage of excess air (by weight) when the flue-gas analysis was 22.1% CO_2, 3.9% O_2, and 0.6% CO?

5–48. When California crude oil was burned with air, the Orsat analysis was 11.6% CO_2, 4.2% O_2, 0.7% CO. Calculate the weight of dry flue gases per lb of oil and the excess air.

5–49. An Orsat analysis of the products of combustion from North Dakota lignite was improved from 10.1% CO_2, 8.3% O_2, and 0.1% CO to 14.9% CO_2, 3.1% O_2, and 0.1% CO by improved operation of the boiler. For 7700 hr per yr of operation at an average load of 5.3 tons of coal per hr and a coal cost 18¢ per million Btu, what was the annual saving in loss due to dry flue gas? In both cases the gas temperature was 355 F, the refuse was 0.1062 lb per lb of coal, and air was at 75 F.

5–50. What is the annual cost of loss due to dry flue gases when burning Washington coal at a cost of 22¢ per million Btu for the following data: 844 Btu per lb heating value of the refuse, 0.0871 lb of refuse per lb fuel, 13.6% CO_2, 5.1% O_2, 0.2% CO, 5800 hr per yr operation, 9.4 tons per hr average consumption.

5–51. Analysis of the products of combustion from burning wood was: 16.6% CO_2, 3.6% O_2, 0.1% CO. Analysis of the dry wood was 50.7% C, 5.9% H_2, 42.8% O_2, 0.6% ash, and the dry heating value was 9175 Btu per lb. What was the weight of dry gases and the loss due to incomplete combustion if there were 0.0086 lb of ash per lb dry wood and the heating value of the refuse was 5114 Btu per lb?

5–52. Calculate a boiler heat balance on the as-fired basis from the following data:

Fuel: Sebastian County, Arkansas
Duration of test: 24 hr
Fuel fired during test: 421.27 tons
Refuse collected during test: 46.3 tons
Heating value of refuse: 805 Btu per lb

Water supplied to boiler during test: 8,985,300 lb
Water entering: 374.6 F
Steam: 840 psig and 750 F
Fuel and air temperature: 91 F
Gas temperature: 439 F
Gas analysis: 14.9% CO_2, 3.7% O_2, 0.2% CO

5-53. Boiler-test data:

Fuel: Webster County, Kentucky
Refuse: 10.4% of fuel
Heating value of refuse: 1076 Btu per lb
Water: 9.17 lb per lb fuel
Water temperature: 256.1 F
Steam: 214 psig and 519 F
Fuel and air temperature: 77 F
Gas temperature: 591 F
Orsat readings: 12.8% CO_2, 5.2% O_2, 0.4% CO

Calculate the boiler heat balance.

5-54. A California fuel oil having essentially the characteristics of the California crude listed in Table 5-4 is used in a boiler test. The resultant data are:

Water: 11.32 lb per lb fuel
Water temperature: 67 F
Steam: 164 psig and 0.8% moisture
Fuel temperature: 62 F
Air: 87 F
Flue gas: 491 F
Gas analysis: CO_2, 13.3%; O_2, 2.1%; CO, 2.2%

Prepare a boiler heat balance.

5-55. How much should a fuel oil similar to Texas crude cost per gallon to be comparable to Latimer County, Oklahoma, coal at $11.75 a ton?

5-56. What is the equivalent that can be paid for 1000 cu ft of Pennsylvania natural gas if a power plant can purchase Allegheny County, Pennsylvania, coal at $4.75 a ton delivered?

5-57. Using the data of Prob. 5-52, calculate the cost of energy lost with the dry flue gases per year of operation. Assume that the boiler operates 5000 hr during the year and that coal costs $5.75 per ton delivered at the plant and $0.75 per ton for handling, crushing, etc., in the plant.

5-58. Calculate the cost of fuel per million Btu for the following: (a) Latimer County, Oklahoma, coal at $6.35 per ton; (b) South Dakota lignite at $3.76 per ton; (c) Texas crude at 12¢ per gal; (d) Ohio natural gas at 7.5¢ per 1000 cu ft; (e) Fayette County, West Virginia, coal at $6.45 per ton.

5-59. Use the data in Example 5-5 and find the annual cost of each of the losses when the plant operates 7600 hr per yr, the fuel costs 26¢ per million Btu, and the plant uses an average of 4.1 tons of coal per hr.

5-60. Calculate a boiler heat balance for the following averaged data:

Volumetric analysis of natural gas:

CO_2	0.20%	C_2H_6	14.63%
CH_4	76.82%	N_2	8.35%

HHV of the fuel: 1034 Btu per cu ft at 60 F and 30 in. Hg

Orsat analysis: 10.8% CO_2, 13.3% O_2, 0.1% CO

Barometer: 28.90 in. Hg

Air temperature: 74 F

Water entered at 53.5 F

Steam left superheater at 167.4 psig and 524 F

Flue gas temperature: 487 F

2293 lb per hr of water

3715 cu ft of natural gas measured at 59 F and 11.2 in. water gage

BIBLIOGRAPHY

ASME Test Code for Steam-Generating Units, 1946.

ASTM Fuel Test Codes.

Bituminous Coal Consumers' Council Data Book. Vols. 1 to 5.

Bureau of Mines. Bulletin Nos. 22, 85, and 193.

Bureau of Mines. Technical Papers Nos. 93 and 158.

CAMPBELL, MARIUS R. "Our Coal Supply: Its Quantity, Quality and Distribution." Presented at International Conference of Bituminous Coal, Carnegie Institute of Technology.

EVANS, FREDERIC C. *Empirical Relations for Coals in the United States.* Cornell University Engineering Experiment Station Bulletin No. 3, 1925.

PRATT, A. D. *Principles of Combustion in the Steam Boiler Furnace.* New York: Babcock & Wilcox Co., 1920.

CHAPTER 6

STEAM GENERATORS

6-1. General. The term *steam generator* has come into acceptance to replace the rather loose term *boiler* and indicates the furnace, boiler, waterwalls, water floor, water screen, superheater, reheater, economizer, air preheater, and fuel-burning equipment. The definition of a steam generator, according to the ASME codes, is a "combination of apparatus for producing, furnishing, or recovering heat, together with apparatus for transferring to a working fluid the heat thus made available." However, the term *boiler* has been used for such a long period of time that the two terms are used interchangeably.

Energy input to the power-plant cycle is from the combustion of a fuel in the steam generator. Some of the energy derived from the combustion is transferred to the water in the furnace through the heating surfaces provided. If there are superheaters, economizers, or air preheaters in the installation, they will also receive energy from the gases.

Large single-unit installations, composed of one boiler serving one turbine, are common because of the saving in piping and because of operating advantages. Units having capacities of 3,750,000 lb per hr are being built. With present economic conditions, flexibility of fuel-burning and heat-absorbing equipment is stressed. Many public utilities have added oil burners to their boilers or made other provisions for the use of an alternate fuel. Many modern plants can now burn several grades of coal. Liberal furnace volumes are an aid in meeting changing coal conditions. Gas is a popular fuel in the Southwest and where economic considerations make its use possible.

Increasing fuel costs have caused more attention to be paid to the efficiency of the steam generator. High feedwater temperatures, cleaner heat-transfer surfaces, and air preheaters capable of operating at low flue-gas temperatures without undue corrosion are some of the improvements.

6-2. Fire-Tube Boilers. Of the many ways of classifying boilers, the distinction based on the fluid inside the tubes is one of the most important. *Fire-tube* boilers are those having the hot gases of combustion inside the tubes. Their opposite are *water-tube* boilers. Fire-tube units are usually limited to steam pressures of about 250-psig working pressure because of the thickness of the large diameter shell for higher pressures. A design pressure of 150 psig is very common for these units and capacities range up to 18,000 lb per hr of steam. Their principal use today in the stationary field is for very small industrial plants or for heating systems. Railroads still use the fire-tube units for their steam engines.

Fig. 6-1 shows one of the more popular types of fire-tube boilers—the *horizontal-return tubular* (HRT) boiler. Coal is fired into the boiler

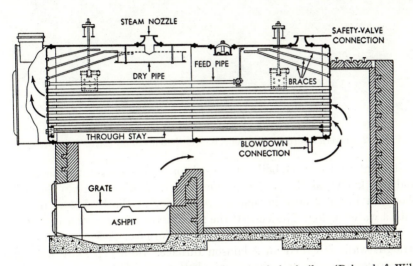

FIG. 6-1. Sectional sketch of a horizontal-return tubular boiler. (Babcock & Wilcox Co.)

through the fire door in the front and is burned on the grate. Products of combustion travel upward over the brick bridge wall that supports the rear of the grate and pass to the rear of the unit. The direction of the gas stream is then reversed so that it passes through the tubes of the shell and leaves at the front of the boiler. The fire tubes are rolled into the tube sheets at each end of the drum and are submerged in the water contained in the shell. Heat transfer through the tubes evaporates the feedwater.

The shell is supported from overhead beams and has a manhole in the top. Cleanout doors are provided in the front and back of the setting for soot removal. Ashes are removed from the ash-pit door directly below the fire door.

A boiler similar in principle to the HRT boiler is shown in Fig. 6-2. While the HRT unit has a so-called *external furnace,* an *internal-furnace* unit has the fire box inside the shell. No brick setting is required for this boiler as is necessary with the HRT type. Stays for the support of the shell heads may be seen in both pictures.

Fɪɢ. 6–2. Internal-furnace, fire-tube boiler. (Combustion Engineering Co.)

The internal-furnace boiler was used for many years in marine service and was known as the *Scotch marine* boiler. It differed from the boiler shown in Fig. 6–2 only in that the refractory at the back was replaced with a water jacket. Notice that the furnace tube for the unit of Fig. 6–2 is corrugated; it is known as the Morrison furnace. A furnace tube with straight walls is called an Adamson furnace. Both types are used, but the corrugated walls are said to make better provision for relative expansion and to shed soot deposits as the walls expand and contract.

With increased steam pressures and capacities, the Scotch marine boiler has not been used for marine applications for many years. Recently, however, this style of boiler has been converted to the use of light and heavy oil and gas fuels, and with the addition of automatic controls, is known as a *packaged* fire-tube boiler, Fig. 6–3. It has found wide acceptance for both heating and process steam applications. Riveted construction has been replaced by welding.

When delivered to the plant site, a packaged boiler needs only a concrete slab on which to rest, and fuel, electrical, breeching, steam, and safety valve connections. Fuel oil pump or gas regulator, forced-draft blower, and all controls are a part of the boiler. The controls will follow a pre-arranged sequence of starting and purging, will automatically adjust the fuel during operation, and will shut the unit down on low steam demand, low water level, low voltage, flame failure, etc.

The boiler shown in Fig. 6–3 is arranged with four passes and can burn either oil or gas. The furnace tube is the first pass; the tubes on each side of the furnace tube constitute the second pass; the tubes imme-

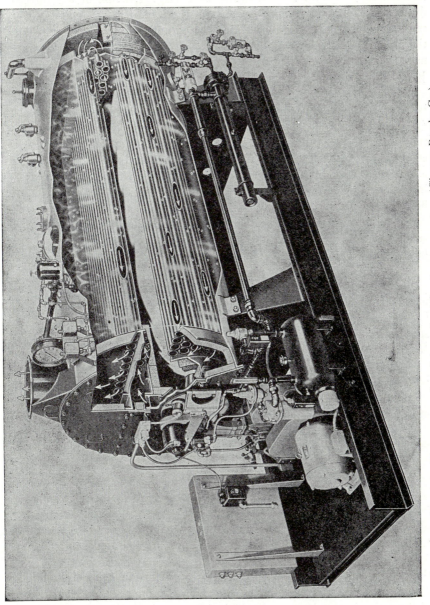

FIG. 6-3. Four-pass, packaged-type, fire-tube steam generator. (Cleaver-Brooks Co.)

diately above the furnace tube form the third pass; and the top row of tubes compose the fourth pass. The number of tubes in each pass is selected to give similar velocities in each pass. Because of the forced-draft fan, the boiler is pressurized, and there may be a slight positive pressure at the flue gas outlet. This also permits higher gas velocities in the unit. Temperatures of the products of combustion leaving the furnace range from 1800 F to 2000 F, and the gas pressure drop ranges from $\frac{1}{2}$ to 1 in. of water per pass.

Packaged fire-tube boilers are cheaper, in first cost, than water-tube boilers until the capacity exceeds some 15,000 to 18,000 lb per hr of steam (packaged water-tube boilers cost 25 to 30% more than packaged fire-tube boilers). At maximum load, these units should attain better than 80% efficiency with about 13.5% CO_2 in the flue gas when using oil, will have a heat release of about 100,000 Btu per (hr) (sq ft), and will have a heating surface of 5 sq ft per boiler horsepower (see Art. 6–11 for definition of boiler horsepower).

An advantage of the fire-tube boiler is its ability to meet wide and sudden load fluctuations with only slight pressure changes because of the large volume of water stored in the shell.

6–3. Water-Tube Boilers—Circulation. Demands for increased capacity and pressure over that economically obtained from the fire-tube unit led to the development of the water-tube unit. Although available for smaller capacities, the field of the water-tube boiler begins at about 15,000 lb per hr of steam.

Water-tube steam generators may be classified in several ways: according to the shape of the tubes (straight or bent), drum position (longitudinal or cross), method of water circulation (forced or natural), number of drums, service (marine or stationary), capacity, and the thermal conditions. Forced-circulation boilers are such a special case that this article will deal with natural-circulation units only.

Circulation of the feedwater within the tubes is one of the most important problems of natural-circulation boiler design. An elementary boiler is shown in Fig. 6–4, in which we shall assume that all the heat transfer from the gases to the water takes place in the riser. Water at essentially saturated conditions will flow into the downcomer from the drum. Since the water flows through the riser on its return to the drum, heat transfer will cause part of the water to evaporate into steam, with the result that the fluid in the riser will be composed of a mixture of water and steam. The density of the mixture in the riser will then be less than the density of the water in the downcomer. This difference in density in the U-tube provides the pressure to overcome the friction loss occurring in the system.

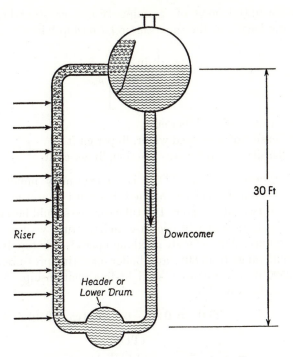

FIG. 6–4. An elementary water-tube boiler.

Thus, when the system is in equilibrium,

$$\Delta p = \Delta p_D + \Delta p_R = \frac{Z}{144} \, (\rho_D - \rho_m) \qquad (6\text{–}1)$$

where Δp = pressure differential due to differences in density, psi

Δp_D = pressure loss in the downcomers, psi

Δp_R = pressure loss in the risers, psi

Z = system height, ft

ρ_D = density of water in the downcomers, usually saturated, lb per cu ft

ρ_m = mean density of water-steam mixture in the risers, lb per cu ft

A mixture of steam and water leaves the risers and enters the drum. The term *dryness fraction* refers to the amount of steam in this mixture of steam and water and may be expressed as a fraction by volume or by weight. When dryness fraction is expressed as a weight ratio it represents the quality of the mixture. *Top dryness fraction* applies to the mixture leaving the tubes.

Mean density (ρ_m) of the mixture in the steaming section of the tubes may be found by integration of the mixture density with respect to elevation above the initial point of steaming. For a constant heat transfer throughout the length of the tube, this relationship * is

$$\rho_m = \left[\frac{\ln \frac{\rho_w}{\rho_t}}{\frac{\rho_w}{\rho_t} - 1} \right] \rho_w \qquad (6\text{-}2)$$

where ρ_m = mean density, lb per cu ft

ρ_w = density of saturated water, lb per cu ft

ρ_t = density at top dryness fraction, lb per cu ft

The top dryness fraction (TDF) is an extremely important design factor because the inside surface of the tubes must be wet at all times to ensure satisfactory heat transfer. If a tube surface should become dry, the coefficient of heat transfer will decrease and the tube wall will overheat, blister, and rupture. Approximate maximum permissible top dryness fractions by volume range from 80% at a boiler operating pressure of 250 psia to 50% at 2500 psia in almost a straight-line relationship. An equation for this relationship would be

$$\text{TDF}_v = 0.80 - 0.000133\,(p - 250) \qquad (6\text{-}3)$$

also

$$\text{TDF}_w = \frac{(\text{TDF}_v)\,v_f}{v_g - (\text{TDF}_v)\,(v_{fg})} \qquad (6\text{-}4)$$

and

$$\text{Circulation ratio} = \frac{1}{\text{TDF}_w} \qquad (6\text{-}5)$$

$$\rho_t = \frac{\text{TDF}_v}{v_g(\text{TDF}_w)} \qquad (6\text{-}6)$$

where TDF_v = top dryness fraction by volume, cu ft steam per cu ft mixture

TDF_w = top dryness fraction by weight, lb steam per lb mixture

p = steam generator operating pressure, psia

and v_f, v_{fg}, and v_g are as given in the steam tables.

---

* The basic equation is

$$L\rho_m = \int_0^L \rho_m{}' \, dl$$

by taking

$$\rho_m{}' = \frac{\rho_w}{\frac{KS}{M} + 1}$$

In these equations, L is the total vertical height of the tubes in ft, $\rho_m{}'$ is the density of the mixture in lb per cu ft at height l, S is the steam generated in lb per hr, M is the total flow of mixture in lb per hr, and $K = \frac{\rho_w}{\rho_s} - 1$, where ρ_s is the density of saturated steam in lb per cu ft. Other symbols are as in Eq. (6-2).

The circulation ratio represents the water flow through the down-comers for each pound of steam generated in the tubes.

EXAMPLE 6–1.   How many pounds of water must enter the downcomers of a boiler operating at 1000 psia to produce 1 lb of steam leaving the risers?

SOLUTION.   From Eq. (6–3),

$$\text{TDF}_v = 0.80 - 0.000133(1000 - 250) = 0.70$$

From the steam tables, $v_f = 0.0216$, $v_{fg} = 0.4240$, and $v_g = 0.4456$ cu ft per lb, and from Eq. 6–4,

$$\text{TDF}_w = \frac{0.70 \times 0.0216}{0.4456 - 0.70(0.4240)} = 0.1018$$

From Eq. (6–5)

$$\text{Circulation ratio} = \frac{1}{0.1018} = 9.83 \text{ lb water per lb steam}$$

Pressure losses in the downcomers consist of losses at entrance and exit ($\frac{1}{2}$ and 1 elbow equivalent, respectively), loss due to changes in tube flow area ($\frac{1}{4}$ to $\frac{1}{2}$ equivalent elbow), loss due to bends ($\frac{3}{4}$ equivalent elbow if 90-deg bend), and the loss due to straight tube. Equivalent lengths may be determined from Fig. 2–22 and the pressure drop in pounds per square inch may be determined from Eq. (2–16). The friction factor to be used in this equation should be determined from Reynolds' number and Fig. 2–19. However, a friction factor of 0.02 will give sufficiently accurate results for many purposes.

Resistances in the risers will include all those listed for the downcomers; in addition there will be a loss through the steam separator in the drum (see Art. 6–5). For problem solution, this loss can be taken as 1 psi, which is a realistic value for a large steam generator.

It is very common to have two or more riser circuits of a modern boiler fed by one set of downcomers. This represents a problem in flow through parallel circuits.

EXAMPLE 6–2.   In a 1000-psia boiler, one set of 50 risers is to be served by two downcomers. Each riser will be 60 ft long, have a difference in elevation of 40 ft, have two 90-deg bends, be 3-in. OD and 2.56 in. ID, and have a heat transfer rate of 50,000 Btu per (hr)(sq ft of projected surface). Water entering the downcomers is saturated. Determine the size of the two downcomers if they are to have 50 ft of length and 3 bends.

SOLUTION.   In Example 6–1 it was determined that the circulation ratio should be 9.83 lb of water per lb of steam at a TDF_w of 0.1018.

The steam evaporated in the risers will be determined from the rate of heat transfer times the projected tube area and divided by the enthalpy of evaporation. The projected tube area = $\frac{3}{12}$ ft tube width × 40 ft length × 50 tubes = 500 sq ft. Therefore, the steam evaporated = 50,000 Btu per hr × 500 sq ft ÷ 649.4

Btu per lb = 38,500 lb per hr. Total circulation will be the steam flow times the circulation ratio, or $38,500 \times 9.83 = 378,000$ lb per hr, with the flow per tube being $378,000 \div 50 \times 60 = 126$ lb per min, and the flow per downcomer being $378,000 \div 2 \times 60 = 3150$ lb per min.

The equivalent length of the risers, at 7 ft per elbow, will be $60 + 7(0.5 + 1 + 2 \times 0.75) = 81$ ft for the straight pipe, the entrance, the exit, and the two elbows. From the steam tables, $\rho_w = \dfrac{1}{v_f} = \dfrac{1}{0.0216} = 46.3$ lb per cu ft and $v_g = 0.4456$ lb per cu ft. Substituting in Eq. (6–6),

$$\rho_t = \frac{0.70}{0.4456(0.1018)} = 15.43 \text{ lb per cu ft}$$

Use Eq. (6–2) to obtain the mean density,

$$\rho_m = \left[\frac{\ln \dfrac{46.3}{15.43}}{\dfrac{46.3}{15.43} - 1} \right] 46.3 = 25.4 \text{ lb per cu ft}$$

Substitute into Eq. (2–16) and add the loss through the separator to obtain the loss in the riser circuit,

$$\Delta p_R = \frac{0.01214 \times 0.02 \times 81(126)^2}{25.4(2.56)^5} + 1 = 1.11 \text{ psi}$$

From Eq. (6–1), the allowable pressure drop in the downcomers can be

$$\Delta p_D = \frac{40}{144}(46.3 - 25.4) - 1.11 = 4.7 \text{ psi}$$

The size of downcomer tubes to give this pressure loss can be determined by trial and error. Try two 5-in. OD tubes with 0.340-in. wall thickness for an ID of 4.32 in. The equivalent length of an elbow will be 12 ft, and the equivalent length of each downcomer will be $50 + 12(0.5 + 1.0 + 3 \times 0.75) = 95$ ft. The friction loss will be

$$\Delta p = \frac{0.01214 \times 0.02 \times 95(3150)^2}{46.3(4.32)^5} = 3.3 \text{ psi}$$

This is as near to a balance as can be obtained with commercial sizes of tubing. Using these downcomers would mean that the TDF would be less than the maximum permissible.

6–4. Water-Tube Boilers—Descriptions.

Water-tube boilers were developed to permit increases in boiler pressure and capacity with reasonable metal stresses. The first designs employed *straight tubes* rolled into headers at each end, Fig. 6–5. In the *longitudinal-drum* design, the water leaves the drum at the rear of the boiler and flows to the sinuous rear headers. The tubes are inclined upward at an angle of 15 deg, or slightly more, to the horizontal. The mixture of steam and water leaves the tubes at the front of the boiler and returns to the drum by way of the front headers. In some designs the drum is inclined at the same angle as the tubes. At the bottom of the rear headers is a square (sometimes round)

FIG. 6-5. Longitudinal-drum boiler. (Babcock & Wilcox Co.)

horizontal header that is known as a *mud drum*. Solids in the boiler water collect in this drum and are discharged to waste through the *blowoff* connection.

Note that the water circulation path forms a U-tube. Hot gases generated in the furnace traverse the tube bank in three passes and are directed by vertical *baffles* to give cross-flow of the gases. In some units the baffles are slightly inclined; in other units they are horizontal.

Because the number of tubes that can be accommodated by a single, longitudinal drum are limited, even when two or three drums are used, the *cross-drum* boiler was developed, Fig. 6-6. The drum of the cross-drum boiler has a more uniform temperature than the longitudinal-drum

Fɪɢ. 6–6.   Cross-drum boiler.   (Babcock & Wilcox Co.)

boiler. While longitudinal-drum boilers have a capacity range of 5000 to 80,000 lb per hr of steam, the cross-drum style has been built up to 525,000 lb per hr.

Sinuous headers, Fig. 6–7, permit staggered tubes and are the same for both styles. A handhole is provided opposite each tube end for installing, rolling, and cleaning the tube. Also, the gas and water flows are the same for both.

Low draft loss and simple field erection methods are advantages of these units. However, they have a serious disadvantage in that they are subject to faulty circulation at high capacity when the upper parts of the tubes may become dry, i.e., water may flow in the bottom part of the tube and steam may flow in the upper part. This may cause corrosion along the water line or blistering and rupture of the upper part of the tube.

Longitudinal-drum boilers are seldom, if ever, installed in recent years and new cross-drum boilers are very rare.

The next step in the development of the steam generator was the *bent-tube* boiler. The first of these units, frequently called the *Stirling type*,

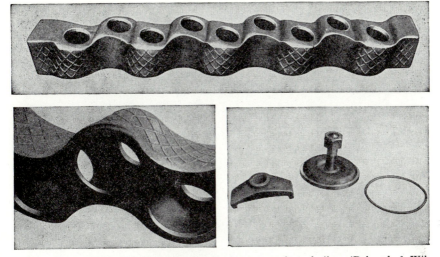

FIG. 6–7. Sinuous header for longitudinal- or cross-drum boiler. (Babcock & Wilcox Co.)

contained four drums, Fig. 6–8, consisting of three upper drums and one lower drum called the *mud drum*. Note that the tubes enter directly into the drum. They are bent so that all tubes enter a drum wall on a radius of the drum.

The lower drum is completely filled with water and is provided with a blowoff connection. All the upper drums contain steam and water and are connected by tubes both above and below the water line. Feed enters and steam leaves from the rear drum. Tubes in the back of the rear pass are downcomers, while those in the front pass are risers. In most cases the tubes in the middle pass are also risers. However, the exact number of downcomers may change with load, possibly causing circulation difficulties or nearly stagnant tubes. Note the gas flow and the baffles to divide the gas passages. Baffles may be made of metal or, as in this case, of ceramic material.

A three-drum, bent-tube boiler, Fig. 6–9, which is especially designed for spaces with low head room, is very common for pressures to 450 psig and capacities to 40,000 lb per hr. The gas space is baffled for three passes. Water flows downward in the rear tubes and up through the front drum to the top drum. Tubes in this boiler, as in nearly all well-designed bent-tube boilers, are spaced on alternately large and small centers so that any one tube may be removed and replaced.

Another design of three-drum boiler that has been successfully used in marine applications, Fig. 6–10, is known as an **A** type of boiler. This unit has the advantages of low weight per pound of steam, compactness, rapid

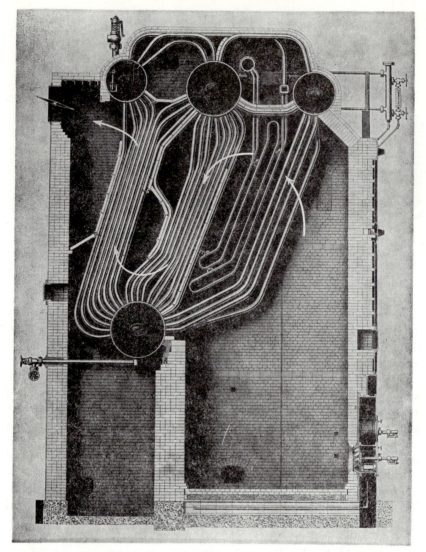

Fɪɢ. 6–8. Bent-tube boiler of the Stirling type. (Babcock & Wilcox Co.)

steaming due to the use of many small tubes, and an ability to provide rapid variations in steam output because of the large drum capacity. Three oil or gas burners are shown in the center of the picture. Gas flow divides, and there are two gas outlets that can contain either economizers or air preheaters. A superheater is shown in the left tube bank. The front and rear walls are of refractory and of double-casing construction. Combustion air flows inside the double casing to be preheated.

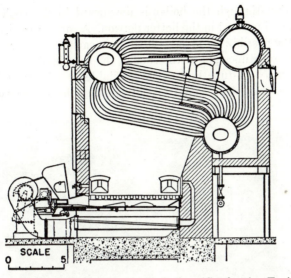

FIG. 6-9. Low-head, three-drum, bent-tube boiler. (Combustion Engineering Co.)

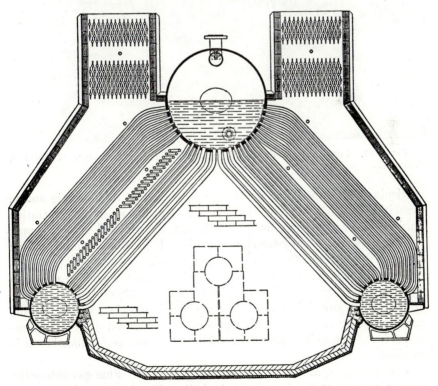

FIG. 6-10. Three-drum marine boiler   (Combustion Engineering Co.)

Circulation through the boiler is downward in the cooler portion of each bank and upward in the hotter portion. As an aid to circulation, and to reduce the possibility of stagnant tubes in each bank, a few large downcomers are sometimes put at the ends of the tube banks.

Two-drum, bent-tube boilers are available in a large variety of sizes and designs. Fig. 6–11 shows one design of a low-capacity unit. This

FIG. 6–11.   Flat-floor, two-drum, oil- or gas-fired steam generator. (Babcock & Wilcox Co.)

particular unit is available for pressures ranging up to 1000 psig and in capacities up to 350,000 lb per hr. The flat furnace floor in the illustration signifies oil or gas firing, but the unit may be arranged with a hopper bottom (to be discussed later) for coal firing. If the upper drum is moved forward, the height of the boiler may be reduced.

With slight variations, this boiler can be factory-assembled and then becomes a packaged unit in capacities up to about 30,000 lb of steam per hr and with a forced-draft fan. As a packaged unit, the gases in the furnace are under a slight positive pressure (no induced-draft fan is required), and the unit is placed on skids so that no special foundation is required.

The gas passes of the unit shown are horizontal. Flue gas enters the convection zone at the far end of the boiler. The location of the baffles shows that the gases flow horizontally across the convection zone in three passes.

Large high-pressure steam generators are usually of either the two-drum or three-drum, bent-tube design. In the two-drum design, the lower drum may be either a drum or a header. Fig. 6–12 shows one such unit.

FIG. 6–12.   Open-pass, pulverized-coal steam generator for 1,330,000 lb per hr of steam at 2000 psig and 1050 F with 1050 F reheat. (Babcock & Wilcox Co.)

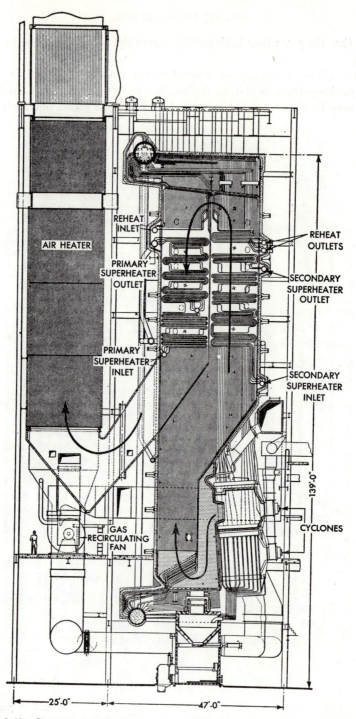

REHEAT INLET

AIR HEATER

REHEAT OUTLETS

PRIMARY SUPERHEATER OUTLET

SECONDARY SUPERHEATER OUTLET

PRIMARY SUPERHEATER INLET

SECONDARY SUPERHEATER INLET

GAS RECIRCULATING FAN

CYCLONES

139'-0"

25'-0"

47'-0"

FIG. 6–13. Steam generator with cyclone burner for 850,000 lb per hr at 1485 psig and 1000 F and 1000 F reheat. (Babcock & Wilcox Co.)

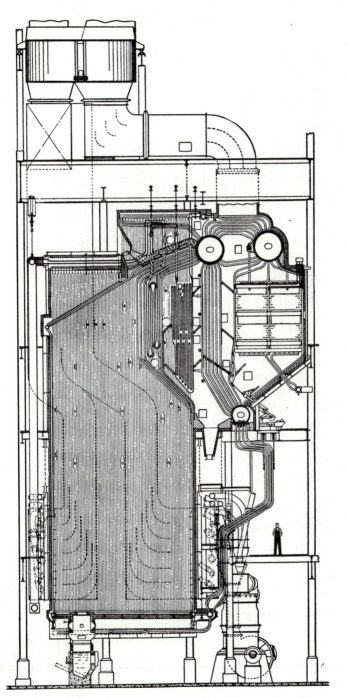

FIG. 6–14. Three-drum steam generator: 1,000,000 lb per hr, 1625 psig, 925 F.
(Combustion Engineering Co.)

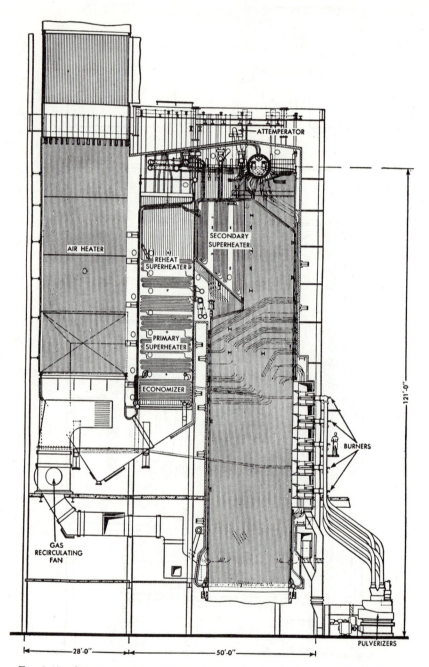

Fig. 6–15. Steam generator: 1,050,000 lb per hr, 2300 psig, 1050 F, 1000 F reheat. (Babcock & Wilcox Co.)

Notice that there is a large header or small drum at the bottom of the boiler that is served by external downcomers. This style of unit may be called an *open-pass* boiler because of the large middle pass that is unobstructed with tubes. The only true boiler tubes in this unit are a few rows at the bottom of the last pass.

A different arrangement of this two-drum design is shown in Fig. 6–13 where cyclone burners have been installed (cyclone burners will be discussed in Chap. 7).

Two large steam generators that evaporate nearly all the water in the radiant furnace are shown in Figs. 6–14 and 6–15. The first of these two units is a three-drum style with a small amount of boiler tube surface. Observe that the downcomers are largely outside the casing. The second of these two units is a two-drum style wherein the lower drum is a header rather than a drum. Other features of both these units, such as superheaters, reheaters, economizers, air preheaters, etc., will be discussed later.

Observe that the gases leaving the furnaces of these steam generators pass through a few (frequently only two) rows of tubes. These tubes are known as the *slag screen.* It is the purpose of this screen to reduce the gases to such a sufficiently low temperature that there will not be excessive slag accumulations on the following tubes. Slag-screen tubes are spaced so that there will be 9 to 12 in. of clear space between the tubes; thus, any slag accumulations on the screen cannot bridge-over and close the gas passage. Slag-screen tubes may be either staggered or in-line. The maximum permissible gas temperature leaving the slag screen will depend upon the ash-softening temperature.

Observe that there are two types of furnace bottoms illustrated; the furnace of Fig. 6–14 has a floor that is nearly horizontal, while the furnaces of Figs. 6–15 and 6–28 have hopper bottoms. These latter units are called *dry-bottom* furnaces because all the ash falling to the bottom of the furnace has been cooled to a solid state and is not fluid. Gas temperatures in a dry-bottom furnace must be below the ash-softening temperature at the point in the furnace where the ash is deposited. This means that the heat release within the furnace must be conservative. Furnace heat release should be calculated as follows:

$$\text{Heat release} = \left[\text{HHV} + 0.24 W_{aa}(t_{ah} - t_a) \right.$$
$$\left. - 1040(9H_2) - 14{,}600(C - C_{ab}) \right] \frac{W_f}{V} \qquad (6\text{–}7)$$

where HHV = higher heating value of the fuel, Btu per lb

W_{aa} = actual air, lb per lb fuel

t_a = temperature of air surrounding the boiler, F

t_{ah} = temperature of the air leaving the air heater or entering the burners, F

and H_2 = hydrogen in the fuel (including the hydrogen in the moisture in the fuel) from the ultimate analysis, lb per lb of
fuel

C = carbon in the fuel from the ultimate analysis, lb per lb of
fuel

C_{ab} = carbon actually burned, lb per lb of fuel

W_f = weight of fuel burned, lb per hr

V = furnace volume, cu ft

Observe that the hydrogen term in this equation includes the hydrogen in the moisture in the fuel and would be the value listed in Table 5–1 if the fuel were coal. The constant of 1040 is the assumed value of h_{fg} for the water. The term $(C - C_{ab})$ represents the unburned carbon.

A flat, nearly horizontal furnace floor may be used for oil or gas firing, or as in Fig. 6–14, the unit discharges liquid or molten ash when burning pulverized coal. In the latter event, the furnace is known as a *slag-tap* or *wet-bottom* furnace. The molten slag may be removed *continuously* or *intermittently*; when removed intermittently, the slag may be collected for periods as long as 8 to 12 hr. Slag-tap furnaces are not considered suitable for coals having an ash fluid temperature in excess of 2600 F or an ash viscosity of more than 250 poises at the fluid temperature; usually they are used only on coals having a much lower fluid temperature.

Wet-bottom furnaces must be designed for a high temperature at the point of ash deposit. This can be accomplished by high furnace-heat release values and by having the flames near to the furnace floor. Frequently it is difficult to maintain the ash in a molten state at less than one-half to one-third load.

Molten slag will dissolve refractory and attack metal tubes. Therefore, the furnace floor tubes must be covered with a protective layer that may consist of tight cast-iron blocks over the tubes; these blocks in turn are covered with plastic chrome ore.

Dry-bottom furnaces have increased in popularity over wet-bottom furnaces (exclusive of cyclone furnaces) because the troublesome pool of molten slag in the furnace bottom necessitates an extremely tight furnace floor. Also, the dry-bottom furnace, if properly designed with a low heat release, can burn a wide variety of coals with ease.

Heat release rates at maximum load of about 21,000 Btu per (hr) (cu ft) are suitable for slag-tap furnaces, while for dry-bottom furnaces the rates vary from 15,000 to 21,000 Btu per (hr) (cu ft), depending on the ash-fusion temperature.

A boiler *setting* is construed to mean the enclosure that surrounds the pressure parts and the furnace. Originally this term applied to the brickwork around the boiler. Brick settings are now used only for the smaller boilers. Spalling and erosion of refractories in furnaces were encountered in the early days of pulverized coal firing due to the high heat release.

A *tube-and-brick* waterwall suitable for the furnace of an oil- or gas-fired steam generator or for the cool convection zones of a stoker or pulverized coal unit is shown in Fig. 6–16. The tubes may be partially

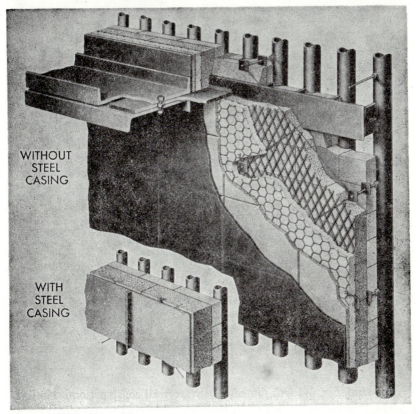

WITHOUT
STEEL
CASING

WITH
STEEL
CASING

FIG. 6–16. Construction of a tube-and-brick waterwall. (Babcock & Wilcox Co.)

imbedded in the refractory tile. Usual tube spacing allows 3 to 8 in. between tubes. The tile is followed by block insulation or castable refractory that is held in place by expanded metal lath. Next are several coats of plastic insulation reinforced by galvanized chicken wire; the last layer of material is an outer coat of sealing compound. Note that the insulation is held in place by studs welded to the tubes.

Closely spaced tubes, or *tangent* tubes, are shown in Fig. 6–17. This construction may be used for hopper-bottom furnaces and for all but the extremely high-temperature portions of wet-bottom furnaces.

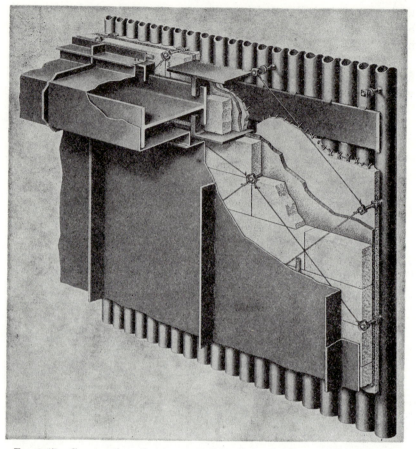

Fig. 6–17. Construction of a tangent-tube waterwall with a steel casing. (Babcock & Wilcox.)

Flat-stud, Fig. 6–18, or *finned-tube* waterwall construction is similar to tangent-tube construction in that both provide a continuous metal surface. The backing materials for the tubes are the same as for the tangent tubes.

Where high temperatures and resistance to slag erosion are required, as in the lower portions of slag-tap furnaces, *partially studded* or *fully studded* tubes are used in the waterwalls, Fig. 6–19. The studs are welded to the tubes and are packed with plastic chrome ore refractory that is pounded into place. This protects the tubes but also reduces the heat transfer rate, thus producing a higher gas temperature. The chrome ore is supported by the tube studs and is also cooled by the studs. The backing for the tubes is the same as for the tangent tubes.

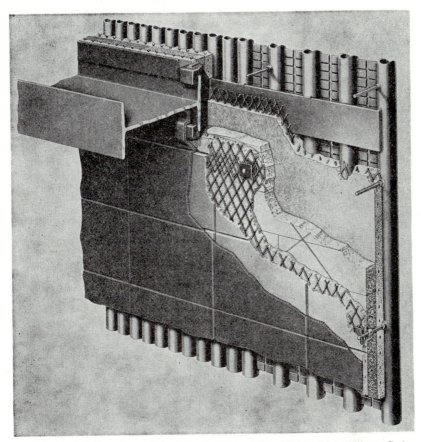

FIG. 6–18. Flat-stud and tangent-tube waterwall. (Babcock & Wilcox Co.)

Notice in Figs. 6–16 through 6–19 that the tubes are tied together by flat, horizontal *tiebars* that are welded to the tubes and that are at tube temperature to prevent any temperature differential. In addition, horizontal *buckstays* are fastened to the tiebars. The buckstays move up and down with the wall as the wall expands and contracts. The buckstays are of sufficient strength to limit deflection of the furnace wall due to the differential pressure across the wall.

In some of the preceding pictures of wall construction, a section is shown with a welded or bolted steel outer casing. Metal casings represent the best construction and are mandatory for large outdoor steam generators.

Some recent large steam generators have been installed with forced-draft blowers and no induced-draft fans. This places the furnace under

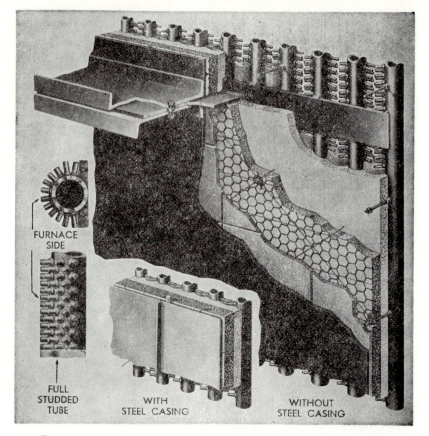

FURNACE
SIDE

FULL
STUDDED   WITH   WITHOUT
TUBE   STEEL CASING   STEEL CASING

FIG. 6–19.   Partial and full stud tube wall.   (Babcock & Wilcox Co.)

a pressure of some 10 in. of water (rather than the conventional slight vacuum) and gives rise to the term *pressurized* boiler. A pressure setting must have an all-steel, welded inner casing fastened to the tubes, Fig. 6–20, to prevent gas leakage from the furnace to the boiler room. Such leakage would not only be noxious to the plant personnel but would also overheat the setting.

Advantages of a pressurized unit are: (1) reduced fan power due to pumping cold rather than hot air or gas, (2) reduced maintenance because of elimination of the induced-draft fan, (3) improved boiler efficiency because of a tighter casing and no air infiltration. The boiler shown in Fig. 6–13 is pressurized.

Insulation thicknesses for the setting should be determined by the desired outside casing-wall temperature to give an installation that will

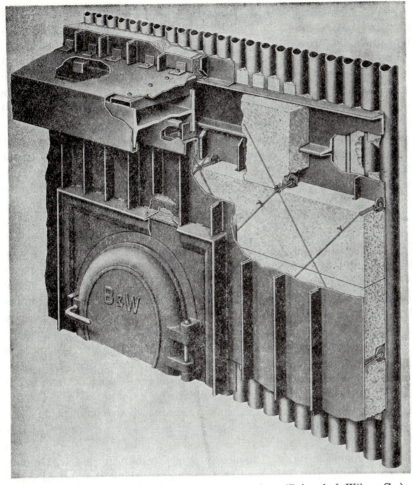

Fig. 6-20.   Tangent-tube wall with steel inner casing.   (Babcock & Wilcox Co.)

not be so hot as to be dangerous and also to give a wall temperature that will not cause too hot a boiler room. Such a casing temperature will usually be lower than would be determined purely by the economics of the problem. Outside-casing temperatures of 125 F to 135 F are recommended, while 150 F would be a maximum. Fortunately for outdoor installations, the ambient temperature and the air velocity over the casing will not materially affect heat loss through the casing.

6–5. The Steam Drum.   Steam leaving the drum of a modern boiler is one of the purest substances produced commercially. If moisture leaves the drum with the steam, the moisture will contain impurities that are particularly troublesome in the turbine by causing turbine blade deposits. Some of the impurities in the steam may be vaporized silica.

Impurities in the water and steam are measured in parts per million (ppm). Maximum boiler-water concentration recommended by the American Boiler Manufacturers' Association varies with pressure. The values range from a maximum of 3500 ppm for boiler pressures up to 300 psi to 1000 ppm for pressures of 1000 to 1500 psi.

Present standards of boiler operation require steam purity of about 1 ppm for central-station units. If a boiler-water concentration of 1000 ppm is assumed, then for a steam impurity of 1 ppm the moisture in the steam should be $(1 \times 100)/1000 = 0.1\%$. Reduction of boiler-water concentration can be accomplished by *blowing down* the boiler. During this operation, boiler water from a low point in the boiler is discharged to atmosphere and wasted. The blowdown lost is replaced by feedwater of a higher purity, and the boiler concentration is reduced. Blowdown may be either continuous or intermittent, depending largely on the rate of increase in boiler water contamination. In order to save some of the energy in the blowdown water, it may be partially flashed to vapor at atmospheric pressure and the vapor returned to the system, or the energy may be saved by using a blowdown heat exchanger.

Moisture traveling with the steam is known as *carry-over* when due to imperfect drying of the steam. An excessive quantity of moisture due to surging of the water level in the drum or due to high water level is referred to as *priming*. One other phenomenon encountered in the delivery of impure steam from the drum is *foaming*. This condition is characterized by the formation of bubbles on the water surface, i.e., water films surrounding the steam, and is caused by grease, oil, certain dissolved salts, and high alkalinity. Frequently a *surface blowdown* or *continuous blowdown* connection is located on the drum at about the normal water level to allow some of the foam to be drawn off the water surface and wasted.

Some present-day small boilers and most old boilers were provided with a *dry pipe* in the boiler drum to assist in the removal of moisture from the steam. A dry pipe is a pipe connected to the steam outlet nozzle, which extends about one-third or one-half the length of the drum and is perforated with many small holes. Steam entering the pipe through the holes must make a 90-deg turn, and the centrifugal action causes some of the moisture to impinge on the pipe wall. The moisture can be drained back to the drum from the dry pipe while relatively dry steam leaves the boiler.

Such methods seldom provide a steam quality of better than 99.5%; a value that is intolerable in a high-pressure plant. For high-steam purity, modern separators use three steps: initial separation, washing, and drying. Apparatus to perform all these steps must be located above the water level. Sometimes the drum becomes so crowded that the second step may be omitted.

We have previously noted that a mixture of water and steam enters the drum from the risers. This mixture may contain 1 part of steam with 5 to 15 parts of water by weight. The first operation in purifying the steam is to increase the steam quality by removing the major portion of the water. Initial separation is often accomplished by bringing the risers into the drum behind a baffle plate and forcing the steam through a restricted opening in the baffle plate to increase its velocity. By proper arrangement of baffles, the steam is then reversed in direction. Centrifugal action causes separation of the heavier water from the lighter steam particles.

When washing is included in the purifying process, it consists of replacing the concentrated boiler water in the moist steam with relatively pure incoming feedwater. Passing the steam through a rain of feedwater will accomplish this phase.

The final phase of purification should remove the last fraction of water that may be in the form of a bulk liquid, atomized spray, or foam bubble films. Passing the steam through narrow lanes formed by corrugated plates or screens provides a centrifugal action to assist in the water removal and provides a surface to collect the water.

Drum size may well be determined largely by the space requirements for all the internals to perform the steps of purification. The pressure drop of steam flowing through a separator is from 0.5 to 2 psi.

One design of separator is shown in Fig. 6–21. Steam and water from the risers enter the drum behind the baffle shown. The mixture is then guided in a vortex path through the primary separators consisting of cylinders. Water is forced to the cylinder walls and returns to the water space through the annulus at the bottom of each cylinder. Steam leaves each cylinder through the corrugated scrubber section at the top. For further purifying, more corrugated scrubbers are located at the top of the drum as a means of secondary separation or drying.

Feedwater is introduced into the drum of Fig. 6–21 by a distributor at the bottom of the drum. Incoming feedwater must be distributed along the drum to prevent the cold water from coming into contact with the drum shell or the tube ends. Observe that downcomers leave the drum through tubes between the two baffled riser sections.

6–6. Superheaters and Reheaters. Superheating the steam supplied to prime movers increases efficiency of a plant both because it increases the cycle efficiency and because it increases the efficiency of the steam engines or turbines. Each 100 F increase in steam superheat temperature

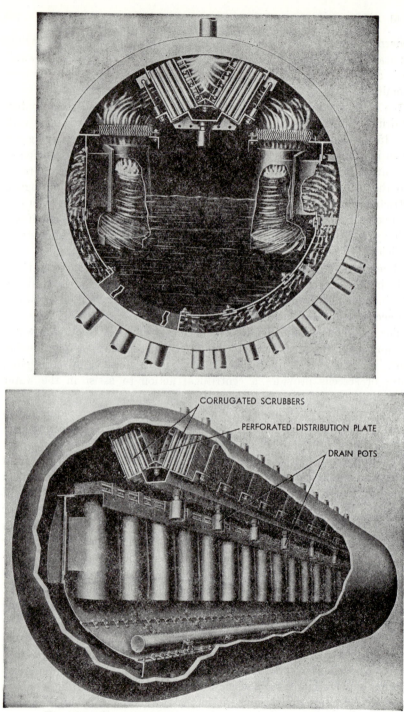

CORRUGATED SCRUBBERS

PERFORATED DISTRIBUTION PLATE

DRAIN POTS

Fig. 6–21. Steam drum with internals. (Babcock & Wilcox Co.)

should decrease the station heat rate some 3%. Reheating the steam to its initial temperature after it has passed part way through the turbine will reduce the heat rate as much as 5% (see Chap. 11 for a discussion of steam cycles).

Since the construction, operation, and control of reheaters is essentially the same as superheaters, the following discussion of superheaters will be taken to apply equally to reheaters.

Except for some special applications, superheaters are located within the steam generator setting (*internal* versus *external*) at a point where the gas temperatures are high enough to produce the desired steam temperature with a reasonable amount of tube surface and without excessive tube metal temperatures. Superheaters may be classified as *convection* or *radiant* units. A superheater that can "see" the furnace flames is a radiant unit and receives its energy by radiant heat transfer. A superheater that cannot "see" the flames receives much of its heat transfer by convection and is known as a convection superheater; however, there is some heat transfer by radiation from the nonluminous gases, as will be shown subsequently by calculations in this chapter.

A radiant superheater installed in the furnace walls, similar to waterwalls, is shown in Fig. 6-22. Radiant superheaters must be carefully designed, since the tubes may become overheated, causing reduced life. High mass-steam flows (pounds per hour of steam per square foot of steam-flow area) are used for radiant superheaters to assure proper cooling of the tubes. Furnace heat release for a radiant superheater is usually low to assist in prolonging the tube life.

Another form of radiant superheater, known as a *platten* superheater, is shown in Fig. 6-23. The tubes of each platten are spaced on close centers, and the plattens are suspended in the top of the furnace.

Most superheaters are of one of the many forms of convection units. When located over the deck of a straight-tube boiler, as in Fig. 6-5, they are called *overdeck* units. In order to increase the steam temperature, the tube deck may be separated part way up and the superheater located in this space; this is an *interdeck* superheater. *Intertube* superheaters have the steam tubes between the vertical rows of boiler tubes, while *interbank* superheaters have the tubes between adjacent banks of boiler tubes, as in Fig. 6-8.

Large steam generators usually employ two styles of convection superheaters, the *pendent* style and the *horizontal* style. The latter, or horizontal superheater, is shown in Fig. 6-12, while a pendent superheater is shown in Fig. 6-24. Both types are shown in Figs. 6-15 and 6-28. The distinctions between the pendent and platten superheaters are that platten superheaters are radiant and the tubes are closely spaced, while pendent superheaters are convection and the tubes are on much larger centers.

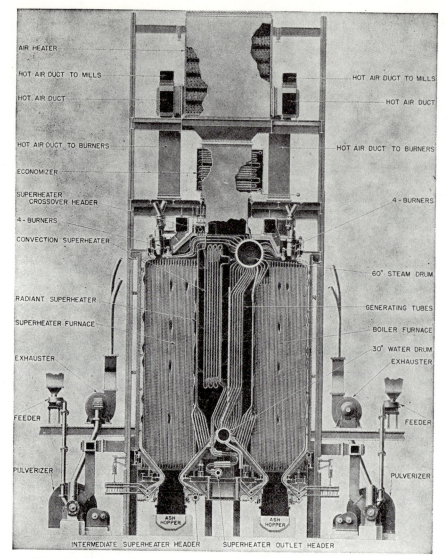

FIG. 6–22. Twin-furnace boiler with radiant and convection superheaters. (Foster Wheeler Corp.)

Observe that platten and pendent superheaters are *nondrainable,* i.e., water in the tubes from steam condensed during a boiler shutdown cannot be drained out of the tubes. When the boiler is fired, the water in the tubes must be boiled out; this necessitates a careful starting procedure. Horizontal superheaters, and some others, have the distinct advantage of being drainable.

Fig. 6–23. Platten superheater. (Combustion Engineering, Inc.)

Fig. 6–24. Pendent superheater. (Combustion Engineering, Inc.)

The illustrations show that modern superheaters have many passes (ten passes for Fig. 6–24) and that the tubes are arranged in-line rather than staggered.

6–7. Superheat and Reheat Temperature Control. Typical characteristic temperature curves for radiant and convection superheaters are shown in Fig. 6–25. Superheat temperature decreases with increased load

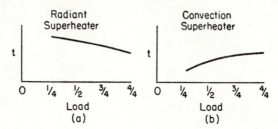

Fig. 6–25. Characteristic temperature curves for convection and radiant superheaters.

for a radiant superheater because, as the load is increased, the furnace temperature increases at a much less rapid rate than the load. Since radiant heat transfer is a function of the temperatures, the heat transfer increases slowly with steam flow; or the steam temperature decreases.

Convection superheaters perform in the opposite manner, since steam temperature rises with increased load. As boiler load increases, both gas flow over the superheater tubes and steam flow within the tubes increase; also, the temperature of the gases entering the superheater bank increases moderately, as indicated for a radiant superheater. These factors increase the rate of heat transfer and the mean temperature difference. The result is an increase in steam temperature with increase in boiler load.

Effects of inlet water temperature and boiler cleanliness on the steam temperature of a convection superheater are important. Reduced boiler-inlet water temperature may occur when a feedwater heater, particularly the final or highest pressure heater, must be taken out of service. More fuel and air must be supplied to the boiler to evaporate the colder water. Thus the mass flow of gases over the convection superheater surface is increased with a corresponding increase in final steam temperature.

Slag deposits on the furnace waterwall or the screen tubes increase the gas temperature to the convection superheater. This reduces the heat transfer to the waterwalls and increases the gas temperature entering the superheater. Thus the steam temperature is increased.

Since a power plant should be designed to operate at a wide range of load, most economical operation at all loads would indicate that a constant steam temperature at all steam flows would be desirable. However, a constant steam temperature at all steam flows would usually require a large investment in the superheater and accessories. When considering investment cost, fuel cost, and typical load curves (see Chap. 15 where a load curve is indicated to be a curve of plant load throughout the day), a constant steam temperature can be justified economically for a steam flow of from one-half to three-quarters steam flow up to full steam flow. The following methods of obtaining this constant steam temperature are used:

(a) *Divided Furnace.* Divided or twin-furnace boilers, Fig. 6–22, or a separate superheater external to the boiler provide a means of superheat temperature control, since the fuel input to the burners in the superheater section can be regulated to provide the desired steam temperature.

(b) *Tilting Burners.* Tilting or vertically adjustable burners, Fig. 7–16, change the gas temperature entering the superheater section. At low steam flow, the burners are tilted upward (the range is ±24 deg) so that the lower portion of the furnace becomes less effective in absorbing energy. Thus the gases enter the superheater at a higher temperature than if the burners were fixed, and the steam temperature leaving a convection can be maintained constant from less than one-half load to full load. This is a very satisfactory and economical method of temperature control.

(c) *Burner Selection.* Large steam generators have several burners that are arranged in horizontal rows, Fig. 6–15. Since only part of the burners are needed at partial load, use of the upper row of burners will increase steam temperature in the same way as tilting the burners upward. Of course the tilting burners permit a continual control of steam temperature that is not permissible with the burner selection method.

(d) *Combined Radiant and Convection Superheater.* With a dropping-temperature characteristic with increased steam flow for a radiant superheater and a rising-temperature characteristic for a convection superheater, it is possible to obtain a nearly constant superheat temperature by using both types of superheater. Proper proportioning of the surfaces in each section produces a curve, as shown in Fig. 6–26. The steam generator of Fig. 6–22 has a combination radiant-convection superheater.

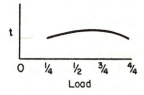

FIG. 6–26. Characteristic temperature curve for combination radiant-convection superheater.

A convection superheater located near to the furnace and not shielded by screen tubes, so that a few front rows of tubes can receive radiant energy, will show a smaller steam temperature change with load than a shielded convection superheater.

(e) *Gas By-pass.* If at any time the flow of gases through a convection superheater is reduced without changing the steam flow, then the final steam temperature will be reduced. Gas by-passing of the super-heater or *damper control* utilizes this principle. If it is desired to have constant steam temperature for three-fourths steam flow and higher, then the superheater is designed to give the proper temperature at three-fourths steam flow. At greater steam flows, some of the flue gases are by-passed around the superheater to maintain the desired steam temperature. Fig. 6–27 shows a temperature curve for a superheater, with and without by-pass damper control, while Fig. 6–28 shows the by-pass dampers in a boiler that may be used to control either superheat or reheat temperature.

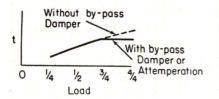

Fig. 6–27. Characteristic temperature curve for by-pass damper or attemperation control.

(f) *Attemperation.* Control by attemperation means that the steam temperature is reduced by removing energy from the steam. There are two basic types of attemperator, the tubular type and the water injection type. In the first type, superheated steam is passed through the tubes of a tubular heat exchanger. The tubes may be located below the water level in the steam drum or in a separate shell-and-tube heat exchanger. In either case, boiler water is used as the cooling medium.

A spray type of attemperator is shown in Fig. 6–29 wherein water is sprayed directly into the steam; the water is evaporated into steam and the temperature of the mixture is reduced. Care must be given to be sure that the spray water does not contain impurities. Usually, boiler feed-water has sufficient purity. The venturi section of the attemperator creates high velocity and turbulence to promote mixing of the steam and water. The thermal sleeve, usually of a high chrome steel, protects the pipe from thermal shock from the cold water. A temperature curve for spray attemperation is the same as one for gas by-pass control, Fig. 6–27.

When convection superheaters are arranged in two stages, the primary superheater and secondary superheater, the attemperator should be located between the two stages to reduce the tube metal temperatures for the high-temperature section.

Spray attemperators are also called desuperheaters and are used at many points within the plant where superheated steam temperature must be reduced.

Calculations to determine the water requirements of a mixing type of desuperheater are the same as for a mixing type of feedwater heater (see Chap. 8) and are based on the Law of Conservation of Energy and the Law of Conservation of Mass.

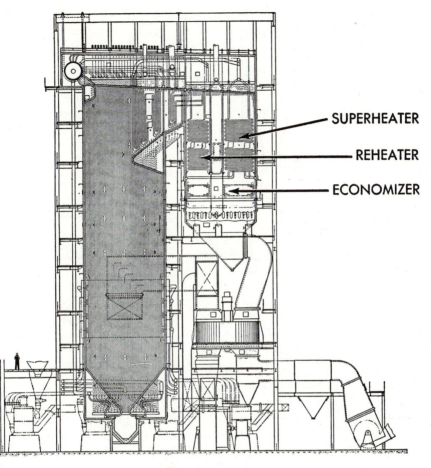

FIG. 6–28. Triple by-pass damper, radiant, reheat, outdoor steam generator for 1,000,000 lb per hr, 1615 psig, 1005 F, 1005 F. (Combustion Engineering Co.)

Equating the energy entering the system per unit of time to that leaving the system,

$$W_1 h_1 + W_2 h_2 = W_3 h_3 \qquad \text{(a)}$$

But from the second of these two laws,

$$W_1 + W_2 = W_3 \qquad \text{(b)}$$

Combining Eqs. (a) and (b),

$$W_2 = \frac{W_1(h_1 - h_3)}{h_3 - h_2} \qquad \text{(6–8)}$$

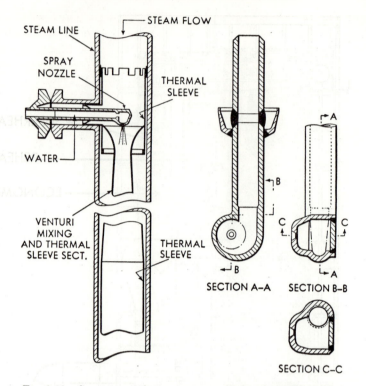

Fig. 6-29. Spray type of attemperator. (Babcock & Wilcox Co.)

in which W_1 = steam entering the desuperheater, lb per hr
h_1 = enthalpy of entering steam, Btu per lb
W_2 = water entering desuperheater, lb per hr
h_2 = enthalpy of wter, Btu per lb
W_3 = steam leaving desuperheater, lb per hr
h_3 = enthalpy of steam leaving, Btu per lb

(g) *Excess Air.* Increased excess combustion air has the effect of lowering furnace temperature. Since radiant heat transfer is a function of the difference in the fourth power of the temperatures, furnace heat absorption will be materially reduced when furnace temperatures are lowered by the use of excess air. Thus, the gas temperature entering the superheater is actually increased. This increase in gas temperature, combined with the increase in mass flow, raises steam temperature.

Although increased excess air is a means of steam temperature control, the increased mass of stack gas materially lowers the boiler efficiency.

(h) *Flue-Gas Recirculation.* The use of flue-gas recirculation is a

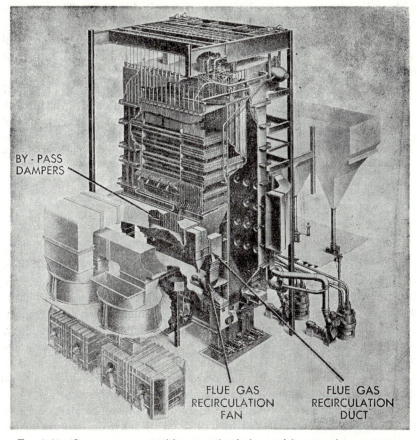

BY-PASS
DAMPERS

FLUE GAS
RECIRCULATION
FAN

FLUE GAS
RECIRCULATION
DUCT

Fig. 6-30. Steam generator with gas recirculation and by-pass damper temperature control. (Babcock & Wilcox Co.)

means of steam temperature control by the same principles employed by
using increased excess air but without the adverse effect on efficiency.
Fig. 6-30 shows a boiler with the ductwork and fan used to extract flue
gas from the entrance to the air heater and to reintroduce the gas into the
bottom of the furnace. Recirculated gas tends to blanket some of the
furnace wall with a layer of cooler gas. This reduces the rate of heat
absorption. Since this gas continually recirculates through the furnace at
partial load, the mass of flue gas leaving the stack is not increased so
much as for the use of increased excess air. Thus, the efficiency is much
better than when steam temperature is controlled by increased excess air.
Gas-recirculation fan power for this type of control is appreciable and
must be considered in any economic comparison. Temperature curves

with and without gas recirculation are shown in Fig. 6–31. These curves are essentially the same as for excess air control.

In a reheat unit, such as Fig. 6–30, two types of temperature control are necessary; one to control reheat temperature and one to control the temperature of the main superheated steam.

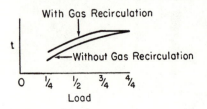

FIG. 6–31. Characteristic temperature curve with and without gas recirculation.

Although the curves of Figs. 6–27 and 6–31 have been drawn for full steam temperature in the range of three-fourths to full load, the systems can be designed for a much larger range.

6–8. Secondary Surface. A very approximate rule-of-thumb for the temperature of the exhaust gases leaving the boiler proper for economically proportioned surface is the saturated-steam temperature plus 100 F. Referring to the steam tables, this figure would indicate that the gas temperature for a boiler pressure of 200 psig would be in the order of 490 F, while for 600 psig it would be about 590 F. Similarly, for a boiler pressure of 1200 psig, the temperature would be some 670 F. Although in many cases the actual gas temperature will be somewhat lower than these figures, they do illustrate the need for some means of recovering part of the energy of these gases. Either *economizers* or *air preheaters* or both may be added to the steam-generating unit for this purpose.

Cast-iron tubes were used in the early designs of economizers because of the ability of this material to withstand attack from oxygen in the feedwater. An operating pressure limit of 250 psig is usually placed on cast-iron economizers.

Steel tubes are required for the economizers of all modern boilers to withstand the pressures encountered. Also the internal corrosion problem has been reduced by the use of deaerated water and by maintaining proper feedwater alkalinity, pH of 8 to 9 (see Chap. 7 for definition of pH). External tube corrosion due to the flue gases acting on the steel tubes has been reduced in modern boilers by the use of higher incoming-water temperatures. External corrosion will be mentioned again in connection with air preheaters.

Although *extended surface* or *finned* tubes may be used for economizers, most modern installations employ the *continuous tube* design as shown in many of the previous illustrations. Economizers have been arranged as a bank of bent tubes connecting two horizontal drums that are arranged one above the other; this is an *integral* or *bent tube* economizer. Older economizer designs provided for removable tube ends (at one end of the unit) for cleaning the internal tube surfaces. Continuous-tube units have no provision for accessibility. Treated feedwater, chemical cleaning of the tubes, and gas leakage from the casing at accessibility points have promoted the use of continuous-tube units.

Water flow upward through the economizer section is preferred (but not universally employed) because it reduces the possibility of vapor binding some of the tubes, provides more stable flow conditions, and makes the economizer drainable. Although economizers have been operated satisfactorily when as much as 20% of the water is evaporated into steam in the economizer (called a *steaming economizer*), the practice is not favored by many engineers because of possible tube deposits and because of the possibility of vapor binding. Steaming economizer surface is cheaper than boiler surface. In order to prevent steaming, the feedwater leaving the economizer should be at least 50 F below saturation temperature.

Since counterflow of gas and water is desirable to reduce heat transfer surface and draft loss, it is advisable to have downward flow of flue gases over the economizer with the upward flow of water. Because of special conditions, these flow patterns cannot always be obtained. When both air heaters and economizers are used, the economizer employs the higher temperature gas because the entering water temperature is higher than the temperature of the air entering the preheater.

Where the feedwater temperature leaving the feedwater heaters and entering the economizer is high, economical boiler operation necessitates an air preheater. The efficiency of most steam generators can be changed several percentage points by increasing or decreasing the air heater surface. In addition, preheated air improves combustion particularly at low loads, produces higher furnace temperature, shortens the time required for ignition, and is necessary in a pulverized fuel unit to dry the coal. Air supplied to a stoker-fired furnace should not exceed 300 F, or 350 F at the most; otherwise, stoker warpage and maintenance will be too high. Pulverized units can employ 500 F to nearly 700 F air if the coal is wet.

Although steam coils and separately fired furnaces have been used to preheat air (the steam coils are economical and rather common when using steam extracted from the turbine), our discussion will be limited to those using flue gas. These may be classified as either *recuperative* (*tubular* or *plate* types) or the *regenerative* type known as the *rotary* or *Ljungstrom* unit (diphenyl with heat exchangers have been used; pebbles have also been used).

The plate type of preheater, using No. 12 U.S. gage steel plate, provides alternate passages for cool air and for flue gases. The plates are welded to reduce leakage, and the air and gas passages are of different widths to account for the greater volume of flue gas. Because of difficulties of warpage and buckling of the plates, and because of the difficulty of replacing corroded plates, this type of unit is seldom used for modern plants.

Tubular air preheaters are shown in several of the illustrations in this chapter and in Fig. 6–32. Heaters with vertical tubes and with the gases

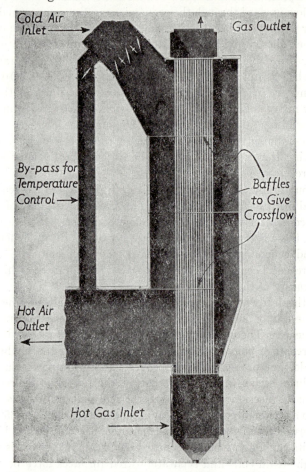

FIG. 6–32. Tubular air preheater. (Babcock & Wilcox Co.)

inside the tubes are preferred; otherwise, soot accumulations on the tubes and on the baffles would be troublesome. The heater of Fig. 6–32 has four air passes and the one gas pass. The tubes, of about No. 14 U.S. gage steel, are rolled into tube sheets at the top and bottom of the heater, and the tubes are usually staggered to improve heat transfer. Counterflow of air and gases is preferable.

Ljungstrom heaters, Fig. 6–33, have become very popular for modern steam generators. They contain a grid or basket of corrugated steel plates formed into a welded drum and attached to a shaft that rotates at 2 to 4 rpm. As the plates are alternately heated and cooled by passing from the hot gas side to the cool air side, they transfer energy from the flue gas to the air. The shaft may be horizontal or vertical. A horizontal shaft is not recommended for coal firing, as accumulations of soot on the plates may cause an unbalance.

As with any rotating element in a stationary system, seals must be provided to reduce air leakage into the flue-gas system. The amount of leakage will depend on the pressure differential between the gas and air ducts on the seals. The seals should be maintained in good condition to reduce leakage. In addition to leakage, there is a small amount of carry-over of air into the gas passage, and vice versa, due to air and gas trapped

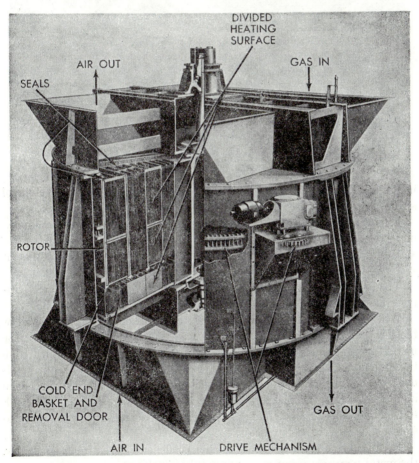

FIG. 6–33(a). Ljungstrom air preheater. (Air Preheater Corp.)

FIG. 6–33(b).   Ljungstrom air preheater.   (Air Preheater Corp.)

in the rotor. Total air loss should not exceed 8 to 10% of the weight of air flowing.

This air loss reduces the flue-gas temperature entering the chimney, increases the power and capacity of the forced- and induced-draft fans, and permits dust to be transferred from the flue-gas system to the air system. The leakage for a tubular heater is essentially negligible.

The height of the rotor of a Ljungstrom heater determines the temperature rise of the air.

Whenever the metal of an air preheater or an economizer falls below the dew-point temperature of the flue gas, there will be corrosion of the metal due to sulfuric acid. The gas dew point cannot be determined from the amount of moisture in the gas because of the presence of sulfur trioxide. The sulfur content of the fuel and the type of firing have an important bearing on the dew point, which may range from about 165 F to well over 250 F. Since flue-gas temperatures decrease with load, corrosion is more severe at low loads.

It can be seen from the heat transfer coefficients given in Chap. 4, that the metal temperature of economizer tubes is close to the water temperature, while the metal temperature of air preheater tubes would be about midway between the air and gas temperatures. Thus, economizer tubes will not be subject to severe external corrosion except when there is a low entering-water temperature. Air preheaters are subject to frequent corrosion conditions.

Usually, the cold ends of air preheaters are designed for easy removal of the metal. The lower part of the baskets in the Ljungstrom heater, Fig. 6–33 (a), are so arranged. It is not uncommon to make the tubes of a tubular heater in two sections; the short, cold-end section may be replaced easily, Fig. 6–13. The use of expensive corrosion-resistant alloys has not proved economical, but glass tubes may be suitable.

Several methods of increasing the cold-end metal temperature of air preheaters have been used: (a) increase the inlet air temperature by using a steam coil before the air heater, (b) recirculate some of the hot air from the heater outlet back to the heater inlet by using a fan or to the forced-draft fan inlet to increase the heater inlet air temperature, (c) by-pass part of the air around the heater. This last method, illustrated in Fig. 6–32, is also used on Ljungstrom heaters. Notice that two sets of dampers are necessary to obtain regulation of the air flows.

6–9. Forced-Circulation Boilers. As the pressure in the boiler becomes higher and higher, the difference in density between the downcomers and the risers decreases. Since the difference in density times the height is the head causing water circulation, the drums of extremely high-pressure, natural-circulation boilers must be at a high elevation. Forced-circulation boilers overcome this difficulty by employing pumps to force the water through the tubes. The *La Mont* boiler uses pumps to provide the circulation of water through the unit, while the *Benson* boiler passes the water through the tubes only once and therefore is known as a *once-through* unit.

Fig. 6–34 shows a controlled-circulation boiler. Observe that the pumps take the water from the drum and supply it to the headers at the bottom of the boiler. Each tube leaving the header contains an orifice that establishes the flow through that tube. Thus, the flow in each tube can be established independently of the other tubes. The circulation pumps develop about 50-psi pressure and consume some 5% of the normal station auxiliary power. Depending on average load and fuel cost, the break-even point for natural versus controlled circulation is about 1800 psig for large units.

A controlled-circulation boiler has several advantages in addition to control of the circulation: smaller and thinner-walled tubes can be used, reduced boiler weight and reduced weight of water in boiler, lighter sup-

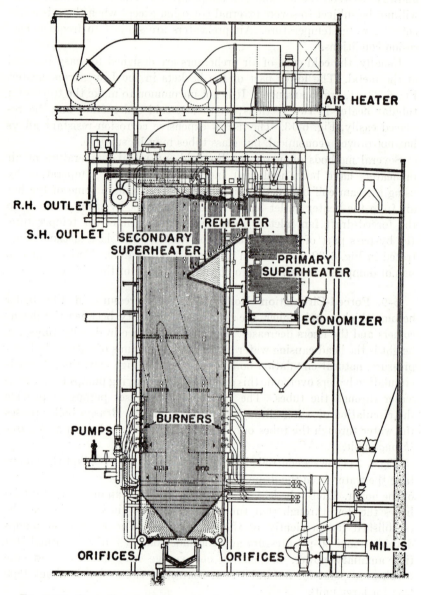

FIG. 6–34. Controlled-circulation, reheat steam generator for 900,000 lb per hr, 1850 psig, 1000 F, 1000 F. (Combustion Engineering, Inc.)

porting structural members and foundations, and greater safety in the event of a tube rupture.

The once-through boiler shown in Fig. 6–35 was designed for supercritical pressure. Once-through boilers can be designed to operate at subcritical pressures.

Fig. 6–35. Supercritical-pressure steam generator to deliver 675,000 lb per hr of steam at 4500 psig and 1150 F with reheat to 1050 F and 1000 F. (Babcock & Wilcox Co.)

Water enters this unit at 5500 psig and is first heated in the walls of the cyclone furnace and in the front wall of the primary furnace. The fluid then flows to the rear of the unit where the primary superheater is located and where the fluid temperature is increased above the critical temperature. Secondary superheat sections and two reheat sections are also shown. The low-pressure reheat section is in parallel, gaswise, with the primary superheater. Dampers are used to control the gas flow over the low-pressure reheater to control temperature.

Observe that there are no drums or steam-separating equipment because the unit is of the once-through style.

6–10. Chemical Cleaning. Internal boiler surfaces of a new unit contain considerable grease and oil remaining from the manufacturing processes and from rolling the tubes into the drums and headers. This grease and oil is removed in a process known as *boiling-out* the unit, wherein a solution of caustic soda and either soda ash or trisodium phosphate (the latter is preferred by many) is put into the boiler. These chemicals are used at the rate of about 1 lb of each chemical per 1000 lb of water. For excessive grease, higher concentrations may be used.

The boiler containing this solution is fired for a period of from 4 to 24 hr at a pressure of not less than 50 psig for low-pressure boilers to a pressure as high as 1000 psig for high-pressure boilers. After the boiler has cooled, the chemical solution is drained, and the boiler is flushed once or twice with clean water at a temperature of about 180 F.

Steam generators that have undergone extensive repairs should also be boiled-out.

New boilers also contain mill scale, scale formed during erection and storage, and scale from welding processes. Old boilers contain scale deposits from water impurities. It is difficult and sometimes impossible to remove these deposits by mechanical cleaning; chemical methods are more economical and faster.

Chemicals used for acid cleaning are 3 to 5% solutions of hydrochloric acid, phosphoric acid, and sulfuric acid. To prevent serious attack on the metals, these acids are inhibited by the addition of such chemicals as arsenic compounds, barium salts, starch, formaldehyde, etc. The particular inhibitor chosen will depend on the chemical and any alloying elements in the steel.

The acids are introduced into the boiler at temperatures ranging from 140 F to 160 F by heating the water external to the boiler. At higher temperatures the inhibitors may not operate satisfactorily. The acids remain in the boiler from 4 to 8 hr, depending on the amount of scale. Periodic tests of the acid strength show when cleaning has been completed because acid strength will decrease during scale removal but will remain essentially constant after all scale has been dissolved.

Two methods of cleaning are the *continuous-circulation* method and the *soaking* method. Phosphoric acid is suitable for the continuous-circulation method, but inhibited hydrochloric acid may react with the metal if the solution velocity becomes too high. The circulation method does not always assure that all surfaces will be cleaned, since some tubes may be by-passed.

Hydrochloric acid is frequently used in the soaking method, the acid being allowed to remain in the tubes without circulation. Sulfuric acid is not used extensively because it is more expensive, is not so effective on some scales, is more difficult to inhibit, and forms relatively insoluble sulfates rather than soluble chlorides.

Acid-cleaning generates hydrogen gas that is dangerous and may cause hydrogen embrittlement of the steel if it is cold-worked immediately after cleaning. To reduce the danger of explosion, nitrogen is used to force the acid out of the boiler and to displace the acid.

The boiler should be flushed twice with warm water (160 F) after acid-cleaning, followed by a caustic boil-out of up to 40-hr duration, using the same procedure and chemicals as given for the degreasing boil-out. This caustic boil-out not only neutralizes the residual acid but, if carried out at several hundred pounds pressure, also forms an even oxide scale on the surface that is beneficial to the unit.

6–11. Boiler Ratings and Performance. There is no generally accepted method of rating the large, modern high-pressure boilers except by giving the pounds of steam per hour together with the steam and water conditions for which the unit is designed. Units of *millions of Btu per hr* and *kBtu per hr* have been proposed but are not generally used.

A term used for many years, and still used to some extent for small boilers, is *boiler horsepower*. Originally, this term was intended to relate the output of the steam generator to the steam requirements of a prime mover. Improvements in the design of both the prime mover and the boilers have made this misnomer obsolete. One boiler horsepower is equivalent to the generation of 34.5 lb of steam per hr from water at 212 F to saturated steam also at 212 F, i.e., *from and at 212 F*. The energy of evaporation for steam at 212 F was taken at 970.3 Btu per lb, making the product equal to $34.5 \times 970.3 = 33,475$ Btu per hr. Usually, this figure is approximated at 33,500 Btu per hr. Thus

$$\text{Boiler hp} = \frac{W(h_2 - h_1)}{33,500} \qquad (6\text{--}9)$$

in which $W =$ steam generated by boiler tubes, economizer, and superheater, lb per hr

$h_1 =$ enthalpy of water entering economizer, Btu per lb

$h_2 =$ enthalpy of steam leaving boiler or superheater, Btu per lb

Because power boilers do not generate steam at 212 F, the actual energy change of the water in passing through the boiler compared with the latent energy of evaporation at 212 F is the *factor of evaporation*. Thus,

$$\text{Factor of evaporation} = f = \frac{(h_2 - h_1)}{970.3} \qquad (6\text{--}10)$$

and   Equivalent evaporation $= f \times W$ (6–11)

and   Boiler hp $= \dfrac{f \times W}{34.5} = \dfrac{\text{equivalent evaporation}}{34.5}$ (6–12)

At the time these terms were standardized, boilers could develop 1 boiler hp with 10 sq ft of boiler heating surface. Therefore, the *manufacturers' boiler horsepower* is defined as 10 sq ft of boiler heating surface. Surface areas for superheaters, economizers, or air heaters are not included as boiler heating surface. Therefore, the terms of boiler horsepower, etc., are meaningless for large, modern steam generators.

It is common for even small, modern boilers to generate more than 1 boiler hp with 10 sq ft of heating surface. Most boilers can develop 150% of rating—200% is common and some units have exceeded 400%—when *percentage of boiler rating* is defined as:

$$\text{Boiler rating} = \frac{\text{boiler horsepower} \times 10}{\text{boiler heating surface}} \qquad (6\text{–}13)$$

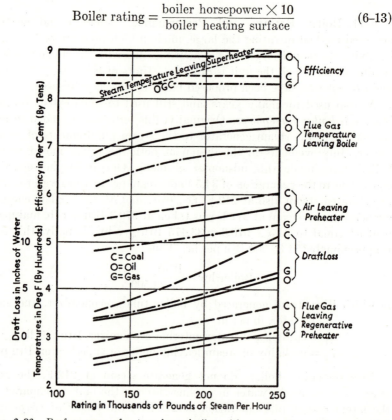

Fig. 6–36. Performance of a two-drum boiler with superheater and regenerative air preheater. Designed for: oil, gas, or bituminous coal; 925-psig working pressure; 1025-psig design pressure; dust collector when coal-fired; forced- and induced-draft fans; dry bottom; waterwalls; 80 F air entering preheater; 245 F feedwater temp.

Steam-generator performance is dependent on many factors, such as type of unit, fuel, size and type of secondary surfaces, and space limitations. Some of the more general characteristics may be visualized from the performance curves given in Fig. 6–36, which are for a recent two-drum boiler designed for operation with pulverized bituminous coal, oil, or gas composed largely of methane and ethane. Other conditions of design are given on the curve sheet. It is interesting to note that this boiler operates at over 400% of rating for the design output of 250,000 lb per hr.

As might be expected, the draft loss from the furnace to the air-heater gas outlet varies as the square of the gas flow, which is essentially the same as the square of the steam output.

The maximum efficiency of most boilers is reached at a very low load. Therefore, the efficiency will decrease with increased load in the range for which performance is predicted. In this case the efficiency drops 1.0% from one-half to full load. One reason that coal firing is less efficient in this unit is because more excess air is used for coal combustion, which increases the dry gas loss. Increased excess air reduces the temperature in the furnace, which is desirable when the ash softens at a low temperature and there is a dry bottom. Unburned combustible for coal firing is another contributing factor to lower efficiency. The large amount of hydrogen in the gaseous fuel increases the loss due to moisture formed by its combustion over that for either oil or coal.

Boiler Design

6–12. General. A rigorous description of the calculations required for the design of a boiler, together with the necessary underlying theory, would consume more space than is available here. Therefore, the purpose of this section is to present some of the problems encountered in the actual design and to offer approximate solutions.

As a demonstration problem we shall assume that it is desired to find the efficiency of an oil-fired boiler whose plan view is shown in Fig. 6–37. An elevation view would be similar to the boiler shown in Fig. 6–11 except that the tube size and spacing are different.

Other data will be assumed to be:

Furnace waterwalls and roof: 3-in. bare tubes
Convection section: 2-in. tubes on 4-in. centers, diamond staggered arrangement
Working pressure: 250 psig
Water entering: 200 F
Air entering and room temperature: 80 F

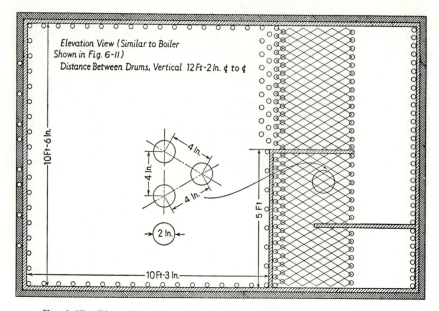

Fig. 6–37.   Plan view of boiler. (Drawing is schematic and not to scale.)

Steam leaving: saturated
Radiation and unaccounted-for loss: 3%
Oil heating value: 18,500 Btu per lb
Oil analysis: 84% C, 13% H_2, 2% S, 1% N
Flue gas: 12.5% CO_2 on dry basis, corresponding to 20% excess air
Convection section: 32 tubes wide, 126 in. between walls
Furnace section: 18 tubes on front wall, sidewalls, and roofs; water screen in path of gases to convection zone has tubes staggered.
Load: 22,000 lb per hr of steam

An efficiency of **75.7%** will be assumed and calculations will be made to check this efficiency.

6–13. Furnace Calculations.   The first step will be to determine the wall areas and the tube surface. To simplify the problem, the heat transfer to the floor tubes will be neglected and the average tube height for the wall will be taken as 12 ft. Then the furnace envelope surfaces will be as follows:

$$\text{Front wall area} = 12 \times 10.5 = 126 \text{ sq ft}$$
$$\text{Side wall area} = 2 \times 12 \times 10.25 = 246 \text{ sq ft}$$
$$\text{Rear wall area} = 12 \times 5 = 60 \text{ sq ft}$$
$$\text{Roof area} = 10.5 \times 10.25 = 108 \text{ sq ft}$$
$$\text{Screen tube area} = 12(10.5 - 5) = 66 \text{ sq ft}$$

The area of the furnace envelope backed by refractory is $126 + 246 + 60 + 108 = 540$ sq ft, and the envelope area represented by the screen is 66 sq ft.

Since the waterwall tubes are set immediately in front of the refractory, called *tangent refractory*, there is radiation to the backs of the tubes from the refractory. Adjustment factors for the furnace envelope area to account for the tube arrangement and spacing are given in Fig. 6–38. For

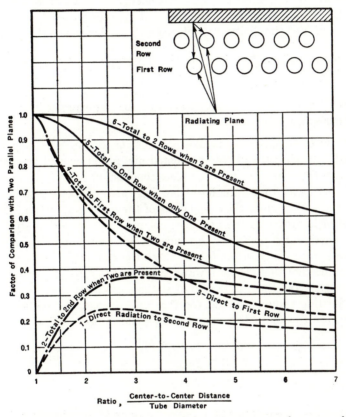

FIG. 6–38. Adjusted-surface correction factors. For tangent tubes, use a factor of 1.0; use curve 5 for tangent refractory; use curve 3 for tubes half-embedded in refractory or for water-screen tubes. [Hottel, *Mechanical Engineering* (July, 1930).]

the portion of the furnace envelope backed by refractory, the tube spacing (using data for the front wall) is the distance between the centers of the end tubes (10.5 ft $- 0.25$ ft) divided by the tube diameter in feet and divided by the number of spaces between the end tubes, or

$$\frac{(10.5 - 0.25)\,12}{3 \times 17} = 2.41 \text{ diam}$$

The adjustment factor is 0.81 from curve 5, and the adjusted surface for refractory-backed tubes is $0.81 \times 540 = 437$ sq ft. The water screen contains 18 tubes in a distance of 5.5 ft; therefore, these tubes are on approximately 1.2-diam projected centers (assuming 17.5 spaces). Using curve 3, the factor is 0.96 and the adjusted surface is $66 \times 0.96 = 63$ sq ft. Then the adjusted envelope surface is $437 + 63 = 500$ sq ft.

Before the furnace exit gas surface can be obtained from Fig. 6–39, it is necessary to determine the available energy by using an equation similar to Eq. (6–7) except that radiation and unaccounted-for loss should be recognized. For simplicity, these losses will be charged to the furnace.

$$\text{Available energy} = [\text{HHV}(1 - Q_7) + 0.24 W_{aa}(t_{ah} - t_a) - 1040(9\text{H}_2) \\ - 14{,}600(\text{C} - \text{C}_{ab})](W_f) \quad (6\text{--}14)$$

where Q_7 is the radiation and unaccounted-for loss expressed as a decimal, and other symbols are as in Eq. (6–7).

Observe that Eq. (6–14) is an expression for the lower heating value of the fuel with adjustments for preheated air, radiation and unaccounted-for loss, and the loss due to incomplete combustion.

It is necessary to use the lower heating value of the fuel to determine gas temperatures, since the latent heat of the water vapor from the combustion of hydrogen in the fuel does not increase gas temperature. Also, it is impossible for the boiler to condense this vapor and to make use of it to evaporate water in the tubes. From the combustion analysis of the fuel we find that there is 6.5% water vapor in the gases by weight and on the wet basis and there is 18.1 lb of wet gas per lb of fuel. By using the higher heating value of the fuel together with the assumed boiler efficiency and Eq. (5–16), the fuel consumption is found to be

$$W_f = \frac{22{,}000(1201.7 - 168.0)}{0.757 \times 18{,}500} = 1622 \text{ lb per hr}$$

From Eq. (6–14),

$$\text{Available energy} = [18{,}500(1 - 0.03) + 0.24 \times 17.1(80 - 80) \\ - 1040 \times 9 \times 0.13 - 14{,}600(0.84 - 0.84)]1622 \\ = 27{,}150{,}000 \text{ Btu per hr}$$

Then the available energy, the abscissa of Fig. 6–39, is $27{,}150{,}000 \div 500 = 54{,}300$ Btu per sq ft per hr of adjusted envelope surface. From Fig. 6–39, the gas temperature leaving the water screen and entering the convection section is 1770 F.

In order to determine the sensible energy of the gases leaving the furnace, the values of constant pressure instantaneous specific heats are

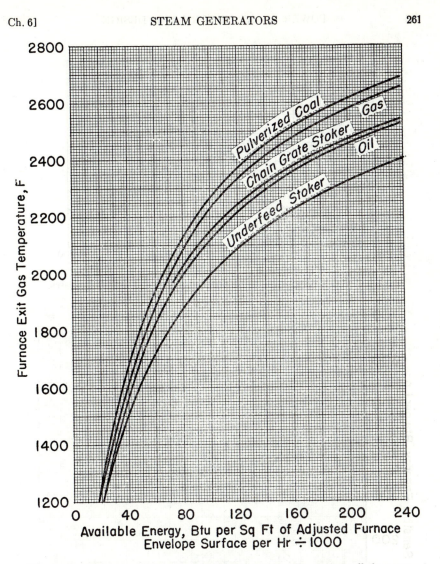

FIG. 6–39. Furnace-exit gas temperatures for bare-tube, waterwall furnaces.

plotted in Fig. 6–40(a) by using the flue-gas analysis.* With these data, a curve of sensible energy above 80 F can be plotted, Fig. 6–40(b). While these curves are for the particular flue gas analysis of this problem, they would not be materially different for many other flue gas analyses.

From Fig. 6–40(b), the sensible energy of the flue gas leaving the water

* R. L. Sweigert and M. W. Beardsley, "Empirical Specific Heat Equations Based upon Spectroscopic Data," *Georgia School of Technology Engineering Experiment Station Bulletin*, No. 2, 1938.

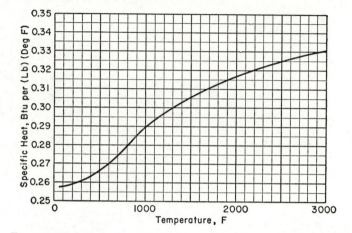

Fig. 6-40(a). Approximate instantaneous specific heat of flue gases.

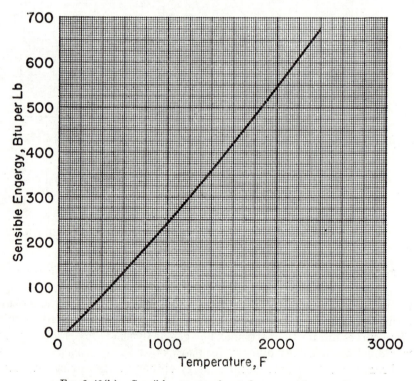

Fig. 6-40(b). Sensible energy of wet flue gases above 80 F.

screen is 478 Btu per lb of gas and the energy transferred in the furnace (Q_f) is

$$Q_f = 27,150,000 - 18.1 \times 1622 \times 478 = 13,130,000 \text{ Btu per hr}$$

The amount of water evaporated in the waterwalls is $13,130,000 \div (1201.7 - 168.0) = 12,700$ lb per hr, based on the inlet feedwater temperature.

The total projected area of the furnace envelope tubes is:

$$\text{Front wall} = 18 \text{ tubes} \times \frac{3\text{-in. diam}}{12} \times 12 \text{ ft long} = 54 \text{ sq ft}$$

$$\text{Side walls} = 17 \text{ tubes} \times \frac{3\text{-in. diam}}{12} \times 12 \text{ ft long} \times 2 \text{ walls} = 102 \text{ sq ft}$$

$$\text{Rear wall} = 7 \text{ tubes} \times \frac{3\text{-in. diam}}{12} \times 12 \text{ ft long} = 21 \text{ sq ft}$$

$$\text{Roof} = 18 \text{ tubes} \times \frac{3\text{-in. diam}}{12} \times 12 \text{ ft long} = 54 \text{ sq ft}$$

$$\text{Screen} = 18 \text{ tubes} \times \frac{3\text{-in. diam}}{12} \times 10.25 \text{ ft long} = 46 \text{ sq ft}$$

$$\text{Total projected tube area} = 277 \text{ sq ft}$$

Therefore the radiant-heat transfer rate is $13,130,000 \div 277 = 47,400$ Btu per (hr)(sq ft of projected surface), or $47,400 \div \pi = 15,100$ Btu per (hr)(sq ft of outside tube surface).

The assumption of boiler efficiency cannot be checked until the convection-surface calculations have been completed.

6–14. Convection-Surface Calculations. Inspection of Fig. 6–37 shows that there are 401 tubes in the convection zone. Assuming an average tube length of approximately 12.5 ft to account for bent tubes and the longer tubes at the rear of the cavity, the convection zone tube surface will be

$$A_0 = 401\pi \frac{2}{12} \times 12.5 = 2625 \text{ sq ft}$$

Heat transfer in the convection zone takes place both by convection and radiation. For the convection heat transfer for boiler tubes it is not unusual to neglect the resistance of the water film inside the tubes and the resistance of the metal; U_0 then becomes equal to h from Table 4–1, Eq. F.

The area between tubes that is available for flow of the gases varies throughout the convection zone, but at any point the area is the number of tube spaces times the width and height of each space. Because of the tube bends, the length of the area will be smaller than the tube length; take 12 ft as the length of the area. The first convection row is 5 ft 5 in. long and must contain 17 tubes or 16 spaces. The area is

$$16 \text{ spaces} \times \frac{2 \text{ in. width}}{12} \times 12 \text{ ft high} = 32 \text{ sq ft}$$

For the next to the last row, the area is

$$7 \text{ spaces} \times \frac{2 \text{ in. width}}{12} \times 12 \text{ ft high} = 14 \text{ sq ft}$$

The average area of 23 sq ft will be used. A more exact method would be to design each section separately.

The gas flow is $1622 \times 18.1 = 29,400$ lb per hr and the density is

$$\rho_0 = \frac{P}{RT} = \frac{144 \times 14.7}{53.3 \times 492} = 0.0808 \text{ lb per cu ft}$$

Then the mass flow based on the average flow area is

$$G = \frac{29,400}{3600 \times 23} = 0.356 \text{ lb per (sq ft) (sec)}$$

$$V_0 = \frac{G}{\rho_0} = \frac{0.356}{0.0808} = 4.4 \text{ fps}$$

$$U_0 = h = \frac{0.91 V_0^{0.69}}{D^{0.31}} = \frac{0.91 (4.4)^{0.69}}{\left(\dfrac{2}{12}\right)^{0.31}} = 4.41 \text{ Btu per (hr) (sq ft) (F)}$$

The next step is to estimate the final gas temperature so that the LMTD may be calculated from Eq. (4–15). Since the furnace was calculated to evaporate 12,700 lb per hr, the convection zone should evaporate $22,000 - 12,700 = 9300$ lb per hr. Each pound of flue gases should then transfer to the tubes in the convection zone by radiation and convection.

$$\frac{9300 (1201.7 - 168.0)}{1622 \times 18.1} = 328 \text{ Btu}$$

The energy of the flue gases entering the convection zone was calculated to be 478 Btu per lb. Then the energy leaving the convection zone will be $478 - 328 = 150$ Btu per lb. From Fig. 6–40(b), this would represent a temperature of 670 F.

This flue gas temperature is too high for an economical steam generator as it is 264 F above the saturated steam temperature of 406 F. For typical conditions, the exit flue gas temperature should be roughly 100 above the saturation temperature.

If the tube temperature is taken to be the same as the water temperature (406 F), in accordance with previous assumptions of negligible resistance through the water film and the metal, LMTD from Eq. (4–15) is

$$\theta_m = \frac{1770 - 670}{\ln \dfrac{1770 - 406}{670 - 406}} = 669 \text{ F}$$

Convection heat transfer is

$$Q_c = U_0 A_0 \theta_m = 4.41 \times 2625 \times 669 = 7{,}740{,}000 \text{ Btu per hr}$$

Gases radiate and absorb energy at intermittent wave-length bands. Radiation in the infrared band from gases has been recognized as important to the design of some heat-transfer apparatus. While the radiation considered for the furnace is from luminous flames and suspended particles, the convection-zone radiation is from inactive nonluminous gases that are not undergoing a chemical change and that carry very little, if any, suspended solids.

Of the constituents in the flue gases, carbon dioxide and water vapor are the only ones that have sufficiently strong radiating characteristics to merit consideration. Sulfur dioxide and carbon monoxide have strong radiating tendencies but usually are present in flue gas in such small quantities that they need not be considered.

The radiation from gases containing carbon dioxide and water vapor may be approximated by

$$Q_r = 0.1723 A \epsilon_s \left[\epsilon_g \left(\frac{T_g}{100} \right)^4 - a \left(\frac{T_s}{100} \right)_4 \right] \qquad (6\text{–}15)$$

in which Q_r = heat transfer by radiation from gases, Btu per hr
$\quad A$ = outside tube surface area, sq ft
$\quad \epsilon_s$ = tube emissivity, 0.80 for boiler and superheater tubes
$\quad \epsilon_g$ = emissivity of the gases at temperature T_g
$\quad T_g$ = absolute gas temperature, R
$\quad a$ = emissivity of the gases at temperature T_s
$\quad T_s$ = absolute tube surface temperature, R

When the gases are at standard atmospheric pressure, as is the case in nearly all boilers, the gas emissivities can be evaluated * from Eq. (6–16).

* Other assumptions for these equations are that $P_c L + P_w L < 0.3$ and that $\dfrac{T_g}{T_s} > 1.25$.

$$\epsilon_g = \epsilon_{cg} + \epsilon_{wg} C_w \qquad\qquad (6\text{-}16)$$

and
$$a = \epsilon_{cs} + \epsilon_{ws} C_w \qquad\qquad (6\text{-}17)$$

in which ϵ_{cg} = emissivity of carbon dioxide at temperature T_g from Fig. 6-41

$\qquad \epsilon_{wg}$ = emissivity of water vapor at temperature T_g from Fig. 6-42

$\qquad \epsilon_{cs}$ = emissivity of carbon dioxide at temperature T_s from Fig. 6-41

$\qquad \epsilon_{ws}$ = emissivity of water vapor at temperature T_s from Fig. 6-42

$\qquad C_w$ = correction factor for water vapor emissivity from Fig. 6-43

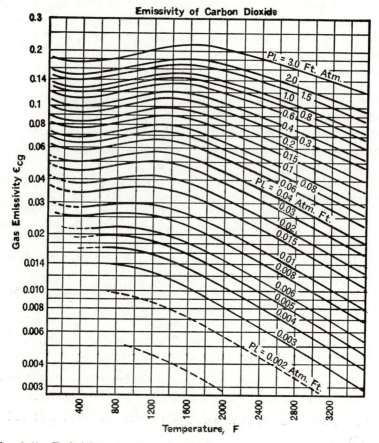

FIG. 6-41. Emissivity of carbon dioxide. (Reproduced with permission of the A.I.Ch.E., *Transactions*, Vol. 38, No. 3.)

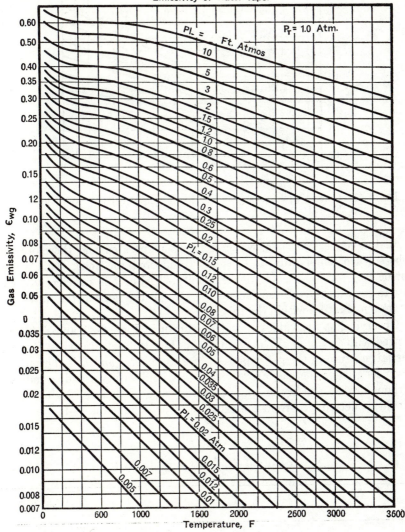

Emissivity of Water Vapor

Gas Emissivity, ϵ_{wg}

$P_T = 1.0$ Atm.

$PL = $ Ft. Atmos

Temperature, F

FIG. 6-42. Emissivity of water vapor. (Reproduced with permission of the A.I.Ch.E., *Transactions*, Vol. 38, No. 3.)

The average gas temperature may be estimated from the equation

$$T_g = 460 + \theta_m + \frac{t_1 + t_2}{2} \qquad (6\text{-}18)$$

where θ_m = log mean temperature difference between gas and surface, F
t_1 and t_2 = surface temperatures at sections where fluid enters and leaves tubes, respectively, F

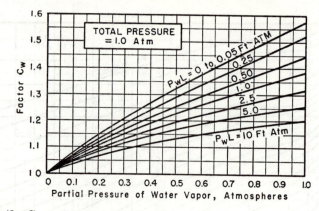

Fig. 6–43. Correction factor for emissivity of water vapor. (Reproduced with permission of the A. I. Ch. E., *Transactions*, Vol. 38, No. 3.)

In the convection zone, water is being evaporated and therefore the tube-surface temperature is constant throughout the zone. This would not be true for superheaters or economizers. Also, it is sufficiently accurate to say that the tube-surface temperature is the same as the water and steam temperature within the tube.

Observe that the values ϵ_c and ϵ_w, shown in Figs. 6–41 and 6–42, are plotted with values of PL as parameters. For each set of curves, P is the partial pressure of the gas expressed in atmospheres and L is the radiant beam length for the gas, expressed in feet. Subscripts c and w indicate carbon dioxide and water vapor, respectively, as before. Values of L should be determined from the expressions given in Table 6–1.

TABLE 6–1

BEAM LENGTHS FOR GAS RADIATION

Application	Length, Ft
Space between infinite parallel planes	1.8 × clearance
Space outside of infinite bank of tubes with centers on equilateral triangles; clearance between tubes equal to tube diam	2.8 × clearance
Same, except clearance of twice the tube diam	3.8 × clearance
Same, except tube centers on squares; clearance equal to tube diam	3.5 × clearance
Infinite cylinder	0.90 × diameter

Parameters of PL are also used in Fig. 6–43. This graph accounts for the effect of the water-vapor partial pressure on radiation.

The tube-surface temperature for our problem is 406 F and the mean gas temperature, from Eq. (6–18) and the previous evaluation of θ_m, is

$$T_g = 460 + 669 + 406 = 1535 \text{ R} \quad \text{or} \quad t_g = 1075 \text{ F}$$

Partial pressures of gases are proportional to the volumetric analysis of the *wet* gas. In this case we have 11.2% carbon dioxide and 10.4% water vapor. Therefore, evaluating L from Table 6–1 as $\frac{2}{12} \times 2.8$, we get

$$P_c L = 0.112 \times \frac{2}{12} \times 2.8 = 0.0523$$

and

$$P_w L = 0.104 \times \frac{2}{12} \times 2.8 = 0.0486$$

From Fig. 6–43, $C_w = 1.08$. For a temperature t_s of 406 F and $P_w L = 0.0486$, ϵ_{ws} is 0.064 from Fig. 6–42. Finding other values in a similar manner, we get, from Eq. (6–17),

$$a = 0.055 + 0.064 \times 1.08 = 0.124$$

and from Eq. (6–16),

$$\epsilon_g = 0.062 + 0.039 \times 1.08 = 0.104$$

Using these values in Eq. (6–15),

$$Q_r = 0.1723 \times 2625 \times 0.80 \left[0.104 \left(\frac{1535}{100} \right)^4 - 0.124 \left(\frac{866}{100} \right)^4 \right]$$
$$= 1{,}835{,}000 \text{ Btu per hr}$$

The total energy transferred for the entire boiler is

$$Q = Q_f + Q_c + Q_r = 13{,}130{,}000 + 7{,}740{,}000 + 1{,}835{,}000$$
$$= 22{,}705{,}000 \text{ Btu per hr}$$

and the evaporation is

$$\frac{22{,}705{,}000}{1201.7 - 168.0} = 21{,}950 \text{ lb per hr}$$

This shows that the assumed efficiency was correct. However, the temperature of the gases entering the convection zone is low. A more economical unit would have less waterwall surface and more convection surface to reduce the final flue-gas temperature.

Observe that the energy transferred by radiation from nonluminous gases in the convection zone amounts to about 20% of the heat transfer in this zone.

6–15. Secondary Surface. Calculations for surfaces of the convection type of superheater, economizer, and tubular air heater follow the same procedures that were used for the convection zone in Art. 6–14. Gas, air, and steam film coefficients may be determined from Table 4–1. Water film coefficients for economizers offer only token resistance to the flow of heat and may be neglected for economizers. Similarly, the metal in the tube walls may be disregarded in calculating the heat flow.

Nonluminous radiant heat transfer from the water vapor and carbon dioxide in the flue gases will amount to a small percentage of the total heat transfer for economizers and air preheaters.

For secondary surface, the flow of water, steam, air, or gas is customarily given as the *mass flow* in units of pound per (hour) (square feet of flow area).

Superheaters, air heaters, and economizers use 2- or 2½-in. OD tubes. These tubes may be placed on approximately 3- to 9-in. centers in superheaters. The wider spacing is to reduce the possibility of slag bridging across the space. With either pendent or horizontal superheater designs, the tubes are in-line and form several passes, as shown in Fig. 6–15. Mass gas flows range from 1500 to 3000 lb per (hr) (sq ft) while mass steam flows are from 200,000 to 300,000 lb per (hr) (sq ft). Mass steam flows may be higher, up to 700,000 or more, for very high-temperature superheaters.

Air preheater tubes should have the air on the outside of the tubes to prevent plugging from soot in the gases. In this way staggered tubes may be used effectively. The tubes are of either No. 12 or No. 14 BWG (0.109 in. or 0.083 in., respectively) thickness and are arranged for at least ½ in. space between tubes. Mass gas flows are from 5000 to 10,000 and mass air flows are from 3000 to 5000 lb per (hr) (sq ft).

Economizers, being of the continuous tube design, are arranged with tubes in-line, and there are many water passes, as shown in the illustrations in this chapter. The tubes are on centers that provide 1½- to 2-in. lanes for gas flow. The spacing parallel to the gas flow ranges from 1 to 3 in. Water velocities in the tubes range from 3 to 8 fps and the mass flow of gases is about 4000 to 7000 lb per (hr) (sq ft).

PROBLEMS

6–1. Find the minimum circulation ratio for the following boiler pressures: (a) 700 psia, (b) 1400 psia, (c) 1950 psia, (d) 2300 psia.

6–2 to 6–4. For each of the following problems, calculate the internal diameters of the downcomers. In each of the three cases the downcomer carries saturated water.

Problem No.	6-2	6-3	6-4
Boiler pressure, psia	1100	2000	2400
Number of risers	100	120	160
OD of risers, in.	3	3	3
Riser wall thickness, in.	0.222	0.360	0.440
Length of risers, ft	100	120	140
Number of 90-deg bends in risers	6	6	8
Circulation ratio	14	8.5	7.5
Steam per tube, lb per hr	1400	1650	2200
Number of downcomers	1	1	1
Length, ft	70	90	100
Number of 90-deg bends in downcomers	2	3	4

6-5 to 6-7. Calculate the required furnace volume for each of the following steam generators if the air surrounding the boiler is at 100 F. [Use Eq. (5-8).] Assume that there is 0.1% unburned carbon.

Problem No.	6-5	6-6	6-7
Heat release, Btu per (hr) (cu ft)....	16,000	18,000	20,000
Coal	Ky.-Letcher	Pa.-Allegheny	Ohio-Jefferson
Excess air, %	20	22	21
Output, lb steam per hr.............	1,000,000	650,000	2,000,000
Efficiency, %	89	90	91
Heat added, Btu per lb steam.......	1,220	1,200	1,180
Air temperature leaving air heater, F.	600	660	620

6-8. A boiler generates steam at 99.4% quality when the drum concentration is 3500 ppm. Determine the steam purity.

6-9. It is desired to have not more than 1 ppm in the steam from a boiler when the boiler concentration is 1500 ppm. What will be the maximum moisture in the steam?

6-10. A boiler has a concentration of 1250 ppm. What percentage of the water in the boiler must be wasted through the blowdown to reduce the concentration to 1000 ppm when the feedwater concentration is 11 ppm?

6-11 to 6-13. Calculate the water flow to the attemperator for each of the following sets of conditions:

Problem No.	6-11	6-12	6-13
Steam flow leaving attemperator, lb per hr.......	690,000	800,000	1,000,000
Steam entering, F...	1,000	880	950
Steam leaving, F	950	800	860
Steam pressure, psia	1,500	1,900	2,100
Water entering, F.............................	270	310	290

6-14 to 6-16. Determine the necessary entering-air temperature to an air preheater to maintain the cold-end metal temperature listed at the top of page 272.

Problem No.	6–14	6–15	6–16
Flue gas leaving, F............................	320	270	250
Metal temperature, F.........................	223	180	190

6–17 to 6–21. Determine the factor of evaporation, equivalent evaporation, boiler horsepower, and the boiler rating, for each of the following:

Problem No.	6–17	6–18	6–19	6–20	6–21
Steam output, lb per hr................	8000	15,000	12,000	2000	700,000
Steam: pressure, psia	135	165	220	120	1,480
moisture, %	1	0.5	0.8	0.6	....
temperature, F	....	....	....	....	1,000
Feedwater temperature, F	120	210	223	215	450
Surface (boiler and waterwall, if any) sq ft	1330	2310	1720	342	15,000
Superheater, sq ft	....	....	....	....	60.000
Economizer, sq ft	....	....	....	....	6,500
Air heater, sq ft	....	....	....	....	227,000

6–22, 6–23. A furnace has the same dimensions and the same water screen as in Fig. 6–34. Neglect the furnace floor and calculate the temperature of the gases entering the convection zone, the heat release per cubic foot of furnace volume, the heat transfer per square foot of projected tube area per hour, and the amount of steam produced if the downcomers carry saturated water.

Problem No.	6–22	6–23
Tubes	3-in. on 5-in. centers	3-in. on 6-in. centers
Fuel, kind	Mo., Boone	Va., Tazewell
quantity, lb per hr	4000	3500
Excess air, %......................	25%	30%
Radiation and unaccounted-for.......	2%	2½%
Method of firing	Underfeed stoker	Chain grate
t_a, F	90	100
t_{ah}, F	90	200
Unburned carbon, lb per lb fuel.......	0.05	0.08
Steam pressure, psia	190	325

6–24. Recalculate the boiler of Fig. 6–34, using 79% efficiency and 30,000 lb per hr of steam.

6–25, 6–26. Assume that the pulverized coal furnace is rectangular in plan and elevation views. The waterwalls are of 3-in. tangent tubes. Include the floor surface (assume flat, horizontal floor) and calculate temperature of the gas leaving the furnace, the heat transfer per hour for each square foot of projected surface, the heat release per cubic foot of furnace volume, and the steam produced if the downcomers carry saturated water.

Problem No.	6–25	6–26
Steam pressure, psia	1650	2300
Fuel, quantity, tons per hr	49	60
kind	Ill., Christian	Ind., Sullivan
Furnace, height, ft	70	80
depth, ft	27	21
width, ft	31	42
t_a, F	80	90
t_{ah}, F	550	620
Radiation and unaccounted-for	1.8%	2.0%
Excess air, %	19	18

6–27, 6–28. Calculate the number of tubes, the surface, and the number of steam passes for in-line tubes in a pendent superheater. Steam enters the tubes saturated. Neglect the nonluminous radiation and the conductivity of the metal, but include both steam and gas films and assume counterflow design.

Problem No.	6–27	6–28
Tube OD, in.	2½	2
Tube-wall thickness, in.	0.25	0.20
Tube spacing, c-to-c, in.	7	5½
Steam pressure, psia	1,300	875
Steam temperature, F	1,070	970
Steam flow, lb per hr	900,000	600.000
Steam mass flow, lb per (sq ft) (hr)	500,000	400,000
Gas mass flow, lb per (sq ft) (hr)	1,700	1,500
Gas flow, lb per hr	1,100,000	800,000
Gas temperature entering, F	1,900	2,050

6–29 to 6–31. Calculate the number of tubes, the surface, and the number of water passes for a counterflow economizer with in-line continuous tubes. Neglect the resistance of the metal and the water film and also neglect the nonluminous radiation.

Problem No.	6–29	6–30	6–31
Tube OD, in.	2	2½	2
Tube-wall thickness, in.	0.15	0.25	0.18
Tube spacing, c-to-c, in.	3¼	4	3
Entering-water temperature, F	450	500	405
Leaving-water temperature, F	505	535	530
Water flow, lb per hr	900,000	1,050,000	380,000
Water velocity, fps	5	5	6
Gas temperature entering, F	925	775	1,150
Gas flow, lb per hr	1,200,000	1,400,000	506,000
Gas mass flow, lb per (sq ft) (hr)	8,000	11,000	10,000

6–32 to 6–34. Calculate the number of No. 14 BWG tubes and the surface for a counterflow air preheater and staggered tubes. Neglect nonluminous radiation and the resistance of the metal. Air is on the outside of the tubes.

Problem No.	6–32	6–33	6–34
Number of air passes	5	5	5
Tube OD, in.	2	2	2½
Tube spacing, c-to-c, in.	2½	3	4
Air temperature entering, F	100	100	100
Air temperature leaving, F	600	528	575
Air flow, lb per hr	414,000	990,000	1,185,000
Air mass flow, lb per (sq ft) (hr)......	9,000	10,000	9,500
Gas temperature entering, F..........	720	605	650
Gas flow, lb per hr	506,000	1,200,000	1,430,000
Gas mass flow, lb per (sq. ft) (hr).....	9,000	10,000	9,500

6–35. If fuel costs 28¢ per 10^6 Btu and a steam generator will operate 7000 hr per yr at a steam flow of 1,200,000 lb per hr with 1255 Btu added to each pound of water, how much increase in air preheater cost can be justified to increase the steam generator efficiency from 89 to 90%. Investment charges are 12%.

BIBLIOGRAPHY

ASME Test Code for Stationary Steam-Generating Units.

DE LORENZI, OTTO. *Combustion Engineering.* New York: Combustion Engineering Co., Inc., 1947.

HOTTEL, H. C., and EGBERT, R. B. "The Radiation of Furnace Gases," *Trans. ASME,* Vol. 63, No. 4, p. 297.

HOTTEL, H. C., and SMITH, V. C. "Radiation from Nonluminous Flames," *Trans. ASME,* Vol. 57, No. 8, p. 463.

LATHAM, ROBERT F. "Naval Boilers." Annapolis, Md.: United States Naval Academy, 1956.

Steam, Its Generation and Use. New York: Babcock & Wilcox Co., 1955.

CHAPTER 7

STEAM GENERATOR AUXILIARIES

It is the purpose of this chapter to discuss some of the more important auxiliaries and accessories required for satisfactory operation of the boiler. The material may be divided into three categories: fuel-burning equipment, equipment for creating furnace draft, and feedwater treatment.

Fuel-burning Equipment

7-1. Pulverizers.   Large coal-fired steam generators have been made possible by the development of pulverized coal firing. The only other method of firing coal in the amounts required by high-capacity boilers is the recently developed cyclone burner that will be discussed later. Pulverizers with their burners can handle a wide variety of coals and can be combined with liquid and gaseous fuels. In addition, boiler efficiency has been improved, due to more efficient combustion with pulverized coal, and maintenance costs are less than with some styles of stokers. The principal disadvantage of pulverized coal is the fly-ash emitted from the boiler, which amounts to 80 to 90% of the ash in the fuel for a dry-bottom furnace and 40 to 60% for a wet-bottom furnace. Unless the plant is located in a remote section, the fly-ash nuisance will require the application of mechanical or electrostatic precipitators. Even when these collecting devices are installed, usually before the induced-draft fan, the fly-ash causes erosion of the fan blades and housing.

Since 1921, when the first plant designed to use pulverized coal was put into operation, this method of coal firing has become popular for steam generators of more than 30,000 lb per hr capacity. Pulverizers or cyclone burners are used almost exclusively for coal-fired units of 150,000 lb per hr capacity or larger.

Like other types of coal equipment, certain preparation of the coal is necessary before feeding it to the pulverizer. Cloth, wood, and straw should be removed so that they will not collect in the mill and create a fire hazard. Magnetic separators are usually installed in the coal conveying system to remove the larger pieces of tramp iron; most pulverizers can reject the smaller pieces without damage. Then the coal should be crushed to a size that is suitable for the pulverizer; even screenings may contain lumps of coal. A uniform coal size is desirable, both to reduce segregation in the bunker and to enable the pulverizer to produce a more uniform product. The size of coal to be fed to the pulverizer depends on the capacity of the unit and ranges from $\frac{3}{4} \times 0$ in. for the smaller sizes to $1\frac{1}{4} \times 0$ in. for the larger pulverizers.

The capacity of a pulverizer is stated in terms of the weight of wet coal entering per hour. Mill capacity is dependent upon several factors, including grindability, fineness of the coal leaving the mill, and moisture of the coal entering the mill.

Coal *grindability* is an indication of the ease of grinding a coal, and it affects the capacity of a mill in that mill capacity reduces as the grindability number decreases. On the Hardgrove * basis, grindability is determined by pulverizing an air-dried sample of coal that has been screened to minus 16- and plus 30-mesh (all the sample will pass through a 16-mesh screen but none will pass a 30-mesh screen) in a standardized mill. The grindability number is determined from the amount of the sample that is minus 200-mesh after grinding. Grindability numbers and ash-softening temperatures for some coals are given in Table 7–1. In most cases, pulverizers should be specified for 45 to 60 grindability coal unless the poorest coal likely to be used in the plant has a lower grindability.

Factors that influence the required fineness of a coal leaving a pulverizer include the size of the furnace, type of coal, furnace temperature, and furnace and burner design. However, the 50-mesh and the 200-mesh screens are important. If 5% or more of the coal is plus 50-mesh, trouble may be encountered from slagging and loss of combustible with the flue gases. In order to promote good combustion, it is important to have a high percentage of very fine coal. Therefore it is customary to specify that the coal leaving the pulverizer shall have 98 to 99.5% minus 50-mesh and 65 to 85% minus 200-mesh. Reducing the minus 200-mesh coal by 5%, other conditions remaining the same, will increase mill capacity about 9%.

Recent difficulties in acquiring fuel have emphasized the importance of flexibility. The present trend of thought is that the most economical plant is one that can burn the lowest and cheapest grade of fuel available to the plant even though that might increase the initial investment. Specifying the pulverizer grindability and fineness for the lowest grade of coal available to the plant may mean that the pulverizer will be oversized when burning better coals, but that is to be preferred when compared to high maintenance or reduced load resulting from burning inferior coal, during emergencies, in a pulverizer designed for a good grade of coal.

Moisture present in coal may be considered as surface moisture and inherent moisture. Hard coals from the eastern portion of the country are dense, and usually the moisture is predominantly **surface moisture**.

* See *Trans. ASME,* Vol. 55, No. 5.

TABLE 7-1

Grindability Numbers for Some Coal with Ash Data

State	County	Rank	Grindability	Ash Temperatures, F		
				Initial Deformation	Softening	Fluid
Alabama	Bibb	Bituminous	56	....	....	13,800
	Jefferson	Bituminous	87	....	....	13,030
Arkansas	Sebastian	Semibituminous	100	2110	2200	2,360
Colorado	Boulder	Subbituminous	47	....	....	....
	Los Animos	Bituminous	54	2410	2520	2,690
Illinois	Christian	Bituminous	57	1925	2050	2,100
	Franklin	Bituminous	58	2150	2210	2,670
Indiana	Sullivan	Bituminous	63	....	....	....
Kansas	Cherokee	Bituminous	61	....	....	....
Kentucky	Letcher	Bituminous	53	....	....	....
	Webster	Bituminous	63	....	....	10,032
Missouri	Randolph	Bituminous	75	....	....	....
North Dakota	Ward	Lignite	50	....	2560	....
Ohio	Jefferson	Bituminous	60	1940	2020	2,580
Oklahoma	Pittsburg	Bituminous	62	....	2400	2,470
Pennsylvania	Allegheny	Bituminous	54	2350	2680	....
	Cambria	Semibituminous	103	....	2900	....
	Lackawanna	Anthracite	26	....	3000	....
	Luzerne	Anthracite	24	....	2860	....
	Somerset *	Semibituminous	100	....	2100	....
Texas	Webb	Bituminous	32	....	....	....
Virginia	Montgomery	Semianthracite	83	....	....	....
	Tazewell	Semibituminous	102	....	2270	2,650
Washington	Pierce	Subbituminous	100	2240	2410	....
West Virginia	McDowell	Semibituminous	104	....	....	....

* Reference coal.

Western coals are more porous and much of their moisture is contained within the coal structure. In order to remove the inherent moisture, there must be prolonged heating of the coal. Pulverizing the coal exposes the inner structure and aids in the drying process. Fortunately, the surface moisture has the most effect on the mill capacity and is also the easiest to remove. The reduction in mill capacity due to coal moisture is a variable depending on the size of mill, type of coal, grindability, type of pulverizer, and the fineness. Roughly, it may be considered that an increase of 1% in coal moisture will reduce the mill capacity about 1%.

Coal leaving the pulverizer should be dry enough to permit the coal temperature to be raised rapidly to the kindling point in the furnace so that there will be good ignition. Also, the coal must be dry enough to form a dust cloud when leaving the mill. The amount of moisture permissible in the coal leaving the mill will depend on the type of coal but is usually considered to be from 1 to 2% for bituminous coals.

In modern plants, the hot air for drying coal in the pulverizer is supplied from the forced-draft fan and the air preheater, as shown schematically in Fig. 7–1. Most of the air leaving the air heater goes directly

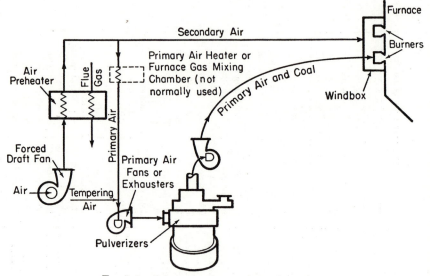

FIG. 7–1. Primary and secondary air system.

to the burner windbox on the boiler and is known as *secondary air*. A small part of the air, known as the *primary air,* is used to dry the air in the pulverizer and to act as the conveying medium to transport the coal from the pulverizer to the burners or storage bins. Burners provide the opportunity for mixing of the coal, primary air, and secondary air.

In rare instances, the temperature of the air leaving the air preheater is not high enough to dry the coal properly. In that circumstance, a *primary air heater* using steam coils will raise the air temperature, or very hot flue gases from the steam generator may be mixed with the air to raise its temperature. The flue gases must be taken from a point in the steam generator where the concentration of CO_2 is low and must be well diluted with air in the mixing chamber. Otherwise, the presence of the CO_2 will be detrimental to the combustion process.

An advantage of coal pulverizers over stokers is the ability of pulverizers to use hot air at temperatures ranging from 500 F to 700 F. These high air temperatures promote good combustion and permit lower flue-gas temperatures, particularly where there are no economizers on the boiler.

Air temperatures leaving the pulverizer should be maintained at 150 F to 180 F (110 F to 130 F when the pulverized coal is to be stored before burning). In order to maintain the desired temperature leaving the pulverizer with variation in coal moisture and air temperature (the temperature of the air from the air preheater will vary with boiler load), *tempering air* from the boiler room is mixed with the primary air at the

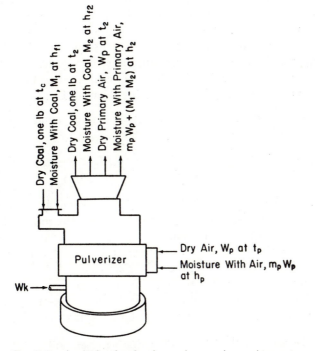

Fig. 7–2. Analysis of pulverizer primary air requirements.

pulverizer inlet to maintain the necessary outlet temperature. Tempering air flow is a minimum when the coal is wet and is a maximum when the coal is dry.

An equation for the primary air flow at maximum coal moisture and zero-tempering air flow may be derived by referring to Fig. 7-2. According to the Law of Conservation of Energy, all the energy for the materials entering the mill must equal the energy for the materials leaving the mill. Taking the specific heat of air as 0.24, the specific heat of water as 1.0, and the specific heat of coal as 0.3,

$$0.24 W_p t_p + m_p W_p h_p + 0.3 t_c + M_1 h_{f1} + Wk$$
$$= 0.24 W_p t_2 + [m_p W_p + (M_1 - M_2)] h_2 + 0.3 t_2 + M_2 h_{f2}$$

where W_p = weight of *dry* primary air per lb of dry coal, lb
m_p = moisture with 1 lb of dry primary air, lb
M_1 = weight of moisture entering with each lb of dry coal, lb
M_2 = weight of moisture leaving per lb of dry coal entering, lb
t_p = temperature of primary air, F
t_c = temperature of entering coal, F
t_2 = temperature of coal-air mixture leaving, F
h_p = enthalpy of moisture in primary air (taken from the steam tables at an assumed partial pressure of 1 psia), Btu per lb
h_2 = enthalpy of moisture in leaving air (taken from the steam tables at an assumed partial pressure of 1 psia), Btu per lb
h_{f1} = enthalpy of liquid moisture in coal entering, Btu per lb
h_{f2} = enthalpy of liquid moisture in coal leaving, Btu per lb
Wk = work expended in pulverizing the coal, Btu per lb coal

About 13 Btu of work are required to pulverize a pound of coal in the more efficient mills.

$$W_p = \frac{0.3(t_2 - t_c) + M_1(h_2 - h_{f1}) + M_2(h_{f2} - h_2) - Wk}{0.24(t_p - t_2) + m_p(h_p - h_2)} \qquad (7\text{-}1)$$

It has already been stated that tempering air will be required when the coal moisture is less than maximum. However, the total weight of air leaving the pulverizer at maximum load will not change because the capacity of the pulverizer fans will remain the same; * nor will the temperature of the mixture leaving the pulverizer change because the amount of tempering air is controlled to maintain a constant outlet temperature. The equation for the amount of tempering air may be derived in the same manner as Eq. (7-1). The total weight of air leaving the pulverizer is the

---

* When the fan is at the inlet to the pulverizer, there will a slight change in capacity due to the change in mixture temperature.

sum of the weights of tempering air and the new quantity of entering primary air, and is taken to be the same as W_p in Eq. (7-1). Then

$$W_t = [0.3(t_2 - t_c) + M_1(h_2 - h_{f1}) + M_2(h_{f2} - h_2) + 0.24W_p(t_2 - t_p)$$
$$- Wk + m_p W_p(h_2 - h_p)] \div [0.24(t_t - t_p)$$
$$+ m_t(h_t - h_2) + m_p(h_2 - h_p)] \quad (7\text{-}2)$$

where W_t = weight of dry tempering air per lb of dry coal, lb

m_t = weight of moisture with 1 lb of dry tempering air, lb

t_t = temperature of the tempering air, F

h_t = enthalpy of the moisture with the tempering air (taken from the steam tables as saturated steam at t_t) Btu per lb

When both the tempering air and the air to the air heater are taken from the boiler room, m_t and m_p will be equal.

Pulverized-fuel systems may be arranged in the *bin* or *unit system*. In the bin system, fuel is fed from the bin to the burners as required. When the supply of pulverized coal in the bin reaches a predetermined minimum, the pulverizers are operated at full capacity until the bin is filled. The advantages of this system are that the power requirement of the pulverizers is less than for the unit system, since the pulverizers are operated at full load only (about 20% less power for a bin system than for a unit system for an average of half-load on the plant) and that load on the plant need not be reduced for a short outage of 1 mill. The bin system has not been used on recent installations because it requires the additional bin (plus building space) for the pulverized coal and because of the very serious fire hazard. Pulverized coal is extremely explosive in the presence of air.

Power requirements at full pulverizer load for the unit system range from 10 to 26 kwhr per ton of coal, depending on the type of pulverizer, the coal, and the fineness. For this system each pulverizer supplies coal directly to its particular group of burners. The load on the pulverizer will depend on the boiler load and on whether those particular burners are in operation. Outage of a mill, on the unit system, will reduce the maximum boiler load.

EXAMPLE 7-1. Primary air is supplied to a pulverizer from an air heater at 560 F. If all the surface moisture (varying up to 15%) is removed in the pulverizer, find the pounds of primary air required per pound of wet coal for 150 F temperature of the mixture leaving the pulverizer. Assume air entering the air heater to have 0.013 lb of water vapor per lb of dry air, the coal to be entering at 60 F and to contain 2% inherent moisture.

SOLUTION. Use Eq. (7-1) and take $M_1 = \dfrac{0.17}{1 - 0.17} = 0.205$, $M_2 = \dfrac{0.02}{1 - 0.17} = 0.024$, and $m_p = 0.013$.

$$W_p = \frac{0.3(150 - 60) + 0.205(1127.9 - 28.1) + 0.024(117.9 - 1127.9) - 13}{0.24(560 - 150) + 0.013(1316.7 - 1127.9)}$$

$$= 2.13 \text{ lb of dry primary air per lb of dry coal}$$

or

$$2.13(1 - 0.17) = 1.77 \text{ lb of dry primary air per lb of wet coal}$$

EXAMPLE 7-2.   If the driest coal to be used in the pulverizer of the previous example will have a surface moisture of 4%, determine the amount of tempering air at 80 F that will be required. All other data remain the same.

SOLUTION.   The weight of dry primary air leaving the pulverizer will be $W_p = 2.13$ lb per lb of dry coal, from Example 7-1. Thus,

$$M_1 = \frac{0.06}{1 - 0.06} = 0.0638, \quad M_2 = \frac{0.02}{1 - 0.06} = 0.0213$$

and m_t may be taken as equal to m_p.

$$\begin{aligned} W_t = [&0.3(150 - 60) + 0.0638(1127.9 - 28.1) + 0.0213(117.9 - 1127.9) \\ &+ 0.24 \times 2.13(150 - 560) - 13 + 0.013 \times 2.13(1127.9 - 1316.7)] \\ \div [&0.24(80 - 560) + 0.013(1096.6 - 1127.9) + 0.013(1127.9 \\ &- 1316.7)] \end{aligned}$$

$$= 1.29 \text{ lb of dry tempering air per lb dry coal}$$

Tube mills, as shown in Fig. 7-3, are sometimes called Hardinge mills and are simple, sturdy, and dependable. They have very low maintenance costs. They are particularly suitable where space, noise, and power costs are not important. Coal is fed by gravity from the coal feeder at the upper portion of the picture to the two large classifiers at each end of the drum and thence into the drum. As the drum rotates at a speed of about 18 to 27 rpm, the coal is pulverized by attrition between the tumbling balls. The conical ends of the drum promote size segregation of the balls and the coal. It is not necessary to remove any tramp iron that may enter with the coal.

Hot primary air enters through the inner ducts at each end of the drum and carries the ground coal through the classifiers. Coarse coal returns by gravity from the classifiers to the tube for regrinding. The exhauster at the upper left of Fig. 7-3 maintains a slight vacuum in the revolving tube and transmits the coal with its primary air from the pulverizer to the burners.

The large drum or tube contains an appreciable amount of coal and, during a normal shut-down, the pulverizer must be operated until all the coal has been ground and removed from the mill so that it will not ignite.

A ball-and-race type of mill, Fig. 7-4, is a medium-speed mill, whereas the tube mill is a slow-speed mill. Built similar to a ball bearing, the ball-and-race mill contains balls riding between an upper, stationary, horizontal race and a lower rotating race. Some of these mills use two

FIG. 7–3. Ball type of coal pulverizer. (Foster Wheeler Corp.)

rows of balls and three races. The springs attached to the upper race prevent the race from turning and also provide the necessary pressure between the balls and races. Rotation of the lower race is provided from the spiral bevel gears in the base and from a separate driving motor.

Raw coal is supplied to the mill from the *rotating table type of feeder* that is driven by a separate two-speed motor. The rotating table is much like a phonograph record table. Coal falls onto the rotating table from the coal bunker or the coal scales. A fixed plow blade above the table scoops the coal from the table and into the pulverizer. The amount of coal entering the mill is controlled by the table speed. The coal is fed from the table feeder into the inside of the grinding rings and onto a table

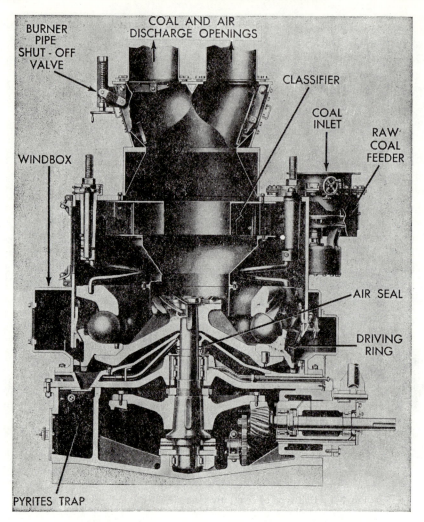

FIG. 7–4. Ball-race type of coal pulverizer. (Babcock & Wilcox Co.)

that is attached to the revolving race. Centrifugal action carries the coal into the path of the balls. Primary air picks up the ground coal and carries it into the classifier where the larger particles are separated from the mixture by inertia and then returned to the pulverizing zone. The flat blades in the classifier are arranged to provide openings counter to rotation of the air leaving the pulverizing zone.

Tramp iron and pyrites are detrimental to this type of mill and are collected in the pyrites trap and removed intermittently.

The ball-race mills may be operated under vacuum by placing an exhauster at the outlet or, more commonly, operated under pressure by using a primary air fan on the inlet. A primary air fan may require less power than an exhauster and will have less erosion on the fan blades. However, the exhauster blades are flat and can be lined with inexpensive checkered floor plate. If there is any leakage at the pulverizer, a unit under vacuum will have air leakage into the unit rather than coal dust leakage out of it.

Ring-roll (bowl) mills, Fig. 7–5, are also of medium speed. The rolls are spring-loaded and are inside the bowl. However, they are adjusted so that they do not touch the bowl. The peripheral speed of the bowl is about 1200 fpm, which corresponds to about 200 rpm for the small mills, to 70 rpm for the larger ones. The rolls rotate about their axes as the bowl revolves, but the roll assemblies are stationary. As the rate of coal grinding increases, the rolls are forced farther away from the bowl, and the force of the springs increases.

Tramp iron and pyrites collect in the bowl and are carried to a tramp-iron spout.

Fig. 7–5. Bowl mill. (Combustion Engineering Co.)

Primary air conveys the fines from the pulverizing chamber to the classifier where the angularity of the adjustable vanes creates a vortex. Coarse particles return to the bowl. The position of the vanes determines the fineness of the coal leaving.

Coal fines leave the mill with the primary air at the top by the exhauster. One motor drives both the mill and the exhauster, but a separate motor is provided for the feeder. Since an exhauster is used, the mill operates under vacuum.

An example of pulverization by impact and also by attrition is shown in Fig. 7-6. This is a double mill, i.e., coal enters at each end and leaves

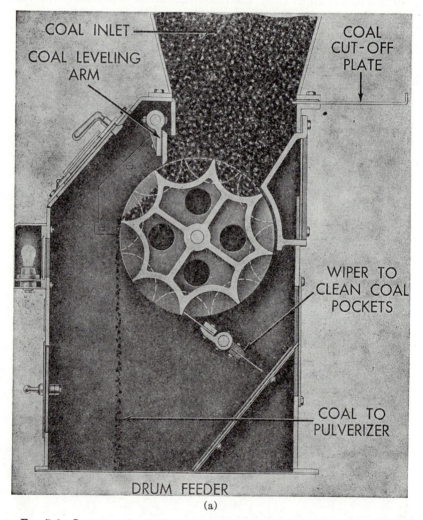

(a)

FIG. 7-6. Impact and attrition type of pulverizer with sections through the

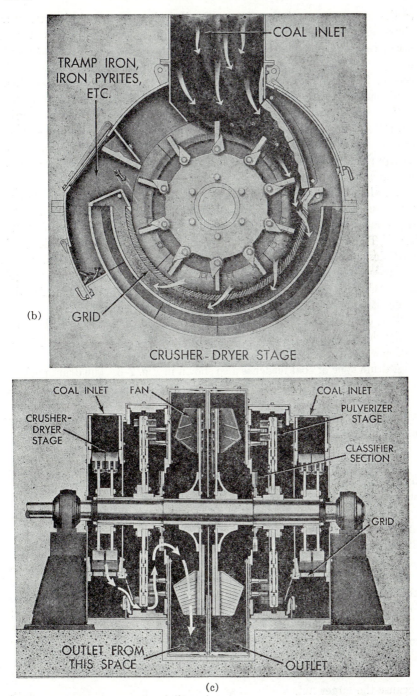

feeder mechanism (a) and the crusher-dryer element (b, c). (Riley Stoker Corp.)

at the center; the pulverizer is symmetrical about the vertical centerline. Considering the right-hand portion, coal and primary air enter the crusher-dryer section at the top. In this section, the coal is crushed so that 95% passes through an 8-mesh screen and 40% passes through a 50-mesh screen. The evaporation of surface moisture takes place in this section. Since the evaporation occurs before pulverization, the performance of this mill is not affected by coal moisture. Crushed coal passes through a grid with ¼-in. openings and into the pulverizing section. The cross-section of the crusher section shows the grid, the means of disposing of tramp iron and pyrites, and the impact crusher hammers. The hammers are faced with a chilled chrome-nickel alloy casting.

Final pulverization of the coal is accomplished by attrition between tungsten-carbide-faced moving and stationary pegs. Upon leaving the pulverizer section, coal must pass through the small rotating rejector arms. These arms impart sufficient centrifugal force to the heavier coal particles so that they are thrown back into the pulverizing section. Fines can pass through and into the fan section where the primary air and fuel are discharged to the burners. These mills operate at 900 rpm for the large sizes and up to 1800 rpm for the small sizes.

It is claimed that the coal in these pulverizers cannot explode because coal velocity through the unit is faster than the speed of flame propagation.

A drum type of coal feeder is also shown in Fig. 7-6. The drum pockets are offset on the two halves to provide more even coal supply. A leveling arm provides uniform measurement of the coal, and wiper blades clean sticky coal from the pockets. The feeder is driven by a constant-speed motor and a variable-speed device to control the flow of coal to the boiler.

7-2. Stokers. The stoker manufacturers have classified stokers as overfeed and underfeed. Chain-grate, traveling-grate, and spreader stokers are types of overfeed stoker because in all three cases the coal is fed onto the grates above the point of air admission. Underfeed stokers are those in which the coal is admitted to the stoker below the point of air admission.

Spreader stokers (Fig. 7-7) are popular for steam generator capacities up to 150,000 lb per hr of steam but may be obtained for units up to about 300,000 lb per hr. The reason for their popularity is their ability to burn any coal ranging from lignite to semi-anthracite and regardless of whether the coal is free-burning or coking. They are especially suitable for lower grades of coals having low-ash fusion temperatures and high ash content. No other stoker can accommodate such a wide range of coal.

The coal size may be up to 1½ in., although ¾-in. or less is preferred because the larger lumps will do much of their burning on the grate with

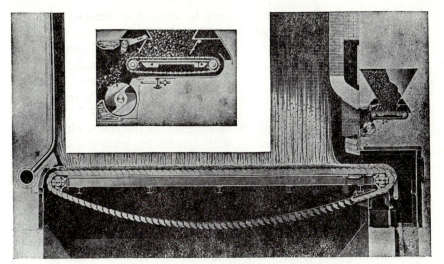

Fig. 7-7. Spreader stoker. (American Engineerng Co.)

an accompanying high loss due to unburned carbon. Minus $\frac{1}{16}$-in. coal will burn satisfactorily but will have a high cinder carry-over to the stack.

Fuel feed rates from the hopper into the unit are controlled by the feeder. Distributor blades pick up the coal as it falls from the small endless chain and literally throw the coal into the furnace. The distributor blades are shaped to distribute the coal evenly over the grate area. From 10 to 45% of the coal is burned in suspension, depending on the rank of the coal, its size, and the burning rate. Most of the volatile matter is distilled off before the coal reaches the grate.

Combustion air is supplied by a forced-draft fan through the ash pit and the holes in the grate. The air, plus the ash bed on the grate, tends to keep the grates cool and maintenance low. Air at 300 F to 350 F may be used. With these temperatures the grate will be about 50 F higher than the air temperature. The ash pit is laterally sectionalized to form plenum chambers under the grate, one chamber for each feeder. Each air zone is dampered to control the air supply for each feeder section.

Unburned fuel and ash are deposited on a stationary or moving grate. Stationary grates may be of the type that requires hand cleaning of the ashes by manually raking them out through doors in the front of the boiler. This style of grate is the most inexpensive but is used only on small installations. Spreader stokers may be equipped with hand dumping grates or with power-operated dumping grates using steam or air as the power medium. Stationary-grate spreader stokers are usually arranged in two or more zones with a feeder for each zone. When the grates need cleaning (every two hours for a high-ash coal and less frequently for a low-ash coal), one section may be cleaned or dumped while the remaining sections are in operation. After cleaning a section, it is necessary only to start the feeder, since the fuel will be ignited automatically by the adjacent section that has been in operation. Usually the grate should not be allowed to accumulate more than a 5-in. layer of ash.

For steam generators of at least 75,000 lb of steam per hr or more, the grate should be of the traveling type, as shown in Fig. 7-7. The grate usually travels from the rear of the furnace to the front, although the reverse direction has been used. With a traveling grate the ash is continuously dumped into the ash pit. These grates are similar to the traveling-grate stokers to be discussed later. Spreader stokers with traveling grates will have a furnace depth of at least 12 ft and may have depths up to 16 ft or more.

Combustion rates of 30 to 40 lb of coal per sq ft of grate per hr are common and as much as 70 lb has been attained. A more satisfactory measure of performance is for a normal heat release of 450,000 Btu per sq ft per hr on dumping grates, up to 900,000 Btu per sq ft per hr on traveling-grate spreaders. However, the higher figure may cause excessive maintenance. Therefore a maximum of 600,000 to 650,000 Btu per sq ft per hr is recommended.

Spreader stokers usually require 25% to 40% excess air.

Because much of the fuel is burned in suspension, there is more carry-over of ash and unburned coal with a spreader stoker than with any other kind of stoker. Therefore, a suitable dust-collecting system is mandatory if high fly-ash emission from the stack is objectionable. Since the fly ash may contain 50% or more unburned carbon, the fly ash with its unburned

FIG. 7-8.   Construction of a chain-grate stoker. (Babcock & Wilcox Co.)

carbon is frequently returned to the furnace to be reburned. This places an extra load on the dust-collecting system but can reduce the heat loss from unburned carbon from as high as 12% to as low as 2 to 3%.

Chain-grate stokers consist of staggered links mounted on bars, Fig. 7-8. The result is a continuous chain that is placed over sprocket wheels at each end of the stoker, Fig. 7-9, and that receives coal from the hopper

Fig. 7-9. Chain-grate stoker installation. (Babcock & Wilcox Co.)

at the front of the boiler. The thickness of the fuel bed, 3 to 8 in., is controlled by the gate at the hopper outlet. Constant-speed a-c motors with variable-speed transmissions, variable-speed d-c motors, or steam turbines may be used to drive the grate at speeds ranging from 5 to 20 in. per min.

Primary air is supplied from below the stoker to each zone obtained from the use of intermediate vertical baffles. Free air space between the links amounts to about 10% of the grate area. The draft loss through the grate and fuel bed for the primary air is usually 1 to 2 in. Thus, forced-draft fans are usually necessary. Permissible air temperatures depend on the fuel but range from 250 F to an extreme of 450 F. While preheated air increases boiler efficiency and promotes ignition of the fuel, high air temperatures will cause the grate to overheat and warp and may cause the coal to cake.

Since there is practically no agitation of the fuel bed, noncaking coals are best for chain-grate stokers. Likewise, the dust loading of the flue gas is very low. An appreciable ash content of the coal (say, 7% or more) is necessary to assure enough ash on the grate to protect it from overheating. If the fuel is bituminous coal, it should be less than 1 in. in size, with most of the coal passing through a ¼-in. screen.

Combustion rates for chain grate stokers should not exceed about 500,000 Btu per sq ft per hr.

Arch design over the stoker is dependent on the type of fuel to be burned. High-moisture coals are slow to ignite and therefore need arches close to the grate to keep the fuel temperature high. The arches may be farther away from the grate and shortened when high-volatile coals are

FIG. 7-10. Details of a traveling-grate stoker. (Riley Stoker Corp.)

used. A front arch reflects radiant energy onto the fresh green coal to promote rapid ignition. Special rear arches are required when fine coal, such as anthracite fines, is burned.

Secondary or *overfire air* is frequently beneficial to promote good combustion of bituminous coal. Proper combustion of any fuel requires *time, turbulence,* and *temperature,* sometimes called the three "T's" of combustion. Introduction of secondary air at the proper locations and in the amount of some 10% of the total air with 6- to 10-in. water pressure increases turbulence and reduces soot in the flue gas.

Traveling-grate stokers, Fig. 7-10, use cast-iron blocks, called *grate clips,* which are supported on rock angles extending across the furnace. These, in turn, are driven by longitudinal chains. Because there is less relative movement between adjacent grate clips than between the links of the chain-grate stoker, the traveling-grate stoker is better able to burn anthracite fines and coke breeze. Otherwise, the discussion of the chain-grate stoker applies to the traveling-grate stoker.

Underfeed stokers introduce coal into the furnace below the point of air admission. Fuel leaving the hopper, Fig. 7-11, is pushed into the *retort*

Fig. 7-11. Longitudinal section of an underfeed stoker. (American Engineering Co.)

trough by the *ram.* After being heated and having the volatiles distilled by the energy penetrating downward from the combustion zone, the coal reaches the region of active combustion in the form of coke and ash. By the time the fuel has been pushed down onto the reciprocating ash-discharge plates at the back of the furnace, combustion has been completed. Agitation of the fuel is not only from the ram but also from the three pusher plates along the bottom of the retort.

A rear view of an underfeed stoker having seven retorts is shown in Fig. 7-12. Air is admitted into the combustion zone through the *tuyeres* (holes

Fig. 7–12.   Multiple-retort underfeed stoker.   (Riley Stoker Corp.)

or slots) at each side of the stoker and also through the tuyeres that form the sides of each retort. The retort sides containing the tuyeres are split in the vertical plane and reciprocate to aid in the movement of the coal. At the end of the retorts there are reciprocating grates that constitute an over-feed stoker, since air is admitted between the grates and coal is pushed over the grates. Rocker plates at the end of the stoker move horizontally and vertically to crush and eject the refuse.

Continuous agitation of the fuel bed by all the moving parts keeps the fuel bed porous and free from clinkers. Thus, this type of stoker is well adapted to burning the caking varieties of coal.

Underfeed stokers are built with only one retort for small boilers, but then the refuse is usually dumped sideways rather than from the end. Combustion rates up to 600,000 Btu per sq ft per hr may be used but they should be reduced in cases where the fuel has a low ash-softening temperature.

Combustion air should be zoned to permit air control at various parts of the grate when changes in the fuel bed occur. Air temperatures should not exceed 350 F because of maintenance costs and might well be kept in the range of 250 to 300 F. Overfire air is beneficial to underfeed stokers. Coal size may vary from 2 in. to slack.

7–3. Oil, Gas, and Coal Burners.   Fuel oil has found wide application in marine installations and has completely replaced coal for all naval ships. Oil is easier to handle, more flexible in operation than stokers, and requires much less auxiliary equipment. A number of pulverized-coal boilers are arranged for firing up to full load with either coal or oil. Natural gas is an even more desirable fuel than oil, from the standpoint of auxiliary equipment, since a meter, pressure regulator, and burner with piping are all that are needed; no tanks, pumps, or bins are required.

Highly volatile petroleum fuels are too expensive for normal operation; therefore, the heavy fuel oils must be used. A schematic diagram, Fig. 7–13, shows the auxiliary equipment included in a fuel-oil system. Because of the high viscosity of the fuel oil, steam-heating coils are installed throughout the oil tank to warm the oil and to facilitate pump-

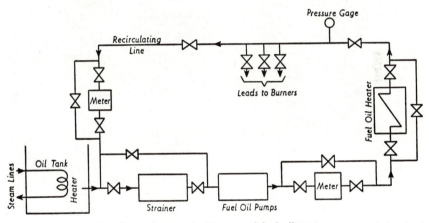

Fig. 7–13. Schematic diagram of fuel-oil system.

ing. Steam smothering lines, consisting of perforated steam pipe, may be located at the top of the tank. In the event of a fire in the tank, steam admitted to the lines will fill the tank and smother the fire.

Oil pumps, often of the reciprocating type, receive the fluid from the strainers and discharge it to the burners through heaters and a meter. Fuel-oil heaters differ little from the feedwater heaters and raise the oil temperature to some 150 F or more to promote combustion.

Recirculating lines are necessary to prevent stagnant oil from collecting in unused sections of pipe and cooling to the point where it becomes solid. Usually, a light oil is used for a half-hour or so when starting and stopping the boiler to wash the pipes clean of the heavy oil.

The major parts of an oil burner are the atomizer, air register, and necessary oil lines with strainers and valves. Atomization of the liquid promotes ignition and complete combustion and is accomplished mechanically or with steam or air. In the last two cases, a low-pressure jet of

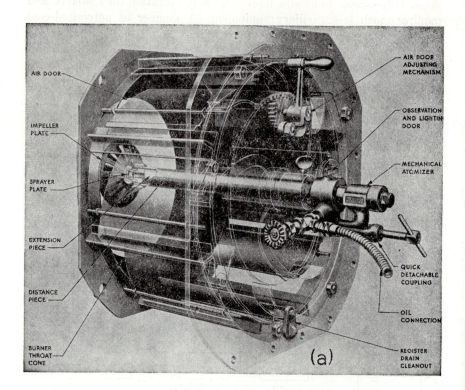

(a)

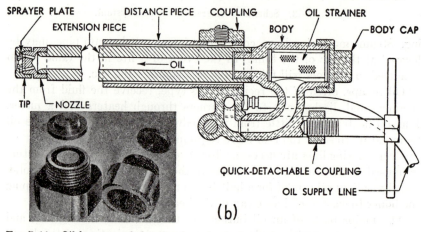

(b)

Fig. 7–14. Oil burner and details of mechanical atomizer. (Babcock & Wilcox Co.)

oil impinges on a jet of air or steam to break up the oil into minute particles and to enable better mixing with the air for combustion. Although steam atomization heats the oil, it is necessary to use 1 to $1\frac{1}{2}\%$ of the boiler output for this purpose. Air atomization is also undesirable for marine use because of the air compressors needed.

Mechanical atomization employs oil at high pressure, 100 to 300 psi, while air or steam atomization requires a lower oil pressure. In the mechanical unit the oil flows through an extension piece to the *nozzle* and *sprayer plate*, Fig. 7-14. Small passages in the sprayer plate guide the oil through a right-angle turn before it leaves through the orifice at the center of the plate. Changing direction creates a vortex that assists in developing a fine mist beyond the impeller plate. Air enters the burner through adjustable *air doors* and mixes with the oil beyond the burner throat cone. Air for combustion may have been preheated from 300 to 500 F before entering the burner.

Fig. 7-15. Circular type of pulverized-coal burners. (Babcock & Wilcox Co.)

Small load variations can be met by changing the oil pressure entering the burner, but pressures considerably lower than the design pressure will prevent good atomization. For extreme load variations, smaller atomizers are inserted in the burner while the boiler is in operation, or some burners are shut off. This latter practice may have a detrimental effect on turbulence in the furnace. Combustion air flow can be regulated by the air doors.

When gas is to be burned in a combination oil and gas burner, a perforated gas pipe is placed in the burner throat cone. Then either fuel may be burned separately or both fuels burned simultaneously.

A *circular type* of pulverized coal burner is shown in Fig. 7-15. The burners and windbox are shown as they would be installed on a boiler. Primary air and pulverized coal are fed to the burner from the pulverizer through the flanged pipe connections at a velocity of 3000 fpm to 5500 fpm. The high velocity prevents coal from settling out of the air stream. At the end of the coal tube is an impeller that deflects the primary air and coal into the secondary air stream to obtain good mixing. Inside the coal tube is the oil ignition torch. It is used at very low loads and for starting the unit. Larger oil burners may be used if the boiler is to operate with oil only at higher loads.

Circular burners may be obtained in capacities up to 125×10^6 Btu per hr each when burning pulverized coal. Frequently, these burners are

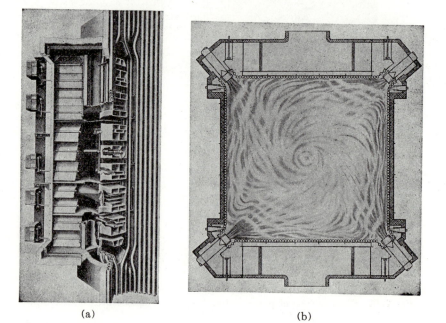

(a)   (b)

FIG. 7-16. A tangential burner. (Combustion Engineering Co.)

called *horizontal* burners because of their typical location in a furnace, as shown in the illustrations of Chap. 6. However, they have been used in vertical positions and have been used in both hopper-bottom and wet-bottom furnaces. Although the former application is more common, the lowest row of the burners may be aimed slightly downward to keep the ash in a molten state on the bottom of a wet-bottom furnace.

A burner arranged for *tangential firing* of pulverized coal, shown in Fig. 7–16, consists of an assembly containing dampers for the control of the secondary air, two coal nozzles, auxiliary air compartments, a lighting-off torch, and an ignition electrode to light the oil. Additional coal nozzles may be added. An assembly is installed in each of the four corners of the furnace and is aimed slightly off the furnace center. In this way, great turbulence is created in the furnace.

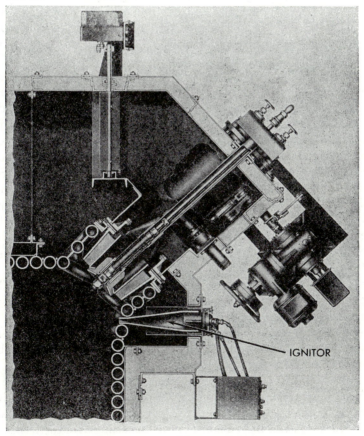

IGNITOR

(c)

Fig. 7–16 (cont'd.). Cutaway view of tangential burner. (Combustion Engineering Co.)

Note that the coal burner tips and the auxiliary nozzles are adjustable in the vertical direction. When operated by either electric motors or hydraulic cylinders, the burners may be adjusted to ±24 deg from horizontal. This corresponds to about ±12 ft at the center of a large furnace. When the flames are directed upward, the gas temperature leaving the furnace will be increased because less of the waterwalls are exposed to the flame. When the burners are directed downward, the waterwalls are more effective and the temperature of the gases leaving the furnace is reduced. The change in furnace outlet-gas temperature can be as much as 160 F for maximum burner adjustment. Observe that the heat transfer is due not only to radiation but also to convection as the turbulent gases sweep the waterwalls.

Since the temperature of the gases entering the superheater will directly affect the superheated steam temperature, tilting burners provide an excellent means of steam temperature control either with changes in load or with changes in slag accumulation on the tubes. For the 160 F change in furnace exit-gas temperature mentioned, the change in superheat temperature will be 80 F.

Tangential firing may be used for gas or oil as the fuel.

One of the most serious problems encountered in burning pulverized coal is the amount of fly ash carried through the boiler and induced-draft fans and into the stack. Costly fly-ash erosion of the induced-draft fans, the difficulty and cost of disposing of the fly ash, and the dust nuisance in the area surrounding the plant require the use of mechanical or electrostatic dust collectors between the air heaters and the fans. The ASME Code recommends a maximum fly-ash discharge of 0.85 lb of fly ash per 1000 lb of flue gas with 50% of excess air. This corresponds to approximately 1.06 lb per 1000 lb of flue gas at 20% excess air. About 80 to 90% of the ash in the coal is discharged out of the boiler with the flue gas of a hopper-bottom boiler. For a liquid slag-tap furnace, the figure is 40 to 60%.

Some efforts have been made to use fly ash as a by-product. Three applications that show some promise are (1) cinder pellets for use in cinder blocks, (2) replacement for cement in concrete (up to 20%), and (3) as a base for asphalt roads when mixed with dirt and lime. Some published figures for plants in metropolitan areas indicate a cost of about $1.00 per ton for disposing of the fly ash.

Cyclone furnaces, Fig. 7–17, were developed in an effort to reduce the fly-ash difficulties and to reduce the cost of maintenance of the pulverizing equipment by the elimination of the pulverizers. Hot primary air mixes with the coal (crushed to $\frac{1}{4}$ in. or less) as the coal leaves the feeder and enters the cyclone with a vortex motion at the center of the cyclone. Sec-

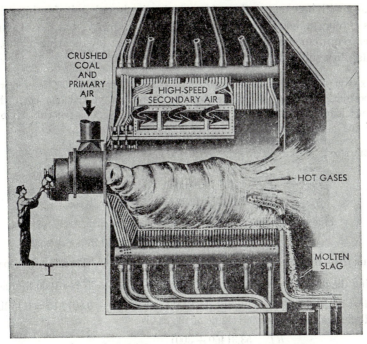

CRUSHED
COAL
AND
PRIMARY
AIR

HIGH-SPEED
SECONDARY AIR

HOT GASES

MOLTEN
SLAG

Fig. 7-17. Cyclone furnace. (Babcock & Wilcox Co.)

ondary air enters with a velocity of 350 fps on the side of the cyclone through the ports shown at the top and aids in the vortex motion. The velocity of the gases in the cyclone is about 200 mph, while their temperature is in the order of 3000 F. Hot gases carrying 10 to 20% of the ash in the coal leave through the center cone while molten ash flows to the side of the cyclone due to the centrifugal action and leaves through the opening below the cone. Molten slag is continuously drained off the main furnace floor (see Fig. 6-13) into the water-filled slag tank. Note the water tubes surrounding the cyclone. The cyclone may operate with about 12% excess air; the carbon loss with the ash should be less than 0.1%. Cyclones are from 5 to 9 ft in diameter with capacities of 85×10^6 to 400×10^6 Btu per hr. Most coals, particularly lignite through bituminous rank coal, can be used in a cyclone. Slag viscosities and ash-softening temperatures of less than 2500 F are important characteristics of a coal to be used in a cyclone. Oil and gas may be substituted for coal in the cyclone furnace.

While there is a saving in pulverizer power and primary fan power when using a cyclone furnace, these furnaces have an air-pressure loss of 20 to 40 in. of water. The net result is that auxiliary power for a cyclone furnace will be greater than for a pulverized-coal unit for bituminous coals having a good heating value and grindability, but less than the power requirements of pulverized-coal units for low-heating value, low grindability subbituminous and lignite coal.

EXAMPLE 7–3. If a dust collector having an efficiency of 80% is used with a wet-bottom pulverized-coal furnace burning Bibb County, Alabama, coal, what is the dust loading in pounds per 1000 lb of gas and in grains per cubic ft of the 300 F flue gas going up the chimney for 25% excess air?

SOLUTION. Using Eq. (5–8),

$$W_{ta} = 7.65 \times \frac{14,157}{10,000} = 10.83 \text{ lb air per lb coal}$$

$$W_{aa} = 1.25 \times 10.83 = 13.54 \text{ lb per lb coal}$$

Flue gas weight $= W_{aa} + 1$ lb coal $-$ ash in coal

$$= 13.54 + 1 - 0.0519 = 14.49 \text{ lb per lb coal}$$

Assume that 60% of the ash is carried over by the flue gas. Then the dust-loading entering the dust collector is $0.60 \times 0.0519 \times \frac{1000}{14.49} = 2.15$ lb per 1000 lb flue gas. The dust in the flue gas leaving the dust collector is $(1.0 - 0.80)2.15 = 0.43$ lb per 1000 lb flue gas. The specific volume of the gas is

$$v = \frac{RT}{P} = \frac{53.3(460 + 300)}{144 \times 14.7} = 19.14 \text{ cu ft per lb}$$

Since there are 7000 grains per lb, the dust-loading may be expressed as

$$\frac{7000 \times 0.43}{19.14 \times 1000} = 0.157 \text{ grains per cu ft of 300 F gas}$$

Draft Equipment

7–4. Chimneys. At times a distinction is made between *chimneys* and *stacks,* chimneys indicating brick or concrete construction and stacks designating steel construction. The two names will be considered synonymous in this text.

The type of chimney construction is a function of economics and plant arrangement. Concrete and brick chimneys must be located on the ground because of their weight. Steel stacks may be lined with a thin layer of sprayed-on concrete and may be located on the plant roof. Steel stacks are usually presumed to be the least expensive up to about 150 ft over-all height. From that height to about 250 ft, radial brick or precast concrete is most economical, while monolithic concrete is preferred for higher chimneys.

Large stacks, as used in the large, modern power plants are constructed to the nearest ½ ft in diameter.

An idea of the proportions and other details of construction of a reinforced concrete chimney may be had from Fig. 7–18. Locations of reinforcing bars are indicated. A lightning rod, not shown on the drawing, was included with the installation and the top 25 ft of the rod were lead-covered to reduce corrosion from acids formed by the flue-gas components. The chimney is tapered to promote stability.

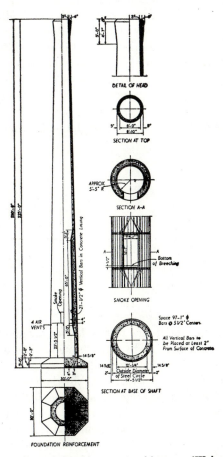

Fig. 7–18.   Details of a reinforced concrete chimney. (Weber Chimney Co.)

A combination of three factors will determine the height of a chimney. Obviously, the chimney must be high enough to provide the necessary draft, but the maximum height may be limited by aeronautical regulations if the plant is in the vicinity of an airport.

Perhaps one of the most difficult problems encountered in chimney design is that of smoke annoyance. So many factors affect the smoke path from the chimney that no definite rules can be given. Model testing has been found most useful in predicting actual conditions. Modern plants that are located in congested areas use electrostatic precipitators or other forms of dust-recovery systems. However, these systems do not completely eliminate the annoyance or remove SO_2 from the gas.

Surrounding buildings or other obstructions to the wind can cause turbulence around the stack that may result in objectionable downdrafts. In the same way, the direction of the wind, i.e., in line with a row of chimneys or across the row, may change the flow pattern. The downwash from high chimneys will generally strike the ground farther away from the plant than it will from low chimneys when other conditions are the same. Some tests have shown that high gas velocities leaving the chimney, about 60 fps, are most advantageous in dissipating the gases.

Observe in Fig. 7-19 that air enters the furnace through the grate and after supporting combustion passes over the boiler tubes to the chimney.

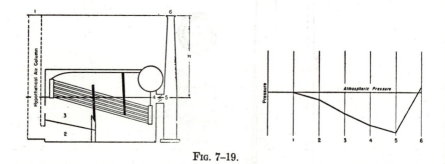

Fig. 7-19.

All these passages offer resistance to the flow of air and products of combustion that must be overcome by the chimney. The air enters the system at atmospheric pressure and immediately some of this pressure must be converted into velocity pressure or kinetic energy to permit flow. Losses are encountered in passing through the fuel bed and grate, the boiler tubes, dampers, and the duct work leading from the boiler to the chimney. Although the gases will be discharged into the atmosphere, they will have a velocity leaving the chimney. Therefore, the total gas pressure leaving the chimney will be higher than atmospheric pressure by the amount of the kinetic energy.

The pressure differential created by the chimney is due to the difference between the densities of the air column and the gas column in the chimney. Any draft created in the boiler between the air inlet and the flue gas outlet will be credited to the boiler and accounted for in the manufacturer's estimate of draft loss through the boiler. Therefore, in calculating chimney draft, it is customary to measure the height above the centerline of the gas entrance to the chimney. Since pressure, in pounds per square foot, is equal to the height of a fluid column times its density,

$$\Delta P = H(\rho_a - \rho_g) \qquad (a)$$

where ΔP = pressure differential created by the two columns, psf
H = height of the stack above centerline of gas inlet, ft
ρ_a and ρ_g = air and gas densities, lb per cu ft

Since the pressure differential is very small, it is customarily measured in *inches of water* rather than in pounds per square foot. Therefore,

$$\Delta P = \frac{p_t \times \rho_w}{12} \qquad (b)$$

in which p_t = theoretical pressure differential or draft, in. of water
ρ_w = density of water used for measuring, lb per cu ft

The abbreviation WG, meaning water gage, is frequently used to indicate pressure in inches of water.

The characteristic gas equation $(PV = WRT)$ gives us a convenient means of calculating the air and gas densities. We can make the assumptions that the gas constant (R) is the same for both air and gas and that both are under a pressure very nearly approximating atmospheric. Combining Eqs. (a) and (b), and using these assumptions, we get

$$p_t = \frac{12HP_a}{\rho_w R}\left(\frac{1}{T_a} - \frac{1}{T_g}\right) \qquad (7\text{-}3)$$

where P_a = atmospheric pressure, psf
T_a and T_g = average absolute air and gas temperatures, R

More convenient units may be used in Eq. (7-1), including the use of 62.4 as the density of the water in the draft gage if it is at 60 F. Then

$$p_t = 0.256BH\left(\frac{1}{T_a} - \frac{1}{T_g}\right) \qquad (7\text{-}4)$$

in which B is the barometric pressure in inches of Hg.

Since the gas inside the chimney is at a higher temperature than the surrounding atmosphere, the gas will raise the chimney temperature with a corresponding reduction in gas temperature. Also, any leaks in the duct work or in the chimney itself will allow infiltration of cool air and further reduce the gas temperature. Estimates of average gas temperature are doubtful at best, but some investigations * indicate that the mean temperature may be taken as shown in Eq. (7-5).

$$t_g = t_a + C(t'_g - t_a) \qquad (7\text{-}5)$$

where t_g = average gas temperature, F
t'_g = gas temperature leaving boiler, F
t_a = air temperature, F
C = empirical constant, ranging from 0.85 to 0.95 for chimney heights and diameters normally used for power plants

Note from Eq. (7-4) that the theoretical draft from a chimney is directly proportional to the barometric pressure and that the draft decreases as the atmospheric temperatures increase. Therefore, when designing a chimney, it is advisable to use the lowest re-occurring barometric pressure (actual barometer, not corrected to sea level) and the highest re-occurring atmospheric temperature encountered during plant operation. For purposes of stack design it is customary to assume that the barometric pressure decreases 0.1 in. Hg for each one hundred-foot rise in elevation. Barometric pressures corrected to sea level usually vary from about 29.5 to 30.2 in. Hg.

Stack outlet velocities will depend on the installation. When the chimney is the sole draft-producing equipment, outlet velocities must be in the range of 15 to 30 fps; otherwise, the friction losses and the velocity head at the chimney outlet will be high. If the installation includes induced-draft fans, the primary purpose of the stack is to disperse the gases over a huge area. Then the stack need do little more than overcome its own friction loss. Outlet velocities of 50 to 60 fps represent good practice and help disperse the gases. However, stack diameter is not only a function of velocity but is also determined by the stack structural stability. The gas-velocity pressure at the chimney outlet must be included in the system losses. If there are fans in the system, they will provide the velocity pressure loss for the system, but the chimney must supply this loss if the chimney provides all draft for the system.

Rearrangement of Eq. (2-26) into units that are more suitable for stack calculations produces

$$p_v = \frac{B\left(\dfrac{Q}{1000}\right)^2}{0.5602 D^4 T_g} \qquad (7\text{-}6)$$

or

$$p_v = \frac{T_g\left(\dfrac{W}{100,000}\right)^2}{0.3535 B D^4} \qquad (7\text{-}7)$$

* Cotton, *Mechanical Engineering*, September, 1923.

where p_v = velocity pressure, in. of water
  B = barometric pressure, in. Hg.
  Q = gas flow, cfm
  W = gas flow, lb per hr
  D = stack diameter, ft
  T_g = absolute gas temperature, R

Stack losses due to friction will be small (say, 5 to 10%) for a natural draft boiler where the stack velocity is low. However, the losses in a high-velocity stack would be much greater. Friction factors for determining the stack loss are quoted to be from 0.02 to 0.065. Construction materials and features will affect the loss. Unlined steel stacks should have a lower loss than concrete-lined steel, concrete, brick, or brick-lined chimneys. Conservative values of friction factors that have proven satisfactory are 0.065 for all stacks except unlined steel, where a value of 0.050 is recommended. Eq. (2–27) may be used with Eqs. (7–6) and (7–7) to obtain

$$p_f = \frac{fHp_v}{D} = \frac{fHB\left(\dfrac{Q}{1000}\right)^2}{0.5602D^5T_g} \tag{7–8}$$

and

$$p_f = \frac{fHT_g\left(\dfrac{W}{100{,}000}\right)^2}{0.3535BD^5} \tag{7–9}$$

where p_f = stack friction loss, in. of water
  f = dimensionless friction factor
  H = stack height above centerline of gas inlet, ft, and other symbols are as given for Eq. (7–7).

Eq. (5–8) is convenient and usually sufficiently accurate for stack and fan calculations.

EXAMPLE 7–4. A concrete-lined steel stack is to be 90 ft high (above the centerline of the gas inlet) and is to handle 250,000 cfm of flue gas at a stack inlet temperature of 300 F. The velocity should be 53 fps. Assume a 28.5 in. Hg. barometer and 90 F air, and find the theoretical draft, the friction loss, and the velocity pressure loss at the stack outlet.

SOLUTION. From Eq. (7–5) for a constant of 0.90,

$$t_g = 90 + 0.90(300 - 90) = \text{about 280 F}$$

From Eq. (7–4),

$$p_t = 0.256 \times 28.5 \times 90 \left(\frac{1}{550} - \frac{1}{740}\right) = 0.307 \text{ in. WG}$$

$$D = \sqrt{\frac{Q}{(\pi/4) \times \text{velocity}}} = \sqrt{\frac{250{,}000}{(\pi/4) \times 60 \times 53}} = 10 \text{ ft}$$

From Eq. (7–6),

$$p_v = \frac{28.5\left(\dfrac{250.000}{1000}\right)^2}{0.5602(10)^4\,740} = 0.430 \text{ in. WG}$$

From Eq. (7–8),

$$p_f = \frac{0.065 \times 90 \times 0.430}{10} = 0.252 \text{ in. WG}$$

7–5. Mechanical Draft. From the foregoing it can be seen that stacks cannot be counted upon to produce sufficient draft to overcome the losses in a large steam generator. For very small units, a stack will produce the necessary draft if the stack velocity is low, the stack is high, and the gas temperature is high. For the location of *forced-* and *induced*-draft fans in the system, see Fig. 1–2 and the illustrations in Chap. 6.

A *forced-draft* fan is required to supply sufficient static pressure to overcome the resistance of some combination of the following equipment: (1) duct on fan inlet, (2) duct from fan to air heater, (3) air heater, (4) duct from air heater to windbox, (5) air-cooled furnace walls, (6) stack effect due to changes in elevation, and (7) the required windbox pressure to provide flow through the burners or stokers. The resistances for an *induced-draft* fan system consist of a combination of some of the following: (1) boiler, (2) superheater, (3) reheater, (4) economizer, (5) dampers, (6) ducts from the economizer to the air heater, (7) ducts from air heater to dust collector, (8) dust collector, (9) ducts or flues to the fan and from the fan to the stack, and (10) any stack effect in the system, exclusive of the chimney. Obviously, all these items will not occur in every system.

It is not necessary to have both forced- and induced-draft fans. In small installations it is not uncommon to provide a forced-draft fan to overcome the air resistance of a stoker or oil burner. But without superheaters, economizers, or air heaters, the loss through the remaining portions of the boiler and flues may be within the limits of a chimney without any assistance from an induced-draft fan.

In the pressurized steam generators previously mentioned, a forced-draft blower that develops from 40 to 70 in. WG (water gage) overcomes all the resistances in the air and gas system from the blower to the stack entrance. In this case the furnace is under a pressure rather than under the customary 0.1 to 0.2 in. WG of suction. No induced-draft fans are used on a pressurized steam generator.

Fans for power-plant service may be driven by variable-speed motors, constant-speed motors, constant-speed motors with hydraulic couplings, steam turbines, and in rare instances, by d-c motors. Forced-draft fans, because of the lower air temperature, usually operate at a full-load synchronous speed of 900 or 1200 rpm, and occasionally small fans operate at 1800 rpm. Induced-draft fans operate at synchronous speeds ranging from 400 to 900 rpm.

There are many similarities between fans and pumps. Both may be classified as either *radial flow* or *axial flow*. Power-plant fans of the latter type are essentially the same as the common house fan, and the air flow through the impeller is substantially parallel to the shaft. They are used for small boiler installations as forced-draft fans and are located immediately adjacent to or in the boiler wall. They are suitable for such an application because they will develop only a small static pressure.

Radial-flow fans compress the gas by the action of the *rotor* revolving within the *scroll-shaped casing*. Basic vane or blade shapes for the rotors, as in the case of pumps, are *straight* or *radial, backward curved*, or *forward curved*. There are many variations of these basic shapes. The vector diagrams included in the chapter on pumps (Chap. 3) apply to fans as well as to pumps. Although the effect of blade shapes will be discussed further in connection with fan performance, it may be seen from the vector diagrams of Chap. 3 that the backward-curved blade fan would be a higher-speed fan that the straight-blade fan; and the straight-blade fan is in turn usually a higher-speed fan than the forward-curved blade fan.

The optimum number of blades on a rotor must be determined by experiment. Generally, however, too many blades increase the friction through the fan and also increase the noise frequency. Radial-bladed fans have from 6 to 12 blades, while backward-curved blade fans have from 16 to 24 blades. Forward-curved blade fans are frequently called *multiblade* fans because of the large number of blades, 32 to 66.

Backward-curved blade fans are not suitable for induced-draft service where the gas contains soot and fly ash because these particles will be thrown from the face of one blade to the back of the next blade. These particles will cling to the blade, thus causing unbalance. Forward-curved and radial-tip blades are recommended for induced-draft service.

A side view of a fan is shown in Fig. 7–20. The *scroll* or *volute* theoretically is shaped like a spiral curve, but in practice it is actually formed of several arcs of circles. The scroll is made of sheet steel. Steel, renewable *wearing plates* or *liners* may be located inside the back plate of the scroll to absorb the abrasion of fly ash when the fan is used for induced-draft service. Wearing plates are also used on the side sheets of the scroll for erosion protection.

In addition to collecting the air discharged from the rotor, the scroll may assist in the conversion of kinetic energy to pressure energy by gradually reducing the air velocity. In that case it is known as an *expanding scroll* and is frequently used on ventilating fans. *Nonexpanding scrolls* are designed on the assumption that the air leaves the rotor uni-

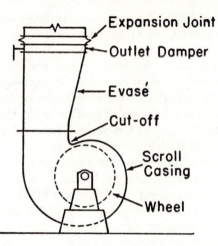

FIG. 7–20.   Outline of a fan and evasé.

formly, and therefore the scroll areas are computed to accommodate the increasing volume from the *cut-off* to the outlet.

Since the velocity of the air leaving the wheel is much higher than the economical duct velocity, either an expanding scroll or an expanding nozzle outlet, or both, is required for efficient operation of the fan. An *evasé* on the fan outlet will assist in efficiently converting the high outlet velocity of the fan into static pressure.

The distance from the wheel periphery to the cut-off is usually from 5 to 10% of the wheel diameter. A closer cut-off would be more efficient but also would have a higher noise intensity.

The fan shown in Fig. 7–20 is *full housed* and *full set*. The latter term indicates that the fan is completely above the floor. When the fan is sunk into the floor, it is seven-eighths set or three-quarters set. This has the effect of conserving headroom and lowering the shaft. Thus, the bearing pedestals and the motor would be lower and more accessible. When seven-eighths or three-quarters set, the concrete floor may be used to form part of the scroll. Since this is not good practice for power-plant service because of leakage, it is recommended that the fans be full-housed.

A view of a fan with a portion of the housing raised is shown in Fig. 7–21. Air or gas enters the fan at the bottom on both sides, *double inlet,* through *inlet boxes.* These inlet boxes are used on fans that receive the air or gas from a duct and provide the means of connecting the fan to the duct. Constant-speed fans with inlet dampers that handle flue gas containing fly ash should have renewable liners in the inlet boxes.

The erosion of inlet boxes is greatly reduced when a variable-speed control is used, in which case inlet box liners are not necessary. A double-

Fɪɢ. 7–21.　Forward-curved blade, induced-draft fan. (American Blower Corp.)

inlet fan usually is approximately *double width* (1½ to 2). A double-width fan has twice the capacity of a single-width fan.

Smaller power-plant fans may be *single inlet* rather than double inlet. There will be some unbalance in a single-inlet fan as there is for a single-inlet pump impeller; however, much of the unbalance may be eliminated by special vanes on the back side of the *back plate*.

From the inlet boxes the air or gas flows through the *inlet cones* into the rotor. The inlet cones provide a smooth path for the air and are an efficient means of increasing the air velocity.

When handling erosive gases, the *center plate* of a double-inlet rotor is frequently undercut at the face of the blade because erosion is excessive at this point. Also, the blades have wearing pads either welded or riveted to either the face or the back of the blade for protection.

Fixed inlet-guide vanes are frequently employed at the entrance to the inlet cones of backward-curved blade fans. These vanes are shaped so that the air will be turned in the direction of fan rotation. The resultant forced vortex reduces the shock losses of the gas entering the wheel, with a corresponding improvement in fan efficiency. This same principle of a forced vortex is used as one means of fan control, and of variable inlet-vane control, which will be discussed later.

7–6. Fan Performance. Air horsepower for a fan may be determined from either the product of pounds per minute times developed pressure in feet or from the cubic feet per minute times developed pressure in pounds per square foot. While the first combination is most satisfactory for pumps, the second combination is the more suitable for fans. Converting WG to psf,

$$\text{Developed pressure in psf} = \frac{p \times \rho_w}{12}$$

where p = developed pressure, in. of water
  ρ_w = density of the measuring fluid, lb per cu ft

Using 62.4 as the density of water in the gage,

$$\text{Air hp} = \frac{\text{cfm} \times p \times 62.4}{12 \times 33,000} = \frac{\text{cfm} \times p}{6350} \qquad (7\text{–}10)$$

If E is used to indicate fan efficiency, the brake horsepower input is

$$\text{Shaft hp} = \frac{\text{air hp}}{E} \qquad (7\text{–}11)$$

Note that the fan flow is the cubic feet per minute of gas measured at fan outlet conditions.

As previously stated, the gas leaving the fan wheel has had both its velocity pressure and static pressure increased over those at inlet conditions. The velocity of the air leaving the wheel is far higher than that permissible in the ducts. Therefore, a conversion from kinetic energy to pressure energy is indicated. For this reason, and because the resistances to be overcome in the duct system are static pressure resistances, it is customary to specify the fan developed pressure as *static pressure* and the fan efficiency as *static efficiency*. Air horsepower calculated from static pressure is known as the *static air horsepower*; when based on total pressure it is called the *power output*. Similarly, *static efficiency* is associated with static air horsepower, and *mechanical efficiency* is associated with power output. Eqs. (7–10) and (7–11) become:

$$\text{Static air hp} = \frac{\text{cfm} \times p_s}{6350} \qquad (7\text{–}10a)$$

$$\text{Power output} = \frac{\text{cfm} \times p_t}{6350} \qquad (7\text{–}10b)$$

$$\text{Static efficiency} = E_s = \frac{\text{static air hp}}{\text{shaft hp}} \qquad (7\text{–}11a)$$

$$\text{Mechanical efficiency} = E_t = \frac{\text{power output}}{\text{shaft hp}} \qquad (7\text{–}11b)$$

$$p_t = p_s + p_v \qquad (7\text{–}12)$$

Observe that the ratio of static and mechanical efficiencies is the same as the ratio of static and total pressures.

EXAMPLE 7–5. A fan develops 5.0-in. static pressure and 0.7-in. velocity pressure when the flow is 8000 cfm of 100 F air, the shaft hp is 7.6, and the speed is 1800 rpm. Calculate the static and mechanical efficiencies.

SOLUTION. From Eq. (7–10a),

$$\text{Static air hp} = \frac{8000 \times 5.0}{6350} = 6.3$$

From Eq. (7–11a),　　　　$E_s = \dfrac{6.3}{7.6} = 82.9\%$

From Eq. (7–12),　　　　$p_t = 5.0 + 0.7 = 5.7$ in.

From Eq. (7–10b),　　Power output $= \dfrac{8000 \times 5.7}{6350} = 7.18$

From Eq. (7–11b),　　　　$E_t = \dfrac{7.18}{7.6} = 94.5\%$

It might seem from Eq. (7–10) that the gas temperature or pressure would have no effect on fan performance. Such is not the case. *Any centrifugal pump or fan displaces a volume of the fluid against the feet-head of the fluid being compressed.* Let us consider two columns of air: both have the same height and are composed of the same fluid, but column A has a higher temperature than column B. The feet-head of both columns is the same, but the unit pressure at the base of the high-temperature column A would be less than the pressure at the base of the low-temperature column B. Therefore, the inches of water pressure for column A, as used in Eq. (7–10), would be less than the pressure for column B. In a similar manner, variations in the gas pressure would affect the inches of water pressure. We may say from this discussion that *when the fluid temperature for a fan is changed, the developed static pressure and the horsepower vary inversely as the absolute fluid temperature and directly as the absolute fluid pressure; or the developed pressure and the horsepower vary directly as the fluid density.* This statement assumes that efficiency remains constant—an assumption that is correct for practical purposes. Remember that air density varies directly as the absolute pressure and inversely as the absolute temperature. The characteristic gas equation, $PV = WRT$ (using $R = 53.3$ for air or flue gas), is applicable to most power-plant fan calculations. Thus,

$$p_f = p_i \left(\frac{P_f T_i}{P_i T_f} \right) \tag{7–13}$$

$$\text{hp}_f = \text{hp}_i \left(\frac{P_f T_i}{P_i T_f} \right) \tag{7–14}$$

where P is air pressure, psia, T is air absolute temperature, R, and the subscript i indicates the condition before correction, and the subscript f indicates the condition after correction.

Variable-speed fan calculations are identical to variable-speed pump calculations and are, for *constant efficiency* and *constant temperature*,

$$\frac{\text{rpm}_1}{\text{rpm}_2} = \frac{Q_1}{Q_2} = \sqrt{\frac{p_1}{p_2}} = \sqrt[3]{\frac{\text{hp}_1}{\text{hp}_2}} \tag{7-15}$$

Values based on either static or total pressure may be used in Eq. (7-15), provided the same base is used for both points 1 and 2.

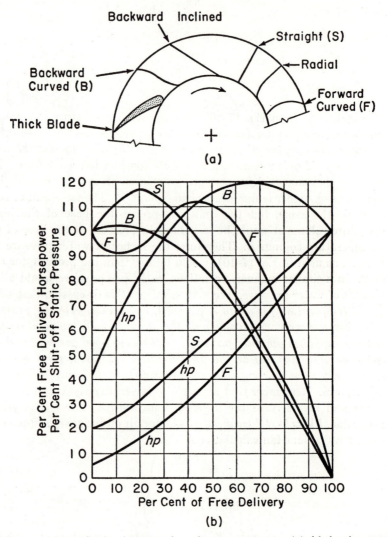

FIG. 7-22. Fan blade shapes and performance curves: (a) blade shapes; (b)

EXAMPLE 7-6. A manufacturer rates his fan at 4.43 in. of water static pressure for 6000 cfm of 70 F air, 1500 rpm, 29.92 in. Hg barometric pressure, 71% static efficiency. What will be the flow, static pressure, and shaft hp at 1800 rpm, 200 F air, and 26.0 in. Hg barometric pressure?

SOLUTION. From Eq. (7-15), for 70 F,

$$Q_2 = 6000 \left(\frac{1800}{1500}\right) = 7200 \text{ cfm}$$

$$p_2 = 4.43 \left(\frac{1800}{1500}\right)^2 = 6.38 \text{ in.}$$

Correcting for density will not affect the volume flow but will affect the developed pressure. For 200 F air and 26.0 in. Hg barometer, use Eq. (7-13).

$$\text{Corrected } p_2 = 6.38 \left(\frac{26.0 \times 530}{29.92 \times 660}\right) = 4.45 \text{ in.}$$

From Eqs. (7-10a) and (7-11a),

$$\text{Shaft hp} = \frac{7200 \times 4.45}{0.71 \times 6350} = 7.11$$

As previously mentioned, there are three basic fan-blade shapes: forward-curved, straight, and backward-curved. In addition, there are many variations of these basic shapes, Fig. 7-22; i.e., radial blade with

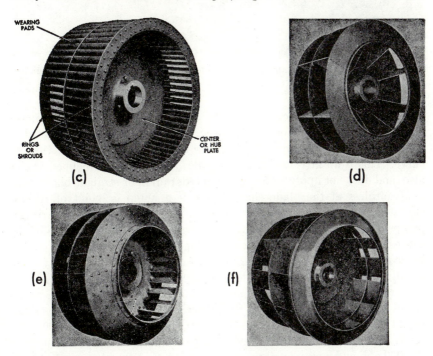

characteristic performance curves. (c to f. American Blower Corp.)

curved inlet, backward-inclined, the streamlined backward-curved blade (sometimes called a *thick blade*), and a large number of variations of the S-shaped, backward-curved blade. Although first introduced many years ago, the thick, streamlined blade, Fig. 7–23, has been favored in many recent installations for forced-draft service because its peak static efficiency is about five percentage points better than the other forms of backward blades. Because of possible difficulties from fly-ash erosion and from unbalance due to fly ash, the thick blade has not been applied to induced-draft service.

Typical performance curves for the three basic blade shapes are given in Fig. 7–22. The design point or rated capacity for these fans would be at 50 to 60% of the free delivery capacity. Observe that the backward-curved blade has a non-overloading horsepower characteristic; i.e., its

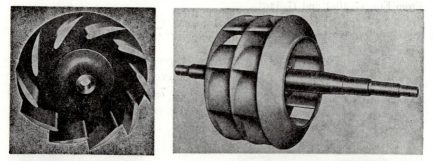

FIG. 7–23. Thick or streamlined fan blade. (Sturtevant Division, Westinghouse Electric Corp.)

maximum horsepower occurs near the design point. Therefore, a motor selected for the power required at design point could not be overloaded to a dangerously high power at any other operating condition on the fan curve. This is an attribute of thick blades and backward-inclined blades also, and does not apply to forward-curved, radial, or straight blades.

Regardless of the blade shape, there is a hump or point of reverse curvature in the static pressure curve. As in the case of centrifugal pumps, two fans operating in parallel at capacities near the hump of the curve would be unstable, as shown in Fig. 7–24 for two forward-curved blade fans. Curve *AC* of Fig. 7–24 represents the curve for the two fans in parallel with the capacity divided equally between the two fans. However, curve *ABC* represents the performance for the two fans with a different flow for each fan; i.e., at 12-in. static pressure the flow for one fan would be 450,000 cfm, while the flow for the other would be 165,000 cfm for a total flow of 515,000 cfm. Therefore, at a total system flow of 515,000 cfm and a system static pressure of 12 in., either fan could operate

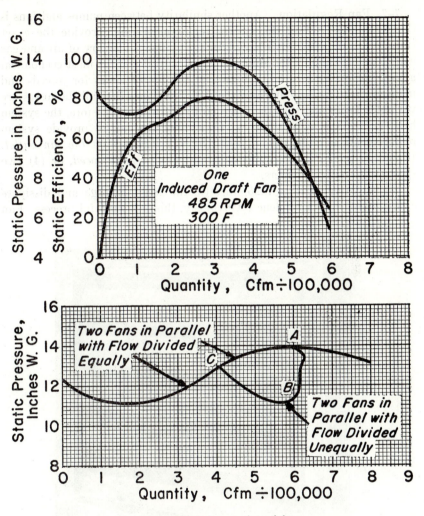

FIG. 7–24. Parallel operation of fans.

at either flow and still satisfy the system. The two fans may *oscillate*, or *hunt*, and be unstable. These fans could be unstable for any system head curve that would intersect curve *ABC*. This instability would exist for any pair of fans or pumps that do not have a continuously drooping head curve from shut-off to wide-open delivery. Any fans operating in parallel should be carefully selected so that all system head curves for the fans intersect the characteristic curve at a greater capacity than the capacity at maximum pressure (or to the right of any hump in the characteristic curve).

7-7. Fan Regulation. Another similarity between pumps and fans is that both must follow the system requirements and provide the exact static pressure needed by the system at any flow. The flow of air and gas will be determined by the amount of fuel burned and the most economical amount of excess air. Examination of the list of losses for forced- and induced-draft fan systems given in Art. 7-5 shows that all the losses are variables losses and that there are no fixed losses. Therefore, the system head curves are square curves. Control of the fan to provide system requirements may be accomplished by (1) *dampers on the fan outlet,* (2) either *inlet dampers* or *inlet vanes,* (3) *variable speed,* or (4) two speeds with outlet dampers.

Outlet dampers, Fig. 7-25, throttle the fan discharge and dissipate excess pressure by turbulence. Obviously, this is a wasteful process. Fan

Fig. 7-25. Outlet fan dampers. (American Blower Corp.)

performance curves, such as those of Figs. 7-22 and 7-24 are determined by a standardized test procedure in which the discharge is throttled.

When dampers are applied to a fan at the inlet to the inlet boxes, Fig. 7-26, the air or gas will have a spinning action in the form of a helix as it enters the wheel. If the direction of spin is in the direction of rota-

Fig. 7-26. Inlet dampers for fan regulation. (Sturtevant Division, Westinghouse Electric Corp.)

tion of the wheel, the high tangential velocity of the gas at the wheel inlet lessens the fan developed pressure.*

Inlet vanes, Fig. 7-27, have the same action as inlet dampers but are somewhat more efficient because they are located at the cone inlet. For power-plant applications, inlet vanes are limited to forced-draft fans where the fluid is clean. The fly ash associated with flue gas causes much more erosion on inlet vanes than on inlet dampers.

Power reduction by the use of inlet dampers and inlet vanes is indicated in Fig. 7-28. Applications of these curves will be demonstrated in an example.

Variable-speed control is most efficient, from the standpoint of power input, at low flows. However, the cost of equipment is much greater than for other methods of regulation. The families of curves obtained for variable-speed fans are the same as those for pumps. Reference should be made to Chap. 3. Hydraulic couplings have from 2 to 7% losses at

---

* Developed pressure for a fan is a function, among other things, of the difference in fluid tangential velocity at the wheel outlet and inlet. Increasing the inlet tangential velocity decreases the tangential velocity difference and therefore decreases the developed pressure.

FIG. 7–27.   Inlet vanes. (Sturtevant Division, Westinghouse Electric Corp.)

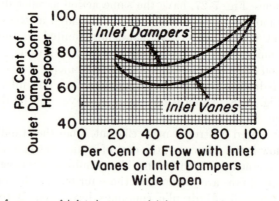

FIG. 7–28.   Performance of inlet damper and inlet vanes (data based on pressure varying as the square of the flow).

full speed. Hydraulic coupling efficiencies may be approximated by Eq. (3–14).

Outlet dampers are usually installed on forced- and induced-draft fans even though some other form of control may be used. However, the outlet dampers are not connected to the automatic control equipment but are positioned by the operator. There are several reasons why the outlet dampers are installed with other forms of control. If there are two or more fans in parallel, outlet dampers permit the isolation of some of the fans while the remainder are operating during low loads on the boiler. Even with only one forced-draft fan and one induced-draft fan, there is a very objectionable draft created when the boiler is not in service. This draft makes maintenance work on the fans more difficult. If a fan handling hot air or gas can be isolated from its system, it can be cooled much more quickly and the maintenance or outage time reduced.

One of the most important reasons for outlet dampers is to provide more positive control at very low loads. Hydraulic couplings become difficult to control accurately at speeds below 20 or 25% of full-load speed. Inlet dampers or vanes do not give satisfactory control below 20 or 25% of design flow because of the leakage between closed blades and because a very slight movement of the blades means a large percentage change in flow. Thus, at low boiler loads, partial closing of the outlet dampers provides better regulation of the air or gas flow.

The fourth method of fan control is a combination of outlet dampers and either a multispeed motor or two motors direct-connected to the fan shaft, with each motor designed for a different speed. Recent installations seem to favor the use of two direct-connected motors as being more economical than a multispeed motor. Selection of the proper speed for the lower speed motor is a matter of economic analysis giving due consideration to the hours of operation at various boiler loads. However, a very common practice is for the lower speed to be approximately 80% of the full-load speed. Since the horsepower varies as the cube of the speed, this would mean that the lower-speed motor would have one-half the horsepower of the higher-speed motor.

A comparison of the power required by these several types of control is given in Fig. 7–29. At 100% volume, the hydraulic coupling requires more power than the other methods because of coupling losses. The comparison of inlet vanes and inlet dampers on the same fan application is hypothetical because, as previously explained, inlet vanes are used on forced-draft fans and inlet dampers on induced-draft fans.

Refined calculations for induced-draft fans will include the effect of decreasing flue-gas temperatures with decreasing loads. However, such refinement is not necessary for many calculations.

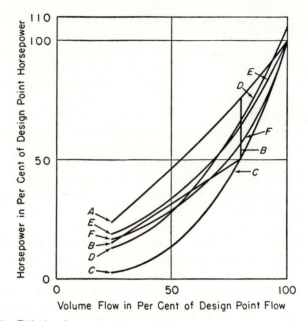

FIG. 7–29. Relative fan power for (a) damper control, (b) two motors, (c) variable speed without coupling losses, (d) hydraulic coupling input, (e) inlet dampers, (f) inlet vanes.

EXAMPLE 7–7. For the induced-draft fan data of Fig. 7–24, calculate the annual savings for 4000 hr per yr of operation at 50,000 kw (11,200 Btu per kwhr heat rate) by using inlet dampers rather than outlet damper control. Take the gas flow as 190,000 cfm, the fan design point as 410,000 cfm, the fuel cost as 25¢ per 10^6 Btu, and the motor efficiency as 90%.

SOLUTION. From Fig. 7–24, at 190,000 cfm the static pressure is 12.6 in. and the efficiency is 70%. From Eqs. (7–10a) and (7–11a), for outlet damper control,

$$\text{Shaft hp} = \frac{190{,}000 \times 12.6}{6350 \times 0.70} = 539$$

With inlet damper control, the fan will operate at $190{,}000/410{,}000 = 46.3\%$ of flow and the power, from Fig. 7–28, will be 72.5% of outlet damper power, or $0.725 \times 539 = 391$ shaft hp. The fan developed pressure will vary as the square of the flow from the design point; or $13.0(190{,}000/410{,}000)^2 = 2.8$ in.

$$\text{Annual saving} = \frac{(539 - 391)0.746 \times 4000 \times 11{,}200 \times 0.25}{10^6 \times 0.90} = \$1373$$

EXAMPLE 7–8. For the data of Example 7–7, find the annual savings for variable-speed operation of the fan by using a fluid coupling. Use 500 rpm as the motor speed.

SOLUTION. From the previous calculations, the static pressure at 190,000 cfm will be 2.8 in. WG.

Using Eq. (7–15), $\text{rpm} = 485 \left(\dfrac{190{,}000}{410{,}000} \right) = 225$

From Eqs. (7–10a) and (7–11a), and by using the static efficiency of 68% at design point (since it is on the same square curve as this operating point) from Fig. 7–24,

$$\text{Shaft hp} = \frac{190{,}000 \times 2.8}{6350 \times 0.68} = 123$$

From Eq. (3–14),

$$\text{Coupling efficiency} = \frac{225}{500} = 45\%$$

$$\text{Coupling input} = \text{motor output} = \frac{123}{0.45} = 274 \text{ hp}$$

Compared with outlet damper control,

$$\text{Annual saving} = \frac{(539 - 274) \times 0.746 \times 4000 \times 11{,}200 \times 0.25}{10^6 \times 0.90}$$
$$= \$2460$$

Feedwater Treatment

7–8. General.   Treatment of boiler feedwater is a most important and technical subject. Details of any treating process for a plant should be determined by a specialist in that field, since they may materially affect the plant design. Deposits of scale-forming water impurities on the boiler tubes can not only restrict the water flow but can raise the metal temperature with resultant tube rupture. We have found that there is a temperature difference across a tube wall and the stagnant films on each side of the wall. As boiler scale is a poor conductor of heat, the temperature drop across a tube wall with scale will be greater than the drop across the tube wall without scale. Since the temperature of boiler water is constant for a given pressure, the tube outside-surface temperature will be higher when scale is present on the inside of the tube. If the metal temperature is high enough, the tube will blister and eventually rupture. Even if the scale is not thick enough to cause failure of the tube, it will reduce the heat transfer and reduce the boiler efficiency.

Most scale-forming substances have one property in common, i.e., the solubility of the substance varies inversely as the temperature. Thus, the impurities in the boiler feedwater will not precipitate from the solution at the temperatures existing in the feed cycle. But when the solution enters the boiler, its temperature is raised, and the concentration is increased until precipitation may occur.

The mechanics of scale formation may be considered as two separate phases. An impurity may precipitate from the entire solution to deposit eventually on boiler surface and to bake into a scale.

However, in watertube boilers this precipitation is not the major source of trouble. We have already observed that there is always a laminar film adjacent to the confining wall regardless of whether the main body of the flow is laminar or turbulent. Also, this stagnant film is at a higher temperature than the rest of the fluid. Since the solubility of the scale-forming substance decreases with higher temperatures, and since there is a tendency for higher concentrations in the film, precipitation of the salts from the film may occur when the concentration of the total volume does not indicate precipitation. Most scale formations have been attributed to this film precipitation.

Scale-forming impurities may be roughly divided into two types. In the first group are impurities that are in suspension in the water and those that are emulsified or in colloidal form. This group includes mud, silt, clay, and silica (SiO_2), either in suspension or in colloidal form. *Turbidity*, lack of clearness in a water, is due to these impurities. Oil may be either emulsified or suspended. Mud may cause water starvation of the tubes or it may be occluded in boiler sludge and scale. Silica may form an extremely hard glasslike scale that is very difficult to remove. Oil may carbonize and form an asphaltic scale or may cause foaming. All these impurities have one characteristic in common in that they must be removed external to the boiler. The usual method of removal is by coagulation with iron sulfate, sodium aluminate, aluminum sulfate, or lime in a large basin, or by filtration in sand filters. Filtration by itself has the disadvantages of (1) the possibility of increasing the silica content of the water and (2) the removal of only the larger of the suspended solids. The capacity of sand filters is approximately 3 gpm of water per sq ft of bed area.

A second group of impurities comprises solids in a dissolved form such as calcium bicarbonate [$Ca(HCO_3)_2$], calcium carbonate [$CaCO_3$], calcium sulfate [$CaSO_4$], magnesium bicarbonate [$Mg(HCO_3)_2$], magnesium carbonate [$MgCO_3$], magnesium sulfate [$MgSO_4$], and sodium silicate [Na_2SiO_3]. Calcium salts are the most common cause of boiler scale, of which calcium is usually one of the chief components. While magnesium salts have important scale-forming characteristics, they produce a scale slightly softer than the calcium compounds. Silica, particularly troublesome in boiler waters, may be in the form of calcium and magnesium silicates or in complex structures in combination with aluminum ions. Silica scales are very hard. The treatment of silica should be accomplished outside the boiler (external treatment). Internal treatment consists of injecting chemicals directly into the boiler to precipitate scale-forming salts.

Oxides of copper and iron and metallic copper are frequently deposited in boilers. The iron pick-up is due to corrosion from dissolved oxygen and carbon dioxide in the feedwater, while the copper pick-up is a result of the corrosive action of oxygen, carbon dioxide, and ammonia.

The hardness of a water is expressed in units of *grains per gallon* or *ppm* (parts per million) in terms of calcium carbonate, $CaCO_3$, regardless of what salts of calcium or magnesium are in the water. Grains per gallon times 17.1 equals parts per million. Therefore, the hardness of a water is indicative of its scale-forming quality and is measured by the soap test. The cubic centimeters of a standard soap that are required under certain prescribed conditions to form a lather lasting for 3 min are equal to the grains per gallon of hardness. Gravimetric determinations are more accurate than the soap test.

A particularly convenient method of determining steam and water purity is by measurement of the electrical conductivity (conductivity is inversely proportional to electrical resistance). The steam is condensed before the measurement is made. The conductivity measurement, being an electrical measurement, is particularly adaptable to automatic recording and control. Pure water is extremely resistant to electrical current and therefore has a low conductivity. The presence of ionizable material in a water increases the conductivity of the water. Thus, the conductance of a water is a measure of the total ion concentration and a measure of the water purity. The unit of measurement is the micromho, which is one millionth of a mho and corresponds to 10^{-6} ohm. The exact relationship between micromhos and parts per million depends on the particular water and on the presence of ammonia, carbon dioxide, and the hydroxide ion. For feedwater, 1 micromho equals about 0.9 ppm, while for condensed steam, 1 micromho equals 0.5 to 0.6 ppm. Pure water has a conductance of about 0.04 micromho or a resistance of 25,000,000 ohms.

Conductivity meters do not measure the silica content of a water.

Hydrogen ion concentration in a water indicates the acidity or alkalinity of the water. There is a continual association and dissociation in water of HOH and H plus OH. The hydrogen ion concentration is expressed as the pH value, which is the logarithm of the reciprocal of the fraction of hydrogen ions present. Thus, a pH value of 7 means a neutral water at 60 F with a concentration of 10^{-7} hydrogen ion. A pH value of more than 7 indicates alkalinity, while less than 7 indicates acidity at 60 F. Pure distilled water is neutral and has pH values of approximately 6.1 at 212 F, 5.8 at 300 F, and 5.6 at 400 F. pH values determine the intensity of corrosion.

Caustic embrittlement is a form of intercrystalline corrosion that attacks the boiler-drum plate at the rivet holes and tube ends. Metal so attacked is weakened to the extent of failure, and many boilers have exploded from this cause. The process of caustic embrittlement is thought to take place when minute leaks permit concentration of the water impurities at points of high stress. Also, the water must have embrittling characteristics. Unfortunately, there is at present no chemical analysis that will indicate whether a water has embrittling characteristics. However, the presence of sodium hydroxide in the boiler water is known to create embrittling tendencies.

Current practice is to arrange a metal test specimen so that it can be stressed while in contact with concentrated boiler water. Cracking of the specimen indicates embrittling characteristics of the water.

Thorough calking of riveted joints and welding of drum seams has reduced caustic embrittlement in modern boilers. However, intercrystalline cracking can still be encountered where tubes are rolled into the drums of headers. Embrittlement inhibitors, such as trisodium phosphate, sodium nitrate, tannins, and others are used in the boiler water.

One method of reducing the impurities in the feedwater—the use of evaporators to produce a make-up water that is low in impurities—will be discussed in Chap. 8. However, treatment of the water supplied to the evaporators is frequently advisable to reduce scale formations in the evaporators. Because of feedwater contamination due to condenser leakage and because the boiler water is concentrated by the constant evaporation of the water in the boiler, internal treatment of water in the boiler is usually necessary when using evaporators as well as with other forms of make-up water treatment.

There are many treatments for boiler water and feedwater, each with a specific purpose and application. Only a very few of these can be described briefly.

7–9. Zeolite Treatment. Zeolite is either a natural or an artificial material that has the property of being able to exchange its base. In the natural form, zeolite is called *greensand* and is found very near the earth's surface in New Jersey and other places. The material is granular with a varying size which is often that of extremely coarse sand.

Zeolite treatment is one of the ion-exchange processes wherein cations * of two or more positive charges are exchanged for sodium or hydrogen.

Natural zeolites have a sodium base that can be exchanged for the calcium or magnesium in raw make-up water. For example, water containing $Ca(HCO_3)_2$ may be passed through a bed of natural zeolite with the following reaction:

$$Na_2Z + Ca(HCO_3)_2 \rightarrow CaZ + 2NaHCO_3 \qquad (7\text{–}16)$$

Greensand weighs about 90 lb per cu ft and that amount will exchange about 3000 grains of equivalent calcium carbonate hardness. The zeolite

---

* Cations are positively charged atoms of which H, K, Na, and NH_4 have one charge; Ca, Cu, Fe, Mg, and Mn have two charges; and Al and Fe have three charges.

will then have exchanged its sodium for calcium or magnesium. The zeolite must then be regenerated by taking it out of service and passing NaCl brine through the bed. A regeneration ritual requires about ¾ hr. Salt for regeneration weighs about 52 lb per cu ft, and roughly 0.5 lb is needed per 1000 grains of hardness.

Sodium zeolite systems have the advantages of being very easy to operate, require little maintenance, and produce an effluent of practically zero hardness. Their disadvantages are that there is no reduction in the alkalinity or total dissolved solids, there must be coagulation and filtration of the influent before entering the zeolite bed if the turbidity is more than 10 ppm, or if there is ferric iron present, there is a possibility of silica pick-up, and the effluent will contain dissolved corrosive gases such as oxygen and carbon dioxide. Ferric iron and turbid waters coat the zeolite grains and reduce their exchange abilities.

Carbonaceous zeolites may be made of lignite coal, or they may be sulfonated coal or coke (treated with sulfuric acid), or they may be synthetic resins. These zeolites may be operated on the sodium cycle like greensand, or they may be regenerated with sulfuric acid (H_2SO_4), producing a *hydrogen zeolite* cation exchange. The use of hydrogen zeolite provides a means of eliminating not only calcium and magnesium salts from the raw water but also sodium.

A reaction for acid-regenerated zeolite would be

$$H_2Z + Ca(HCO_3)_2 \rightarrow 2H_2CO_3 + CaZ \qquad (7\text{-}17)$$

H_2CO_3 will decompose into H_2O and CO_2.

Synthetic zeolites weigh about 50 lb per cu ft, a quantity which will have an exchange value of up to 9000 grains of hardness.

Advantages and disadvantages of the hydrogen system are the same as those of the sodium system except for the added disadvantage of handling the acid and except for the added advantage that the hydrogen system removes sodium and the bicarbonates become acids.

7–10. Lime-Soda Treatment.   This form of treatment may be carried out on either cold or hot raw water. The same chemicals are used in either case, but the hot process supposedly is more compact and provides better treatment. The water is, of course, at saturation temperature where reaction is many times faster than that for cold water.

Fig. 7–30 shows the equipment used in a hot lime-soda softener. At the extreme left is a phosphate mixing tank and next to it is the lime (calcium hydroxide) and soda ash (sodium carbonate) mixing tank and proportioner. The chemicals are fed to the contact heater where they are introduced into the hot water.

FIG. 7-30.   Hot-process lime-soda water softener. (Cochrane Corp.)

The following reactions with the lime, $Ca(OH)_2$, take place in the softener:

$$Ca(HCO_3)_2 + Ca(OH)_2 = 2CaCO_3 + 2H_2O \qquad (7\text{-}18)$$

$$Mg(HCO_3)_2 + 2Ca(OH)_2 = Mg(OH)_2 + 2CaCO_3 + 2H_2O \ (7\text{-}19)$$

$$MgSO_4 + Ca(OH)_2 = Mg(OH)_2 + CaSO_4 \qquad (7\text{-}20)$$

$$MgCl_2 + Ca(OH)_2 = Mg(OH)_2 + CaCl_2 \qquad (7\text{-}21)$$

For the soda ash, Na_2CO_3, the reactions are:

$$CaSO_4 + Na_2CO_3 = CaCO_3 + Na_2SO_4 \qquad (7\text{-}22)$$

$$CaCl_2 + Na_2CO_3 = CaCO_3 + 2NaCl \qquad (7\text{-}23)$$

Magnesium hydroxide formed in the above reactions is gelatinous, serves as a coagulate, and assists in the precipitation process. When there is insufficient magnesium in the raw water to form magnesium hydroxide, magnesium oxide may be added to form the coagulating agent.

A second and extremely important function of the magnesium hydroxide is its ability to absorb soluble silica from the solution.

Use of the phosphate in the softener of Fig. 7-30 assists in the precipitation of the calcium salts and the silica.

Mixing of the steam with the water at the top of the softener to produce hot water (212 F or higher) not only reduces the reaction time but also aids in the precipitation process by reducing the water density.

Effluent leaves the softener, Fig. 7-30, through the collecting cone and passes through the pressure filter where the precipitates and floc are removed. Anthracite or other nonsilicious materials should be used in the filter to prevent silica pick-up in the effluent.

Hot lime-soda softening of water in power plants has been popular because the process removes calcium and magnesium salts and because it removes troublesome silica. Filtration at the end of the process removes turbidity, and deaeration can be accomplished in the hot softener. Note that the process has several advantages over the zeolite process. However, hot lime-soda softening does not produce so soft a water as zeolite because the effluent from the lime-soda process cannot be reduced to less than 10 to 90 ppm of hardness as calcium carbonate. Therefore it should not be used on low-hardness waters.

7-11. Internal Treatment. Even though the feed to a boiler is low in impurities, continued input of water and continued evaporation of the water into steam cause a build-up of impurities in the boiler. Therefore the concentration of impurities in a boiler is always greater than in the water fed to the boiler.

Such unrelated substances as potato peelings, molasses, manure, and even kerosene have been used for internal treatment of scale-forming impurities.

Internal treatment may be applied in two ways: (1) conversion of the soluble scale-forming and magnesium salts into insoluble compounds and (2) maintenance of the precipitated sludge in a fluid form that can be removed from the boiler with the boiler blowdown. Boiler water alkalinity will cause the magnesium salts to form flocculent magnesium hydroxide.

Phosphate injection is currently common practice to precipitate the calcium salts. These phosphates are the same as those discussed in connection with Fig. 7-30, namely, trisodium phosphate, disodium phosphate, sodium metaphosphate, and monosodium phosphate. Monosodium phosphate and sodium metaphosphate are acid and are used when it is desired to reduce the boiler water alkalinity. Trisodium phosphate increases the alkalinity.

Any of these phosphates converts into trisodium phosphate after injection into the drum. Trisodium phosphate reacts with the calcium salts to form tricalcium phosphate, which is a finely divided, flocculent precipitate. This reaction takes place most readily when the alkalinity corresponds to a pH of 9.5 or more.

The phosphates act so rapidly that they should be pumped directly into the boiler drum, either intermittently or continuously. If the phosphates are introduced into the feedwater lines, the floc that is formed may clog valves, etc.

For the second phase of internal treatment, organic substances are injected into the drum to keep the sludge fluid so that it can be removed from the boiler by the blowdown. It is thought that the organic compounds tend to coat the precipitates so that there is less tendency for them to adhere to the metal surfaces.

Organic agents for internal treatment are seaweed derivatives, starches, tannins, and lignins. Cornstarch is used for its antifoaming action as well as for its sludge fluidity.

7–12. Demineralization. Purification of water for boiler purposes by demineralization is a very recent development. Demineralization is another ion-exchange process which is reversible. Only materials that will ionize in water can be removed by demineralization. The process involves the use of a hydrogen cation-exchange material that displays a greater affinity for the metallic cations * than the hydrogen cation. Reversal of the process occurs during regeneration and only if the hydrogen ion is present in high concentration. There is also an anion-exchange material that may be either weakly basic or strongly basic. The strongly basic material is used for power-plant purposes because it removes silica (SiO_2).

Removal of silica and the high purity of demineralized water are two reasons why this method of water treatment has been adopted so rapidly. Silica will not only create a very hard boiler scale, but will also vaporize and leave the boiler with the steam. It will then be deposited on the turbine blades. The purity of the demineralized water may be as high as 0.06 to 1.0 micromho.

Anion- and cation-exchange materials are usually organic resins. Equations for a cation exchange involving calicum sulfate and a strongly basic anion exchange for silicic acid are:

$$2RSO_3H + CaSO_4 = (RSO_3)_2Ca + H_2SO_4 \qquad (7\text{-}24)$$

$$R_4NOH + H_2SiO_3 = R_4NHSiO_3 + H_2O \qquad (7\text{-}25)$$

Regeneration of the cation-exchange material is usually accomplished by the use of sulfuric acid (H_2SO_4). Hydrochloric acid may be used but it is more difficult to handle. Caustic soda (NaOH) is used to regenerate

---

* Cations are calcium, hydrogen, magnesium, sodium, etc. Anions include the negative ions, such as bicarbonate, carbonate, chloride, sulfate, etc.

a strongly basic anion-exchange resin. After regeneration, the resin is washed with water.

Maximum operating temperature for a demineralizing system is determined by the strongly basic anion-exchange material at about 110 F.

Whenever the influent to the demineralizing system contains appreciable silt, organic matter, or iron, the water should be treated by a coagulation system. Silt and iron will coat the exchange material and reduce its effectiveness. Organic material may cause a slime that will also reduce the exchange value. At times, cold process softening may be used in place of coagulation.

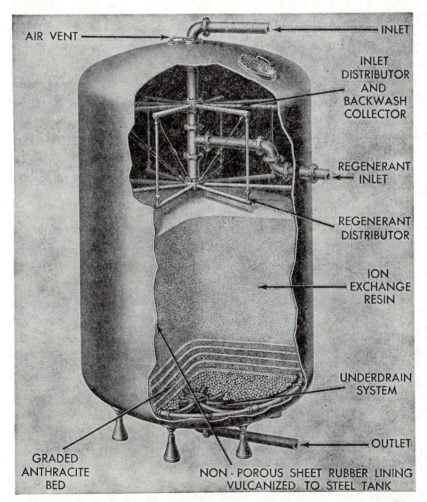

AIR VENT

INLET

INLET DISTRIBUTOR AND BACKWASH COLLECTOR

REGENERANT INLET

REGENERANT DISTRIBUTOR

ION EXCHANGE RESIN

UNDERDRAIN SYSTEM

OUTLET

GRADED ANTHRACITE BED

NON - POROUS SHEET RUBBER LINING VULCANIZED TO STEEL TANK

Fig. 7–31.   Cation- or anion-exchange unit for a multibed system. (Graver Water Conditioning Co.)

Effluent from a demineralizing unit contains dissolved oxygen. This, together with an approximately neutral (pH) condition, will cause the effluent to be extremely corrosive. Plastic piping is necessary to contain the water without excessive iron pick-up.

A *multibed* demineralizing system will contain a minimum of two tanks in addition to the tanks for the regenerating agents; one tank will be for the cation-exchange material and the other will be for the anion-exchange material. The two exchange tanks are identical, Fig. 7–31. Raw water enters at the top, passes through the material, and leaves at the bottom. The regenerating ritual consists of first backwashing the unit by admitting water at the bottom and removing it at the top. Regenerant solution, usually sulfuric acid for the cation unit and caustic soda for the anion unit, enters at the side of the unit and is distributed over the bed. The regenerant leaves at the bottom of the unit. The unit is then rinsed.

Many arrangements of cation exchangers with weakly basic and highly basic anion exchangers have been used as multibed systems. One arrangement that has found considerable acceptance is shown in Fig. 7–32. The

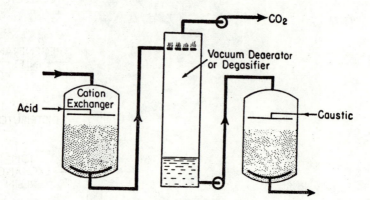

FIG. 7–32. Multibed demineralizer system.

degasifier between the cation and anion exchangers removes carbon dioxide. A vacuum deaerator may be used in place of a degasifier to remove nearly all the dissolved gases, including oxygen. These units are included in the system when bicarbonate alkalinity is high because it is more expensive to remove the carbon dioxide in the anion exchangers.

A multibed system using two exchanger tanks can deliver an effluent containing 0.5 to 5 ppm of totally dissolved solids, including 0.05 to 0.1 ppm of silica. If four beds are used, there will be still better results.

Another way of increasing the effluent purity is to use both cation- and anion-exchange resins in one tank which is known as a *mixed-bed* or

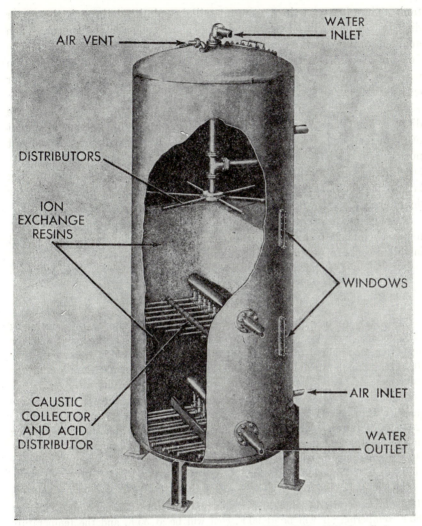

FIG. 7-33. Monobed exchange unit. (Graver Water Conditioning Co.)

monobed unit, Fig. 7-33. The mixing of the two types of resins resembles an infinite series of beds. Monobed units will provide an effluent of from 0.1 to 1.0 ppm of totally dissolved solids, including 0.0 to 0.05 ppm of silica.

Cation- and anion-exchange resins are mixed when the water is being purified. When the units are backwashed, the two resins will separate into two layers. The anion resin, having a lower density, will occupy the upper layer, and the cation exchange resin will be at the lower part of the tank. Therefore caustic regenerant will be admitted through the upper distributors and will leave through the center collector-distributor. Then acid enters through the center collector-distributor and leaves through the bottom collector. After rinsing, the two layers are mixed by agitation with air. Windows to inspect the separation of the layers are provided for each layer.

Some advantages of demineralizers over evaporators (evaporators are discussed in Chap. 8) are: (1) demineralizers are completely independent of plant load or boiler output; (2) demineralizers provide greater effluent purity; (3) there is less energy degradation with demineralizers; (4) greater flexibility in plant arrangement is made possible by eliminating large steam and vapor pipes; (5) lower operating costs are effected for demineralizers with medium or low solids influents.

Advantages for evaporators include: (1) less complex operation, (2) fewer chemicals or no chemicals to handle, (3) lower operating and installed costs for waters having high solids.

PROBLEMS

7-1. If a coal is expected to have a maximum moisture of 13%, including 2.5% surface moisture, determine (a) the primary air required and (b) the tempering air required for a minimum total moisture of 4.5%. Other data are: 100 F tempering air, 620 F primary air, 40 F coal, 180 F temperature leaving the pulverizers, and 0.013 lb of water vapor with each pound of dry air.

7-2. A steam generator will require 50 tons per hr of dry coal. Calculate the quantity of primary air and tempering air if the maximum coal moisture is to be 12% (total) and the minimum is 3%. There is 1% inherent moisture. Tempering air will be at 90 F; mill outlet temperature will be 170 F; primary air will be at 540 F; tempering air will be at 110 F; coal will be at 50 F; moisture in the air will be at 0.006 lb per lb of dry air.

7-3. A power plant is to be designed to use a coal that will vary from 7 to 21% total moisture, of which 3% is inherent moisture. Primary air will be at 660 F, tempering air will be at 90 F, coal will be at 35 F, and the temperature leaving the mills will be 160 F. Moisture with the air will amount to 0.010 lb per lb of dry air. Calculate the amount of primary and tempering air required.

7-4. A plant is scheduled to operate 2000 hr per yr at 45 tons of coal per hr, 3000 hr per yr at 30 tons per hr, and 2000 hr per yr at 15 tons per hr. If the pulverizers require 16 kwhr per ton, 18.9 kwhr per ton, and 21.6 kwhr per ton respectively at the above loads, what will be the annual savings for a bin system over a unit system? Motor efficiency is 91% and electrical energy averages 4 mills per kwhr.

7-5. Calculate the annual savings for a bin compared with a unit pulverized-coal system for the data given in the table on page 335, at a fuel cost of 34¢ per 10^6 Btu.

	Load A	Load B	Load C
Coal, tons per hr................	20	14	7
Operation, hr per yr.............	1000	3000	4000
Power, kwhr per ton..............	26.0	31.3	53.6
Motor efficiency, %	90	88	80
Heat rate, Btu per kwhr	13,100	13,800	14,900

7-6. Determine the most economical mill, on a present worth basis, for the following bids at a fuel cost of 19¢ per 10^6 Btu and at 14% investment charges:

	Load A	Load B	Load C
Coal, tons per hr....................	30	20	10
Operation, hr per yr.................	2000	2000	3500
Motor efficiency, %..................	92	90	83
Heat rate, Btu per kwhr	10,200	10,600	11,700
Bid A: Power, kwhr per ton...........	16.5	20.2	24.6
Maintenance, cents per ton..........	5	5	5
Bid B: Power, kwhr per ton...........	16.0	19.7	23.7
Maintenance, cents per ton..........	8	8	8

7-7. Determine the most economical pulverizing equipment for a fuel cost of 23¢ per 10^6 Btu (use annual cost basis):

	Load A	Load B	Load C
Coal, tons per hr....................	80	60	40
Operation, hr per yr.................	4000	3000	1000
Motor efficiency, %..................	91	89	85
Heat rate, Btu per kwhr.............	9200	9300	10,700
Bid A: Power, kwhr per ton...........	26	31.3	38.2
Maintenance, cents per ton..........	3	3	3
Bid B: Power, kwhr per ton	13.6	18.5	27.7
Maintenance, cents per ton..........	6	6	6

7-8. What size (area) of spreader stoker would you recommend for a steam generator whose output is 90,000 lb per hr of 175 psig, 450 F steam? Water enters at 220 F, the efficiency is 76%, and the coal has 11,900 Btu per lb heating value.

7-9. Use the data of Prob. 7-8 for a chain-grate stoker.

7-10. Estimate the grate area for an underfeed stoker for a steam generating capacity of 32,000 lb per hr of 150 psig, 400 F steam. Water enters at 210 F, the efficiency is about 69%, and the heating value of the coal is 13,050 Btu per lb.

7-11. Calculate the dust loading (in grains per cubic feet) for a dry-bottom boiler that uses Christian County, Illinois, coal with 21% excess air if there is a 86% efficient collector. Flue gas is at 280 F.

7-12. What minimum dust-collector efficiency would be required to meet the ASME Code for a slag-tap boiler designed for 15% excess air and a flue-gas temperature of 310 F when burning Morgan County, Texas, coal?

7-13. Estimate the minimum dust-collector efficiency required to meet the ASME Code for a cyclone furnace burning Jefferson County, Ohio, coal at 25% excess air and 300 F flue-gas temperature.

7-14. A boiler burns 1 ton per hr of Carbon County, Montana, coal with 50% excess air. The plant elevation is 2500 ft, and the flue-gas temperature entering the brick stack is expected to be 420 F. Find the stack diameter and height for an average velocity of 15 fps, a draft of 1.15 in. of water plus the velocity pressure, and 80 F air temperature.

7–15. A natural-draft stack (unlined steel) is to be designed for 6 tons per hr of Jefferson County, Alabama, coal at 35% excess air, a temperature of 390 F entering the stack, and 0.95 in. of water plus velocity pressure. Atmospheric temperature will be as high as 95 F; the plant elevation is 750 ft. Calculate the stack height.

7–16. Calculate the average velocity for a lined stack that will be 140 ft high and that will serve a boiler to be equipped with induced-draft fans. Sixty tons per hour of 12,000 Btu per lb coal will be consumed with 21% excess air. The temperature of the flue gas entering the stack will be 275 F. The plant elevation will be 1200 ft; atmospheric temperature should be taken as 90 F.

7–17. A steam generator will be equipped with an induced-draft fan with a capacity of 420,000 cfm at 280 F. Calculate the stack height (lined) to overcome friction at an average velocity of 60 fps when the air temperature is 80 F and the elevation is 6000 ft.

7–18. A fan develops 4.6-in. static pressure and 0.85-in. velocity head when delivering 11,200 cfm of 85 F air. Static efficiency is 71%. Find: (a) static air horsepower; (b) total air horsepower; (c) brake horsepower; (d) efficiency based on total or dynamic head.

7–19. A fan manufacturer rates his fan at 6.0-in. static pressure for 20,000 cfm of 70 F air, 1200 rpm, and static efficiency of 69%. At what speed should this fan operate to develop 5.1 in. when the temperature is 600 F? Also find the cubic feet per minute and brake horsepower.

7–20. What is the horsepower input required by an induced-draft fan developing 15-in. static pressure when delivering 75,000 cfm of 475 F gas with 74.5% static efficiency?

7–21. A fan is driven by a belt and pulley so that its speed can be changed by changing the pulley diameter. The fan is rated at 8000 cfm and 5-in. static pressure for 1700 rpm and 60 F air. The static efficiency is 73%. What is the maximum speed at which the fan can operate without overloading a 10-hp motor if the air temperature is 200 F?

7–22. Calculate the speed and shaft power for the fan of Fig. 7–24 to displace 350,000 cfm of 270 F flue gas at 10.5-in. static pressure.

7–23. Determine the speed and shaft power for two fans operating in parallel with a total flow of 600,000 cfm and a static pressure of 11.7 in. Gas temperature will be 325 F and the characteristics are given in Fig. 7–24.

7–24. A power plant will have an induced-draft fan whose characteristics are as shown in Fig. 7–24 and whose design point is 450,000 cfm. The fan will operate 5000 hr per yr at 400,000 cfm, 1500 hr per yr at 300,000 cfm, and 1000 hr per yr at 100,000 cfm. The system head curve is a square curve from the design point to zero flow. Fuel costs 18¢ per 10^6 Btu, and the heat rates are 11,200, 11,700, and 13,600 Btu per kwhr, respectively. A hydraulic coupling for this fan would cost $20,000 installed, would have a motor speed of 500 rpm, and would have an annual maintenance cost of $200. Inlet and outlet dampers will be provided on the fan regardless of the type of control. Calculate the annual cost for outlet damper control, inlet damper control, and variable-speed control with 12% investment charges. Assume 87% motor efficiency.

7-25. A forced-draft fan is to be controlled by outlet damper and is to be driven by two motors, one at 875 rpm and the other at 700 rpm. The fan performance at 875 rpm and 100 F air is:

Delivery, cfm	Static Pressure, in.	Static Efficiency, %
280,000	13.4	74
200,000	14.1	68
150,000	14.3	60.2
100,000	14.4	48
50,000	14.4	28.5
0	14.2	0

Piot the characteristic curves for 875 rpm and the shaft power at 875 and 700 rpm.

7-26. A manufacturer offers the following fans for forced-draft service with a design point of 270,000 cfm, 100 F air, 13.3-in. static pressure, 850-rpm fan speed, 875-rpm motor speed.

	Fan A	Fan B
Blade shape	Backward	Streamlined
Design point static efficiency	78%	83%
Cost of fan	$10,000	$11,200
Cost of hydraulic coupling	$12,000	$10,000
Motor cost	$ 9,000	$ 8,000

The system head curve is a square curve, and the plant is expected to operate at 230,000 cfm for 4500 hr per yr and at 130,000 cfm for 2500 hr per yr with heat rates of 10,500 and 11,700 Btu per kwhr. Investment charges are 5%, and fuel costs 41¢ per 10^6 Btu. Motor efficiency averages 89%.

7-27. A manufacturer proposes the following fans for induced-draft service. The design point is 410,000 cfm at 13.0-in. static pressure and 300 F. The system head curve is a square curve.

	Fan A	Fan B
Fan speed	480	420
Motor speed	500	435
Cost of fan	$23,500	$26,500
Cost of motor	$14,000	$16,000
Cost of hydraulic coupling........................	$18,000	$18,000
Design point static efficiency	70%	76%

If the investment charges are 13% and the fan will operate at 370,000 cfm for 3700 hr per yr and at 130,000 cfm for 4000 hr per yr, determine the present worth of each fan. Fuel costs 22¢ per 10^6 Btu; the heat rates are 10,300 and 11,900 Btu per kwhr; motor efficiencies are 91% and 85%.

7-28. Calculate the annual savings for a fan using inlet dampers as compared with outlet dampers when electrical energy costs 4 mills per kwhr. The fan performance at 300 F is given in the tabulation on page 338.

Delivery, cfm	Static Pressure, in.	Static Efficiency, %
200,000	11.1	88
175.000	11.6	85.5
150.000	11.7	80.5
100,000	11.65	66.5
50,000	11.3	46
0	10.7	0

The fan will operate for 3000 hr per yr at 175,000 cfm, 3000 hr per yr at 125,000 cfm, and 1500 hr per yr at 75,000 cfm. The system head curve is a square curve from the design point of 200,000 cfm; the motor efficiency is 88%.

7–29. What size of tank would be required to hold sufficient salt to regenerate three times a system that will soften 50,000 gal of water (from 8 grains per gal hardness to zero hardness) between regenerations? Allow 15% extra capacity in the tank.

7–30. Write the chemical equations for the reactions of the following water impurities with greensand: (a) $CaSO_4$, (b) $CaCl_2$, (c) $Ca(NO_3)_2$, (d) $Mg(HCO_3)_2$, (e) $MgSO_4$, (f) $Mg(NO_3)_2$.

7–31. Write the chemical equations representing the reactions occurring when greensand is regenerated with salt (NaCl). Calcium and magnesium salts have combined with the zeolite.

7–32. Write the equations for the regeneration of carbonaceous zeolite with H_2SO_4. Calcium, magnesium, and sodium salts have combined with the zeolite.

7–33. Write the chemical equations for the reactions of the following water impurities with hydrogen zeolite (H_2Z): (a) $Ca(HCO_3)_2$, (b) $CaSO_4$, and (c) $CaCl_2$.

7–34. Estimate the present worth of the cost of fly-ash disposal from the dust collector in the example of Art. 7–3 if investment charges are 12.5%, the average heat rate for the plant is 9.800 Btu per kwhr at an average generation of 500,000 kw for 8000 hr per yr.

BIBLIOGRAPHY

Air Conditioning and Engineering. Detroit: American Blower Corp., 1955.
CALISE, V. J. "Status of Demineralizing for Treatment of Boiler Feedwater in Today's Power Plants," ASME Convention, 1955.
Combustion Engineering. New York: Combustion Engineering, Inc., 1947.
Fan Engineering. Buffalo: Buffalo Forge Co., 1949.
Handbook of Industrial Water Conditioning. Philadelphia: W. H. & L. D. Betz, 1953.
KENT, R. T. *Kent's Mechanical Engineer's Handbook.* New York: John Wiley & Sons, Inc., 1950.
MARKS, L. S. *Mechanical Engineer's Handbook.* New York: McGraw-Hill Book Co., Inc., 1951.
Steam. New York: Babcock & Wilcox, Co., 1955.

CHAPTER 8

HEAT EXCHANGERS

8-1. General. Steam from the prime movers in many of the older steam power plants was exhausted to atmosphere. Although the steam was thus lost to the system, atmosphere was really the condenser for the cycle. Some plants still operate on this method, but these plants are usually small or have special requirements.

Recovery of the exhaust steam in a condenser reduces the make-up water that must be added to the system from 100%, when exhausting to atmosphere, to 1 to 5%. Make-up cannot be reduced to zero because there are always some losses from traps, boiler blowdown, leaks at packing glands, and contaminated returns from steam-heating systems or oil-consuming equipment.

Reduction of make-up is not the only purpose of the condenser. By using a condenser, the exhaust pressure may be lowered from standard atmospheric pressure to 1 in. of mercury absolute, and the efficiency of the cycle may be doubled. Such a gain is well worth the cost of the additional equipment required to reduce the exhaust pressure. Further improvement may be made in the cycle efficiency by adding feedwater heaters to raise the temperature of the condensed steam before inducting it into the boiler. Although the regenerative cycle, i.e., a cycle using steam extracted from the prime mover to heat the feedwater, is economically important, it is a necessity for modern high-pressure boilers. It often requires several hours to heat a boiler from dead cold to operating temperatures, and once these temperatures are obtained, tremendous stresses would be developed by introducing cold condensate into the boiler. Sudden chilling of the hot surfaces would cause local contraction of the metals that would easily rupture the boiler parts.

The purpose of this chapter is to discuss the design and operation of the heat-exchange equipment used in the feed cycle of a modern plant.

Condensers

8-2. Barometric and Low-Level Jet Condensers. Two methods of condensing a vapor are: (1) mixing the vapor with a liquid so that the vapor can reject its latent energy to the liquid, with a consequent increase in the temperature of the liquid; and (2) transferring the latent energy of the vapor through a surface to another fluid that is at a lower temperature. Condensers as well as feedwater heaters are constructed to operate on both principles.

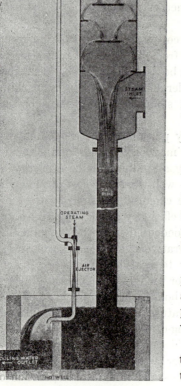

FIG. 8–1. Barometric condenser. (Ingersoll-Rand Co.)

Mixing or direct-contact condensers are either of the *barometric* or *low-level-jet* type. Fig. 8–1 shows one design of a barometric condenser. Steam enters the condenser shell at the lower right-hand corner and travels upward toward the top of the shell. Circulating water, entering at the upper right-hand part of the shell, impinges on the top baffle and cascades downward by gravity over the remaining baffles and baffle rings, forming successive curtains of water. In order to travel upward, the steam must pass through these curtains of water. The steam is condensed by mixing with the water.

Noncondensable gases are drawn out of the shell at the top by the air ejector.

Compression of the mixture of circulating water and condensed steam from the low shell pressure to atmospheric pressure in the *hotwell* is accomplished by the column of water in the *tail pipe*.

Theoretically, the height of the tail pipe is equal to the difference between atmospheric pressure and shell pressure plus the tail-pipe friction loss and the velocity head.

Low-level jet condensers are similar to barometric condensers except that the tail pipe is eliminated, i.e., the water is compressed from condenser pressure to essentially atmospheric pressure either by a pump or by making use of the kinetic energy of the water jets, as shown in Fig. 8–2. Elimination of the tail pipe reduces the headroom required by the condenser.

Mixing condensers are seldom used in modern power stations in spite of their low initial cost because the condensed steam is mixed with the circulating water. Neither the contamination of the boiler feedwater nor the cost of treating all the circulating water is permissible. Modern boilers must have water that is free of scale-forming impurities.

Calculations for water quantities for mixing condensers are the same as for mixing feedwater heaters, Eqs. (8–11), (8–12), and (8–13).

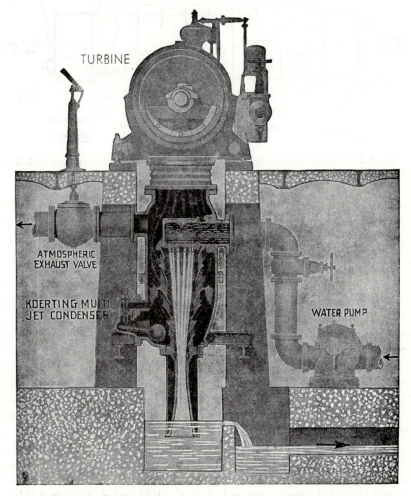

TURBINE

ATMOSPHERIC
EXHAUST VALVE

KOERTING MULTI
JET CONDENSER

WATER PUMP

FIG. 8–2. Low-level jet condenser. (Schutte & Koerting Co.)

8–3. Surface Condensers. Contamination of the feedwater by mixture with the circulating water is eliminated in the surface type of condenser. At the right side of Fig. 8–3, circulating water enters the cold-end *water box*, flows through the tubes to the warm-end water box, and leaves at the top of the warm-end water box. Circulating-water outlet connections are usually (but not always) located at the top of the water box.

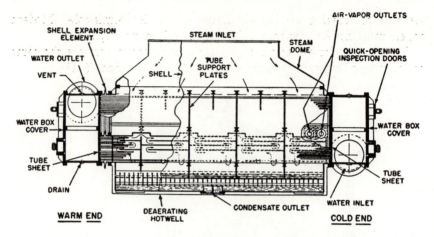

Fig. 8-3. Longitudinal section of a single-pass surface condenser (Ingersoll-Rand Co.)

and the inlet is located at the bottom of the water box. Thus, the water passages of the condenser form the top of an inverted U-tube circuit. At the outlet, the water is normally less than atmospheric pressure. Therefore, placing the water outlet at the top permits scavenging air and gases out of the water passage to prevent air binding.

In a single-pass condenser, as shown in Fig. 8-3, the circulating water makes one passage through the unit. Two-pass condensers have the inlet and outlet on the same water box. If the unit of Fig. 8-3 were a two-pass unit, the inlet and the outlet connections would be on the same water box, and the horizontal baffle in that water box would be solid to force the water through the tubes. The water box at the end opposite to the inlet and outlet connections is then known as the *return water box*. Usually, the first pass, having the coldest water, is at the bottom of the condenser. Condensers may have one, two, three, or four passes; one- and two-pass condensers are the most common.

Water boxes may be divided by a vertical partition into two equal parts. Then there are two inlet and two outlet circulating-water connections. *Divided water boxes* permit the operators to open the water boxes for one side of the condenser and clean the tubes while the other side of the condenser is in operation. Of course, the turbine cannot carry full load while half of the condenser is out of operation.

In passing through the tubes, the circulating water receives energy from the steam by convection and conduction through the tube walls, and its temperature is increased from the initial t_1 to the final t_2. Since the steam enters the condenser in a moist state, its temperature remains constant throughout the condenser, but its latent energy is removed. Condensed steam, called *condensate,* collects in the *hotwell,* whose capacity should be at least equal to the maximum condensate produced in 1 min, and is removed by the *condensate pump.*

Certain constructional details are worthy of mention. Water boxes and covers are made of cast iron and are held to the *shell* by bolts through the flanges. *Tube sheets* are inserted between the water-box flange and the shell flange. These sheets are usually made of rolled Muntz metal (60% copper, 40% zinc) or rolled Naval brass (60% copper, 39.25% zinc, 0.75% tin), while the shell is welded copper-bearing steel or cast iron. Shells vary from $\frac{3}{8}$ to $\frac{7}{8}$ in. thickness, while tube sheets vary from $\frac{7}{8}$ to $1\frac{1}{2}$ in. thickness. Welded construction has become popular because of improvements in welding technics that allow reduced costs. In any event, the copper alloy tubes and the steel or iron shell may give an electrolytic action when there is salt in the circulating water. It is common practice to suspend zinc plates in the water boxes of marine installations so that any electrolytic action will attack the zinc rather than an important part of the condenser.

Support plates are located at periodic intervals along the length of the tubes to steady them and to prevent vibration from impingement of high-velocity steam. Support plates may be of steel, Muntz metal, etc. Muntz metal has the advantage that it does not cut the tubes when there is vibration. However, Muntz metal is subject to electrolysis.

When the entering steam contains a large percentage of moisture, it is possible to have considerable erosion on the upper tubes. A hard metal shield may be placed over each tube of the top one or two rows of tubes for protection. Or, if erosion is severe, the tubes in the top of the condenser may be metalized or a piece of grating may be installed over the tubes.

Admiralty metal (primarily 70 to 73% copper, 1% tin, 26 to 29% zinc) is the most popular tube material, although arsenical copper, cupronickel, Muntz metal, red brass, and aluminum brass are used. Addition of very small amounts of arsenic, antimony, and phosphorus to the Admiralty metal inhibits the metal to prevent dezincification.

The different metals used for the tubes and shell of a surface condenser will cause a different expansion for each part as the temperature of the unit changes. Fig. 8–4(a) shows the method used in past years to provide for tube expansion, but which at the present is used only on small condensers. The inlet end of the tube is rolled into the tube-sheet hole and the end is flared to offer the least possible resistance to the water entering the tube. However, the outlet end is packed into its hole. Relative expansion of the tube is compensated by slippage of the tube in the

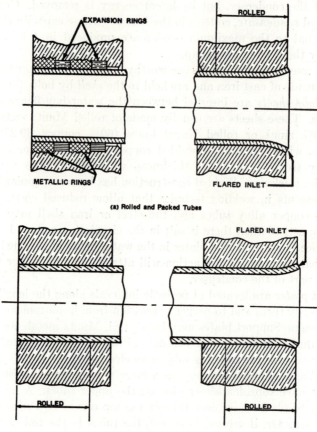

FIG. 8-4. Two methods of fastening tubes to condenser tube sheets. (Westinghouse Electric Corp.)

packed joint. But, since the tube can slip in the packing, the joint will not be absolutely tight. Leakage of water past the packing and into the shell space may be a serious point of condensate contamination.

In recent years, condenser tubes have been rolled into both tube sheets, forming a tight joint at both ends and reducing circulating-water leakage, Fig. 8-4(b). With this system, expansion joints have been used to take care of the relative expansion, Fig. 8-5. The tubes may expand or contract and the shell may change its length by expansion or contraction of the flexible expansion joint.

A new development is the technique of welding the tubes into the tube sheets of large condensers.

FIG. 8–5. Expansion joint for condenser shell. (Allis-Chalmers Manufacturing Co.)

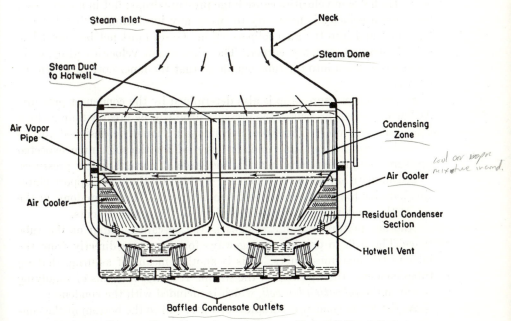

FIG. 8–6. Cross-section of a surface condenser. (Ingersoll-Rand Co.)

The cross-section of the surface condenser in Fig. 8–6 shows the arrangement of tubes and steam and gas passages. The vertical and inclined lines in the center of the unit are the lines of tube centers. Of course the tubes would be perpendicular to the page. Steam entering the condenser finds the first row of tubes widely spaced to permit access to the succeeding tubes. One of the features of modern condensers is the ample tube spacing in contrast with the old condensers which had as many tubes inside the shell as space would permit. In some cases these old condensers have had their performance improved by removing enough tubes to allow steam to reach the inner tubes.

As steam progresses through the condenser, some of it is condensed, the water droplets falling from the tubes by gravity. With the total volume of steam reducing as it progresses through the condenser, the steam space may be reduced both by narrowing the condenser shell and by locating the tubes on smaller center distances. The theoretical condenser cross-section would approximate a **V**. When condensing, the steam contracts in the volume ratio of about 40,000:1 in a fraction of a second. However, because of construction features and steam lanes at the sides of the condensers, the condensers may have other shapes.

The pressure drop of the steam in passing over the tube bundle seldom exceeds 0.05 in. Hg. As in any other flow problem, the pressure drop is a function of the weight of fluid, the specific volume of the fluid, and the area. High steam velocities cause large pressure drops; but in a condenser, high steam velocity is necessary to sweep any insulating air blankets off the tubes and into the air-cooling section. Gases must not be allowed to collect in some stagnant corner. The high steam velocities also aid in forcing the steam into the tube bundle so that the innermost tubes will be effective.

Steam side pressure drop is also determined by the condenser pressure. At low condenser pressures, the pressure drop will be greater than at high condenser pressures. Therefore, a condenser should be designed for its average operating pressure; if it is designed for the maximum expected pressure, the pressure drop may be too large at normal or low pressures.

Steam distribution in the condenser is aided by the tube-support plates which divide the bundle into small compartments. The steam dome at the top of the condenser distributes the steam to the compartments.

Another feature of these designs is the so-called *laning* of the tube space, Fig. 8–6, to provide passageways for the steam directly from the inlet to the innermost tubes. This is another means of assuring that all tubes condense their portion of the steam. It is also a means of supplying steam to the *reheating hotwell*, which is included with the condenser.

As the steam condenses, droplets of water fall to the bottom of the condenser and cascade from one tube to another. Since the drops are at a higher temperature than the tube walls enclosing the cold circulating water, the rain of water is cooled below its saturation temperature. It has not been uncommon to find condensate leaving an old condenser at from 5 to 10 deg less than the saturation temperature corresponding to the condenser pressure. This temperature difference is known as *subcooling*.

Proper laning will provide a passage for the steam directly from the inlet to the hotwell. The condensate flows through a **U** trap into a trough in the hotwell. From there it flows into the hotwell proper through small holes and by spilling over the edge of the troughs so that the water is broken up into small drops that mix with the steam. Turbine exhaust flows directly from the steam dome to the hotwell and mixes with the condensate to reheat it to saturation temperature. The combination of breaking the water up into droplets and heating it to saturation temperature deaerates the condensate. Thus, a reheating hotwell is also a *deaerating hotwell*. The spray should fall a distance of at least 16 to 18 in. in the deaerating section to assure that it will be broken up into small enough droplets.

Whenever steam is condensed, regardless of the pressure, there will always be some noncondensable gases present. With condenser pressure less than atmospheric, the gases will enter the condenser with the steam and also will leak into the shell through any joints that are not absolutely tight. These gases cannot leave with the condensate because water at saturation temperature cannot theoretically absorb gas. If gas is left in the shell, the condenser will become air bound and the pressure will rise.

Noncondensable gases that are vented from the hotwell and that have been swept from the tubes are carried into the *air-cooler section*. This section is always located where it will receive the coolest circulating water (in a multipass condenser). The noncondensable gases flow through the multipass sections of the air coolers at velocities in the order of 50 fps. Much of the steam is condensed. The gases leave at the *air outlets* and flow to air-removal equipment. Tubes in the air-cooler sections are concentrated on close center-to-center distances to assist in condensing as much steam as possible.

As operating temperatures and atmospheric conditions change, a condenser will expand in a transverse direction as well as in a longitudinal direction. Four procedures of condenser mounting are available to care for diametral expansion. The simplest method, Fig. 8–7(a), consists of mounting the condenser rigidly on its foundation. A flexible expansion joint is used between the turbine and the condenser. The joints may be made of stainless steel, copper, or rubber. Stainless steel joints are expensive, and copper joints corrode. Rubber joints crack with age and fail at the high temperatures obtained if the turbine goes noncondensing (due to circulating-water failure, etc.). Another disadvantage is that it is extremely difficult to renew an expansion joint on a large unit because of the lack of room in the turbine foundation.

If the condenser is rigidly connected to and supported wholly from the turbine, Fig. 8-7(b), all piping connected to the condenser must be provided with expansion joints. Certain precautions must be observed with this style of connection. Although the shape of the turbine-exhaust connection may be other than circular, the center of gravity of the condenser must be below the center of the turbine exhaust connection when the condenser is loaded with water under operating conditions. Vibration and the reactions caused by the change of direction of the circulating water in the condenser may be overcome by locating a stabilizer under the hotwell. To restrain motion in any direction except vertical, ribs or blocks are welded

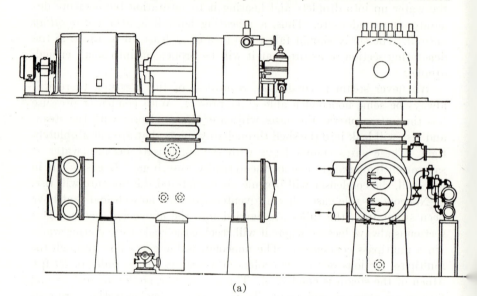

(a)

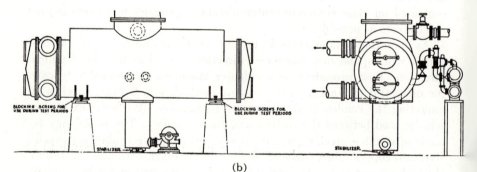

(b)

Fig. 8-7. Methods of supporting surface

onto the hotwell so that they butt against corresponding ribs which are bolted and grouted to the floor. Blocking screws must be installed for use during erection and at times when the condenser is unbalanced, i.e., when the condenser steam space is flooded to test for leakage or when the circulating-water spaces have been drained.

Spring supports, as shown in Fig. 8–7(c) and (d), are used for nearly all large units. Turbine and condenser are bolted rigidly together, and piping for all auxiliaries contains expansion joints to permit free movement of the condenser. Each spring nest must be designed to carry the load existing at its location and to prevent any unbalance on the turbine exhaust. The proper tension on each nest is obtained by adjustment of the leveling screws. Screw jacks or hydraulic jacks are provided to support the condenser during erection, etc. The precautions noted above for the type of connection illustrated in Fig. 8–7(b) also apply to spring supports.

A fourth system, which eliminates the tedious adjustment of the springs by substituting a weight with lever and fulcrum at each of the four condenser footings, Fig. 8–8, has been used on a very few large condensers.

Copper or rubber expansion joints cost about twice as much as spring

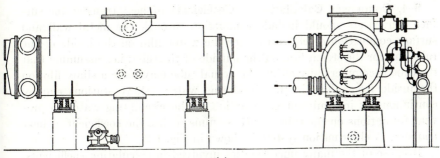

(c)

(d)

condensers. (Allis-Chalmers Manufacturing Co.)

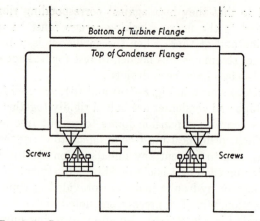

FIG. 8-8. Lever and counterweight condenser support.

supports, and counterweights cost about ten times as much as spring supports. Spring supports are used on condensers of 5000 sq ft and larger, while fulcrum supports are not used below 35,000 sq ft of surface.

8-4. Condenser Calculations. Coefficients of heat transfer for surface condensers should include conduction through the metal wall and surface conduction or convection for the water film on the inside of the tube and for the steam film on the outside of the tube; i.e., assuming that the tube is commercially clean. An actual tube may have a slimy film on the inside from algae in the circulating water. Many stations employ continuous or intermittent chlorination of the circulating water to prevent formations on the tubes, as these films reduce the coefficient of heat transfer. A chlorination system is shown in Fig. 1-2.

Because of the many uncertainties involved in calculating coefficients, the Heat Exchange Institute, which is composed of the major manufacturers of heat-exchange equipment, has standardized the over-all coefficients of heat transfer (U) to be used in the design of surface condensers, Fig. 8-9. Values given on the curves are for 70 F water entering commercially clean tubes made of No. 18 BWG Admiralty metal. Correction factors are given for other metals and thicknesses.

A temperature-surface diagram for a condenser would consist of a constant temperature line for the steam condensing and a rising curved line for the circulating water being heated, Fig. 4-7(b). It was also shown in Chap. 4 that the equation for log mean temperature difference and the quantity of energy transferred are

$$\theta_m = \frac{\theta_2 - \theta_1}{\ln \dfrac{\theta_2}{\theta_1}} = \frac{t_2 - t_1}{\ln \dfrac{t_{sat} - t_1}{t_{sat} - t_2}} \qquad (8\text{-}1)$$

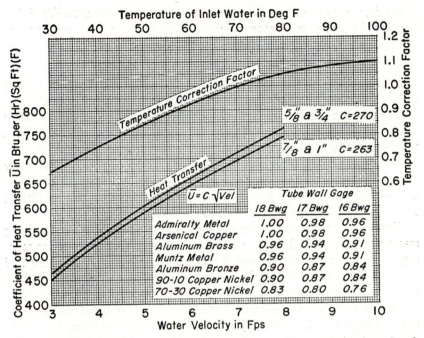

FIG. 8–9. Heat-transfer curves for surface condensers. (By permission from *Standards for Steam Surface Condensers*, 4th ed. Copyright, 1955, by Heat Exchange Institute.)

and

$$Q = UA\theta_m \tag{8-2}$$

where θ_m = log mean temperature difference, F

θ_1 = minimum terminal temperature difference, $t_{sat} - t_2$, F

θ_2 = maximum terminal temperature difference, $t_{sat} - t_1$, F

t_{sat} = saturation temperature of the steam, F

t_1 = inlet circulating water temperature, F

t_2 = outlet circulating water temperature, F

U = over-all coefficient of heat transfer based on outside tube area, Btu per (sq ft) (hr) (F)

A = outside tube area, sq ft

The quantity θ_1 is known simply as the *terminal difference* and is the *difference between the saturation temperature of the steam and the temperature of the water leaving*. The coefficient of heat transfer to be used in Eq. 8–2 cannot be obtained directly from Fig. 8–9 but can be defined by

$$U = \overline{U} \times F_t \times F_m \times F_c \times F_p \tag{8-3}$$

where $\overline{U}$ = coefficient of heat transfer from Fig. 8–9, Btu per (sq ft) (hr) (F)

F_t = temperature correction factor from Fig. 8–9

F_m = tube material and thickness correction factor from Fig. 8–9
F_c = cleanliness factor
F_p = prime mover factors, 1.0 for turbines and 0.75 for steam engines

Coefficients of heat transfer depend on the velocity of the fluid flowing through the tubes and on the film temperatures. The Institute's curves assume a water film temperature corresponding to a water temperature of 70 F. The correction for any other water temperature may be found from the *temperature curve,* whose abscissa scale is at the top of the sheet.

The cleanliness factor is concerned with films formed by algae and dirty water. When circulating water is clean or is chlorinated, a factor of 0.85 is normally used. Lower values should be used for bad water conditions and a value of 0.95 should be used for good water conditions.

Heat transfer coefficients given in Fig. 8–9 are for a condenser receiving steam that is free of oil, such as a turbine exhaust. Condenser tubes serving a steam engine become coated with an oil film that causes drops of water on the tube surface rather than a water film on the surface. This reduces the heat transfer rate.

Condenser tubes are selected from the four sizes listed on the heat transfer curves and are usually of a thickness corresponding to No. 18

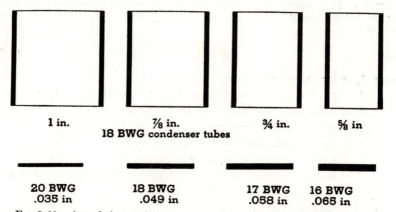

| 1 in. | ⅞ in. | ¾ in. | ⅝ in |

18 BWG condenser tubes

| 20 BWG | 18 BWG | 17 BWG | 16 BWG |
| .035 in | .049 in | .058 in | .065 in |

Fig. 8–10. Actual sizes and thicknesses of condenser and feedwater tubes.

BWG (Birmingham Wire Gauge). Tubes smaller than ⅝-in. OD become clogged too easily, while tubes larger than 1-in. OD require excessive amounts of water to maintain a reasonable velocity. Even ⅝- and ¾-in. tubes clog very easily and for that reason are used only in small condensers. The ⅝-in. tubes are used for marine installations; for stationary plants they are seldom used above 500-sq ft capacity. For condensers of 500 to 2750 sq ft, ¾-in. tubes are used. Other ranges are: ¾- or ⅞-in.

TABLE 8-1

PHYSICAL PROPERTIES OF CONDENSER AND FEEDWATER HEATER TUBES

	Thickness		Tube Size			
	In.	BWG	⅝ in.	¾ in.	⅞ in.	1 in.
Outside diam, in.			0.625	0.750	0.875	1.000
Inside diam, in.	0.134	10	0.357	0.482	0.607	0.732
	0.120	11	0.385	0.510	0.635	0.760
	0.109	12	0.407	0.532	...	0.782
	0.095	13	0.435	0.560	...	0.810
	0.033	14	0.458	0.584	0.709	0.834
	0.072	15	0.481	0.606	...	0.856
	0.065	16	0.495	0.620	0.745	0.870
	0.058	17	0.509	0.634	0.759	0.884
	0.049	18	0.527	0.652	0.777	0.902
Surface (outside), sq ft per lin ft			0.1636	0.1963	0.2297	0.2618
Length in ft for 1 sq ft surface			6.112	5.094	4.367	3.817
Water quantity, gpm at 1 ft per sec velocity		10	0.310	0.568	0.902	1.314
		11	0.364	0.638	0.988	1.420
		12	0.404	0.696	...	1.496
		13	0.464	0.770	...	1.608
		14	0.514	0.836	1.230	1.704
		15	0.566	0.902	...	1.798
		16	0.600	0.942	1.360	1.854
		17	0.630	0.980	1.410	1.910
		18	0.680	1.042	1.480	1.994
Weight of tube, lb per lin ft		10	0.800	1.040	1.240	1.41
		11	0.735	0.918	1.096	1.28
		12	0.683	0.845	...	1.18
		13	0.610	0.752	...	1.04
		14	0.548	0.670	0.798	0.92
		15	0.482	0.592	...	0.81
		16	0.442	0.540	0.636	0.74
		17	0.399	0.494	0.576	0.66
		18	0.342	0.417	0.490	0.57

Weights are theoretical and for copper.

tubes from 3000 to 4750 sq ft; ¾- or ⅞- or 1-in. tubes from 5000 sq ft to 14,000 sq ft; either ⅞- or 1-in. tubes for 15,000 sq ft and above. The proper tube size is a matter of economic selection and engineering judgment as to the possibilities of clogging. Because of the latter factor, many engineers prefer to select the largest possible tube diameter. No. 18 BWG thickness will withstand ordinary water pressures. Unusual conditions may require heavier tubes. Sizes and thicknesses may be visualized from Fig. 8–10.

Useful physical data for condenser or feedwater heater tubes are given in Table 8–1.

Condenser tube length is of extreme importance because of its direct effect on friction loss, Fig. 8–11, and because of steam distribution over the tube bundle. Imagine a small condenser with a steam-inlet nozzle 2 ft in diam and with tubes 25 ft long. The ends of the tubes would not receive sufficient steam to operate successfully. On the other extreme, tubes 10 ft long would be entirely too short for a large condenser with a square steam-inlet nozzle 15 ft on each side. Table 8–2 may be used as a

TABLE 8–2

APPROXIMATE TUBE LENGTHS FOR SURFACE CONDENSERS

Surface Area (Sq Ft)	Approximate Tube Length (Ft)
500 to 1500	8–14
1500 to 3000	10–18
3000 to 6000	14–20
6000 to 10,000	16–22
10,000 to 20,000	18–24
Over 20,000	22–30

rough guide for length of condenser tubes to give a well-proportioned condenser. Tubes longer than 30 ft (actual, not effective, length) are seldom used because there is a cost penalty for excess length.

Inspection of the temperature-surface diagram for surface condensers (Fig. 4–7b) shows that inlet circulating-water temperatures must be materially below the saturation temperature corresponding to the pressure expected in the condenser. Since fresh-water temperatures vary throughout the year, the condenser should be designed for the average inlet temperature expected during the year, but the condenser should always be large enough to accommodate all the steam exhausted from the turbine during its absolute maximum load operation. An oversized condenser will increase the initial cost of the station but will reduce operating costs by permitting the turbine to operate with a low exhaust pressure. The reverse

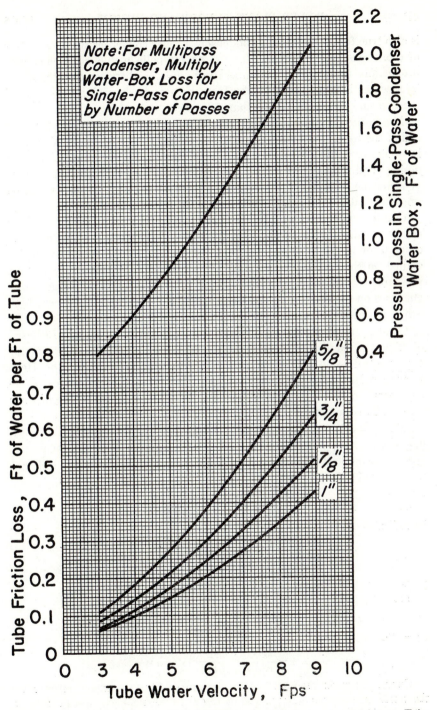

Fig. 8-11. Friction loss for No. 18 BWG condenser tubes and water boxes. Tube loss based on Williams and Hazen formula using $c = 130$. (By permission from *Standards for Steam Surface Condensers,* 4th ed. Copyright, 1955, by Heat Exchange Institute.)

is also true for an undersized condenser. Proper condenser size is a matter of economical justification. Many designers believe that good condenser proportions result when the temperature difference between the steam and inlet water is 20 F to 25 F. The Institute standards prohibit a terminal temperature difference of less than 5 F.

Maximum circulating-water velocity is limited by erosion on the inlet end of the tubes and by the friction loss through the condenser. Velocities in excess of 8 fps are seldom used. The practical, minimum velocity at design conditions is 5 or 6 fps. Velocities in excess of 6 fps help to prevent marine organisms from attaching themselves to the tubes. It is good practice to use 7- to 7½-fps velocity; 8 fps is not uncommon.

Condensers should be protected from excessive steam pressures in the condenser shell in the event of a stoppage of circulating water or some other fault in the system. Formerly, *atmospheric relief valves* were installed on the condensers. These valves opened automatically when the pressure rose above atmospheric. Modern large turbines have lead blowout discs on the exhaust casing of the turbine. Also, there may be a high exhaust pressure trip-out installed on the turbine.

Condenser calculations for non-reheat turbine installations are based on 950 Btu per lb of steam rejected and 1000 Btu per lb of steam for engine applications, all in accordance with the standards of the Heat Exchange Institute. Although not provided for in the standards, some engineers use 1000 Btu per lb rejected for reheat turbines because of their lower exhaust moisture.

EXAMPLE 8–1. Design a surface condenser for a turbine that exhausts 50,000 lb per hr of steam at 2.0 in. Hg abs with circulating water entering at 80 F. Use an 8-degree terminal difference, 7-fps velocity, ¾-in. 18 BWG tubes, and a cleanliness factor of 0.85.

SOLUTION.

$$t_{sat} = 101.1 \text{ F}, \ t_2 = 101.1 - 8.0 = 93.1 \text{ F}$$

From Eq. (8–1),

$$\theta_m = \frac{93.1 - 80}{\ln \dfrac{101.1 - 80}{8}} = 13.54 \text{ deg}$$

From Fig. 8–9, $\overline{U} = 713$, and the temperature correction factor is 1.05.
From Eq. (8–3),

$$U = 713 \times 1.05 \times 1.0 \times 0.85 \times 1.0 = 636 \frac{\text{Btu}}{(\text{sq ft})(\text{hr})(\text{F})}$$

From Eq. (8–2),

$$A = \frac{50,000 \times 950}{636 \times 13.54} = 5516 \text{ sq ft}$$

Circulating water requirements are determined from the energy rejected by the steam and the temperature rise of the water. Assuming a specific heat of unity for the water, we compute the flow to be

$$\frac{50,000 \times 950}{93.1 - 80} = 3,626,000 \text{ lb per hr} \qquad \dot{M}_{cw}\,C_p\,\Delta T = \dot{q}$$

Dividing by 500 gives a flow of 7252 gpm.

Regardless of the number of passes, all this water must flow through *each* pass. Table 8–1 indicates that each ¾-in. No. 18 BWG tube will pass 1.042 gpm at 1 fps. At 7 fps, there will be

$$N_{tubes/pass} = \frac{7252}{1.042 \times 7} = 995 \text{ tubes per pass}$$

The number of passes will be determined by the tube length. Also from Table 8–1, each tube has 0.1963 sq ft of outside surface per foot of length. Assuming one pass with 995 tubes, the effective tube length would be

$$L_{effective} = \frac{5516}{995 \times 0.1963} = 28.2 \text{ ft}$$

Consulting Table 8–2, we find this length to be unreasonable. For a two-pass condenser, the effective length would be $28.2 \div 2 = 14.1$ ft, which is reasonable. The total number of tubes would be $2 \times 995 = 1990$.

A condenser of this size would have ½-in. thick tube sheets, so the actual tube length would be 14.2 ft.

From Fig. 8–11, the friction loss in the condenser would be

$$2 \times 14.2 \times 0.40 + 2 \times 1.41 = 14.18 \text{ ft}$$

8–5. Economic Selection of Condensers. Condenser manufacturers have standardized condenser sizes. Effective tube lengths are in 2-ft increments and four different lengths for each surface area. Tube diameters have already been listed in the preceding article. Surface areas are standardized in 100-sq ft increments from 100 to 1500 sq ft, in 250-sq ft increments from 1500 to 8000 sq ft, in 500-sq ft increments from 8000 to 10,000 sq ft, in 1000-sq ft increments from 10,000 to 20,000 sq ft, in 2500-sq ft increments from 20,000 to 50,000 sq ft, and in 5000-sq ft increments up to 100,000 sq ft. Also, the condensers may have one to four passes.

With such a large variety of sizes to choose from, the engineer must make an economic study to select the proper unit. The factors to be considered in such a study include initial cost of condensing equipment, cost of the effect of condenser vacuum on plant heat rate, cost of circulating pump power, cost of intake and discharge structures, cost of building, and cost of change in turbine capacity because of vacuum.

Some general observations can be made on condenser sizes:

1. For a given steam flow, a single-pass condenser usually requires 50% more water than a two-pass condenser.
2. Average circulating water temperature rises are about 15 F for two-pass and 10 F for single-pass condensers.

3. Single-pass units condense about 10 lb per hr of steam for each square foot of surface, while two-pass units condense about 8 lb per hr for each square foot of surface.

4. For each 1-degree increase in circulating water inlet temperature, the surface and circulating water requirements will increase 5%, all other conditions remaining unchanged.

5. To decrease condenser pressure 0.1 in. Hg, the surface and circulating water requirements must be increased 9%, all other conditions remaining unchanged.

6. An increase of 0.1 fps water velocity decreases the surface by 0.85%, and the circulating water will increase 0.45%, all other conditions remaining unchanged.

7. An increase of five percentage points in cleanliness factor will decrease surface and circulating water requirements 3.5%, all other factors remaining unchanged.

8. If sufficient circulating water is available, the most economical condenser is a single-pass unit with the smallest tube diameter permissible.

No. 18 BWG Admiralty condenser tubes cost about $1.45 per sq ft of surface. Arsenical copper tubes cost about 15% more than Admiralty metal, while aluminum brass costs about 13% more than Admiralty metal.

Condenser prices, exclusive of tubes, but including air ejectors, air meters, hot-well liquid-level control, air piping, condensate piping, supervision of installation, labor of installation, spring supports, and foundation bolts, may be approximated for problem purposes for medium and large condensers with 1-in. tubes as:

Two-pass condenser cost * = $15,000 + $4.85 (surface in sq ft)   (8-4)

Single-pass condenser cost * = $15,000 + $5.25 (surface in sq ft)   (8-5)

For ⅞-in. tubes, deduct 20¢ per sq ft.

Vertical circulating water pumps with motors of medium to large capacities are priced (for problem purposes) approximately as follows:

Mixed flow pump cost = $4,200 + 0.55 (gpm)   (8-6)

These prices for circulating pumps are based on a developed head of 40 ft. If the developed head is greater or less than 40 ft, the price increases 0.5% for each foot or fraction of a foot above or below 40 ft. Combined circulating pump and motor efficiencies (wire-to-water efficiency) may be assumed as 75%.

Of the economic factors previously listed, cost of intake and discharge structures, building, and change in turbine capacity are very difficult to determine accurately and are usually not very large. However, an increase in the maximum power required by the circulating pumps does

---

* These equations neglect price variations due to tube length.

decrease the maximum plant output. Theoretically, the plant capacity could be increased a few kilowatts to account for an increase in pump motor demand. The change for this demand is an incremental value, and as such, would be appreciably less than the plant unit cost. For example, the incremental plant cost might be \$125 per kw on a plant that would have a unit cost of \$175 per kw.

It will be convenient to rearrange Eq. (8–1) to produce:

$$\theta_1 = \frac{t_2 - t_1}{(e)^{(t_2 - t_1)/\theta_m} - 1} \tag{8-7}$$

The following example illustrates the method of economic analysis for a selection of a commercial size of condenser.

EXAMPLE 8–2. Select the most economical of two condensers (one having 15,000 sq ft and the other having 16,000 sq ft of surface) to serve a 22,000-kw turbine generator. The heat rate for the unit (including boiler) will be 11,700 Btu per kwhr at 1½ in. Hg abs. Both condensers are two-pass and should be designed for: 150,000 lb per hr of steam, 70 F inlet water, No. 18 BWG Admiralty metal tubes that are 1 in. in diam, 12.5% investment charges, fuel cost of 27¢ per 10^6 Btu, condenser pressure correction factor for heat rate of 115 Btu per kwhr for a change of ½ in. Hg condenser pressure, effective tube length of 22 ft, 16-ft friction loss in circulating water system external to condenser, 7.5-fps water velocity, demand charge of \$130 per kw, 90% cleanliness factor, 3500 hr of operation per yr at full load.

SOLUTION. See tabulation on page 360. Note the order in which the items were calculated.

8–6. Circulating Pumps. Three kinds of circulating pump that are used for surface condensers—Francis, mixed flow, and axial—have been discussed in Chap. 3. Likewise, calculations of system head curves and friction losses in conduits have been previously discussed. However, it will be to advantage to show the application of these two topics to the circulating-water systems of surface condensers. Fig. 8–12 illustrates schematically a circulating-water system arranged to take advantage of the siphon effect in the intake and discharge pipes. Using the figures from Example 8–1, 7252 gpm are required with a friction loss through the condenser of 14.18 ft of water. From the procedures and formulas given in Chap. 2, friction loss and velocity head for the piping may be calculated. We shall assume that these have been found to be 2.5-ft friction loss for the piping, 0.76-ft velocity head, 1.0-ft loss each through the inlet and discharge tunnels. If a screen has been placed in the entrance to the inlet tunnel to remove any debris that might be in the river water, as shown in Fig. 1–2, there may be an additional loss of 1.0 ft. The total of all these losses would be 20.44 ft.

At low water level, the height of the siphon above the water level is

Solution to Example 8-2.

Item		15,000 sq ft	16,000 sq ft
1	Tubes per pass $= \dfrac{\text{surface}}{\text{passes} \times \text{length} \times \text{surface per ft}}$	$\dfrac{15{,}000}{2 \times 22 \times 0.2618} = 1302$	1390
2	Gpm of water = tubes per pass × vel × gpm @ 1 fps	$1302 \times 7.5 \times 1.994 = 19{,}500$	20,800
3	Water temperature rise $= t_2 - t_1 = \dfrac{\text{Btu per hr}}{500 \times \text{gpm}}$	$\dfrac{150{,}000 \times 950}{500 \times 19{,}500} = 14.60$	13.70
4	Temperature of water leaving $= t_1 + (t_2 - t_1)$, F	$70 + 14.60 = 84.60$	83.70
5	U, from Fig. 8-9 and Eq. (8-3)	$720 \times 1.0 \times 1.0 \times 0.90 \times 1.0 = 648$	648
6	θ_m, from Eq. (8-2)	$\dfrac{150{,}000 \times 950}{648 \times 15{,}000} = 14.65$	13.75
7	Terminal difference $= \theta_t$, from Eq. (8-7)	8.54	8.01
8	$t_{sat} = t_2 + \theta_t$, F	$84.60 + 8.54 = 93.14$	91.71
9	Water temperature range $= \theta_2 = t_{sat} - t_1$	$93.14 - 70 = 23.14$	21.71
10	Condenser pressure corresponding to t, in. Hg abs	1.57	1.50
11	Condenser friction, from Fig. 8-11, ft	$2 \times 1.58 + 22 \times 0.31 \times 2 = 16.80$ ft	16.80 ft
12	Circulating pump developed dynamic head condenser loss + condenser vel head + other losses, ft	$16.80 + \dfrac{(7.5)^2}{2g} + 16 = 33.67$	33.67
13	Circulating pump motor input, from Eq. (3-4), kw	$\dfrac{0.746 \times 19{,}500 \times 33.67}{3960 \times 0.75} = 165$	176
	Equipment Costs		
14	Tubes	$15{,}000 \times 1.45 = \$21{,}750$	\$23,200
15	Condenser and auxiliaries, installed, Eq. (8-4)	$15{,}000 + 4.85 \times 15{,}000 = \$87{,}700$	\$92,500
16	Two circulating water pumps, each at half-capacity and 40-ft head, Eq. (8-6)	$2\left[4200 + 0.55 \times \dfrac{19{,}500}{2}\right] = \$19{,}120$	\$19,840
17	Correction to circulating pump cost for 33.67-ft head	$19{,}120\,(40 - 33)\,0.005 = \$670$	\$690
18	Total equipment cost	\$129,240	\$136,230
	Annual Costs		
19	Investment cost	\$16,180	\$17,020
20	Vacuum penalty above arbitrary datum of 1.5 in. Hg abs	$\dfrac{22{,}000 \times 3500 \times 115 \times 0.07 \times 0.27}{0.5 \times 10^6} = \$340$	0
21	Annual cost of circulating pump demand	$0.125 \times \$130 \times 165 = \$2680$	\$2860
22	Annual cost of circulating pump power	$\dfrac{11{,}700 \times 3500 \times 165 \times 0.27}{10^6} = \$1820$	\$1950
23	Total comparative annual cost	\$21,020	\$21,830

360

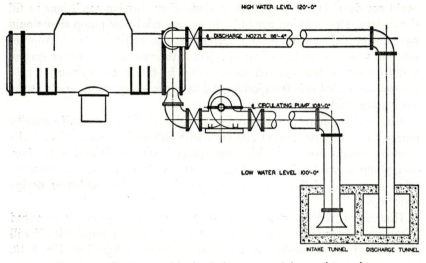

Fᵢɢ. 8–12. Condenser with circulating pump, piping, and tunnels.

16.33 ft. If we assume that the siphon has 10% losses due to entrained gases, water vapor pressure, etc., the total dynamic head required of the pump at low water would be $20.44 + 1.6 = 22.04$ ft at 7252 gpm.

When the river is at high water level, the siphon will be completely submerged. All losses will be the same as those for low water level except for the siphon loss. Therefore, the pump should develop only 20.44 ft at 7252 gpm. The pump should be designed for the higher head so that it will operate satisfactorily at low water level.

Siphon efficiency is a function of: (1) the amount of air entrained and dissolved in the circulating water, (2) the amount of horizontal run at the top of the siphon, and (3) the velocity of the water in the condenser outlet. The long, horizontal run in the discharge pipe of Fig. 8–12 would be detrimental to good siphon efficiency. A velocity at the condenser outlet of 8 fps or more promotes good siphon efficiency.

Now let us investigate the conditions for starting the pump. Obviously, there will be practically no flow at starting conditions, and so we can neglect friction and velocity heads. Also, when the river is at high water level, the system can be filled without starting the pump. At low water level, the pump must develop enough head when starting to raise the water into the discharge pipe and to start the siphon. The static elevation of the siphon above the low water level is 16.3 ft. Since we have specified the pump head to develop 22 ft at 7252 gpm, it will develop at least slightly more than this amount when starting, which will be sufficient to fill the system.

Now suppose that the height of the siphon is 30 ft instead of 16.3 ft. Then it would be entirely possible that the centrifugal circulating pump could not develop sufficient head at shut-off or starting conditions to fill the system. One solution would be to use an axial-flow pump that would have a very high shut-off head, i.e., 200% or more of design head. Another solution would be to install an air ejector to evacuate the water system and to assist a low shut-off head pump in filling the system. These ejectors are called *priming ejectors* and may use either steam or air as the motivating fluid.

Circulating pumps usually are installed in pairs, i.e., two half-capacity pumps to operate in parallel and both with full design head. When half-capacity pumps are used with a condenser having a divided water box, each pump may serve its half of the water box. Then a single pump operating alone will deliver very nearly one-half of the condenser design water flow.

However, if the water boxes are not divided, or the piping is arranged so that the water from both pumps is mixed, then one pump by itself will provide more than one-half of the total condenser design flow, Fig. 8–13.

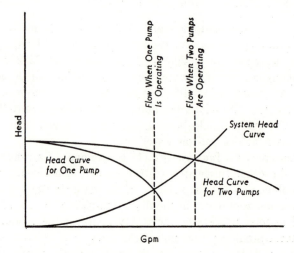

Fig. 8–13 One- and two-pump operation for a condenser circulating-water system.

8–7. Air Ejectors. Air and other noncondensable gases enter the condenser through leaks and along with the steam. The steam is condensed and removed from the condenser by the condensate pump. In older stations the condensate pump removed both water and noncondensable gases. The pump was of the reciprocating type and was called a *wet-vacuum* pump. This practice of removing both water and gases together is objectionable in a modern plant because oxygen in the feedwater may damage high-pressure boilers. Centrifugal condensate pumps cannot accommodate a gas along with a liquid without becoming airbound; and air remaining in the shell would increase the total pressure in the condenser as well as blanket the tubes of a surface condenser. Therefore, some other means must be used to remove the noncondensable gases from the condenser. Reciprocating air pumps, called *reciprocating dry vacuum pumps,* are really motor-driven, reciprocating air compressors. They have been used in some recent high-pressure stations because they have a very high capacity when the condenser is at approximately atmospheric pressure. They have the advantage of not requiring small, hazardous, high-pressure steam lines for their operation. Also, the condenser can be evacuated while the boiler pressure is still low. This permits starting the turbine with a low boiler pressure.

Steam-jet air ejectors have been used extensively and have found acceptance because of small space requirements, simplicity, reliability, low maintenance, and quick starting characteristics.

The principle of ejectors or injectors is discussed in Art. 9–3. Fig. 8–14 illustrates a typical two-stage design of condenser air ejector. Steam enters the first-stage steam nozzle through a strainer to prevent

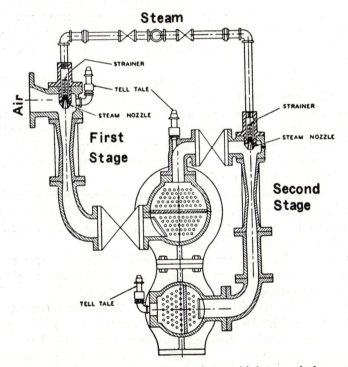

Fig. 8–14. Typical two-stage, steam-jet air ejector with inter- and aftercondensers. (Allis-Chalmers Manufacturing Co.)

clogging of the very small opening in the nozzle. After expansion through the nozzle, the steam enters the air chamber at a very high velocity and entrains air particles. Some of the kinetic energy of the steam is imparted to the air, and the mixture of steam and air is compressed in the diffusor by conversion of some of the kinetic energy to pressure energy. The total pressure at the outlet of the first stage is often 4 to 5 in. of Hg abs. Steam and air flow into the small surface type of condenser, known as the *intercondenser* or *intercooler*, where the steam is condensed on the cold tubes. Air is not condensed in the intercooler and is drawn into the second stage of the ejector. The second stage is practically a duplicate of the first stage and further compresses the air to slightly above atmospheric pressure. The second small condenser, *aftercondenser* or *aftercooler*, liquefies the motivating steam used in the second stage, and the remaining gases are vented to atmosphere.

Condensed steam or drains from the intercondenser are returned to the main condenser steam space by a loop seal or drainer. The minimum height of the loop seal should be about $7\frac{1}{2}$ ft for an intercondenser pressure of 4 or 5 in. Hg abs. Drains from the aftercondenser may be returned to the condenser through a drainer or liquid-level controller, Fig. 3–31. Because the drains from the aftercooler frequently contain considerable ammonia (from water treatment chemicals, etc.), it is preferable to discharge them to waste so that the ammonia will not attack the condenser tubes.

When condensate is used as the cooling medium in the inter- and aftercondensers, nearly all the energy of the steam supplied to the ejector unit and not used for compression is transferred to the feedwater and retained by the power cycle. Even though ejectors are very inefficient, when considering the amount of work done and the energy supplied to them, they are economical to operate because the only energy lost from the power cycle is that from the drains.

Jet condensers may be used as inter- and aftercondensers in place of the surface type of condenser. However, they have the characteristic disadvantage of all jet condensers in that steam is mixed with and contaminated by the condensing water.

Air entering the condenser from small leaks is the major source of noncondensable gases. Very few, if any, central stations, marine plants, or industrial stations are designed for continuous condensing operation at higher than 3 in. Hg abs. Therefore, air leaking into a condenser is a function of the tightness of all joints and connections and is independent of the normal operating pressure.* Air ejectors are operated at constant-steam consumption regardless of turbine load. But the quantity of condensate flowing through the inter- and aftercondensers decreases as the

---

* This is because the pressure drop from atmospheric to condenser pressure is greater than the critical pressure drop.

turbine load decreases. Under these conditions the temperature of water leaving the inter- and aftercondensers will increase as turbine load decreases, until a point is reached where the condensers cannot liquefy all the steam used by the ejectors. Two schemes that may be used to maintain operation of the ejector unit at low turbine loads are: (1) a portion of the water spaces of the inter- and aftercondensers may be connected to the river-water system, or (2) a by-pass connection from the condensate pipe located after the ejector condensers may allow some of the condensate to return to the condenser hotwell. The first scheme has the disadvantage that there is always the possibility of river water leaking into the condensate and contaminating it. The second scheme is popular for both stationary and marine plants. Fig. 8–15 is a schematic drawing

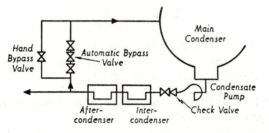

Fig. 8–15. Air-ejector condenser by-pass valve for low-load operation.

of such a by-pass system. The by-pass valve is actuated either by the temperature of the condensate leaving the aftercondenser (130 F to 180 F as specified by the manufacturer) or by a flow meter in the condensate line.

Whenever there is a mixture of a gas and a vapor in one chamber, the total pressure is the sum of the partial pressures of each constituent. The amount of air in the condenser shell should be small compared with the amount of steam, and the total condenser pressure is usually taken as the steam pressure. However, at the air outlet of a well-designed condenser there should be a small percentage of steam and a large percentage of air.

Let us consider a surface condenser operating at a total pressure of 1 in. Hg abs, which corresponds to a saturation temperature of 79.0 F. If the gas-vapor mixture leaving the air outlet were at a temperature of 79.0 F, Dalton's law shows that there would be no air present in the mixture because steam at 79.0 F has a pressure of 1 in. Hg abs and the total pressure is 1 in. Hg abs. Thus, the partial pressure of the gas would be zero or there would be no air in the mixture.

We have shown that the temperature of the gas-vapor mixture leaving the condenser must be lower than the temperature existing in the main portion of the condenser. Manufacturers designate a section of the condenser near the air outlet as the *air-cooler section* and make sure that it receives the coldest portion of the circulating water, i.e., it is located in the first pass of a multipass condenser.

The Heat Exchange Institute specifies that the temperature of the mixture at the air-vapor outlet, regardless of whether the condenser is fitted with a separate or integral air cooler, should be saturated at $7\frac{1}{2}$ deg less than the saturation temperature of the steam in the condenser, for the purposes of selecting vacuum pump capacity. It is customary always to consider the steam to be saturated. The weight of steam per pound of dry air at the condenser outlet may be calculated from an equation derived by assuming steam to be a perfect gas and by equaling $V = WRT/p$ for air to the same equation for steam.

$$W = 0.62 \frac{p_s}{p - p_s} \tag{8-8}$$

where W = weight of steam per lb of dry air, lb
　　　p_s = steam vapor pressure at air off-take temperature
　　　p = pressure at condenser air off-take

The pressures may be in either pounds per square inch absolute or inches Hg absolute, provided that the ratio is dimensionless.

In order to reduce the condenser pressure quickly during start-up periods, *priming ejectors* are also installed on condensers. The units have very large capacities for the rapid initial evacuation but will not reduce the condenser pressure below 4 to 6 in. Hg abs. The evactors may use steam, water, or compressed air as the motivating fluid. Since they are used for only a short time during start-up, no aftercondensers are used for the steam-driven units.

These rapid evactors are also used on the water space of some condensers to assist in starting the circulating-water pumps, as previously mentioned.

Feedwater Heaters

8–8. Surface Type of Feedwater Heater. Fundamentally, there is little difference between a surface condenser and a surface feedwater heater. Both heat water that normally flows inside metal tubes by condensing steam on the outside of the tubes. Temperature-surface diagrams for both are the same. The differences between the two pieces of equipment lie in such physical characteristics as the temperatures of both fluids, pressures of the fluids, and the strengths for which the parts are designed.

Fig. 8–16 illustrates a horizontal, surface type of feedwater heater with the important parts labeled. Water enters the *stationary head* or *water box* from the bottom of this two-pass heater and flows through the first pass to the reversing chamber, called the *floating head*. After its direction is reversed, the water flows back to the stationary head and out at the top, taking advantage of the natural tendency of warm water to flow upward. Of course the water is under pressure and must flow through the path provided, regardless of natural circulation tendencies. Steam enters the top of the shell and impinges on an impact baffle directly below the inlet nozzle. This baffle prevents vibration and erosion of the tubes from direct impingement of the steam and also forces steam to flow to both ends of the heater so that the full length of the tubes may be utilized.

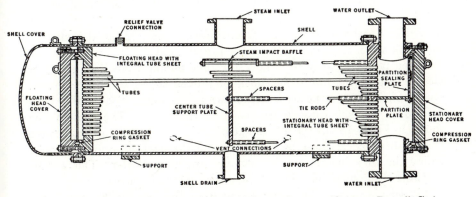

Fig. 8–16. Horizontal, surface type of feedwater heater. (Griscom-Russell Co.)

Tubes are rolled into both the *stationary-head tube sheet* and the *floating-head tube sheet*. The tubes are laid out on a triangular pitch; the pitch being about 1.25 times the nominal tube diameter (see Fig. 8–20). One constructional difference between the heater and condenser is that heater shells must be designed for steam pressures varying from vacuum to 800 psi or more. Expansion joints of the type used on condenser shells would be impractical at these pressures. Therefore, relative tube and shell expansion is taken care of by the floating head which can slide back and forth inside the shell. The tube support plates used in heaters are often semicircular in shape.

Details of the stationary head and floating head will depend on the design pressures. The heater of Fig. 8–16 has both tube sheets made of steel. They are made integral with their respective heads by hogging out billets that were originally a solid piece of metal having the maximum depth and diameter of the head. Such construction, while expensive, ensures a head suitable for high pressures. Heads may also be formed by welding for high-pressure use. Another type of stationary head is shown in Fig. 8–17 where the tube sheet is separate from the *channel* or water

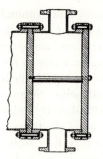

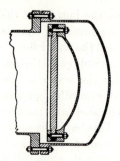

Fig. 8–17. Channel with separate tube sheet. (Griscom-Russell Co.)

Fig. 8–18. Nonpull-through type of floating head. (Griscom-Russell Co.)

box. This construction has the disadvantage of an additional joint that may leak and is therefore used for low pressure only.

When it is necessary to remove the tube bundle from the heater for cleaning, inspection, or replacement of broken tubes, the whole bundle (including stationary head) is withdrawn to the right. The floating head and cover will slip through the inside of the heater. The construction shown in Fig. 8–18 has the advantage of a smaller shell diameter, but the floating-head cover must be removed from the floating-head tube sheet before the tube bundle can be removed from the heater. This entails breaking the flanged joint between the shell and shell cover.

Floating heads may be eliminated by using U tubes, Fig. 8–19, with the advantage that some joints are thereby eliminated, and thus there is less likelihood of leaks developing. U tubes should not be used when there is a scale-forming fluid in the tubes unless the tubes are to be chemically

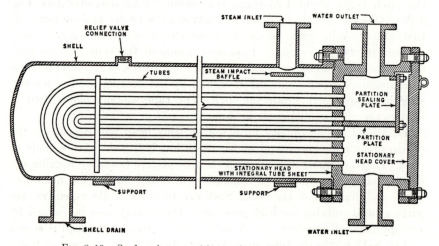

Fig. 8–19. Surface heater with U tubes. (Griscom-Russell Co.)

cleaned with acid. Another disadvantage is that some good tubes may have to be cut out of the bundle in order to reach and replace a faulty tube near the center of the bundle. Generally, the advantages outweigh the disadvantages, since more than three out of four feedwater heaters currently being installed are of the U-tube type.

Horizontal-closed feedwater heaters, as shown in Fig. 8–16, require little headroom but need a clear floor space in front of the heater that is slightly longer than the tube bundle so that the bundle can be withdrawn from the heater. Vertical heaters may be used when headroom is ample and floor space is at a premium. They can be arranged for removal of the shell from the bundle, Fig. 8–20, or the bundle from the shell by invert-

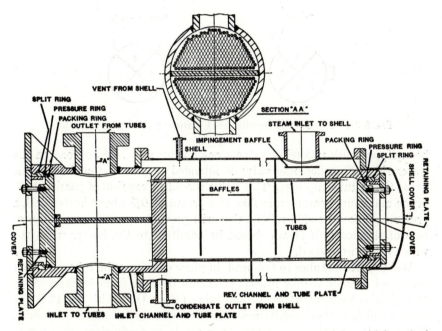

Fig. 8–20. Vertical surface type of feedwater heater arranged for withdrawal of the shell from the tube bundle. (Westinghouse Electric Corp.)

ing the heater. The feet for the heater are on channel at the left of the illustration.

Channels of surface feedwater heaters are partitioned for even numbers of passes, ranging from two to eight passes. For a given surface area and velocity, increasing the number of passes above two will shorten the heater but will increase the shell diameter. Generally, a long heater with a small diameter will cost less than a short but large-diameter heater. Channel partitions may be arranged as indicated in Fig. 8–21. Different temperatures in the first and fourth passes will cause different expansion of the tubes in the two passes. Therefore, the tube sheet has a tendency to warp when the partitions are arranged in quadrants.

Design pressures for heater shells should be at least 10% above the maximum pressure indicated by the turbine performance. Heaters should

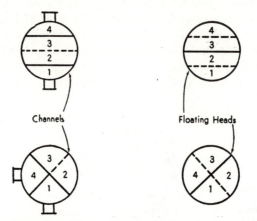

Fɪɢ. 8–21.   Channel and floating head partitions for four-pass heaters.

be equipped with two relief valves; one on the shell side and one on the water side. The shell-side valve is intended to protect the shell in the event that a tube ruptures. This valve should be capable of passing 10% of the maximum heater water flow at a pressure 10% above heater design pressure. The tube-side relief valve will prevent overpressure and rupture of the heater if steam should be admitted to the heater when the water valves are closed.

Welding of the tubes to the head, in place of rolling the tubes, is being employed in some U-tube heaters to ensure freedom from leakage. Also, welded tubes permit the use of small tubes (⅝-in. OD) with No. 12 BWG walls or thicker; these small thick tubes cannot be rolled into a tube sheet. Roughly, this thickness of tube corresponds to feedwater pressures of higher than 3300 psig, depending on the material.

8–9. Heat-Balance Calculations for Surface Heaters. Terminal difference for a surface heater is defined in the same way as that for a surface condenser; *it is the difference between the saturation temperature of the steam in the heater and the temperature of the water leaving the heater.* Saturation temperature is always used, even though the steam may be entering the heater in a superheated state.

Steam requirements for a heater may be determined by the Law of Conservation of Energy; or, all the energy entering a system during a given period of time must equal the energy leaving during the same period when the process is steady flow. The period of time used in heater calculations is customarily 1 hr.

Steam, condensing on the heater tubes, will fall to the bottom of the heater as a rain of drops. Since the drops impinge on successive tubes of a horizontal heater or run down the sides of vertical tubes, additional heat transfer takes place with the result that the drains leave the heater at less than saturation temperaure. The reduction below saturation temperature, known as *subcooling*, may amount to 5 or 10 deg for large heaters. The higher value is associated with vertical heaters. It is common to neglect subcooling in heater calculations. Subcooling is often a benefit to the cycle rather than a detriment.

Losses of energy from the heater because of radiation, convection, and conduction through the heater insulation are minor and should be neglected in calculations, unless it is known that the heater will be uninsulated.

Making the energy balance around the heater, Fig. 8–22, we get:

$$\overset{\text{Entering}}{S \times h_s + W \times h_1} = \overset{\text{Leaving}}{S \times h_d + W \times h_2} \tag{8-9}$$

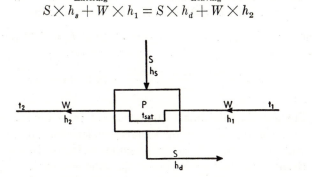

Fig. 8–22.   Heat balance for a surface heater.

With one equation we may have one unknown. Usually, but not always, the unknown is the steam required, S. Solving for S,

$$S = \frac{W(h_2 - h_1)}{h_s - h_d} \tag{8-10}$$

where S = steam flow, lb per hr

W = water flow, lb per hr

h_s = enthalpy of the steam, Btu per lb

h_d = enthalpy of drains corresponding to saturated liquid at steam pressure, Btu per lb

h_1 = enthalpy of water entering corresponding to temperature t_1, Btu per lb

h_2 = enthalpy of water leaving, corresponding to temperature
$t_2 = t_{sat} - $ TD, Btu per lb

p = steam pressure, psia

t_{sat} = *saturation* temperature of steam, F

t_1 = temperature of water entering, F

t_2 = temperature of water leaving, F

TD = terminal difference, F

Note that each item of Eq. (8–9) is a product of pounds per hour times Btu per pound, giving Btu per hour.

Many problems will be encountered where several fluids are entering the heater shell in addition to the steam. Regardless of the number, the energy entering the heater per unit time must equal the energy leaving the heater per unit time.

If the effects of compressed liquid are to be considered, then the feedwater enthalpies should be taken from the table on "Compressed Liquid" in the steam tables.

EXAMPLE 8–3.   Find the amount of steam needed by the heater, Fig. 8–22, at 75 psia, 350 F, with 5 F TD, 350,000 lb per hr of water entering at 260 F.

SOLUTION. $t_{sat} = 307.6$ F, $t_2 = 307.6 - 5 = 302.6$ F, $h_2 = 272.3$ Btu, $h_1 = 228.6$ Btu, $h_d = 277.4$ Btu, $h_s = 1205.3$ Btu. From Eq. (8–10),

$$S = \frac{350,000(272.3 - 228.6)}{1205.3 - 277.4} = 16,480 \text{ lb per hr}$$

8–10. Surface Calculations.   Heat-transfer rates for surface heaters, Fig. 8–23, are similar to those for surface condensers. Since clean water flows through the heater tubes, a cleanliness factor is not used with heaters. Because the water film inside the tube is a controlling factor in the heat transfer, its effect is defined in accordance with a somewhat arbitrary *film temperature*, which is found by subtracting eight-tenths of the log mean temperature difference from the saturated steam temperature. If calculations give a film temperature of more than 250 F, then the curve for 250 F should be used to find the heat-transfer rate.

Observe that the velocity used as the abscissa of Fig. 8–23 is a "cold water velocity" using a density of 62.4 lb per cu ft regardless of the actual water temperature. Cold water velocities in excess of 10 fps are not used for design purposes because of possible erosion. However, this velocity may be exceeded under overload conditions.

Admiralty metal tubes are used for low-pressure heaters where the tube metal temperature (metal temperature for condensing zones is assumed as the saturated steam temperature at the shell design pressure) does not exceed 450 F. Maximum temperatures for other common tube materials

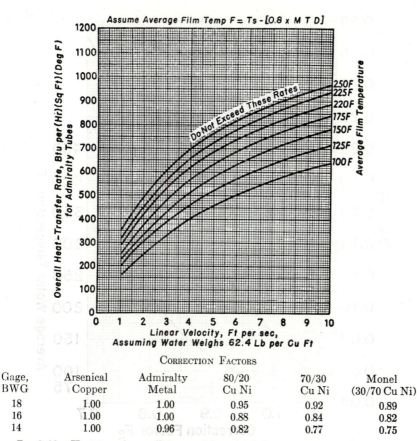

Fig. 8–23. Heat transfer coefficients, with multiplying factors, for feedwater heater tubes. (Bleeder Heater Manufacturers Association, Inc.)

CORRECTION FACTORS

Gage, BWG	Arsenical Copper	Admiralty Metal	80/20 Cu Ni	70/30 Cu Ni	Monel (30/70 Cu Ni)
18	1.00	1.00	0.95	0.92	0.89
16	1.00	1.00	0.88	0.84	0.82
14	1.00	0.96	0.82	0.77	0.75

are: 400 F for arsenical copper, 700 F for 80/20 and 70/30 copper nickel, and 800 F for Monel metal.

Friction loss through the tubes can be calculated from the data of Fig. 8–24. Average water temperature is taken as the saturated steam temperature minus the log mean temperature difference.

EXAMPLE 8–4. A 12,650-kw turbine generator may have either three or four stages of feedwater heating. If the unit will operate at this load for 4000 hr per yr, calculate the return on the investment for the fourth heater for the following data: ⅝ in. No. 16 BWG 80/20 Cu-Ni tubes, water temperature of 275 F entering heater and 350 F leaving, improvement in heat rate is 175 Btu per kwhr, heater costs $14.00 per sq ft of surface plus $6000 for installation and piping, 5 F TD, 120,000 lb per hr of water, fuel cost 31¢ per 10⁶ Btu, pump efficiency is 69% (wire-to-water), additional piping pressure loss is 4 psi, cold water velocity in the tubes to be 8 fps, plant heat rate of 12,500 Btu per kwhr with four heaters.

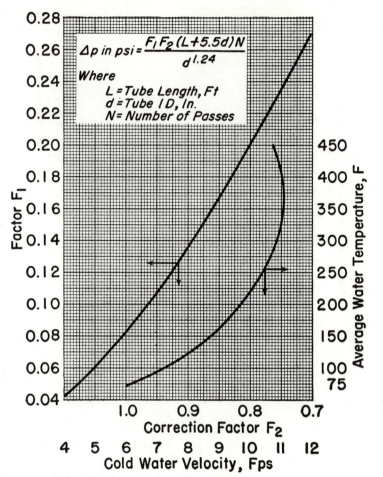

FIG. 8–24. Tube-side pressure drop for surface type of feedwater heaters and drain coolers. (Bleeder Heater Manufacturers Assn., Inc.)

SOLUTION. $t_s = 350 + 5 = 355$ F. From Eq. 8–1,

$$\theta_m = \frac{350 - 275}{\ln \dfrac{355 - 275}{5}} = 27.05 \text{ F}$$

$$Q = W \, \Delta h = 120,000(321.6 - 243.9) = 9,340,000 \text{ Btu per hr}$$

Film temperature $= 355 - 0.8(27.05) = 333.3$ F

Since this film temperature is above 250 F, use the curve for 250 F in Fig. 8–23 to get $U = 910$ and the correction factor equal to 0.88. The corrected value of U is then $910 \times 0.88 = 801$. From Eq. 8–2,

$$A = \frac{9,340,000}{801 \times 27.05} = 431 \text{ sq ft}$$

Using the data of Table 8–1, the tubes per pass will be, for a cold water velocity of 8 fps and a factor of 0.600 gpm at 1 fps,

$$\frac{120,000}{500 \times 0.600 \times 8} = 50 \text{ tubes per pass}$$

If the heater has two passes, the effective tube length will be, using a factor of 0.1636 sq ft of surface per lin ft of tube from Table 8–1,

$$\frac{431}{2 \times 50 \times 0.1636} = 26.4 \text{ ft}$$

From the dimensions of feedwater heaters given in the Appendix, it can be seen that this would be a very long heater of small shell diameter. Therefore, select a four-pass heater with an effective tube length of $(26.4 \times 2)/4 = 13.2$ ft.

From Fig. 8–24, the average water temperature would be $355 - 27.05 = 328$ F, and the friction loss would be

$$\frac{0.135 \times 0.75(13.2 + 5.5 \times 0.495)4}{(0.495)^{1.24}} = 15.43 \text{ psi}$$

The additional pressure loss in the feedwater system will be $4 + 15.43 =$ say, 19.4 psi. The additional input to the boiler feed-pump motor will be, from Eqs. (3–5) and (3–4a) and at a specific gravity of 0.903 for the average water temperature of 328 F,

$$\frac{120,000}{500 \times 0.903} = 266 \text{ gpm}$$

$$\text{Motor input} = \frac{0.746 \times 266 \times 19.4}{0.69 \times 1714} = 3.25 \text{ kw}$$

The annual cost of this additional power can be obtained from the plant heat rate, the fuel cost, and the hours of operation as

$$\frac{3.25(12,500 - 175)4000 \times 0.31}{10^6} = \$49.80$$

Observe that this cost is very small and usually can be neglected. The annual savings from the reduced heat rate is calculated in a similar way—by using the reduction in heat rate, the load in kilowatts, the fuel cost, and the annual hours of operation—to be

$$\frac{175 \times 12,650 \times 4000 \times 0.31}{10^6} = \$2750$$

The net return is $2750 - $49.80 =$ say $2700.

The cost of the heater installed will be 431 sq ft times $14.00 per sq ft plus $6000, or $12,030. The return on the investment will be

$$\frac{\$2700}{\$12,030} = 22.4\%$$

8–11. Desuperheating Zones. Manufacturers will guarantee a surface heater of the type that has been discussed for a terminal difference of 2 deg or more. If a customer wants a heater with less than a 2-deg termi-

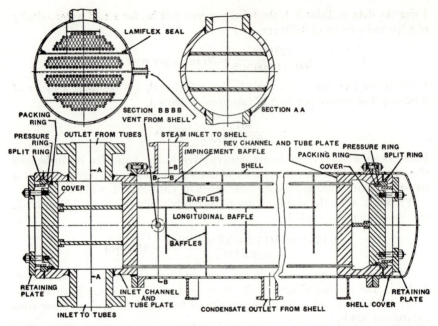

FIG. 8-25. Surface heater with desuperheating zone. (Westinghouse Electric Corp.)

nal difference, the manufacturer will include a *desuperheating zone* in the heater, Fig. 8-25, provided that the heater is supplied with superheated steam. The zone is formed by segregating the tubes of the last pass of the heater with special baffles that force superheated steam over these tubes before it can come in contact with the remainder of the heating surface. In that way, heat is transferred from superheated steam to the feedwater just before the feedwater leaves the heater.

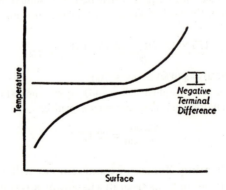

FIG. 8-26. Temperature-surface diagram for a closed heater with desuperheating zone.

A temperature-surface diagram for a heater with desuperheating zone is shown in Fig. 8–26.

Manufacturers will guarantee a minus 3- to 5-deg terminal difference with this construction if the steam is highly superheated.

There is an appreciable pressure drop through the desuperheating zone of heaters so that the steam pressure in the condensing zone is less than the pressure of the steam supplied to the heater. Since the terminal difference is based on the saturation temperature of the steam *supplied* to the heater, the load on the desuperheating zone is increased as the pressure drop through the zone is increased.

Heat transfer through the dry gaseous film on the outside of a tube in the desuperheating zone is less than the wet film of a tube in the condensing zone. Coefficients for desuperheating zones range from 70 to 125 Btu per (sq ft) (hr) (F).

8–12. Contact Heaters and Deaerators. Feedwater heaters that function by mixing steam with the feedwater are sometimes known as *open heaters,* the name being a holdover from the days when the feedwater was heated to only 210 F or 212 F in heaters that were vented to atmosphere to maintain their pressure at approximately 14.7 psia. A preferred name for these units is *contact heater.*

Any contact heater will liberate some of the noncondensable gases, always present to a greater or lesser degree, from the steam and water. These gases enter the system through leaks, by chemical reactions due to water treating, and by dissociation of the steam and chemicals. Of the noncondensable gases present in the water and steam, oxygen, carbon dioxide, and ammonia attack various common power-plant metals and therefore must be removed from the fluids.

A contact heater that is especially designed to remove the noncondensable gases is termed a *deaerator.* This term is a misnomer because it is not the air that is objectionable and must be removed, but the three gases listed above. A modern deaerator will be guaranteed to deliver water containing less than 0.005 cc of oxygen per liter. This corresponds to less than 7 lb of oxygen in a billion pounds of water.

Liberation of the noncondensable gases in a deaerator is based upon Dalton's law (see Art. 8–7) and Henry's law.* The latter law states that the amount of a gas that can be dissolved in a liquid is proportional to the partial pressure of that gas in the atmosphere above the liquid (except for gases that unite chemically with the liquid). Therefore, if the feedwater in a deaerator is at saturation temperature, then theoretically the feedwater cannot contain any dissolved gases because the partial pressure of the steam in the atmosphere above the water will be equal to the total

---

* The process of gas removal by diffusion is also involved.

pressure. Thus, the partial pressures of gases other than steam will be zero, and the gases cannot be absorbed by the water.

In addition to heating the feedwater to its saturation temperature, a good deaerator must break the water into droplets or films to permit release of all gas bubbles that are formed. Agitation of the water also aids in the release of gas bubbles by reducing the surface tension of the liquid.

These principles of deaeration are illustrated in power-plant type of deaerating equipment shown in Figs. 8–27 to 8–29. Refer to Fig. 8–27 in which various parts of the unit have been labeled. Feedwater first flows through the *vent condenser* at the top. This is a U-tube surface feedwater heater. From there, the water flows into the spray distributor in the inner chamber of the deaerator where its temperature is further increased by mixing with the steam that enters the inner chamber through the perforations at the top. The feedwater then cascades over the staggered trays of the heating section. Upon leaving the heating section, the water is practically at saturation temperature. The water passes over another series of trays, known as the *air separating trays.*

Although some noncondensable gases are removed throughout the heating process in the inner shell, the air separating trays provide the final separation of gases from the water, the removal of which permits the low oxygen guarantee. A contact heater, without deaerating features, would not contain the air separating trays but in all other respects would be the same as Fig. 8–27.

Steam containing the liberated noncondensable gases leaves the shell and is conducted to the vent condenser. There, the steam is condensed and the noncondensable gases vented to atmosphere or to a region of lower pressure than the deaerator pressure. Drains from the vent condenser are returned to the air-separating tray section of the deaerator.

Note the downward concurrent flow of water and steam in the inner chamber of the unit. Water leaves the inner chamber through the water-seal pipe. Thus, only fresh steam, containing very little noncondensable gases, comes in contact with the deaerated water in the reservoir. When larger storage capacity is needed, the vertical deaerating tank may be placed on top of a horizontal storage tank (see Plate 9 of the Appendix).

The inner shell of the deaerator is made of stainless steel because of the corrosive gases removed from the water. The trays may be made of either cast iron or stainless steel, the latter being preferred for modern units because of reduced weight as well as corrosion resistance. A view of a stainless steel tray stack also is shown in Fig. 8–27.

Fig. 8–28 shows another deaerator having many of the features of the previous unit but exhibiting several important differences. Observe that the vent condenser has been replaced by an isolated jet preheater section in the top of the deaerator shell. Most of the oxygen is removed in this

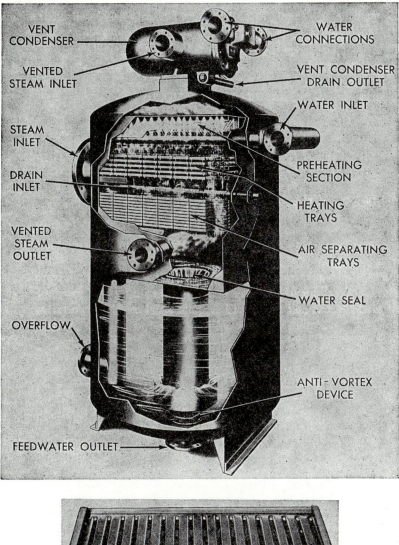

VENT
CONDENSER

WATER
CONNECTIONS

VENTED
STEAM INLET

VENT CONDENSER
DRAIN OUTLET

WATER INLET

STEAM
INLET

PREHEATING
SECTION

DRAIN
INLET

HEATING
TRAYS

VENTED
STEAM
OUTLET

AIR SEPARATING
TRAYS

WATER SEAL

OVERFLOW

ANTI-VORTEX
DEVICE

FEEDWATER OUTLET

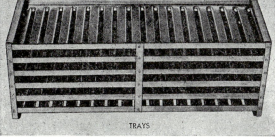

TRAYS

Fig. 8–27.　Tray type of deaerator.　(Cochrane Corp.)

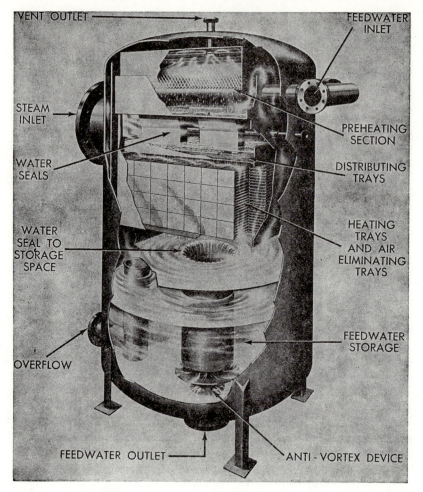

FIG. 8-28. Jet type of deaerator. (Cochrane Corp.)

preheater section. Water leaving the preheater is practically at saturation temperature and passes through the water seal that isolates the preheater section. Remaining noncondensable gases are removed in the air separating trays. Steam flows through the distributing trays and then the air separating trays before entering the preheating section.

Because the steam-pressure loss in this tray type of deaerator is extremely small (a few inches of water), the terminal difference based on the steam pressure at the steam inlet nozzle is practically zero and is guaranteed to be not more than 1 deg. To meet the oxygen guarantee, there must be a minimum temperature rise of 20 F in the vent condenser alone, while a jet unit can meet the guarantee with a total rise of 20 F.

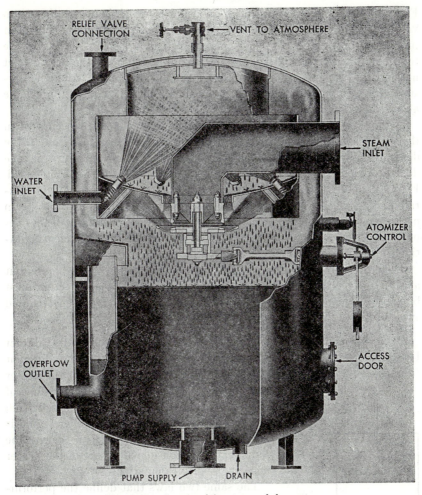

FIG. 8–29.　An atomizing type of deaerator.
(Cochrane Corp.)

Atomizing types of deaerator, Fig. 8–29, were originally developed for marine applications because deaeration will be adversely affected in the tray type due to the pitching and rolling of the vessel (the motion effects the thickness of water film spilling over the edges of the trays). Also, because atomizing deaerators are lighter in weight than tray units (particularly units with cast-iron trays), they can handle waters containing encrusting materials, and corrosive water.

Although there have been some applications of atomizing deaerators to stationary plants, two disadvantages have prevented their general acceptance. One disadvantage is the terminal difference. Observe in

Fig. 8–29 that steam flows through an atomizer which increases the steam velocity to provide agitation. By adjustment of the atomizer control, a pressure drop of about 1 psi is maintained at all steam flows. If the terminal difference is based on the steam pressure at the steam-inlet nozzle, this pressure loss will cause a terminal difference. At high pressures this terminal difference will be negligible, but it will be appreciable at low pressures.

A more serious disadvantage of the atomizing deaerator occurs when there is a load decrease on a turbine supplying extraction steam to the deaerator. When turbine load decreases, the steam pressure in the deaerator decreases (see variation of stage pressures with turbine load, Fig. 9–30 and Art. 9–13). Since the water in the storage reservoir is saturated, it will flash when the pressure drops and will supply the steam necessary for heating the incoming feedwater. Since this flash steam does not pass through the atomizing valve, it will not provide sufficient agitation for proper deaeration.

Deaerators will function at pressures above or below atmospheric, but at subatmospheric pressures, special provisions such as air ejectors must be made to vent the unit properly. Contact heaters have been operated successfully at pressures up to 800 psig.

8–13. Heat-Balance Calculations for Deaerators and Contact Heaters. Heat-balance calculations for a deaerator are no different from those for a contact heater. The calculations assume that the heater

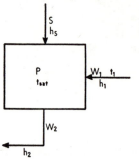

FIG. 8–30. Energy balance for a contact heater.

is perfectly insulated, that the process is one of steady flow, and that the loss of energy from the vent to atmosphere or some other region is negligible. If a vent condenser is used with the equipment, it is considered a part of the heater for heat-balance calculations.

The same two laws that define the performance of all heat-exchange equipment are used for contact-heater calculations. The Law of Conservation of Energy applies to the energy exchange and the Law of Conservation of Mass applies to the weights flowing. Refer to Fig. 8–30.

$$\text{Energy entering} = \text{energy leaving}$$

or
$$Sh_s + W_1 h_1 = W_2 h_2 \tag{8–11}$$

where S = steam flow, lb per hr
  W_1 = water entering, lb per hr
  W_2 = water leaving, lb per hr

h_s = steam enthalpy, Btu per lb
h_1 = enthalpy of entering water, Btu per lb
h_2 = enthalpy of water leaving, Btu per lb

Each product in this equation must equal Btu per hour. Regardless of the number of fluids entering the heater, the energy balance must always hold true. The enthalpy of the water leaving should be taken at zero terminal difference. A typical application of this equation would contain two unknowns, S and W_1. But, by applying the second of the two laws, we get

$$S + W_1 = W_2 \qquad\qquad (8\text{--}12)$$

Solving for the steam quantity,

$$S = \frac{W_2(h_2 - h_1)}{h_s - h_1} \qquad\qquad (8\text{--}13)$$

EXAMPLE 8–5. Steam enters a contact type of heater at 145 psia and 1264.0 Btu per lb. The heater discharges 550,000 lb per hr of water; the entering temperature is 286.0 F. Find the steam required.

SOLUTION.

$$t_s \text{ at } 145 \text{ psia} = 355.8 \text{ F}$$
$$h_1 = 255.2 \text{ Btu per lb}$$
$$h_2 = 327.7 \text{ Btu per lb}$$

Using Eq. (8–13), $S = \dfrac{550,000(327.7 - 255.2)}{1264.0 - 255.2} = 39,530 \text{ lb per hr}$

From Eq. (8–12), $W_1 = 550,000 - 39,530 = 510,470 \text{ lb per hr}$

Evaporators and Miscellaneous Heat Exchangers

8–14. Evaporators. Evaporators are one method of purifying make-up water for the system. On board ship, two sets of evaporators are used. One set purifies water for drinking purposes while the other supplies water for the power plant.

Make-up quantity for well-operated, modern, central stations averages at least 0.5% of the boiler steam output and can easily be 1 to 1½%. For a plant that supplies process steam, the make-up will be 1 to 2% plus the process steam loss. The capacity of the make-up system can well be double the average loss, or from a minimum of 1% up to 4 or 5%.

The two styles of evaporators that have been used are the film type and submerged type. In the film type, a spray of water falls on tubes that are kept at a high temperature by motivating steam on the inside. This design has been replaced by the submerged type, where the tube bundle is submerged in the liquid, Fig. 8–31. Steam entering the unit passes through a U-shaped pipe that provides for expansion and then enters the

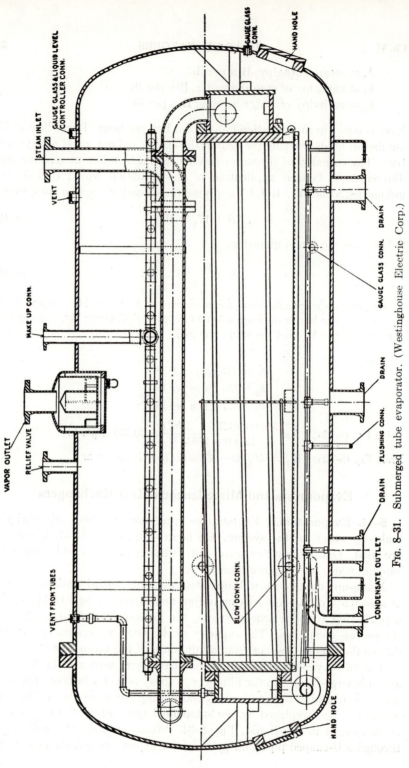

GAUGE GLASS & LIQUID LEVEL
CONTROLLER CONN.

STEAM INLET

VENT

MAKE UP CONN.

VAPOR OUTLET

RELIEF VALVE

VENT FROM TUBES

GAUGE GLASS
CONN.

HAND HOLE

DRAIN

GAUGE GLASS CONN.   DRAIN

DRAIN

FLUSHING CONN.

DRAIN

CONDENSATE OUTLET

BLOW DOWN CONN.

HAND HOLE

Fig. 8-31. Submerged tube evaporator. (Westinghouse Electric Corp.)

steam chest. Each tube receives steam from the chest and discharges condensate from the other end. The level of the raw water is maintained at about the centerline of the shell, so that all tubes are completely submerged. Evaporated water, called *vapor*, leaves the shell through a separator that removes any entrained moisture.

The design of the tube bundle is worthy of more attention. Note that all tubes slope toward the condensate discharge end to assure thorough draining of the tubes. Raw water entering the shell will contain impurities that must be removed. When the vapor leaves the shell, these impurities remain in the shell water. Some of the impurities will form a scale over the tubes which will reduce the heat transfer. The scale, if very soft and spongy or very hard and glasslike, may be removed by hand-cleaning the tubes. The entire tube bundle may be rolled out of the shell on the tracks provided. Another purpose of the U-tube steam pipe is to facilitate disconnecting the pipe for removal of the bundle.

Other scales can be removed by *cracking* the evaporator. The tubes are bowed to assist in the cracking process in which all water is drained from the shell and steam is admitted to the tubes. The tubes will then be heated to higher than their normal operating temperature, which causes the tubes to bend. After the steam is turned off, cold raw water is admitted to the shell as rapidly as possible to chill the tubes and to contract them. This bowing and straightening of the tubes causes the scale to crack off. Scale may then be removed by draining the shell through the blowdown connection.

When steam bubbles form around the tubes and rise to the surface, they break away from the surface with a violent boiling action that may carry some of the shell liquid along with the vapor as carry-over. This moisture, containing a high concentration of impurities, will defeat the sole purpose of the evaporator. This same effect of carry-over may be noted upon opening a bottle of ginger ale. As the gas leaves the bottle with a violent action, some of the fluid is entrained with the gas and may be seen leaving the bottle also.

Carry-over from an evaporator may be controlled by reducing the rate at which vapor leaves the liquid surface, known as *relieving rate* or *disengaging rate*. Relieving rates can be given as the velocity, in feet per second, at which the vapor leaves the water surface; the relieving surface is defined as the width of the water surface in the shell times the distance between the tube sheets. Manufacturers vary their design relieving rates, depending upon the shell-liquor concentration, on the maximum impurity permitted in the vapor and on the shell pressure. Practical values of relieving velocities for 1-ppm impurities in the vapor and 3000-ppm shell-liquor concentration are: 1 fps at 20-psia vapor pressure and 0.2 fps at 300-psia vapor pressure. Presumably, the relieving velocity should be lower at high vapor pressures because of the greater vapor density at the high vapor pressures.

Evaporators have given much trouble in high-pressure central stations and may give a very poor quality of vapor when shell pressures fluctuate. Therefore, it is good practice to install evaporators that are oversized or that have very low relieving velocities. A maximum relieving rate of 0.25 fps is recommended.

The reason for keeping the water surface at the approximate centerline of the shell is that the maximum relieving surface must be obtained for the shell diameter.

Pretreatment of the evaporator raw water will reduce the scale-forming tendencies and thereby reduce maintenance. A common form of pretreatment includes coagulation and zeolite application. Thus, the evaporator influent will contain impurities but not scale-forming impurities.

Any gases, dissolved or entrained, entering the evaporator with influent will pass on through the evaporator and leave with the vapor. If the raw water contains appreciable gases, it is advisable to deaerate the raw water rather than allow the gases to get into the system. Evaporator vapor may be used as the motivating steam in the deaerator.

When the shell-liquor concentration reaches the maximum permissible, then either the water in the shell must be dumped to waste or a part of the shell liquor withdrawn as blowdown. Continuous blowdown of a part of the shell liquor is very common practice. The blowdown water is taken to a heat exchanger where it is used to preheat the incoming raw water in the *blowdown heat exchanger*. The blowdown is then discharged to waste. A blowdown heat exchanger usually shows a very high return on the investment.

Fig. 8–32 shows a schematic arrangement of an evaporator with pretreatment and deaeration. Of course a hot-process softener could have been used.

Steam separators, usually constructed to remove moisture by centrifugal action, are also important in assuring pure vapor.

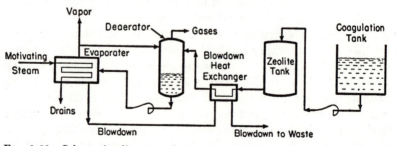

Fig. 8–32. Schematic diagram of evaporator with deaerator, blowdown heat exchanger, zeolite pretreatment, and coagulation.

Flow of raw water into the shell is governed by a float-controlled valve to maintain the desired water level.

8–15. Evaporator Arrangement and Energy Calculations. Steam needed to evaporate a given quantity of water may be calculated by applying the Law of Conservation of Energy in the same manner as for the other heat exchangers discussed in this chapter. Drains from the tubes are assumed to be saturated liquid, and the vapor is also assumed to be saturated. If the vapor were not saturated, then the moisture that it contained would be an indication of impurities present from carry-over.

Evaporators may be arranged in parallel or in series, Fig. 8–33. When in parallel, steam from the same source is supplied to two or more evap-

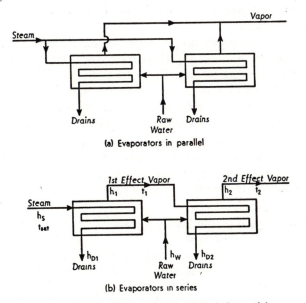

(a) Evaporators in parallel

(b) Evaporators in series

Fig. 8–33. Evaporator arrangements (schematic).

orators. The vapor pressure in each shell is the same, and the vapor from all the shells is condensed in one heater or condenser (not shown on Fig. 8–33). If the vapor is at a high enough pressure, it may be used as process steam. A rough rule for this arrangement is that each pound of motivating steam will evaporate about 0.8 lb of raw water into vapor.

When in series, each evaporator is known as an *effect.* The first evaporator in the chain is the *first effect,* the second evaporator is the *second effect,* etc. Motivating steam enters the first effect only and produces vapor. This first-effect vapor is used as the motivating fluid in the second effect, the second-effect vapor as the motivating fluid in the third, etc. The drains from the second effect are purified raw water. The pressure in each shell is different, the first effect having the highest pressure. Evaporators in series, Fig. 8–33(b), have two advantages over the parallel arrangement: first, series arrangement requires less steam per pound of vapor, and second, less vapor must be condensed in the condenser. A two-effect system will evaporate about 1.5 lb of raw water per lb of steam entering the first effect. The similar value for a three-effect system is about 2 lb of water per lb of steam.

The temperature-surface diagram for any one of the evaporators consists of two horizontal lines; initial steam superheat, if any, is neglected, Fig. 4–7(a). Since log mean temperature difference for a diagram of this type would be indeterminate, the term *heat head* is used to indicate the thermal potential. *Heat head is defined as the difference between the saturation temperature of the motivating fluid and the saturation temperature of the vapor.*

Energy balances for evaporators are based on the Law of Conservation of Energy—the same as for any other heat exchanger. When calculating the steam requirements for a series system, it is usually best to assume 1 lb of motivating steam and to determine the amount of raw water that it will evaporate. A direct proportion may then be used to determine the actual steam necessary to evaporate the required amount of raw water.

EXAMPLE 8–6. Calculate the steam requirements for a double-effect evaporator system, Fig. 8–33(b), when steam enters the system saturated at 50 psia. Vapor leaves the second effect at 16 psia. Raw water at 30,000 lb per hr enters the system at 60 F. Divide the total heat head for the system equally between the two effects.

SOLUTION. $h_s = 1174.1$ Btu per lb, $h_{d1} = 250.1$ Btu per lb, $h_2 = 1152.0$ Btu per lb, $t_{sat} = 281.0$ F, $t_2 = 216.3$ F.

> Total system heat head $= 281.0 - 216.3 = 64.7$ deg
> Heat head for each effect $= 64.7 \div 2 = 32.35$ deg
> $t_1 = 281.0 - 32.4 = 248.6$ F
> Pressure of vapor from first effect $= 29.1$ psia
> $h_1 = 1163.5$ Btu per lb, $\quad h_{d2} = 217.0$

Assume 1 lb of steam entering the first effect, and equate the energy entering to the energy leaving, with W_1 as the raw water. Thus,

$$W_1 = \frac{1(h_s - h_{d1})}{h_1 - h_w} = \frac{1(1174.1 - 250.1)}{1163.5 - 28.1} = 0.8138 \text{ lb}$$

We now have 0.8138 lb of vapor entering the coils of the second effect; it will evaporate W_2 pounds of raw water. Thus,

$$W_2 = \frac{W_1(h_1 - h_{d2})}{h_2 - h_w} = \frac{0.8138(1163.5 - 217.0)}{1152.0 - 28.1} = 0.6853 \text{ lb}$$

Therefore, each pound of steam will evaporate

$$0.8138 + 0.6853 = 1.4991 \text{ lb of raw water}$$

Total steam required $= 30,000 \div 1.4991 = 20,010$ lb per hr

$$\text{First-effect vapor} = 30,000 \left(\frac{0.8138}{1.4991}\right) = 16,300 \text{ lb per hr}$$

$$\text{Second-effect vapor} = 30,000 - 16,300 = 13,700 \text{ lb per hr}$$

EXAMPLE 8–7. Calculate the disengaging area for the first-effect evaporator of Example 8–6.

SOLUTION. Assume a disengaging velocity of 0.25 fps. From the steam tables, v_g at 29.1 psia $= 14.14$ cu ft per lb.

$$\text{Disengaging area} = \frac{\text{weight of vapor per sec} \times \text{sp vol}}{\text{velocity}}$$

$$= \frac{16,300 \times 14.14}{3600 \times 0.25} = 256.5 \text{ sq ft}$$

Extraction of steam from the turbine to provide motivating steam for an evaporator system reduces the turbine-generator output. Additional steam must be supplied to the turbine to restore the reduced turbine-generator output. Empirical equations have been developed for determining the increased steam flow to the turbine when adding an evaporator to the system (see Art. 9–15 for a discussion of the concept of extraction factors). For a single-effect evaporator,

$$Q = \frac{1.35W(\Delta h_1)}{\Delta h_2} \tag{8–14}$$

For a double-effect evaporator,

$$Q = \frac{0.75W(\Delta h_1)}{\Delta h_2} \tag{8–15}$$

The cost of supplying this additional steam for 1 hr of operation is

$$\text{Cost} = \frac{Q(\Delta h_3)(\text{fuel cost})}{\text{boiler efficiency}} \tag{8–16}$$

The investment for a single-effect evaporator system, including the deaerator and fittings and accessories installed in the plant, is

$$\text{Cost} = \frac{27W}{\theta_m} + 13,000 \tag{8–17}$$

where $Q =$ replacement throttle steam, lb per hr

$W =$ total evaporator vapor from all effects, lb per hr

$\Delta h_1 =$ difference in enthalpy of steam at the extraction point supplying the evaporator and enthalpy of the steam at the extraction point where the vapor is returned, Btu per lb

$\Delta h_2 =$ difference in enthalpy of steam at turbine inlet and turbine outlet, Btu per lb

$\Delta h_3 =$ difference in enthalpy of steam leaving boiler and water entering boiler, Btu per lb

$\theta_m =$ total heat head for all effects, F

Multiply Eq. (8–17) by 1.8 to obtain the cost of a double-effect system.

EXAMPLE 8–8. Calculate the return on the increased investment of a double-effect evaporator over a single-effect evaporator for the following data: 12,000 lb per hr of vapor, $\Delta h_1 = 88.2$, $\Delta h_2 = 477.7$, $\Delta h_3 = 1102.5$, 88% boiler efficiency, 3500 hr per yr of operation, 26¢ per 10^6 Btu fuel cost, $\theta_m = 81.8$ (data taken from Fig. 11–10).

SOLUTION. Use Eqs. (8–14) and (8–15).
For a single-effect evaporator system,

$$Q = \frac{1.35(12,000)(88.2)}{477.7} = 2990 \text{ lb per hr}$$

For a double-effect evaporator system,

$$Q = \frac{0.75(12,000)(88.2)}{477.7} = 1660 \text{ lb per hr}$$

Using Eq. (8–16) for the difference in steam flow of $2990 - 1660 = 1330$ lb per hr,

$$\text{Cost} = \frac{1330(1102.5)0.26}{0.88(10)^6} = \$0.434 \text{ per hr}$$

The cost would be 3500 hr × \$0.434 per hr = \$1518 per yr. Cost of a single-effect system, from Eq. (8–17),

$$\text{Cost} = \frac{27 \times 12,000}{81.8} + 13,000 = \$16,960$$

The cost for a double-effect system would be 1.8 × \$16,960 = \$30,500, and the return on the increased investment would be

$$\frac{\$1518}{\$30,500 - \$16,960} = 11.2\%$$

8–16. Surface Calculations. Evaporators do not operate satisfactorily with a heat head of less than 20 deg. When a heat head of about 100 deg is reached, the tubes become blanketed with steam, and the type of evaporation changes from nuclear to film. Nuclear evaporation takes place when bubbles of vapor form on the tube, and film vaporization consists of a film of vapor blanketing the tube and acting as an insulator. Thus, heat-transfer coefficients are much lower for film vaporization than for nuclear vaporization.

Coefficients of heat transfer for evaporator tubes may be calculated from the data given in Chap. 4. Cognizance should be taken of the depth of the tubes below the water surface in the form of a pressure correction. Fig. 8–34 gives some approximate coefficients. Eq. (8–2) applies to the surface calculations, provided that θ_m is taken as the heat head. If the data of Chap. 4 are used, the heat transfer rates should be multiplied by a factor of about 0.60 to account for scale accumulations on the tube.

When predicting the output of a given evaporator system at partial load, it is frequently satisfactory to assume that the vapor output is proportional to the heat head. This assumes that the coefficient of heat trans-

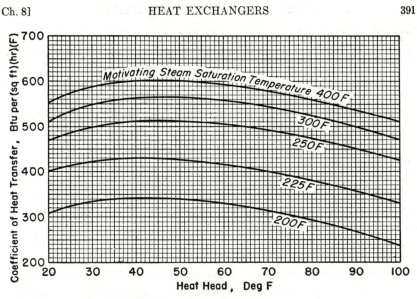

FIG. 8–34.　Heat transfer coefficients for evaporators.

fer is independent of the heat head, an assumption that is not exactly correct (see Fig. 8–34).

Sometimes evaporators in a multiple-effect system are all designed to have the same surface. In that event, each effect will have a different heat head, and the heat head for the first effect of a double-effect unit will be slightly more than one-half the total heat head for both effects. The calculations for that arrangement will be of the trial-and-error variety.

EXAMPLE 8–9. Calculate the tube surface for an evaporator of 12,000 lb per hr capacity that receives steam at 95 psia and 1244.7 Btu per lb; water enters at 60 F; vapor leaves at 26 psia; there is 10% blowdown.

SOLUTION.　Saturation temperature of the motivating steam is 324.1 F and of the vapor is 242.3 F. The enthalpy of the vapor is 1161.3 Btu per lb; of the shell liquor, 210.6 Btu per lb; and of the entering water, 28.1 Btu per lb.

$$\theta_m = 324.1 - 242.3 = 81.8 \text{ F}$$

From Fig. 8–34,　　　　　$U = 530$

and from Eq. (8–2),

$$A = \frac{Q}{U\theta_m} = \frac{12{,}000(1161.3 - 28.1) + 0.10 \times 12{,}000(210.6 - 28.1)}{530(81.8)}$$

$$= 319 \text{ sq ft}$$

8–17. Other Heat Exchangers.　Several other exchangers play an important part in the design and operation of steam plants. Most turbines and many pumps have oil coolers to provide cool lubricant to the bearings. Plants that use heavy oils for fuel must have oil heaters, since heavy oils will not flow in the pipes and will not ignite properly. Other applications of heat transfer are the air or hydrogen coolers for supplying a cool atmosphere to the generator. All this equipment follows the laws of heat transfer, but a detailed discussion would not be advisable here.

One additional piece of equipment that deserves mention is the *drain cooler* because it enters into the station heat balances, which will be discussed in Chap. 11. In general appearance and construction, drain coolers are similar to surface types of feedwater heater, but they differ in that both the hot and cold fluids are water. Drains leaving a surface heater are often at a high temperature and can be used economically to heat the feedwater. A drain cooler is a heat exchanger that has drains on one side of the tube wall and feedwater on the other, with both fluids changing temperature as they traverse the tube surface. Drain coolers may be built with either fluid on the inside of the tubes, friction loss being the determining factor. Internal baffles force the shell fluid to flow in a serpentine fashion over the tube bundle to ensure good heat transfer. With such an arrangement, the flow over the tubes is neither parallel nor counterflow, and therefore corrections should be made to the log mean temperature difference as indicated in the Heat Exchange Institute standards.

Since drain coolers are used only in connection with surface heaters, it is sometimes advantageous to combine the two pieces of equipment in one shell, Fig. 8–35. The drain-cooler section is located in the bottom of the heater and is kept submerged by maintaining a water level in the shell of the heater with a float-controlled valve or drainer. Condensed steam enters the drain-cooler section at the left and flows the length of the heater. Note the many baffles in this section.

Whenever a drain cooler is used in conjunction with a surface heater, the two units should be considered as one for the purposes of heat-balance computations. With this procedure, the trial-and-error calculation for the temperature of the feedwater leaving the drain cooler will be eliminated. After the steam flow is determined, the intermediate temperature may be found.

Terminal difference for a drain cooler is defined as the difference between the temperature of the drains leaving the unit and the temperature of the feedwater entering. This is the least temperature difference found on the temperature-surface diagram, and it is used to simplify the energy-balance calculations.

Heat-transfer coefficients for drain coolers are in the order of 350 Btu per (sq ft) (hr) (F).

In order to operate satisfactorily and not to vapor-bind, the drains in a drain cooler must be in the liquid phase. Therefore, the drainer valve

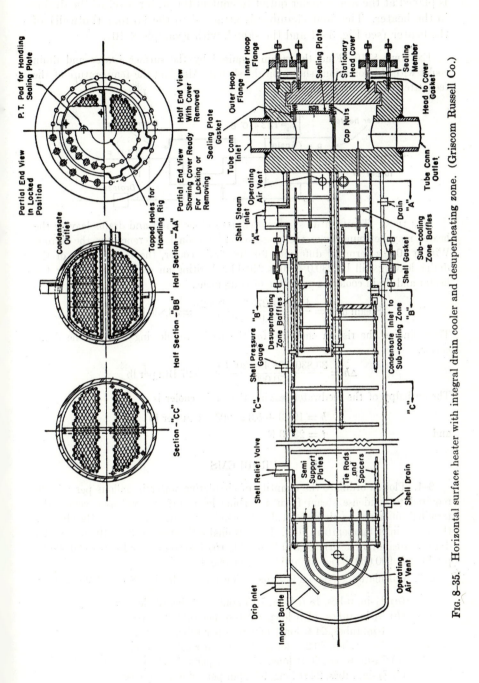

Fig. 8-35.   Horizontal surface heater with integral drain cooler and desuperheating zone.   (Griscom Russell Co.)

is placed at the drain cooler outlet to control the water level of the drains in the heater. The float chamber is attached to the bottom (hotwell) of the heater (see Fig. 3–31 and the sketch with Example 8–10).

EXAMPLE 8–10. Find the steam required by the surface heater and drain cooler with the conditions shown on the sketch. Also find the temperature of the feedwater leaving the drain cooler.

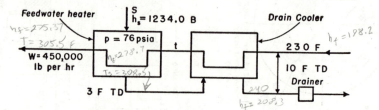

SOLUTION. Temperatures and enthalpies for the water and drains leaving the heater may be found from the pressure and terminal difference for the heater. With a 10-deg terminal difference for the drain cooler, the temperature of the drains leaving will be $230.0 + 10 = 240.0$ F. Writing an energy balance for the heater and drain cooler as one piece of equipment,

$$\text{Steam flow} = \frac{450,000(275.3 - 198.2)}{1234.0 - 208.3} = 33,830 \text{ lb per hr}$$

Then, finding the rise in enthalpy for the feedwater flowing through the drain cooler,

$$\Delta h = \frac{33,830(278.4 - 208.3)}{450,000} = 5.27 \text{ Btu per lb}$$

The enthalpy of the feedwater leaving the drain cooler is

$$h = 198.3 + 5.3 = 203.6 \text{ Btu per lb}$$

and $$t = 235.3 \text{ F}$$

PROBLEMS

8–1. Plot as ordinate the quantity of circulating water in gallons per minute required to condense steam from a turbine in a surface condenser versus the terminal difference as abscissa. Take the condenser pressure as 2 in. Hg abs and the water inlet temperature 70 F. Use terminal differences of 3, 5, 10, 15, and 20 deg and steam flows: (a) 100,000 lb per hr, (b) 300,000 lb per hr, (c) 600,000 lb per hr, (d) 900,000 lb per hr, and (e) 1,200,000 lb per hr.

8–2. Calculate the friction loss for each condenser having No. 18 BWG tubes:

    (a) ¾-in. tubes, 18 ft long, 7.15 gpm per tube, single pass
    (b) ¾-in. tubes, 14 ft long, 8.2 gpm per tube, two pass
    (c) 1-in. tubes, 24 ft long, 14.9 gpm per tube, two pass
    (d) 1-in. tubes, 30.2 ft long, 15.5 gpm per tube, two pass
    (e) ⅞-in. tubes, 26 ft long, 11.3 gpm per tube, two pass
    (f) ⅞-in. tubes, 22 ft long, 9.7 gpm per tube, three pass

8–3. Calculate the surface and friction loss required for each of the following condensers with No. 18 BWG Admiralty tubes to serve turbines:

Steam Flow, Lb/Hr	Condenser Press, In. Hg Abs	Terminal Differ- ence, Deg.	No. of Passes	Cleanli- ness, %	Tube Size	Velocity, Fps	Inlet Water Temp, F
(a) 100,000	2	10	4	80	7/8	6	78
(b) 450,000	2	8	1	70	7/8	7	80
(c) 700,000	1.5	7	2	90	1	7.5	67
(d) 550,000	1	6	1	95	1	8	55
(e) 300,000	1.5	8	2	85	7/8	6.5	70
(f) 200,000	2.0	11	2	80	7/8	6	77
(g) 175,000	2.5	15	1	90	3/4	8	83
(h) 400,000	1.5	13	2	75	1	7.5	65

8–4. A surface condenser is to operate under the following conditions: 150,000-kw turbine, exhaust steam rate 8 lb per kwhr, 1-in. Hg abs, single-pass, 3/4-in. No. 18 BWG Admiralty tubes, 50 F inlet water temperature, 8-fps velocity, 2-ft friction drop in the inlet tunnel, 1 ft for the outlet tunnel, 3-ft drop in the inlet pipe, 1.5 ft in the outlet pipe, top of outlet pipe 18 ft above river level, 85% cleanliness, 15-deg terminal difference, 80% circulating pump efficiency. Find: (a) outlet temperature of circulating water, (b) surface, (c) gpm of circulating water, (d) number of tubes in condenser, (e) effective tube length, (f) total weight of tubes in condenser, allowing 1¼-in. tube sheet thickness, (g) friction drop through condenser, (h) total dynamic head for circulating pump, (i) horse-power of circulating-pump motor.

8–5. The conditions given in a condenser specification were: 50,000-kw turbine, 1.5 in. Hg abs, 9-lb steam per kwhr, 70 F inlet water. A manufacturer proposed the following: single-pass condenser, 1428 tubes, 1-in. No. 18 BWG, 36,300 sq ft of surface, 42,800 gpm of circulating water. Comment on the manu-facturer's proposal, and show calculations to verify your conclusions.

8–6. Design a surface condenser for the following conditions: 25,000-kw tur-bine, exhaust steam rate 8 lb per kwhr, two passes, 7/8-in. No. 18 BWG tubes, 7-fps water velocity, 8-deg terminal difference, 2-ft friction drop in inlet tunnel, 1 ft in outlet tunnel, 3 ft in inlet pipe, 1.5 ft in outlet pipe, outlet pipe 24 ft above river level, 86% circulating-pump efficiency. Find: (a) outlet temperature of cir-culating water, (b) gallons per minute of circulating water, (c) surface area, (d) number of tubes in condenser, (e) total weight of tubes, assuming tube sheet thickness of 1 in., (f) friction drop through condenser, (g) total dynamic head on circulating pump, (h) horsepower of circulating-pump motor.

8–7. It is desired to increase the vacuum of an old condenser, keeping the inlet water temperature and steam flow constant. Suggest two things that might be done to the system.

8–8. For the circulating-pump characteristic given, find the quantity pumped by one pump alone and by two identical pumps operating in parallel. The system head is 16.5 ft at 115,000 gpm and there is no static head on the pump. Characteristics are: 65,000 gpm, 14.0-ft head; 70,000 gpm, 12.0-ft head; 75,000 gpm, 9.7-ft head; 80,000 gpm, 6.9-ft head; 83,000 gpm, 5.0-ft head.

8–9. A manufacturer offers a four-pass condenser having 8700 sq ft of surface for a 40,000-kw turbine that exhausts 6 lb of steam per kwhr at pressure of 1.5 in. Hg abs. Water enters the condenser at 70 F and leaves at 85 F with a velocity of 6 fps. Show by calculations whether the condenser is properly designed.

8–10. A surface condenser is to operate at a load of 10^8 Btu per hr with a pressure of 2 in. Hg abs, an inlet water temperature of 80 F, and a water temperature rise of 14 F. The surface area is 9000 sq ft. Calculate the number of 1-in. tubes per pass at a water velocity of 8 fps.

8–11. A surface condenser has an air leakage of 5.0 cfm of 70 F when the condenser operates at 1.2 in Hg abs. If the temperature at the air outlet is 80 F, how many pounds per hour of steam are lost with the air?

8–12. The temperature at the air outlet of a condenser is 72.7 F when the pressure is 1.05 in. Hg abs. Calculate the pounds of steam leaving per pound of air.

8–13. If 1 lb of steam leaves the air outlet of a condenser with each pound of air when the condenser operates at 1.5 in. Hg abs, what is the temperature at the air outlet?

8–14. If the air leakage into a condenser is 9 cfm of 70 F air at 14.7 psia, how much steam is lost from the condenser if the temperature at the air outlet is 84 F and the pressure is 1.70 in. Hg abs?

8–15. How much would the air leakage increase for the condenser of Prob. 8–14 if the condenser pressure dropped to 1 in. Hg abs.

8–16. Use the data for the example of Art. 8–5 to determine the economic advantage of the following condensers: (*a*) 14,000 sq ft of 1-in. tubes, (*b*) 13,000 sq ft of 1-in. tubes, (*c*) 12,000 sq ft of 1-in. tubes.

8–17. Determine the annual cost for each condenser to serve a 100,000-kw turbine generator for the following data: 9500 Btu per kwhr heat rate at 1.5 in. Hg abs; 100-Btu change in heat rate for each 0.5 in. Hg change in condenser pressure; 14% investment charges; 4000 hr per yr of full load operation; 31¢ per 10^6 Btu fuel cost; demand charge of $120 per kw; 500,000 lb per hr of steam condensed; 30-ft. ⅞-in., No. 18 BWG Admiralty metal tubes; 65 F inlet water; 12-ft friction loss external to condenser; 8-fps water velocity; 85% cleanliness factor. Use condenser surfaces of (*a*) 70,000 sq ft, (*b*) 65,000 sq ft, (*c*) 60,000 sq ft, (*d*) 55,000 sq ft.

8–18. A surface heater operates at 49.2 psia and receives steam at 1300.5 Btu per lb with 1,028,000 lb per hr of water entering at 226.3 F. Drains from another heater enter at 292.1 F and in the amount of 214,300 lb per hr. Calculate the steam flow for a plus 3 F TD.

8–19. The pressure in a surface heater is 1.28 psia when 230,900 lb per hr of water enters at 84.7 F. Drains in the amount of 57,600 lb per hr enter at 116.1 F. Calculate the steam flow at an enthalpy of 1092.5 Btu when the TD is plus 4 F.

8-20. A surface heater has a water flow of 1,100,600 lb per hr when the water enters at 305.2 F and leaves with a 1-deg terminal difference. Steam enters at 222 psig with an enthalpy of 1292.1 Btu per lb. Find the quantity of steam.

8-21. A surface heater has the following steam and water entering: 933,600 lb per hr of feedwater at 155.0 F, 6600 lb per hr of steam at 1382.0 Btu per lb, 25,000 lb per hr of steam at 1152 Btu per lb, 28,700 lb per hr of drains (water) at 242.2 F, 93,000 lb per hr of drains at 287.1 F. The feedwater leaves with a 2-deg terminal difference. Find the additional steam required at 1108.2 Btu per lb to do the heating when the heater operates at 16.4 psia.

8-22. What is the terminal difference of a closed heater that receives 257,500 lb per hr of 387.0 F water and 14,900 lb per hr of 360 psia steam at 537 F?

8-23. A surface heater receives 39,000 lb per hr of water, 2780 lb per hr of 76 psig steam at 98% quality, 3400 lb per hr of drains (water) at 352.2 F, and 750 lb per hr of drains at 302.3 F. If there is a 4-deg terminal difference, what is the temperature of the water entering?

8-24. The feedwater pressure in a surface heater is 1800 psig when the flow is 620,000 lb per hr. There is a 0 F TD, the steam pressure is 417.0 psia, the steam enthalpy is 1360.5 Btu per lb, and the feedwater enters at 380.4 F. Consider compressibility and calculate the steam flow.

8-25. A surface heater operates at 380 psia and receives 670,000 lb per hr of feedwater at 360 F. Calculate the surface and friction drop for No. 14 BWG 70/30 Cu-Ni tubes that are ⅝-in. diameter. Use a cold water velocity of 9 fps and a 2 F TD.

8-26. If the use of the heater in Prob. 8-24 improves the heat rate of a 100,000-kw turbine by 120 Btu per kwhr and costs $24.00 per sq ft installed, is the heater justified for 2900 hr per yr of full-load operation when fuel costs 19¢ per 10^6 Btu and investment charges are 13%?

8-27. The feedwater flow through a heater for a 66,000-kw turbine would be 550,000 lb per hr, with the water entering at 320 F and the steam pressure at 295 psia. Calculate the surface and friction drop for a cold water velocity of 7.5 fps through ¾-in. No. 16 BWG 80/20 tubes at a terminal difference of 4 F.

8-28. The heater of Prob. 8-26 will reduce the heat rate of the turbine by 97 Btu per kwhr. If the unit operates at 5000 hr per yr at full load and if fuel costs 27¢ per 10^6 Btu, what is the return on the investment if the heater costs $19.00 per sq ft installed?

8-29. A heater for a 44,000-kw turbine will have a feedwater flow of 400,000 lb per hr. The steam pressure will be 265 psia, and the entering water temperature will be 300 F. Calculate the surface and friction drop for No. 14 BWG Admiralty metal tubes at a 5 F TD. Use a cold water velocity of 8.5 fps.

8-30. When operating for 3800 hr per yr, the heater of Prob. 8-28 will improve the station heat rate by 100 Btu per kwhr. What is the return on the investment if the heater costs $14.00 per sq ft installed and fuel costs 23¢ per 10^6 Btu?

8-31. A surface heater containing ⅝-in. No. 18 BWG Admiralty metal tubes may be designed for 5 F or 10 F TD when the flow of feedwater is 200,000 lb per hr. Steam pressure is 30 psia, and entering water temperature is 200 F. The differential heater cost is $4.50 per sq ft and the change in heat rate is 5 Btu per kwhr for 20,000-kw generation for 3000 hr per yr. If fuel costs 24¢ per 10^6 Btu, which terminal difference should be used? Use a cold water velocity of 8 fps.

8–32. Determine the most economical terminal difference, 4 F or 6 F, for a surface heater to receive 522,000 lb per hr of water at 108 F when the steam pressure is 9 psia. Use 5⁄8-in. No. 18 BWG arsenical copper tubes. The heater will operate 4100 hr per yr at a generator load of 85,000 kw. Fuel costs 17¢ per 10^6 Btu, the incremental heater cost will be $4.00 per sq ft, and the improvement in heat rate will be 2.5 Btu per kwhr. Use a cold water velocity of 9 fps.

8–33. Compare feedwater heaters designed for 3 F TD and 7 F TD for the following data: 300,000 lb per hr of feedwater entering at 150 F; steam pressure, 18 psia; cold water velocity, 8.5 fps; 3⁄4-in. No. 16 BWG Admiralty metal tubes; generator load, 33,000 kw for 2200 hr per yr; 5 Btu per kwhr heat-rate improvement; $4.75 per sq ft incremental surface cost; 33¢ per 10^6 Btu fuel cost; 7.5-fps cold water velocity.

8–34. A surface heater had the following data used for design: steam at 114 psia; 5-deg terminal difference; 600-psi water pressure; inlet water temperature of 279.9 F; 358,240 lb per hr of water; 17-ft tube length; 189 tubes per pass; 1080 sq ft surface area; two passes; 5⁄8-in. No. 16 BWG Admiralty tubes. When the heater was manufactured, No. 16 BWG 5⁄8-in. low-carbon steel tubes were used. What effect will this have on the heater performance? Low-carbon steel will have 70% of the heat-transfer rate of the Admiralty metal tubes.

8–35. A parallel-flow, low-level jet condenser receives 110,000 lb of steam per hr from a turbine and has a 7-deg terminal difference when operating at 1.5 in. Hg abs. Calculate the circulating-water requirements in gallons per minute when the inlet water temperature is 67 F.

8–36. A barometric condenser operates at 4 in. Hg abs with a 5-deg terminal difference to condense steam from a 1000-bhp engine. Water enters at 80 F, and the engine uses 9.3 lb of steam per bhp-hr. Find the amount of circulating water required in terms of gallons per minute.

8–37. A contact heater operates at 40 psia and receives steam with 50-deg superheat and water at 180 F. Find the quantities of steam and water entering if 77,000 lb per hr of water leaves.

8–38. A deaerating heater receives steam at 6.5 psia and 1150.0 Btu at the rate of 25,000 lb per hr. If the water enters at 84.0 F, what are the water quantities entering and leaving?

8–39. When operating at 3 psig, a deaerating heater receives 10,000 lb per hr of steam at 875 F and other steam at 1% moisture. Water at 600,000 lb per hr enters at 146.5 F and 45,000 lb per hr at 263.0 F. How much of the moist steam is required?

8–40. A contact type of heater operates at 5.2 psia, with water entering at 81.0 F and steam at 1056.2 Btu per lb. Additional water enters at 229.3 F in the amount of 216,800 lb per hr. The total leaving the heater is 1,207,100 lb per hr. Find the amount of steam and water entering.

8–41. A deaerating heater operates at 26.7 psia and receives water at 181.6 F. Drains from higher-pressure heaters enter in the amount of 168,000 lb per hr at 303.7 F. The feedwater flow from the heater is 673,000 lb per hr. Calculate the steam flow for an enthalpy of 1237.2 Btu per lb.

8–42. A double-effect evaporator system is to evaporate 15,000 lb per hr of 60 F raw water when receiving steam saturated at 25 psia. The vapor from the second effect is at 3 psia. Find: (a) the heat head for each evaporator, assuming that the total heat head is divided equally between the two effects, (b) the steam required, (c) the quantity that would be evaporated if the initial steam pressure dropped to 17 psia during partial load operation and the vapor pressure for the second effect stayed the same.

8–43. A 35,000 lb per hr double-effect evaporator system receives steam at 40 psia and 300 F. Raw water enters at 60 F, and the pressure in the second effect is 5 psia. Find: (a) the heat head for each evaporator, assuming that it is equal for both, (b) the steam required, and (c) the quantity that would be evaporated if the initial pressure dropped to 16 psia and the vapor pressure leaving the second effect remained the same.

8–44. Calculate the surface required for each effect from the data given in Prob. 8–42.

8–45. Calculate the surface required for each evaporator in Prob. 8–43.

8–46. Work parts (a) and (b) of Prob. 8–42 by assuming that the surface area of each effect will be equal, rather than the heat heads being equal.

8–47. Repeat parts (a) and (b) of Prob. 8–43, but divide the heat heads so that the surface of each evaporator will be the same, rather than the heat heads being the same.

8–48. Calculate the cost per 1000 gal of evaporated water at 14% investment charges for a single-effect evaporator system that would have a capacity of 10,000 lb per hr. Other data are: 21¢ per 10^6 Btu fuel cost, 89% boiler efficiency, $\theta_m = 51$ F, $\Delta h_1 = 56$, $\Delta h_2 = 453$, $\Delta h_3 = 1136$. Motivating steam is at 89 psia.

8–49. Find the cost per 1000 gal of evaporated water for a double-effect system having a capacity of 20,000 lb per hr. Other data are: 17¢ per 10^6 Btu fuel cost, 87% boiler efficiency, $\theta_m = 64$ F, $\Delta h_1 = 71$, $\Delta h_2 = 470$, $\Delta h_3 = 1071$. Motivating steam is at 37 psia.

8–50. Calculate the return on the increased investment for a double-effect evaporator system versus a single-effect system to evaporate 18,000 lb per hr. The data are as follows: 33¢ per 10^6 Btu fuel cost, $\theta_m = 89$ F, $\Delta h_1 = 101.5$, $\Delta h_2 = 662$, $\Delta h_3 = 1183$. Boiler efficiency is 90% and 50,000,000 lb per yr evaporated steam pressure to the system is 52 psia.

8–51. Calculate the return on the increased investment for a double-effect evaporator to operate 5100 hr per yr with a capacity of 15,000 lb per hr. Other data are: $\theta_m = 55.6$ F, $\Delta h_1 = 81$, $\Delta h_2 = 626$, $\Delta h_3 = 1225$. Boiler efficiency is 88%. Steam pressure to the system is 77 psia.

8–52. Calculate the tube surface and the relieving area for the evaporator of Prob. 8–45.

8–53. Calculate the tube surface and the disengaging area for each evaporator of Prob. 8–46 by assuming equal heat heads.

8–54. Calculate the tube surface and the relieving area for the single-effect evaporator of Prob. 8–47.

8–55. Calculate the relieving area and the tube surface for the single-effect evaporator of Prob. 8–48.

8–56. Feedwater at a pressure of 1800 psia enters a combination heater and drain cooler at 380.4 F in the amount of 620,000 lb per hr. Steam enters at 417 psia and 1360.5 Btu per lb. There is a 0 F TD for the heater and a 10 F TD for the drain-cooler section. Consider compressibility and calculate the steam flow.

8–57. A tubular feedwater heater and drain cooler receives 1,027,000 lb per hr of feedwater at a pressure of 2800 psia and a temperature of 371.2 F. Steam enters at 420 psia and 1344 Btu per lb. Drains from another heater enter at 461.7 F in the amount of 60,100 lb per hr. Terminal differences are −2 F TD for the heater and 10 F TD for the drain cooler. Calculate the steam flow. Consider the compressibility of the water.

8–58. Feedwater enters a combination heater and drain cooler at 2600 psia and 196.3 F at the rate of 230,000 lb per hr. Steam pressure is 21.3 psia. Steam enters as follows: 630 lb per hr at 1399.5 Btu per lb, 1620 lb per hr at 1237.8 Btu per lb, and an undetermined quantity at 1295 Btu per lb. Terminal differences are 1 F TD for the heater and 10 F TD for the drain cooler. Calculate the undetermined steam quantity if 26,800 lb per hr of drains enter at 240.4 F. Consider compressibility of the water.

8–59. The steam pressure in a combination feedwater heater and drain cooler is 153 psia when water enters at 293.5 F and 1600 psia. Drains from another heater enter at 369.8 F and in the amount of 34,700 lb per hr. Terminal differences are 0 F TD for the heater and 10 F TD for the heater and 10 F TD for the drain cooler when the water flow is 507,700 lb per hr. Calculate the steam quantity at 1445.5 Btu per lb by considering the compressibility of the water.

8–60. The vapor from an evaporator is condensed in a feedwater heater that is called an evaporator condenser. Find the pressure in the evaporator condenser for the data given on the sketch.

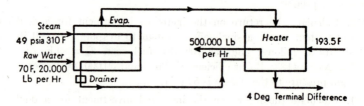

8–61. A surface heater and drain cooler receive 700,000 lb per hr of water entering the drain cooler at 277.0 F. The heater operates at 100 psig and with steam at 390 F. Find the temperature of the feedwater leaving the drain cooler when there is a 5-deg terminal difference for the heater and a 15-deg terminal difference for the drain cooler.

8–62. Two closed heaters operate in parallel and receive water and steam from the same sources. The water first passes through two drain coolers also operating in parallel which receive the drains from their respective heaters. The water enters the drain coolers at 317.7 F with a total flow of 1,240,000 lb per hr. Steam for the heaters is at 215 psig and 1302.0 Btu per lb. There is also a total

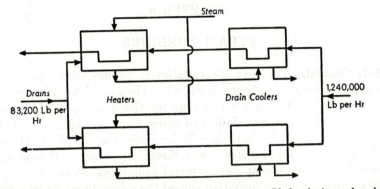

of 83,200 lb per hr of 419.0 F drains entering the heaters. If the drain coolers have a 10 F terminal difference and the heaters a −2 F terminal difference, find the amount of steam supplied to the two heaters. All flows are divided equally between the two heaters. See sketch.

BIBLIOGRAPHY

Proportioning of Surface Condensing Equipment. Milwaukee, Wis.: Allis-Chalmers.

Standards. New York: Bleeder Heater Manufacturers Association, Inc., 1953.

ELLINGEN, WILLIAM E. *Approximate Methods for Sizing and Pricing of Steam Surface Condensers.* ASME Convention, 1954.

Standards for Steam Surface Condensers. New York: Heat Exchange Institute, 1952.

IMPAGLIAZZO, A. M. "Feed Water Conditioning by Evaporation," Proceedings of the American Power Conference, Vol. 15, p. 644.

PUTNAM, G. H. "Condensers Now Handle More Energy in Less Space," *Power,* (January, 1952).

CHAPTER 9

STEAM TURBINES

9-1. General. The first known attempt to use steam to produce mechanical energy was about 150 B.C. when Hero of Alexandria described a steam turbine. The unit was a pure reaction type and did no useful work. In principle it was similar to many lawn sprinklers used today. Steam was generated in the main cylinder, which was spherical. The only means of escape for the steam was through several small bent tubes attached to the sphere at its horizontal centerline. Steam leaving the tubes created a reactive force that caused the sphere to rotate on a vertical shaft.

In 1882 Gustaf de Laval applied the turbine principle to a prime mover for his cream separator. The high rotative speeds obtainable with the turbine were better suited to that purpose than the steam engine. A few years later he brought out a series of small impulse turbines ranging in size from 1 hp at 100,000 rpm to nearly 100 kw at about 6000 rpm. These units were single stage and had only one steam nozzle. At about the same time, 1884, Sir Charles A. Parsons developed a reaction turbine. The American rights to his patent were sold to the Westinghouse Electric Corporation in 1895, and turbines were built and sold by them shortly thereafter. The reaction type of turbine is still referred to as a Parsons turbine. During this same period an American, C. G. Curtis, developed another type of impulse turbine. His designs were bought by the General Electric Company.

All steam turbines operate by converting thermal energy into kinetic energy and then into mechanical energy. The first conversion is accomplished by expansion of the steam through suitable nozzles. In the case of the impulse turbine, the jet of steam issuing from the nozzles impinges on a series of rotating blades or buckets attached to a shaft so that its direction is changed. In so doing, a force is exerted on the blades which causes the shaft to rotate. In the reaction turbine, the force on the blades is created by the reaction due to the pressure drop of steam as it passes through the nozzle-shaped rotating blades.

Impulse turbines operate on the principle that a force is exerted when the *direction* of the fluid stream is changed, in contrast to reaction turbines whose force is due to changing the *speed* as well as the *direction* of the stream relative to the blade. Another distinction between the two turbines is that there is no pressure drop through the moving blades of an impulse turbine, while there is a large pressure drop through the moving blades of a reaction turbine.

From this very recent beginning, the steam turbine has been developed into a highly efficient prime mover. Except in sizes below about 600 kw, the turbine is more efficient than the steam engine at full load. However, many steam engines have a flatter efficiency curve, i.e., they retain good efficiency at a low load. The turbine can be designed for larger capacities and for operation at an exhaust pressure only slightly greater than a perfect vacuum. Many units operate at back pressures of ½ to ¼ in. Hg abs during winter months when cold circulating water may be obtained for the condenser. Very common exhaust pressures for turbines are 1 and 1½ in. Hg abs. Steam engines cannot handle the large volumes of steam encountered at these low pressures because of the limiting sizes of the exhaust ports. Also, there is difficulty in packing a steam engine under such low pressures. A back pressure of 4 in. of Hg abs is considered good for a steam engine. In being able to expand the steam to such low pressures, the turbine takes advantage of the added gains from the complete expansion cycle, with none of the increased friction losses encountered in steam engines. Also of importance is the complete lack of oil contamination of the steam in the turbine.

Turbines for generator drives are usually built to operate at synchronous speeds with 3600 and 1800 rpm most usual for 60-cycle operation. However, when they are used for mechanical applications, the speed will be determined by the driven apparatus, or the better efficiencies of high speed may be utilized by speed reduction with gears. Since the only moving part is the rotor, high speeds are not detrimental to the turbine.

Steam turbines will require less floor space for the same rating than the steam engine, and because of the higher speed, the generator will also be smaller.

Nearly all turbine generators for central stations are arranged to permit the extraction of steam at various pressures for use in processes or for heating feedwater. Thus, the cycle efficiency can be improved and other benefits obtained.

To give an idea of the progress in the design of steam turbines, it is interesting to note that the average steam pressure of turbines sold in 1926 was 350 psig and the maximum was about 1000 psig, although in 1897 Dr. de Laval operated 50- and 100-hp turbines with their boilers and auxiliaries at pressures from 1420 to 3420 psig and at temperatures from 700 to 750 F. Pressures of 2000 psig are common today and the maximum is 5000 psig. Steam temperatures ranged from an average of 650 F and a maximum of 750 F in 1926; today, 1000 F and 1050 F are common and the maximum is 1150 F. The higher pressures and temperatures and the reheat cycle produce higher efficiencies which are due to both the inherent improvement in the theoretical cycle and the better turbine efficiencies. An inspection of the Mollier diagram will show the improvement in the theoretical steam rate by using higher pressures and temperatures with the corresponding decrease in entropy. As the energy that can be converted into useful work per pound of steam is increased, the steam flow required by a unit of a given rating will obviously decrease, thus decreasing the physical size of the unit. Or inversely, the greater will be the output from a unit of a given physical size.

9–2. Classification of Steam Turbines. No one method of classifying turbines is adequate, since all large turbines are built for the particular installation. The following methods are the most common:

(a) *Types of Blades.* As previously mentioned, thermal energy may be converted into mechanical energy by the impulse of the jet or by the reaction due to the jet leaving the nozzle. One stage of an impulse turbine will consist of a stationary nozzle ring attached to the casing and one row of moving blades attached to the shaft. The number of nozzles will depend on the quantity of steam required by the turbine. If the nozzles occupy the entire arc of the ring, the turbine is said to have *full peripheral admission.* With less than this number of nozzles, the turbine would have *partial admission.* One nozzle block and one rotating row of blades forms one *Rateau stage.*

With normal blade speeds, a Rateau stage can handle an enthalpy drop of about 50 Btu per lb of steam with the best efficiency. Small units may use greater energy drops when the efficiency is not important. When high efficiency is desired with a high differential of thermal energy, the stages must be *compounded* by one or both of two methods. *Pressure compounding* consists of several Rateau stages in series, each with its own nozzle block and rotating row of blades. *Velocity compounding* employs one nozzle block with two or more rotating rows. In order to keep the direction of rotation the same for each rotating row, stationary reversing blades are placed between consecutive rotating rows. This type of staging, known as Curtis staging, is often used in small single-stage turbines and as the first stage of large units.

Reaction stages are composed of one stationary row of blades and one rotating row of blades with a pressure drop occurring in each stationary and rotating row. Blades in both rows are shaped so that the area between two adjacent blades of the same row will form a converging nozzle; hence, there is a pressure drop and therefore an increase in the relative velocity across each row. Enthalpy differentials for each stage of a reaction tur-

bine are usually lower than those for impulse stages. Thus, the reaction
turbine requires more stages.

Some manufacturers combine the impulse and reaction principles by
employing a Curtis or Rateau stage as the first stage of a large turbine
and by using reaction stages for the remainder of the expansion. The
advantages of this arrangement over a straight reaction turbine are the
ease of governing during low-load periods and the increased efficiency of
the unit at these times with a simpler steam chest.

(b) *Cylinder Arrangement.* The simplest turbine is one with a *single
cylinder* with all the rotating blades attached to one shaft and the steam
flow all in one direction (see Plate 15 in the Appendix). If operated condensing, present designs would allow a maximum capacity of about
40,000 kw at 3600 rpm (non-reheat). The limiting feature is the maximum allowable length of the low-pressure blading and the allowable bearing span. Single-cylinder units with steam entering in the center and
flowing in two equal quantities, but in opposite directions along the shaft,
are known as *double-flow units* (see Fig. 9–47). This type of construction
is not applied to a single-cylinder turbine because of the difficulty of governing, but it is often used to increase the capacity of a compound unit
when the governing can be accomplished in another section of the turbine.

Tandem-compound units have found wide application. The steam
enters the first section, known as the *high-pressure end,* and expands in
this section to a pressure which is intermediate between the initial and
final pressures for the turbine (see Figs. 9–47 and 9–48). The steam is
then transferred to the *low-pressure end* by a pipe known as the *cross-over*
pipe. The pressure at which this transfer takes place is known as the
cross-over pressure. The cross-over pipe is usually made large, as any
pressure loss in it will decrease the availability of the energy. The low-
pressure end of the turbine is a single-cylinder section with double flow.
Both the high-pressure and the low-pressure ends are on the same shaft.
Tandem-compound units may also have two low-pressure casings that
produce *triple flow*, Figs. 9–49 and 9–50. Many large tandem-compound
units have a high-pressure element, intermediate-pressure element (or
reheat element), and triple-flow low-pressure elements. The limitation on
capacity is the length of the low-pressure blading and the number of low-
pressure ends.

Cross-compound units differ from tandem-compound units only in that
the high- and low-pressure ends are not on the same shaft (see Fig. 9–46).
When used as turbine generators, each end is usually connected to a separate generator, but when used as mechanical drives, as in marine applications, the two ends are connected to the same propeller shaft by reduction
gears. Cross-compound units may be built with several low-pressure sec-

tions receiving steam from one high-pressure section. A cross-compound unit is being constructed for a capacity of 500,000 kw, 638,900 kva, 3600/1800 rpm, with steam conditions of 2400 psig, 1050 F, with reheat to 1000 F, and for 45 psig hydrogen cooling and liquid cooling for both generator stators.

Steeple- or vertical-compound units have been used in a few instances. The high-pressure section is placed directly above the low-pressure turbine. The South Amboy Station has a 25,000-kw steeple-compound unit and the Ford Motor Company plant at Dearborn has another rated at 110,000 kw. This is the largest unit of its kind to date. Because of the obvious disadvantage of vibration and foundation troubles for the upper unit, and because the upper unit is in the way when dismantling the lower unit, steeple-compound units are seldom used. Their only advantage is conservation of floor space.

The space available for the unit, the initial cost, and the efficiency are controlling factors in the selection of the style of turbine. Tandem-compound and single-cylinder units lend themselves to installations where the available space is long and narrow. The tandem-compound units are, of course, longer than single-cylinder units but may be more efficient. Cross-compound units require wide areas whereas a steeple-compound unit needs headroom. Special care must be taken with the design of the high-pressure element supporting structures of steeple-compound units to eliminate objectionable vibration.

In changing from the maximum capacity of one casing arrangement to another casing arrangement, there usually will be an appreciable improvement in efficiency, particularly with high vacuum, due to relieving the crowding of the low-pressure blades. Also, there will probably be a sharp increase in cost.

(c) *Back Pressure.* Turbines are also classified according to the exhaust pressure for which they are designed. Those used for driving pumps, fans, and other auxiliaries in the power plant commonly operate at exhaust pressures approximating atmospheric. Modern turbine generators, known as *condensing units,* make use of the greater efficiencies obtained by high vacuum so that the design exhaust pressures for turbines in most parts of the United States are from 1 to 2 in. Hg abs. During the 1930's, many old plants were improved by the addition of *topping* or *superposed units* that exhaust at high back pressures, about 175 to 250 psig, into old existing units. By this means the plant capacities and efficiencies were increased considerably, due to both the high efficiency of the turbine generator and the better efficiency of the high-pressure cycle without increasing circulating-water requirements and by the addition of only boilers, pumps, feedwater heaters, and a topping unit. In one case a 50,000-kw high-pressure topping unit was installed in a plant in the space formerly occupied by a 14,000-kw vertical unit with an improvement of nearly one-third in the efficiency of the entire plant and with increased plant capacity. Superposed units are customarily single cylinder, 3600 rpm. The maximum capacity for these units is about 75,000 kw, the generator being the limiting factor. Practically no 1800-rpm, noncondensing topping units have been built because of the better efficiency obtained by the higher speed, but their maximum capacity would probably be about 200,000 kw, again limited by the generator.

(d) *Initial Temperature and Pressure.* The exact demarcation between high, intermediate, and low pressures is not definite. Steam pressures in the 1800- to 2400-psig range are considered high pressure, while those above 3206 psig are supercritical pressures; those in the 200- to 400-psig range are obviously low pressure. High temperature is used to signify inlet temperatures above 900 F. The advances made in the increase of throttle temperature and pressure during the last decade have been possible because of the advance in the science of metallurgy.

(e) *Reheat.* When steam is extracted from the turbine and its temperature increased (usually in the steam generator) before being returned to the turbine, the turbine is classed as a reheat turbine. Nearly all turbine generators of 100,000 kw or larger that are currently being installed are for the reheat cycle. Most units in the 50,000 to 100,000-kw range also are for reheat. Usually, the steam is reheated to the same temperature as the high-pressure steam or to within 50 F of that temperature. The most economical pressure for the reheat steam is about 20% of the initial steam pressure. Supercritical-pressure turbines have two stages of reheat, i.e., they are double-reheat units.

(f) *Other Methods.* In addition to identifying turbines by the type of blading used, they may also be divided into *single-stage* and *multistage* units. Single-stage units characteristically have poor efficiencies and are used only for driving plant auxiliaries where the efficiency is of minor importance. There have been some cases where it was more desirable to have an auxiliary driven by a turbine of low efficiency rather than by one of high efficiency. Such peculiar circumstances are due to the requirements of the station heat balance.

Although units may be referred to as large or small units, this nomenclature is even less exact in its meaning than the pressure classification already mentioned. The turbines used to drive the boiler feed pumps in large central stations have higher ratings than many of the main units in small plants.

Because of the particular circumstances caused by the heat balance, it is sometimes advisable to admit steam to the turbine at two or more pressures. Of course the lower the pressure of the steam, the nearer it must be admitted to the exhaust end of the unit. Such turbines are known as *mixed-pressure turbines.* The steam admitted into the low-pressure open-ings may come from old low-pressure boilers during times of high-load periods, or it may be the excess from some auxiliaries or processes.

Another classification that should be used with extreme caution is that of *high speed* or *low speed*. Those associated with the design of central stations are accustomed to think of speeds of 3600 rpm as high speed. However, it is common practice in the merchant marine to operate the high-pressure element of cross-compound geared turbines at a rated speed of 6000 rpm. When a turbine is used to drive an electric a-c generator, the turbine must operate at the synchronous speed of the generator or at higher speeds through reduction gears. This latter method of drive is seldom used for units with a rating greater than about 1500 kw.

Most main units in power plants are provided with extraction openings or bleed belts so that steam may be withdrawn from the turbine for feed-water heating or process. Turbines built with provision for extraction are called *extraction* turbines and those without are called *nonextraction* tur-bines. When heat cycles are discussed, it will be observed that the station efficiency may be improved to a great extent by the use of the regenerative cycle. In cases where the low-pressure element of the turbine is crowded, the extraction of steam will materially improve the efficiency of the tur-bine itself. Also, an extraction opening in the low-pressure stages of the turbine provides an ideal place to withdraw moisture, either by the action of the steam leaving the unit or by using specially designed blades to assist in the removal of the moisture. Blade efficiencies are improved by decreasing the moisture content of the steam.

The use to which the turbine is to be put may help to distinguish it from other units. Common terminology for this method of classification is *sta-tionary, marine, or mechanical-drive turbines*. The marine and stationary turbines are very similar in design, their differences being principally in speed, governing, ability to reverse the direction of rotation for geared marine units, and structural features. As stated previously, mechanical-drive turbines are often smaller in size and poorer in efficiency than marine or stationary units.

9-3. Theory of Nozzles. Since the flow of the steam through a nozzle is a case of steady flow, the GE equation would apply, Eq. (2-1).

Apply this equation to a reversible adiabatic flow process, eliminate all terms that are zero or negligible, and remember that $h = u + Pv/778$ by definition. Then, if subscripts i and o indicate inlet and outlet,

$$\frac{V_i^2 - V_o^2}{2g778} = h_o - h_i$$

If the initial velocity is assumed negligible, then

$$V_o = \sqrt{2g778(h_i - h_o)} = 223.8\sqrt{h_i - h_o} \qquad (9\text{-}1)$$

which is the basic equation for the flow of a fluid through a nozzle or orifice. The last assumption that the initial velocity is negligible will be found to be in error when considering the flow of steam in some turbine nozzles. However, it will be possible to correct for that later.

It is not sufficient to know only the velocity of flow. The continuity equation states that

$$vw = AV \qquad (9\text{--}2)$$

where A = area, sq ft
   w = flow, lb per sec

Or in a more convenient form,

$$a = \frac{Wv}{25V} \qquad (9\text{--}2a)$$

where a = area, sq in.
   W = flow, lb per hr

The results of combining Eqs. (9–1) and (9–2a) for a set of assumed conditions are shown in Fig. 9–1 when they are plotted against the pressure ratio P_o/P_i. Observe that the flow increases as the outlet pressure decreases until a maximum is reached, and then the flow decreases again until it reaches zero for zero outlet pressure. The specific volume and velocity continually increase as the pressure ratio decreases. The point at which the flow reaches the maximum value is known as the *critical point*, and at this condition the velocity of the fluid is the same as the velocity of sound in the fluid. For that reason it is sometimes called the acoustic velocity.

Experiments have shown that the flow will remain constant at the maximum value of critical flow when the pressure ratio is reduced below that corresponding to the *critical pressure ratio*. The critical pressure ratio is not the same for all gases and will vary for the same gas for changes in the pressure and temperature. For steam, the critical pressure ratio is different for the superheated or moist states. The value varies from 0.547 to 0.549 for superheated steam when the initial pressure changes from 0 psia to 1500 psia, respectively, and from 0.580 to 0.594 for saturated or moist steam over the same pressure change. These variations have been expressed in mathematical form by Rattaliata [*] as

Critical pressure ratio (superheated steam) $= 9.417 \times 10^{-7} p_1 + 0.5471$
$$(9\text{--}3)$$

Critical pressure ratio (saturated steam) $= 9.34 \times 10^{-6} p_1 + 0.5800$
$$(9\text{--}3a)$$

where p_1 = steam pressure entering nozzle, psia

[*] J. T. Rattaliata, "Critical Pressure Ratios Applied to Steam Nozzles," *Current Trends*, Allis-Chalmers Mfg. Co., 1939.

These values were determined by using the theoretical formulas for critical pressure (see any text on thermodynamics) and Keenan and Keyes' steam tables.

Further inspection of Fig. 9–1 shows that theoretically the steam flow through the nozzle decreases as the outlet pressure is reduced below the

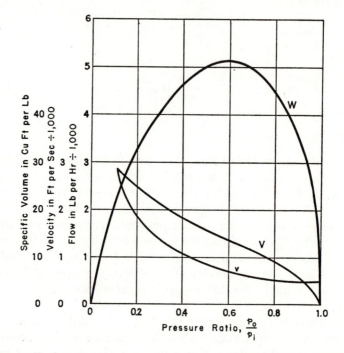

Fig. 9–1. Typical theoretical curves of flow, specific volume, and velocity for steam passing through a nozzle. Initial conditions: 100 psia saturated; throat area, 1 sq in.

critical value until at zero outlet pressure there is no flow. This would mean that the flow through a hole in a steampipe could be stopped by reducing atmospheric pressure to a perfect vacuum! Experience has shown that to be an absurdity and that the *flow through the nozzle will remain constant for all outlet pressures equal to or less than critical pressure*. The pressure at the *throat* or the section of minimum area cannot be reduced below critical pressure regardless of how low the outlet pressure may be. Thus, the section of minimum area limits or determines the flow through the nozzle.

Several empirical formulas have been proposed for determining the flow of steam through nozzles; one such formula is Napier's equation. While this formula is not sufficiently accurate to be used in turbine design, it will be useful in a later article to explain certain phases of turbine performance (see Art. 9–13).

$$\text{Napier's formula,} \quad w = \frac{ap_i}{70} \tag{9-4}$$

in which w = flow, lb per sec
$\quad\quad a$ = area of throat, sq in.
$\quad\quad p_i$ = initial pressure, psia

may be used for conditions where the outlet pressure is less than critical pressure.

So far, we have considered only a nozzle that has a continually decreasing cross-sectional area in the direction of flow. This type of nozzle is called a *converging* nozzle or *nonexpanding* nozzle. The exit area will be the minimum area of the passage, and thus will be the area that should be used in the continuity equation to determine the steam flow. The place of minimum area in a nozzle is called the *throat*. For a nonexpanding nozzle this coincides with the exit, often called the *mouth*.

A converging nozzle may be used when the outlet pressure is less than the critical pressure. Under these conditions the steam will expand up to the throat, as though the outlet pressure were at critical pressure. However, as soon as the steam has left the throat, its pressure will become that of the surroundings, or less than critical pressure, with a corresponding increase in specific volume and velocity. The expansion from the inlet to the throat will be gradual, but the expansion after leaving the nozzle will be in the form of an explosion. Obviously the first part of the expansion, that from the inlet to the throat, would be far more efficient than the latter part where the explosion effect would cause a dissipation of the energy.

When it is necessary to design a nozzle for pressure ratios less than critical, it is customary to add a diverging section after the throat of the nozzle. The extended nozzle passage beyond the throat allows the expansion to take place in an orderly fashion without the loss of energy encountered in the explosion. Such a nozzle is called a *converging-diverging* or an *expanding* nozzle.

In order to understand the several phenomena that take place in a nozzle under different conditions of flow, assume that a converging-diverging nozzle is designed to expand steam from an initial pressure of 100 psia to an exit pressure of 20 psia. The nozzle shown in Fig. 9–2 would correspond to these conditions, provided the converging portion in the sketch is assumed lengthened for the sake of clarity. The critical pressure would be approximately 55 psia at the throat when the exit pressure is 20 psia. Line a would indicate the pressures throughout the nozzle when it operates under design conditions. If the outlet pressure were the

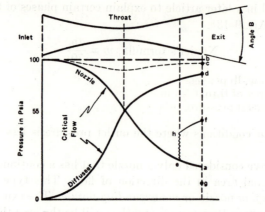

FIG. 9–2. Pressure in a nozzle for various flow conditions.

same as the inlet pressure, 100 psia, there would be no flow and the same pressure would exist at all points in the nozzle, curve b. If the outlet pressure were c, the throat pressure would not decrease to the critical pressure, and therefore the flow would not reach the maximum. This condition exists in a metering venturi where the pressure drop from the inlet to the throat need be only sufficient to give an accurate measurement and should be kept low to reduce the loss as much as possible.

Before discussing the other possibilities for the exit pressure, let us assume that this same nozzle is to be used as a diffusor (see air ejectors, Art. 8–7). Then the steam inlet velocity would be very high and the inlet pressure low. Instead of expanding, the steam will be compressed in the diffusor, and the kinetic energy will be converted into pressure energy. This is the action that takes place in the diffusors of air ejectors used in connection with condensers. The outlet pressure would be at d, and the adjacent line would give the pressure history of the steam during its passage through the diffusor.

Returning again to the nozzle action, if the inlet pressure were 100 psia as previously assumed and the outlet pressure d, the pressure gradient throughout the nozzle would follow the critical flow line starting at 100 psia and decreasing to the critical pressure at the throat. From there on the pressure gradient would follow the diffusor curve to the outlet pressure of d.

So far all the cases considered for exit pressures have involved smooth orderly flow conditions. Such would not be the case, however, if the outlet pressure were at f. If the nozzle should have to operate with that back pressure, the expansion would proceed along the path a until, after passing through the critical pressure point, it reached e, which is a pressure less than the exhaust pressure. At that point a new phenomenon called *com-*

pression shock would occur and the pressure gradient would rise suddenly and then proceed to *f*. The compression from *h* to *f* will take place as in a diffusor and will be gradual. An expansion of this type is called *over-expansion* and is caused by the nozzle mouth area being too large.

One more possible operating condition would be with the outlet pressure as at *g*. The pressure drop from *a* to *g* would occur at the mouth of the nozzle with very rapid expansion of the steam similar to that which would occur if the nozzle did not have the diverging section. This condition is called *underexpansion*.

Note that it is impossible for a converging nozzle to have compression shock.

Conditions such as *a*, *f*, and *g* are common in steam nozzles used in turbines, while conditions such as *b*, *c*, and *d* are very rare.

The ratio of the nozzle mouth area to the nozzle throat area is the nozzle expansion ratio or, briefly, the nozzle ratio. For any set of conditions, it can be calculated by applying the continuity equation to the mouth and throat areas. Theoretically, this would give

$$\text{Nozzle ratio} = \frac{V_t \times v_m}{V_m \times v_t} \qquad (9\text{-}5)$$

The subscripts *m* and *t* indicate the mouth and throat.

EXAMPLE 9–1. Determine the throat and mouth area of a nozzle passing 5000 lb per hr of steam that enters the nozzle at 4 psia and 95% quality and leaves at 1 psia.

SOLUTION. Assume critical pressure ratio as 0.58, or calculate a valve from Eq. (9–3a). Since the outlet pressure is obviously less than critical pressure, the throat pressure is

$$p_t = 0.58 \times 4.0 = 2.32 \text{ psia}$$

Thus
$$h_i = 1077 \text{ Btu per lb} \qquad s_i = 1.781 \text{ Btu per (lb)(F)}$$
$$h_t = 1043 \text{ Btu per lb} \qquad s_t = 1.781 \text{ Btu per (lb)(F)}$$
$$h_m = 995.3 \text{ Btu per lb} \qquad s_m = 1.781 \text{ Btu per (lb)(F)}$$
$$v_t = 140 \text{ ft}^3 \text{ per lb}$$
$$v_m = 297 \text{ ft}^3 \text{ per lb}$$
$$V_t = 223.8\sqrt{1077 - 1043} = 1305 \text{ fps}$$
$$V_m = 223.8\sqrt{1077 - 995.3} = 2023 \text{ fps}$$
$$a_t = \frac{Wv}{25V} = \frac{5000 \times 140}{25 \times 1305} = 21.46 \text{ sq in.} \qquad \text{diam}_t = 5.24 \text{ in.}$$
$$a_m = \frac{5000 \times 297}{25 \times 2023} = 29.4 \text{ sq in.} \qquad \text{diam}_m = 6.11 \text{ in.}$$

9–4. Nozzle Efficiency. So far we have considered that the flow through the nozzle has been a reversible adiabatic process. This can never exist in practice. As mentioned previously when discussing the flow of fluids, a thin film of the fluid will adhere to the nozzle wall, and the friction of the fluid will depend on the height of the protuberances and the thickness of the film. Also, the velocity will not be uniform throughout the fluid cross-section. The nozzle shape and height as well as the length will affect the efficiency of flow. The nozzle height will affect the efficiency only up to about $1\frac{1}{2}$ in. Nozzles less than this have poorer efficiency, but heights greater than this will not materially improve the efficiency. Nozzle efficiency will vary from 94 to 99%, the latter value referring to so-called curved-foil nozzles, when expressed as

E_{nh} = nozzle efficiency based on enthalpy

$$= \frac{\text{actual difference in enthalpy through the nozzle}}{\text{theoretical difference in enthalpy}} \qquad (9\text{-}6)$$

Combining Eqs. (9-1) and (9-6), we get

$$V_o = 223.8\sqrt{E_{nh}(h_i - h_o)} \qquad (9\text{-}7)$$

It is also possible to express nozzle efficiency based on velocity rather than on enthalpy:

$E_{nv}{}^*$ = nozzle efficiency based on velocity

$$= \frac{\text{actual nozzle velocity}}{\text{nozzle velocity based on isentropic expansion}} \qquad (9\text{-}8)$$

Now combining Eq. (9-8) with Eq. (9-1)

$$V_o = 223.8E_{nv}\sqrt{(h_i - h_o)} \qquad (9\text{-}9)$$

Regardless of how we may perform our calculations, the velocity from any given nozzle will be the same. Therefore, we may combine Eqs. (9-7) and (9-9) to get

$$E_{nv} = \sqrt{E_{nh}} \qquad (9\text{-}10)$$

Throughout this text we shall use the efficiency based on enthalpy unless otherwise stated.

It has already been stated that nozzle inefficiency is due to friction and turbulence. But friction is always evidenced as heat, and the most convenient recipient of this heat generated is the steam. Thus, the inefficiency of the nozzle causes an increase in the enthalpy of the steam leaving the nozzle over the enthalpy expected by a reversible adiabatic expansion. The process is illustrated on a Mollier diagram in Fig. 9-3. The ideal expansion would take place from i to o at constant entropy. The actual expansion will not follow this line because of friction, and the steam will actually leave the nozzle at conditions corresponding to point 3. *Reheat*

* Sometimes this is known as the velocity coefficient.

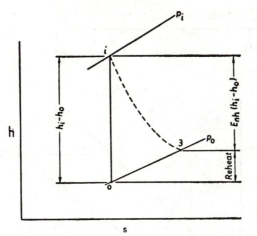

FIG. 9–3. Comparison of actual and ideal expansions of steam through nozzle.

is the difference between the enthalpies at points o and 3. Perhaps, according to strict thermodynamics, we cannot plot the path of the steam in expanding from points i to 3 because the process is irreversible and not in equilibrum. However, we can approximate the path by the dashed line connecting points i and 3 and call it the path of the actual expansion.

In a similar manner, the expansion lines for flow through an expanding nozzle can be analyzed, Fig. 9–4. If the nozzle were ideal, the expansion would be from i to o. Because of friction the expansion actually ends at point m after passing through point t at the throat. If there were a series of turbine blades or something else to absorb the kinetic energy developed by the expansion, then the steam would remain at the conditions of point m. But if there were no turbine blades to transform the kinetic energy into useful work, the high velocity of the steam jet issuing from the nozzle mouth would be dissipated eventually by severe turbulence and friction between sections m and 4. As before, these losses appear as heat that increases the enthalpy of the steam. If we assume that the velocity at the inlet to the nozzle is the same as the velocity at point 4, the energy transformation from kinetic energy to thermal energy from m to 4 will be numerically equal to the energy transformation from thermal energy to kinetic from i to m. Therefore, the enthalpy of the steam at point 4 will be the same as the enthalpy at i. Such a process is called a *throttling* process and is referred to as taking place at constant enthalpy. Although the initial and final enthalpies are the same, the process does not progress from i to 4 at constant enthalpy but detours by way of points t and m.

Converging nozzles will have higher efficiencies than expanding nozzles. Underexpansion and overexpansion cause a decrease in nozzle efficiency from the optimum because the turbulence generated by the explosion or

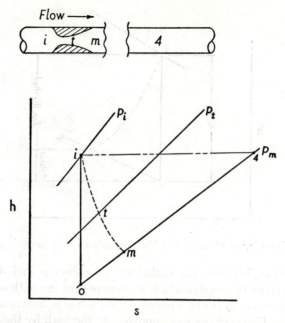

Fig. 9–4. Expansion through a converging-diverging nozzle.

compression shock requires energy that cannot be used for useful work. Since compression shock occurs inside the nozzle, it will reduce efficiency more than the explosion that occurs outside. Overexpansion cannot exist for a nonexpanding nozzle; therefore, the nozzle efficiency will remain at its peak for all pressure ratios equal to or greater than the critical pressure ratio. For a ratio of actual mouth area to ideal mouth area of about 0.7 (underexpansion), the losses will be about 5%, while for a ratio of about 1.3 (overexpansion), the losses will be about 16%.

EXAMPLE 9–2. For the Example 9–1, use a nozzle efficiency of 94% based on enthalpy, and determine the throat and mouth area and the length of the nozzle from the throat to the mouth if the angle formed by the sides is $\beta = 15$ deg (see Fig. 9–7).

SOLUTION. Throat velocity is

$$V_t = 223.8\sqrt{0.94(1077 - 1043)} = 1266 \text{ fps}$$

Enthalpy at the throat is

$$h_t = 1077 - 0.94(1077 - 1043) = 1045 \text{ Btu per lb}$$
$$v_t = 140 \text{ ft}^3 \text{ per lb (no appreciable change)}$$

From Eq. (9–2a),

$$a_t = \frac{5000 \times 140}{25 \times 1266} = 22.15 \text{ sq in.}$$
$$\text{diam}_t = 5.31 \text{ in.}$$

$$V_m = 223.8\sqrt{0.94(1077 - 995.3)} = 1961 \text{ fps}$$
$$h_m = 1077 - 0.94(1077 - 995.3) = 1000.2 \text{ Btu per lb}$$
$$v_m = 300 \text{ ft}^3 \text{ per lb}$$
$$a_m = \frac{5000 \times 300}{25 \times 1961} = 30.6 \text{ sq in.}$$
$$\text{diam}_m = 6.24 \text{ in.}$$

Length of nozzle from throat to mouth is

$$L = \frac{6.24 - 5.31}{2}\left(\cot\frac{15°}{2}\right) = 3.53 \text{ in.}$$

Observe that the change in specific volume of the steam due to nozzle efficiency is very small and can be neglected for high nozzle efficiencies.

9–5. Supersaturation. When steam expands suddenly from the superheat region to the moist region, there is little time for the steam to condense partially, and there are no nuclei present to promote condensation. The point at which condensation of the steam does commence after crossing the saturation line on the Mollier diagram is known as the *Wilson line*. It is quite probable that the point of condensation is not fixed but will be variable and will depend on the velocity or the rate of change of the velocity. However, it may be said that the steam is not in a state of thermal equilibrum, i.e., the condition of the steam is unstable and it can exist under such conditions for only a brief period of time.

For practical purposes it may be assumed that this state of *supersaturation,* or *undercooling,* may continue until the Mollier diagram would indicate about 4% moisture. In effect, there is an extension of the superheat region into the moist region.

Under supersaturated conditions the steam temperature is less than normally would be expected, and consequently the specific volume would be correspondingly lowered. The flow through the nozzle is increased because of the lowered temperature and specific volume. The increase in flow due to supersaturation reaches a maximum of about 3% but is dependent on the moisture.

As soon as droplets of moisture start to form, the supersaturation effect disappears. All authorities, however, do not agree that the restoration of thermal equilibrium takes place as soon as moisture starts to form. Some contend that equilibrium is never complete in the turbine and that some of the effects of supersaturation continue until the steam reaches the condenser.

9–6. Nozzle Designs. About the simplest and most inexpensive nonexpanding nozzle to manufacture is sketched in Fig. 9–5. The cross-section is circular, but since the nozzle is at an angle with the blades, the mouth is elliptical. The hole is drilled and reamed. The angle α is the nozzle angle. To obtain the most energy from the steam jet, the blade must not be shorter than the height of the nozzle; usually, it is made slightly taller than the height of the nozzle.

It will be shown in subsequent text that the angle α should be as small

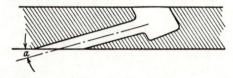

FIG. 9–5. Sketch of a reamed nonexpanding nozzle.

as possible; generally it is about 13 to 20 deg but may range from 10 to about 30 deg. The larger angles are used for low pressure stages. The more that this angle is decreased, the longer will be the length of the nozzle and the greater the friction loss; but the nozzle mouth will also become wider, causing a very bad flow condition. This bad effect is sometimes lessened by reducing the distance between adjacent nozzles until the mouths overlap.

From aerodynamics it can be determined that the foil shape (teardrop) would be the most efficient nozzle design. The inlet is well rounded and the outlet sharp, Fig. 9–6. The angle α is again about 14.5 or 15 deg. The

FIG. 9–6. Sketch of a curved-foil nozzle.

flow characteristics of this type of nozzle are very favorable, and this design is also used for reaction blades.

Expanding nozzles could be designed so that the diverging portion would provide uniform decrease in pressure with respect to length, or uniform drop in enthalpy with respect to length, or uniform increases in specific volume with respect to length. In the first two cases, the mouths would flare rapidly, which would direct the jet of steam over a wide area and would make it ineffective. In all cases the cost of manufacturing such nozzles would be prohibitive, with no particular advantages.

The common design of expanding nozzles, shown in Fig. 9–7, is inex-

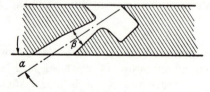

FIG. 9–7. Sketch of an expanding nozzle.

pensive. The passages are reamed after drilling. Angle α is the same as for the reamed nonexpanding nozzle, and angle β, the divergence angle, depends on the nozzle ratio. This angle should not be over 15 deg because the steam will not follow the sidewalls for greater angles. If the steam were to lose contact with the sidewalls, the nozzle would become a non-expanding nozzle with the pressure ratio less than critical and would have the loss of explosion as in the nonexpanding nozzle. The cylindrical extension shown in Fig. 9–7 is to guide the steam. Trapeziform and rectangular nozzles have been used in the past, but they were more inefficient than other designs.

Milled nozzles for high efficiency may be seen in Fig. 9–8. This picture also shows the nozzles assembled in the diaphragm.

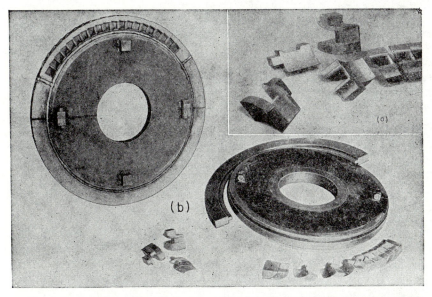

FIG. 9–8. Impulse-turbine diaphragm. At (a), milled nozzle block section which forms the steam passage; at (b) diaphragm parts carefully assembled to maintain calculated nozzle area. (Elliott Co.)

9–7. Theory of Impulse Blades.
The thermal energy of the steam is converted in a turbine first into kinetic energy by the nozzles and then into mechanical energy by the action of the jet on the blades. One stage of an impulse turbine consists of one or more nozzles admitting steam to a row of moving curved blades. This is a *Rateau* stage. A *Curtis* stage differs only in that the steam jet acts on two or three rotating rows of blades before leaving the turbine or before entering another set of nozzles.

The impulse-momentum law provides one method for solving problems on the design of turbine blades. The familiar equation $F = wa/g$, where

F is the force, w/g the mass, and a the acceleration, may be rearranged to become the impulse-momentum relation:

$$F \, \Delta t = \frac{w}{g} \, \Delta V \qquad (9\text{–}11)$$

which may be stated as: an external force F acting on a moving particle which has mass w/g for a period of time Δt will change the velocity of the mass by the amount ΔV. Note that Δt and the mass are scalar quantities, while the force and the velocity are vector quantities. If the direction of either is known, the direction of the other will be established.

In the event that certain components of the force are desired,

$$F_x \, \Delta t = \frac{w}{g} \, (V_1 \rightarrow V_2)_x$$

$$F_y \, \Delta t = \frac{w}{g} \, (V_1 \rightarrow V_2)_y \qquad (9\text{–}12)$$

$$F_z \, \Delta t = \frac{w}{g} \, (V_1 \rightarrow V_2)_z$$

The sign $\rightarrow$ indicates the vector difference between the velocities.

Next, it is necessary to apply these equations to a steam jet striking a moving blade, as in Fig. 9–9. The steam approaches the blade, with the

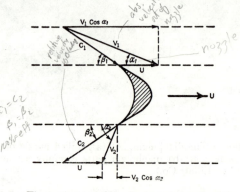

Fig. 9–9. Theoretical velocity diagram for an impulse blade.

absolute velocity represented by vector V_1, at an angle α_1 to the plane of rotation of the blade, which is moving at a velocity represented by vector U. Note that vector V_1 is on the centerline of the nozzle supplying the steam and is the velocity leaving the nozzle that we previously called V_m. Vector V_1 may be resolved into two components, C_1, which is the velocity of the steam relative to the blade, and U, which is the mean velocity of the blade.

If the blade is assumed to be frictionless, then the only effect on the jet of steam will be the change in direction since there will be no change in magnitude. Therefore, the vector representing the jet leaving the blade, C_2, will be in a path along the outlet edge of the blade. Vector C_2 will be the relative velocity of the steam leaving the blade. In order to obtain the absolute velocity of the steam leaving the blade, add vectors C_2 and U to get V_2 at an angle α_2 with the plane of the moving blade. The angles β_1 and β_2 are, respectively, the angles of the relative steam velocity entering and leaving the blade. Obviously, the steam should enter the blade with as little obstruction as possible, and so these angles are also the angles of the two edges of the blade.

The force causing rotation of the wheel will be that component in the x direction, Eq. (9–12), sometimes known as the *whirl* component of the jet velocity. Taking account of the directions and making $w/\Delta t$ equal to W, substituting into Eq. (9–12) gives

$$F_x = \frac{W}{g} \ (C_1 \cos \beta_1 + C_2 \cos \beta_2) \qquad (9\text{–}13)$$

From the geometry of the figure, adding and subtracting the wheel velocity to this equation will result in the formula

$$F_x = \frac{W}{g} \ (V_1 \cos \alpha_1 + V_2 \cos \alpha_2) \qquad (9\text{–}14)$$

But the work done on the blade will be the force times the distance, or the power (P) will be the force times the velocity:

$$P = \frac{WU}{g} \ (V_1 \cos \alpha_1 + V_2 \cos \alpha_2) \qquad (9\text{–}15)$$

$$P = \frac{WU}{g} \ (C_1 \cos \beta_1 + C_2 \cos \beta_2) \qquad (9\text{–}16)$$

Note that the units of P for this equation will depend on the unit of time employed to express the flow rate W, i.e., P has the units of foot-pounds per second if the flow is in pounds per second, or foot-pounds per hour if the flow rate is in pounds per hour, etc.

Because of the friction of the steam passing over the blade, the numerical value of C_2 will be less than that of C_1. Using k * to indicate the velocity efficiency of the blade, Eq. (9–16) will become

$$P = \frac{WU}{g} \ (C_1 \cos \beta_1 + kC_1 \cos \beta_2) \qquad (9\text{–}17)$$

$$P = \frac{WU}{g} \ C_1 (\cos \beta_1 + k \cos \beta_2) \qquad (9\text{–}18)$$

and $\qquad\qquad\qquad C_1 \cos \beta_1 = V_1 \cos \alpha_1 - U$

* Sometimes k is called the bucket velocity coefficient.

When setting the velocity ratio of the blade speed to the steam jet leaving the nozzle (U/V_1) equal to ρ,

$$C_1 \cos \beta_1 = \left(\frac{\cos \alpha_1}{\rho} - 1\right) U$$

$$P = \frac{WU^2}{g} \left(\frac{\cos \alpha_1}{\rho} - 1\right)\left(1 + \frac{k \cos \beta_2}{\cos \beta_1}\right) \qquad (9\text{-}19)$$

Since the energy supplied to the blade is the kinetic energy $WV_1{}^2/2g$, the efficiency of the blade will be Eq. (9-19) divided by the kinetic energy of the jet,

$$E_b = 2\rho (\cos \alpha_1 - \rho)\left(1 + \frac{k \cos \beta_2}{\cos \beta_1}\right) \qquad (9\text{-}20)$$

It is important to note from this equation that the efficiency depends on the velocity ratio of the blade and the jet and not on the absolute values of either. In order to determine the conditions for which the efficiency will be the maximum, differentiate Eq. (9-20) with respect to the velocity ratio ρ, and set it equal to zero. Thus,

$$\frac{d}{d\rho}\left[2\rho (\cos \alpha_1 - \rho)\left(1 + \frac{k \cos \beta_2}{\cos \beta_1}\right)\right] = 0$$

$$\rho = \frac{\cos \alpha_1}{2} \qquad (9\text{-}21)$$

This shows that the best efficiency for the normal nozzle angle of 15 deg will be obtained when the velocity ratio is about 0.48; also that the angle α_2 will be 90 deg. Thus, if the blade velocity were 700 fps, the maximum jet velocity for best efficiency should be only about 1400 fps, which represents a drop of 40 Btu per lb of steam in the nozzle. Mean blade speeds of 350 to 750 fps are common, while the maximum for the largest low-pressure blading is about 1130 fps.

At times ρ is allowed to decrease to about 0.4 in order to accommodate higher enthalpy drops. Eq. (9-21) further indicates that the velocity ratio could be somewhat higher if the nozzle angle were zero and that the efficiency would then be the highest. Obviously, this would be impossible to attain in practice since it would mean a complete reversal of the steam so that the outgoing jet from the blade would interfere with either the incoming jet or the other blades.

Normally, the blades used are unsymmetrical in that the angles β_1 and β_2 are not equal. The blade efficiency, as indicated by Eq. (9-20), can be increased by reducing angle β_2, within limits. However, if the outlet angle is reduced, the blade exit area must be made larger than the nozzle area since the velocity of the steam has decreased considerably. The specific volume can be assumed to be the same across the blade. Only two ways remain to increase the outlet areas: use a large exit angle or increase the blade height. The latter method has limited possibilities, but the exit angle is usually found to be 5 or 10 deg larger than the nozzle angle. In the event that the outlet area of the blade is made smaller than the inlet area, the blade is said to have some reaction effect. It will be noted in the blade-efficiency curves given later that reaction blades are more efficient than impulse blades, and so a small pressure drop through the impulse blades, which is used to give some reaction, is desirable.

The inlet angle must be chosen so that shockless entrance is assured, or else the steam jet will strike either the back of the blade or the concave front portion. If the jet strikes the front of the blade, the impulse will be increased but not efficiently because of the flow disturbances that will be set up. If the jet strikes the back of the blade, it will have a retarding effect that will be still worse. The inlet angle will be about 10 to 16 deg larger than the nozzle angle, as can be seen from the velocity diagram (Fig. 9–9).

Values of k decrease with the increased relative steam velocity, increase with increased blade widths, and decrease with increased total change in steam direction as the steam passes through the blade (called the *angle of deflection*); the values range from about 60 to 95%.

When long blades are used in turbines of higher ratings, there will be some difference between the velocity ratio at the root of the blade and at the tip. In design, the mean velocity is used in the velocity ratio. As a compensation for the variation in blade velocity, the blades are warped or tapered so that the angles change throughout the length of the blade and the blade section decreases toward the tip, Fig. 9–10.

Combined velocity diagrams using different methods of presentation are shown in Fig. 9–11. Parts (a) and (b) are both drawn for an impulse blade with no allowance for friction. The type of diagram used in (a) indicates the velocities by the vector length, but the directions are incorrect. The methods of parts (b) and (d) have the advantage of showing both the magnitude and direction of the vectors. Diagrams such as these are used in practice in preference to the method used in Fig. 9–9. Fig. 9–11(c) is drawn the same as Fig. 9–11(a) but indicates that friction has been taken into consideration.

The velocity diagrams indicate that the steam travels in the direction of the rotating blades. When the nozzle diaphragm has only a part of the arc occupied by nozzles, the second diaphragm for a multistage turbine must be displaced in the direction of rotation to receive the steam with the least turbulence. This *lead* placed on succeeding diaphragms permits efficient use of some of the residual velocity, V_2. Obviously, lead has no meaning when there is full peripheral admission in the first-stage diaphragm.

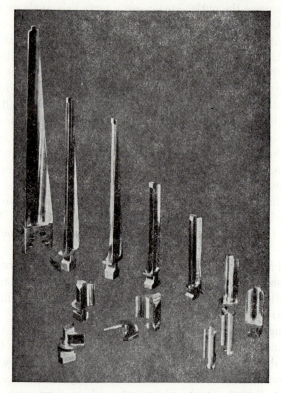

FIG. 9–10. Warped and straight impulse blades. (Elliott Co.)

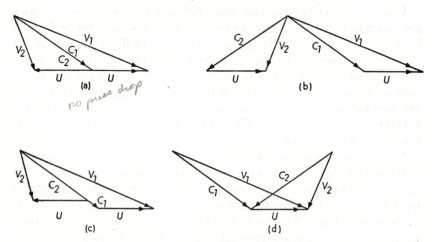

FIG. 9–11. Combined velocity diagrams for no friction loss at (a), (b), and (d); with friction loss at (c).

The energy represented by the velocity V_2 is known as the *leaving loss* and is expressed in Btu per pound. The energy gained by taking advantage of it in the design is known as *carry-over* and will amount to 50 to 75% of the leaving loss. Changing Eq. (9-1) to take care of the carry-over,

$$V_m = 223.8\sqrt{E_{nh}(h_i - h_o) + E_{co}\left(\frac{V_c}{223.8}\right)^2}\qquad(9\text{-}22)$$

where E_{nh} = nozzle efficiency
E_{co} = carry-over efficiency
V_c = carry-over velocity, fps

and other symbols are as for Eq. (9-1) and Eq. (9-5).

EXAMPLE 9-3. The data for one intermediate impulse stage of a turbine are as follows: 500 fps carry-over velocity, 60% carry-over efficiency, 92% nozzle efficiency, steam enters nozzle at 75 psia and 400 F, steam leaves nozzle at 45 psia, 20-deg nozzle exit angle, 650-fps mean blade velocity, symmetrical blading, $k = 0.88$.

Calculate ρ, all the blade angles and velocities, the power developed by each pound of steam flowing per second, the blade efficiency, and the leaving loss for the stage.

SOLUTION. For symbols of blade vectors and angles, refer to Fig. 9-9. From Eq. 9-22,

$$V_m = V_1 = 223.8\sqrt{0.92(1231.4 - 1187.1) + 0.60\left(\frac{500}{223.8}\right)^2} = 1480 \text{ fps}$$

$$\rho = \frac{U}{V_1} = \frac{650}{1480} = 0.439$$

$$V_1 \cos \alpha_1 - U = 1480 \times 0.9397 - 650 = 740$$

$$V_1 \sin \alpha_1 = 1480 \times 0.3420 = 506$$

$$\beta_1 = \tan^{-1}\left(\frac{V_1 \sin \alpha_1}{V_1 \cos \alpha_1 - U}\right) = \tan^{-1}\left(\frac{506}{740}\right) = 34.4 \text{ deg}$$

$$\beta_2 = \beta_1 = 34.4 \text{ deg}$$

$$C_1 = \frac{V_1 \sin \alpha_1}{\sin \beta_1} = \frac{506}{0.5650} = 896 \text{ fps}$$

$$C_2 = kC_1 = 0.88 \times 896 = 788 \text{ fps}$$

$$\alpha_2 = \tan^{-1}\left(\frac{C_2 \sin \beta_2}{C_2 \cos \beta_2 - U}\right) = \tan^{-1}\left(\frac{788 \times 0.5650}{788 \times 0.8251 - 650}\right) = 90 \text{ deg}$$

$$V_2 = \frac{C_2 \sin \beta_2}{\sin \alpha_2} = \frac{788 \times 0.5650}{1} = 445 \text{ fps}$$

Any of the Eqs. (9-15) through (9-19) may be used to obtain the power; use Eq. (9-18):

$$P = \frac{1 \times 650 \times 896}{32.2}(0.8251 + 0.88 \times 0.8251) = 28,056 \text{ ft-lb per sec}$$

From Eq. (9–20), the blade efficiency is

$$E_b = 2 \times 0.439(0.9397 - 0.439)\left(1 + \frac{0.88 \times 0.8251}{0.8251}\right) = 82.6\%$$

$$\text{Leaving loss} = \frac{V_2{}^2}{2g778} = \frac{(445)^2}{2 \times 32.2 \times 778} = 3.95 \text{ Btu per lb}$$

9–8. Velocity and Pressure Compounding. The inability of impulse staging to absorb efficiently more kinetic energy than that represented by an enthalpy drop of about 40 Btu has been discussed. When it is necessary to transform more than this amount, the turbine must be arranged in stages, each of which will transform a part of the total energy.

The stages may be designed on the basis of *velocity compounding* so that the total energy can be used and at the same time the wheel speeds kept relatively low. In velocity-compounding, the energy from the jet of steam issuing from a single nozzle, or group of nozzles, in one diaphragm is imparted to several rows of moving blades. Such a stage, known as a Curtis stage after its inventor, is shown in Fig. 9–12.

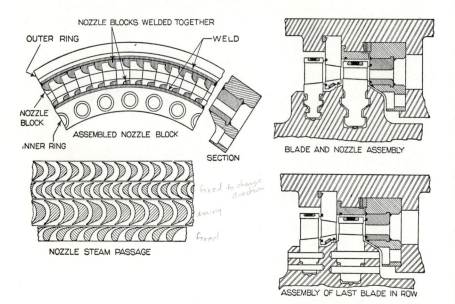

Fig. 9–12. Nozzle block, stationary, and rotating blades of a Curtis stage. (Westinghouse Electric Corp.)

In *pressure compounding*, the steam passes through a nozzle group, then imparts a portion of its energy to a single rotating row of blades, and then passes through the other successive stages, each stage being composed of a group of nozzles and one rotating row of blades. Only a part of the total energy is converted into kinetic energy in each Rateau stage. Under some conditions, velocity compounding is used for the first stage of a turbine, while the remaining stages are pressure-compounded.

Curtis stages have been built with as many as four rotating rows of blades. However, it can be shown that the ratio of work obtained from each row is in the proportion of 7:5:3:1. In other words, the last row does only $\frac{1}{16}$ of the total work of the stage. Two-row Curtis stages are very common.

If the total pressure drop through the turbine is divided among several stages by pressure-compounding to reduce the velocity at each wheel, then the total enthalpy differential for the turbine may be divided equally among the several stages. The velocity for each stage will be, theoretically,

$$V = 223.8\sqrt{(h_1 - h_2)/n}$$

in which V = steam jet velocity for each stage, fps
$\qquad h_1$ = initial enthalpy, Btu per lb
$\qquad h_2$ = final enthalpy, Btu per lb
$\qquad n$ = number of stages

or the original jet velocity for the entire unit will be reduced to $\sqrt{n}$ for each stage if n is the number of stages. This means that for a two-stage turbine with the bucket speed at 700 fpm and the steam-jet velocity twice the bucket speed, the maximum allowable drop in enthalpy for efficient operation will be about 80 Btu per lb

The effect of velocity-compounding in reducing the peripheral velocity of a two-row Curtis stage may be seen by examination of the velocity diagrams, Fig. 9–13. Steam leaves the nozzle with the velocity V_1 and enters the first moving blade with the relative velocity C_1. The jet leaves the first moving blade with the relative velocity C_2, numerically equal to C_1 because friction is neglected, and enters the fixed reversing blade with the absolute velocity V_2. The absolute velocity V_3 of the jet leaving the fixed blade is the same, numerically, as the entering velocity when friction is neglected. Velocities for the second moving blade are determined by the same means as for the first moving blade. The two rows of moving blades may be on separate wheels or on the same wheel. The values of the angles are customarily different from those for a Rateau stage. Note that the horizontal component of the velocity V_1 in Fig. 9–13(b) is four times the bucket velocity, whereas in a Rateau stage the relationship is two to one. In general, if there are n rows of moving blades in a Curtis stage, the velocity of the moving blades will be $1/n$ times the velocity of the single Rateau stage.

Applying this to the same blade speed used for one of the Rateau stages, the allowable enthalpy drop would be almost 160 Btu per lb for a

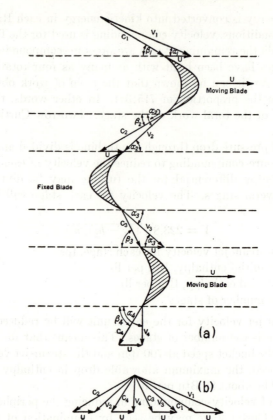

(a)

(b)

Fig. 9-13. Velocity diagrams for a two-row Curtis stage without friction.

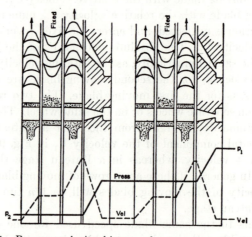

Fig. 9-14. Pressure-velocity history of a two-stage Curtis turbine.

two-row Curtis stage. If friction were taken into account, the results would be slightly different.

The variations in pressure and velocity for flow through a two-stage Curtis turbine and for a two-stage Rateau turbine are shown in Figs. 9–14 and 9–15, respectively.

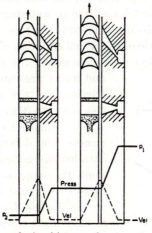

FIG. 9–15.   Pressure-velocity history of a two-stage Rateau turbine.

9–9. Impulse-Blade Efficiencies.   Eq. (9–20) expressed the *bucket* or *blade efficiency* for an impulse stage as the ratio of the energy output of the blade and the kinetic energy of the steam jet entering the blade. Combining this efficiency with the efficiency of the nozzle, Fig. 9–16,

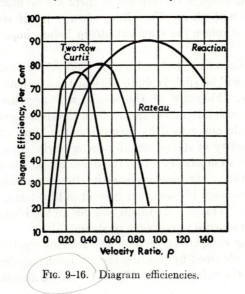

FIG. 9–16.   Diagram efficiencies.

results in a value frequently termed the *diagram efficiency*. This term is understood to include the effects of all losses due to friction in the nozzle and buckets, bucket angles, nozzle angles, velocity ratio, and leaving velocity.

Stage efficiency includes all the losses listed above, and in addition, the losses due to steam leakage through the diaphragm packing, windage or rotational losses, moisture losses due to wet steam, nozzle-end losses for partial peripheral admission, configuration losses, etc. Thus, stage efficiency represents the ratio of the actual enthalpy difference across the stage to the isentropic enthalpy difference across the stage for the same inlet and outlet pressures, Fig. 9–17. Note that friction has the effect of reheating the steam at constant pressure.

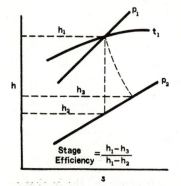

Stage Efficiency $= \dfrac{h_1 - h_3}{h_1 - h_2}$

Fig. 9–17. Stage efficiency shown on Mollier diagram.

The efficiencies used by turbine designers are determined by test, and a detailed discussion of them would be beyond the scope of this text. Briefly, the factors that must be considered are:

(a) *Nozzle Height*. A high nozzle is usually more efficient than a low one.

(b) *Over- and Underexpansion*.

(c) *Steam Displacement*. This factor must be considered when there is partial admission, i.e., when nozzles are used around only part of the nozzle ring. As the blades travel through the arc where there are no nozzles, they carry stagnant steam with them. Therefore, when a jet of steam enters the blades, it must displace the stagnant steam before imparting energy to the blades.

(d) *Mean Diameter*. It has been determined by tests that a large-diameter wheel will be slightly more efficient than a small one.

(e) *Moisture*. Moisture in the steam will decrease the efficiency because (1) the moisture moves slower than the steam and will strike the back of the blade and retard its forward motion, and (2) the moisture reduces the quantity of steam actually doing work. Also, since moisture in the steam has an erosive effect on the blade edges, it is kept to a maximum of about 13%. The moisture is estimated empirically to reduce the efficiency ½% for each 1% moisture present in the steam.

(f) *Windage*. This is a frictional loss due to the blades, shrouds, and discs rotating in a dense atmosphere of steam.

An interesting comparison of the efficiencies of reaction, Rateau, and Curtis stages is obtained by plotting the efficiencies against the enthalpy drop instead of the velocity ratio, Fig. 9–18, for an assumed wheel veloc-

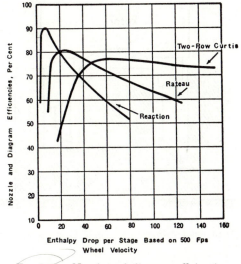

FIG. 9-18. Nozzle and diagram efficiencies.

ity. The reaction stage must operate on a small enthalpy drop, and for this reason reaction turbines have more stages than impulse turbines for the same steam conditions. The Curtis stage, while having a low peak efficiency, maintains better efficiency with large pressure drops than either of the others—an important consideration for the first stage of a multi-stage turbine (see Art. 9-13).

Other losses that must be included in the turbine performance are those due to mechanical losses which include bearing friction, oil pumps, glands, and the governor system. All these losses are determined by test. The turbine is revolved at the desired speeds by some external source, and only sufficient steam is admitted to the casing to maintain the correct density inside. The power required to rotate the turbine will be the sum of the windage, friction, and mechanical losses, which should be deducted from the power generated by the blades to determine unit output. The blade windage losses may be reduced where there are no nozzles by using shields or baffles, Fig. 9-19.

An expression for diagram efficiency would be a ratio of the power output of the blades and the energy supplied to the nozzle.

$$E_d = \frac{P}{778W\left[(h_i - h_o) + \left(\dfrac{V_c}{223.8}\right)^2\right]} \qquad (9\text{-}23)$$

where E_d = diagram efficiency and other symbols are as in Eqs. (9-22) and (9-15) through (9-19).

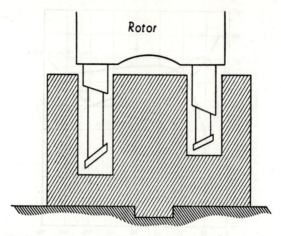

Fig. 9–19. Shields to reduce windage loss of a Curtis stage where there are no nozzles.

EXAMPLE 9–4. For the data in the example of Art. 9–7, calculate the diagram efficiency.

SOLUTION. From Eq. (9–23) and the data in the example of Art. 9–7, $P = 28{,}056$ ft-lb per sec, $h_i = 1231.4B$, $h_o = 1187.1B$, $V_c = 500$ fps, $W = 1$ lb per sec, and

$$E_d = \frac{28{,}056}{778 \times 1 \left[(1231.4 - 1187.1) + \left(\dfrac{500}{223.8} \right)^2 \right]} = 73.1\%$$

EXAMPLE 9–5. Design the blade path for a two-row, single-stage Curtis turbine to develop 100 bhp at 3600 rpm with 97% mechanical efficiency with steam conditions of 200 psig, 500 F, and 5 psig. Mean wheel diam = 33 in.; $\alpha_1 = 20$ deg; nozzle efficiency = 94%; $\beta_2 = 20$ deg; $\beta_4 = \beta_3 - 3$ deg; $C_2 = 0.85C_1$; $\alpha_3 = 22$ deg; $V_3 = 0.93V_2$; $C_4 = 0.91C_3$. Symbols are the same as shown in Fig. 9–13.

SOLUTION. For the nozzle, $\Delta h_s = 1267.3 - 1074.8 = 192.5$ Btu per lb.

Then
$$V_1 = 223.8\sqrt{192.5 \times 0.94} = 3011 \text{ fps}$$

$$U = \pi \times \frac{33}{12} \times \frac{3600}{60} = 519 \text{ fps}$$

$$\rho = \frac{U}{V_1} = 519 \div 3011 = 0.1723$$

$$V_1 \cos \alpha_1 - U = 3011 \times 0.9398 - 519 = 2310$$

$$V_1 \sin \alpha_1 = 3011 \times 0.3420 = 1029$$

$$\beta_1 = \tan^{-1}\left(\frac{V_1 \sin \alpha_1}{V_1 \cos \alpha_1 - U} \right) = \tan^{-1}\left(\frac{1029}{2310} \right) = 24.05 \text{ deg}$$

$$C_1 = \frac{V_1 \sin \alpha_1}{\sin \beta_1} = \frac{1029}{0.4072} = 2530 \text{ fps}$$

Thus $C_2 = 0.85 \times 2530 = 2150$ fps

$$\alpha_2 = \tan^{-1}\left(\frac{\sin \beta_2}{\cos \beta_2 - \dfrac{U}{C_2}}\right) = \tan^{-1}\left(\frac{0.342}{0.9398 - \dfrac{519}{2150}}\right) = 26.1 \text{ deg}$$

$$V_2 = C_2\left(\frac{\sin \beta_2}{\sin \alpha_2}\right) = 2150\left(\frac{0.344}{0.440}\right) = 1680 \text{ fps}$$

and $V_3 = 1680 \times 0.93 = 1563$ fps

$$\beta_3 = \tan^{-1}\left(\frac{\sin \alpha_3}{\cos \alpha_3 - \dfrac{U}{V_3}}\right) = \tan^{-1}\left(\frac{0.3747}{0.9272 - \dfrac{519}{1563}}\right) = 32.21 \text{ deg}$$

$$C_3 = V_3\frac{\sin \alpha_3}{\sin \beta_3} = 1563\left(\frac{0.3747}{0.5333}\right) = 1100 \text{ fps}$$

$$C_4 = 0.91 \times 1100 = 1000 \text{ fps}$$

and $\beta_4 = 32.21 - 3 = 29.21$ deg

$$\alpha_4 = \tan^{-1}\left(\frac{\sin \beta_4}{\cos \beta_4 - \dfrac{U}{C_4}}\right) = \tan^{-1}\left(\frac{0.4882}{0.8728 - \dfrac{519}{1000}}\right) = 54.08 \text{ deg}$$

$$V_4 = C_4\frac{\sin \beta_4}{\sin \alpha_4} = 1000\frac{0.4882}{0.8098} = 603 \text{ fps}$$

The leaving loss will be

$$\frac{AV_4^2}{2g} = \frac{(603)^2}{64.4 \times 778} = 7.27 \text{ Btu per lb}$$

Work for first rotating row [Eq. (9–17) or (9–18)] is

$$\frac{W}{g} U(V_1 \cos \alpha_1 + V_2 \cos \alpha_2) = \frac{1 \times 519}{32.2}\,(3011 \times 0.9398 + 1680 \times 0.898)$$
$$= 69,900 \text{ ft-lb per lb, or } 89.9 \text{ Btu per lb}$$

Work for the second rotating row is

$$\frac{WU}{g}(V_3 \cos \alpha_3 + V_4 \cos \alpha_4) = \frac{1 \times 519}{32.2}\,(1563 \times 0.9272 + 603 \times 0.5869)$$
$$= 29,100 \text{ ft-lb per lb, or } 37.4 \text{ Btu per lb}$$

Ratio of power for first row to the second row is

$$\frac{69,900}{29,100} = 2.4\!:\!1$$

Then $\Delta h = 89.9 + 37.4 = 127.3$ Btu per lb

Diagram efficiency is

$$\frac{127.3}{192.5} = 66.1\%$$

Note that this compares favorably with the data of Fig. 9–16.

Exhaust enthalpy is

$$h_1 - \Delta h = 1267.3 - 127.3 = 1140.0 \text{ Btu per lb}$$

Steam flow is

$$\frac{\text{bhp} \times 2545}{\Delta h \times E_m} = \frac{100 \times 2545}{127.3 \times 0.97} = 2060 \text{ lb per hr}$$

Total nozzle mouth area is

$$\frac{Wv}{25V_1} = \frac{2060 \times 20.0}{25 \times 3011} = 0.547 \text{ sq in.}$$

9–10. Theory of Reaction Blades. Both the underlying principle and the construction of reaction turbines differ materially from those of the impulse unit. One row of stationary and one row of rotating blades constitute a reaction stage. The blades are unsymmetrical, with the outlet angle much larger than the inlet angle, so that the available area for the steam jet is in the shape of a converging nozzle. There is a drop in pressure in both the stationary and rotating rows with a corresponding increase in specific volume. The absolute velocity of the steam increases in the stationary blades and the relative velocity increases in the moving blades, but the absolute velocity of the steam leaving the moving blades is less than the absolute entrance velocity. Since the turbine runs full of steam, the volume flow at any section is the product of the free area and the velocity; but since the volume increases as the steam progresses through the turbine, either the area or the velocity, or both, must also increase. In order to keep the blade lengths within reasonable limits, the area and the velocity are increased toward the exhaust of the turbine.

All turbine blades may be classified according to their *degree of reaction*, which is defined as the ratio of the enthalpy drop in the moving blades to the enthalpy drop in the stage consisting of one stationary row of blades and one rotating row of blades. The degree of reaction may vary from 0 to 100%, zero reaction being pure impulse blades. It is not at all uncommon for impulse blades to have a small amount of reaction (say, 5%) to improve their efficiency.

Parsons referred to the blading he developed as impulse-reaction blading. Since his blades were 50% reaction, they were also 50% impulse. However, blades containing an appreciable percentage of reaction effect, such as 50% reaction, are known by the shortened term of *reaction blades*. Since there is an equal enthalpy drop in the stationary and moving rows of 50% reaction blades, the pressure drop across the moving and stationary rows is equal; also, the blade passages have the same shape. This represents a manufacturing advantage. It also follows that the velocity diagrams for both the stationary and rotating rows are identical, Fig. 9–20. Because of its importance, and because space does not permit a detailed discussion of all varieties of blade, the equations developed for reaction blades will be confined to 50% reaction. Therefore,

$$V_1 = C_2, \quad C_1 = V_2, \quad \alpha_1 = \beta_2, \quad \text{and} \quad \beta_1 = \alpha_2$$

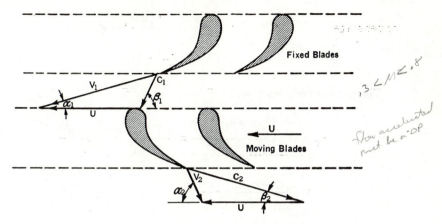

FIG. 9–20. Velocity diagrams for reaction blades.

Then
$$F_x = \frac{W}{g}(C_1 \cos \beta_1 + C_2 \cos \beta_2)$$ (9–24)

$$= \frac{W}{g}(V_1 \cos \alpha_1 + V_2 \cos \alpha_2)$$ (9–25)

Then $V_2 \cos \alpha_2 = C_2 \cos \beta_2 - U$

but $C_2 \cos \beta_2 = V_1 \cos \alpha_1$

Therefore, $V_2 \cos \alpha_2 = V_1 \cos \alpha_1 - U$ (9–26)

Then
$$F_x = \frac{W}{g}(2V_1 \cos \alpha_1 - U)$$

Since the power, P, developed by the blade is UF_x and $\rho = \dfrac{U}{V_1}$,

$$P = \frac{V_1^2 W}{g}(2\rho \cos \alpha_1 - \rho^2)$$ (9–27)

If Δh is the isentropic enthalpy drop per pound for a stage and e_n is the nozzle efficiency, then the amount $(e_n W \, \Delta h)/2$ is the enthalpy converted into kinetic energy per row, since the fixed and moving rows are symmetrical. The exit velocity V_2 will be available as carry-over to the next stage and may be assumed available for the stage under consideration from the previous stage. Since the rows are symmetrical, this carry-over also exists between the stationary and moving rows.

If the carry-over efficiency is e_{co}, the energy is $(e_{co} W V_2^2)/2g$. The energy available per row is, using $A = 1/778$,

$$\frac{e_n W \, \Delta h}{2} = \left(\frac{WV_1^2 - e_{co} W V_2^2}{2g}\right) A$$

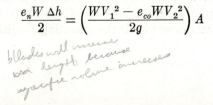

or per stage
$$e_n W \, \Delta h = \left(\frac{W V_1{}^2 - e_{co} W V_2{}^2}{g} \right) A$$

$$\Delta h = \left(\frac{V_1{}^2 - e_{co} V_2{}^2}{e_n g} \right) A$$

The diagram efficiency is

$$E_d = \frac{\dfrac{V_1{}^2}{g} (2\rho \cos \alpha_1 - \rho^2)}{\dfrac{V_1{}^2 - e_{co} V_2{}^2}{e_n g}} = \frac{e_n V_1{}^2 (2\rho \cos \alpha_1 - \rho^2)}{V_1{}^2 - e_{co} V_2{}^2}$$

$$= \frac{e_n (2\rho \cos \alpha_1 - \rho^2)}{1 - \dfrac{e_{co} V_2{}^2}{V_1{}^2}}$$

From the Law of Cosines,

$$V_2{}^2 = C_2{}^2 + U^2 - 2C_2 U \cos \beta_2$$
$$= V_1{}^2 + U^2 - 2V_1 U \cos \alpha_1$$

$$\frac{V_2{}^2}{V_1{}^2} = 1 + \rho^2 - 2\rho \cos \alpha_1$$

$$E_d = \frac{e_n (2\rho \cos \alpha_1 - \rho^2)}{1 - e_{co}(1 + \rho^2 - 2\rho \cos \alpha_1)} \qquad (9\text{-}28)$$

Several interesting conclusions can be obtained from Eq. (9–28). If the carry-over efficiency (e_{co}) is 100%, then the diagram efficiency is the same as the nozzle efficiency and is independent of either the velocity ratio (ρ) or the blade angles. If the nozzle efficiency (e_n) were also 100%, then it would follow that the diagram efficiency would be 100%. With a carry-over efficiency of zero, the diagram efficiency would be $E_d = e_n (2\rho \cos \alpha_1 - \rho^2)$; and also with 100% nozzle efficiency, the diagram efficiency would be $E_d = 2\rho \cos \alpha_1 - \rho^2$. In order to determine the velocity ratio that will give the maximum efficiency, the last equation may be differentiated with respect to ρ and equated to zero to obtain

$$\rho = \cos \alpha_1 \qquad (9\text{-}29)$$

The carry-over efficiency of the blading is very difficult to determine by test and is included with the diagram efficiency as determined from testing. However, it is undoubtedly less than 100% and may be in the neighborhood of 60 to 80%. It can be seen that there would be no carry-over to the first stage of a turbine and that the carry-over from the last stage would be a complete loss. In the latter case it is called the leaving loss and is found in impulse as well as in reaction turbines. Exact values for leaving loss cannot be given because of the many factors involved, such as the design conditions versus the operating conditions, the possible variations in exhaust pressure, and the load. In general, the leaving loss is less for noncondensing turbines than for condensing turbines. It may be in the order of 1% of the total isentropic enthalpy drop for the first case at full load and 4 or 6% in the later case. The difference is due to the tremendous volumes that must be handled at the exhaust of a condensing unit. The specific volume of the exhaust from a noncondensing unit is much smaller than that for a condensing unit.

As in the case of impulse blading, the losses throughout the stage appear as reheat, and so the stage efficiency is also the ratio of the actual enthalpy difference for the stage divided by the isentropic enthalpy difference for the same initial and final pressures, Fig. 9–17.

The more recent designs of reaction blades are of the teardrop shape, Fig. 9–21, as discussed for the curved-foil impulse nozzle. The entrance is thick and well rounded, with the back of the blade as thin as possible to reduce flow disturbances at that point. Because of the rounded shape of the inlet edge of the blade, it is difficult to assign a definite angle for the inlet, although it would probably be some 80 or 90 deg. The exit angle is not often spoken of as such but is called per cent gaging, which is the sine of the exit angle. Gaging indicates the amount of the annular area (about 25 to 60%) that is available as actual blade area for the steam.

It will be noticed that the diagram efficiency curve for reaction blading is flat at the peak so that the best efficiency may be obtained with a rather wide latitude in the velocity ratio, Fig. 9–16. Values of 0.8 to 0.85 for the velocity ratio seem to be customary for constant-speed drives, but at times it is best to favor low-speed operation when the drive is variable speed such as for marine units. Then the velocity ratio for design conditions may be chosen somewhat larger than $\cos \alpha_1$.

Fig. 9–21. A reaction blade. (Westinghouse Electric Corp.)

The factors involved in the efficiency of reaction stages are similar to those for impulse stages. They include, in addition to the diagram efficiency, (1) blade leakage, (2) blade height, (3) blade width, (4) trailing edge thickness, and (5) moisture in the steam. The moisture correction for reaction blades may be higher than that for impulse blades—perhaps 1¼% for each 1% moisture in the steam after isentropic expansion. Blade leakage is a factor that is peculiar to the reaction blades. The pressure drop across the rotating blades of a reaction stage will encourage steam to by-pass the blade by going through the clearance area; thus, the steam does no work. Shrouds (peripheral metal strips that tie the blade tips together) are used for reaction blades. They also act as seals because of the small clearances between them and the stationary portion of the turbine.

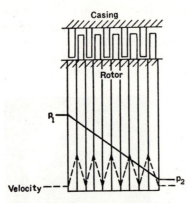

FIG. 9–22. Pressure-velocity history of a group of reaction stages.

The pressure-velocity history for a reaction turbine may be visualized by referring to Fig. 9–22.

Since the reaction turbine runs full of steam at all times, the height of the blades is determined by the continuity equation and must increase as the steam progresses through the turbine. The height may be reduced to some extent by increasing the velocity of the steam in the low-pressure stages, but this cannot completely compensate for the increase in the quantity flowing. When the unit is to operate condensing, the annular area is increased by using a conical drum (shaft). For this arrangement the mean blade velocity will increase towards the low-pressure end of the turbine, but the velocity ratio can be maintained the same throughout the turbine by increasing the steam velocity. As already mentioned, increased steam velocity will decrease the required area.

9–11. Dummy Pistons. In addition to causing leakage across the fixed and moving blades, the pressure differential across the moving blades causes an end thrust on the shaft. If this thrust were to be absorbed entirely by the thrust bearing, it would require bearings that would be out of proportion to the turbine. A balance may be achieved, however, by using a dummy piston or balancing piston similar in principle to that used on a high-pressure pump. A small part of the steam received by the turbine is by-passed through a labyrinth sealing strip, Fig. 9–23, and then to a low-pressure section of the turbine by a pipe shown schematically in Fig. 9–24. Other illustrations of dummy pistons will be shown later in the chapter. By properly proportioning the areas and pressures for the dummy piston to counteract the blading thrust, the load on the thrust bearing may be reduced considerably.

Considering the turbine shown schematically in Fig. 9–24, we see that the pressure drop across the rotating rows of blades will cause a thrust to

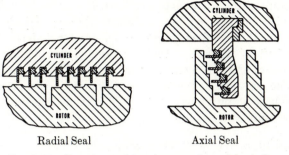

Radial Seal   Axial Seal

FIG. 9-23. Reaction-turbine dummy piston labyrinth.

the right which will be equal to the pressure on the blades times the area.
In order to simplify the calculations, assume that this is only one rotating
row of blades and that its area is a mean for the first and last rows of
reaction blades. A hypothetical mean blade row would have a height of
$(1 + 9)/2 = 5$ in. With 50% reaction, the pressure drop across the rotating rows, and therefore the drop across the hypothetical mean blade row,
would be $(175 - 0.5)0.50 = 87.25$ psi. The area of the annulus represented by the hypothetical mean blade row may be approximated by
multiplying its height by the circumference at the center of the row:
area $= 5(17 + 5)\pi = 346$ sq in. Then

$$\text{Blade thrust} = 87.25 \times 346 = 30{,}189 \text{ lb}$$

$$\text{Thrust for area } A = 0.5 \frac{\pi}{4} (17^2 - 8^2) = 88 \text{ lb}$$

$$\text{Thrust for area } B = 175 \times 3.5 \left(\frac{24 + 17}{2} \right) \pi = 39{,}447 \text{ lb}$$

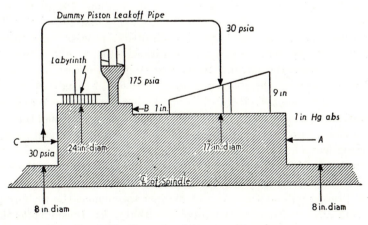

FIG. 9-24. Example of dummy-piston forces.

The pressure of the steam acting on area C will be determined by the point at which the leak-off pipe reinducts the steam into the turbine because the flow through the blading is much greater than the flow through the dummy piston. Therefore, the flow through the blading will be the controlling factor and will determine the pressure at the outlet of the leak-off pipe. The pressure acting on area C must be the pressure at the outlet of the leak-off pipe plus a very small friction drop through the pipe. Using the pressure at C as 30 psia,

Thrust for area $C = 30(24^2 - 8^2)\dfrac{\pi}{4} = 12{,}063$ lb

Total thrust towards the right $= 30{,}189 + 12{,}063 = 42{,}252$ lb

Total thrust towards the left $= 39{,}447 + 88 = 39{,}535$ lb

Net thrust towards the right $= 42{,}252 - 39{,}535 = 2717$ lb

If the dummy piston were not used, the thrust bearing would have a load of 30,189 lb. The Kingsbury type of thrust bearing is designed for bearing loads of from 200 to 400 psi of bearing area when used on turbine generators; the more conservative figure is used for large central-station units. The bearing area needed for the turbine designed without the dummy piston would be 150 sq in. versus 13.5 sq in. with the dummy piston. Of course it would be good engineering to use a thrust bearing somewhat larger than 13.5 sq in.

Since there is practically no pressure drop across the wheels of an im-pulse turbine, dummy pistons are not required for those turbines.

9–12. Performance. Steam entering a turbine must travel through the main steam valve (stop or throttle valve) which includes a strainer, steam chest, and governor valves; all of which cause a pressure drop. Friction loss through this equipment would be about 4% of the initial pressure for a multivalve governor (see Art. 9–13) and 10% for a single-valve turbine. This pressure drop is a throttling process represented by constant enthalpy from the entrance conditions p_1, t_1, h_1, and s_1 to point 2, Fig. 9–25, which is the entrance to the first stage of the turbine.

In the first stage, the steam theoretically would expand isentropically from point 2 to point 3'. Regardless of whether the stages are impulse or reaction, there will be losses occurring in the expansion through the blades and nozzles, as previously outlined, that will appear as thermal energy in the steam leaving the stage. This will result in "reheating" of the steam from enthalpy h'_3 to h_3 at the stage pressure p_3. Upon entering the second stage, the steam will have the enthalpy h_3.

The process will continue in this manner throughout the turbine until the steam emerges from the last stage at enthalpy h_7. Inasmuch as this is

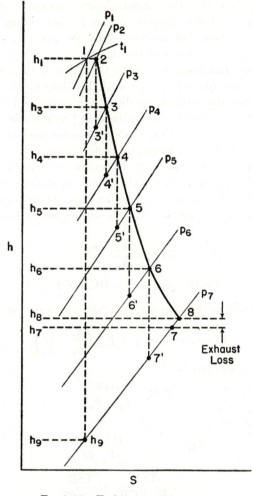

Fɪɢ. 9–25. Turbine condition curve.

the last stage of the turbine, there is no possibility of obtaining useful work from the kinetic energy of the steam going into the condenser where the steam will come to rest. Thus, the kinetic energy known as leaving loss is converted into thermal energy. The leaving loss is represented by the enthalpy difference $h_8 - h_7$, while the energy transferred to the condenser circulating water is represented by the difference between h_8 and the enthalpy of saturated liquid at pressure p_7.

A curved line drawn through points 1, 2, 3, 4, 5, 6, and 8, indicating the condition of the steam throughout the turbine at points where it is possible to extract and measure the steam properties, is known as the *condition curve.* Note that the condition curve does not represent the path of the steam on the Mollier diagram during its flow through any particular stage of the turbine but indicates only the properties of the steam entering and leaving the stages of the turbine.

Stage efficiency has been defined as the ratio of the actual enthalpy drop across the stage to the isentropic enthalpy drop for the same pressure differential. When this ratio is applied to a group of blades or to the entire turbine, it is referred to as the Rankine Cycle Ratio (RCR), or the internal efficiency of the turbine. This expression does not include such mechanical losses as bearing losses, oil pump power, etc.

Likewise, the steam rates may be expressed with or without the mechanical efficiency, and they are the pounds of steam per hour that are required by the turbine to develop 1 kw or 1 hp. These relationships may be developed as follows by defining the enthalpies as indicated in Fig. 9–25:

$$\text{RCR} = \frac{h_1 - h_8}{h_1 - h_9} \tag{9-30}$$

$$\text{Engine efficiency} = \text{RCR} \times E_{mg}$$

in which E_{mg} equals mechanical and generator efficiency; or

$$\text{Engine efficiency} = E_{mg}\left(\frac{h_1 - h_8}{h_1 - h_9}\right) \tag{9-31}$$

The theoretical steam rate is

$$\text{TSR} = \frac{3413}{h_1 - h_9} \tag{9-32}$$

The actual steam rate is

$$\text{SR} = \frac{3413}{(h_1 - h_9)E_{mg}\text{RCR}} \tag{9-33}$$

$$= \frac{\text{TSR}}{E_{mg}\text{RCR}} = \frac{\text{TSR}}{\text{eng eff}} \tag{9-34}$$

These efficiencies have been developed to conform to the generally accepted usage in practice and to include the generator losses. If the turbine is of the mechanical-drive type, then the generator efficiency will have no significance; the expressions would then be based on horsepower and the thermal equivalent of a horsepower (2545 Btu per hr) would be substituted for 3413.

EXAMPLE 9–6. A 3600-rpm, 60,000-kw unit receives steam at 385 psig and 800 F with a back pressure of 1.5 in. Hg abs. Engine efficiency is 78% and the combined mechanical and electrical efficiency is 95%. Find: (a) RCR, (b) theoretical steam rate, (c) actual steam rate, (d) the steam flow, and (e) the exhaust enthalpy.

SOLUTION.

(a) RCR = (engine efficiency) ÷ (mech and elect efficiency)

$$= 78 \div 0.95 = 82.1\%$$

(b) $\text{TSR} = \dfrac{3413}{\Delta h_s} = \dfrac{3413}{1416.4 - 925.8} = 6.95$ lb per kwhr $(s = 1.6842)$

(c) $\text{SR} = \dfrac{3413}{(\text{eng eff}) \times \Delta h_s} = \dfrac{6.95}{0.78} = 8.91$ lb per kwhr

(d) Steam flow = 8.91 × 60,000 = 534,600 lb per hr

(e) Exhaust enthalpy = $h_1 - \Delta h_s(\text{RCR})$

$$= 1416.4 - 0.821(1416.4 - 925.8)$$

$$= 1013.6 \text{ Btu per lb}$$

From the standpoint of efficiency, the condition curve may be divided into three sections: (1) high-pressure section, (2) intermediate-pressure section in the superheat region, and (3) the moist region. The high-pressure section is not too efficient because of the amount of leakage with small blades and nozzles and also because of the low-efficiency Curtis stage commonly used in this region. In the intermediate-pressure section, the blades, either Rateau or reaction, are of good height and the losses from leakage are moderate so that the efficiency is highest in this section. The efficiency of the low-pressure section, or moist region, is the lowest of any part of the turbine because of the losses from moisture and exit velocity.

Accurate determination of the condition curve can be made only by an analysis of the blading by the turbine designer. However, conditions arise in the design of a station where it is necessary to estimate the condition curve. This is particularly true during preliminary calculations. One common method of determining the condition curve under these circumstances is to estimate the RCR for the entire turbine and to use that to establish the exhaust enthalpy. The condition curve is then drawn as a straight line between the inlet conditions and the exhaust conditions.

A more accurate but more difficult solution may be obtained by estimating the efficiency of the three sections of the condition curve. With a certain amount of trials, four points may be located that will give a smooth curve and also the desired over-all Rankine Cycle Ratio (RCR).

TABLE 9–1

TURBINE INTERNAL EFFICIENCIES 25,000 KW AND ABOVE

Curtis element, first 150 Btu for high-pressure units..	Avg. 75% *
High-pressure reaction, above 200 psi................	82%
Intermediate pressure	85%
Low pressure, moist region	78%

* Range 70 to 78%

The values of efficiencies in Table 9–1 may be used for the solution of problems, with results that will be within reasonable limits of those to be found in practice.

Engine efficiencies of small turbines, without generators, may be determined from Fig. 9–26. The values shown on these curves are typical. Actual units may have higher or lower efficiencies, depending on details of design. Mechanical losses of small units vary from 2 to 5%. Losses for reduction gears are 2% or 3% at rated load.

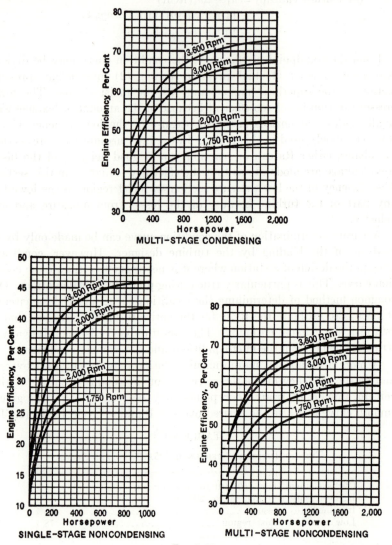

Fig. 9–26.

Warren and Knowlton * have presented curves for engine efficiencies of large turbine generators, Figs. 9-27, 9-28, and 9-29.

Fig. 9-27 indicates the engine efficiencies that may be expected at rated

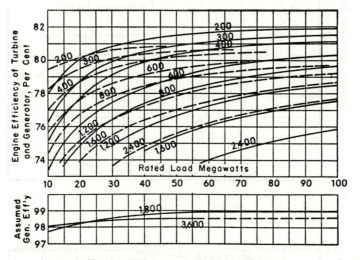

FIG. 9-27.   Engine efficiencies of large condensing turbine generators. Drawn for 300 F superheat, 4% exhaust loss, 1.25% mechanical loss. Full lines, 1800 rpm; dash lines, 3600 rpm. Generator efficiencies assumed as shown; hydrogen cooling. Figures on curves are throttle pressure, psig. (*Trans. ASME,* February, 1941, pp. 125–135.)

load for large condensing turbines at 1800 rpm and 3600 rpm with initial pressures of 200 to 2400 psig with hydrogen-cooled generators and for the exhaust loss and mechanical efficiency of 4% and 98.75%, respectively, all for an initial superheat of 300 F. The engine efficiency should be corrected for other superheat conditions, as indicated on Fig. 9-28. The mechanical efficiency will vary for large units from about 98.7 to 99.25%. Generator efficiencies for rated load and partial loads may be determined from Fig. 9-29.

Although the curves just discussed were derived from the performance data of General Electric turbines, they may be considered representative of all manufacturers for the conditions specified. Actual efficiencies of turbines may vary to some extent from these curves for several reasons; viz., the exhaust losses may differ from the 4% on which the curves were based, the mechanical or generator efficiencies may vary, the control valves may be arranged differently to fit the most efficient load point to the conditions of the expected load duration curve.

* G. B. Warren and P. H. Knowlton, "Relative 'Engine Efficiencies' Realizable from Large Modern Steam Turbine-Generator Units," *Trans. ASME* (February, 1941), pp. 125–135.

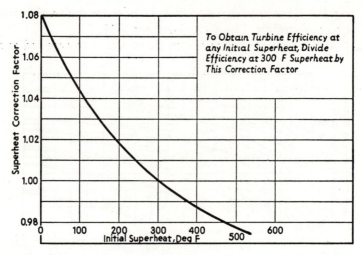

Fɪɢ. 9–28. Correction factors for superheat to be used with Fig. 9–27. (*Trans. ASME*, February, 1941, pp. 125–135.)

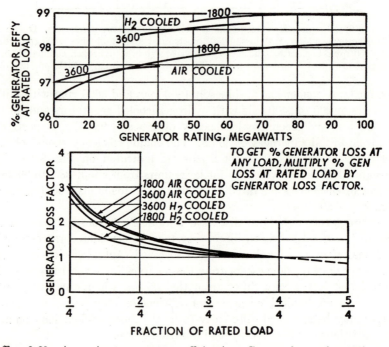

Fɪɢ. 9–29. Approximate generator efficiencies. Curves drawn for 80% power factor; for 90%, add 0.15% to rated-load generator efficiency; for 70%, subtract 0.25% from rated-load generator efficiency. (*Trans. ASME*, February, 1941, pp. 125–135.)

Currently there is a trend toward rating large turbine generator units at their maximum guaranteed capacity. Preferred Standard units (see Art. 11-10) are so rated but have a design tolerance that usually permits an additional 5 to 10% of capacity. In the past, units were designed to deliver approximately 5/4 of their rating.

EXAMPLE 9-7. Estimate the engine efficiency and RCR at rated load for a 50,000-kw turbine generator operating at 1800 rpm, 850 psig, 900 F, 1 in. Hg abs, air cooled.

SOLUTION. Superheat $= 900 - 527.3 = 372.7$ F. Engine efficiency (from Fig. 9-27) $= 77.8\%$ at 300 F superheat, 98.75% mechanical efficiency, and 98.85% electrical efficiency. Engine efficiency corrected for superheat factor (from Fig. 9-28) $= 77.8/0.99 = 78.6\%$.

$$\text{RCR} = \frac{78.6}{0.9875 \times 0.9885} = 80.5\%$$

Actual generator efficiency (Fig. 9-29) $= 97.75\%$ Actual engine efficiency $= 80.5 \times 0.9875 \times 0.9775 = 77.7\%$

9-13. Governing. Before we can discuss the effect of governing and load and blade types on the shape of the condition curve, it is necessary to understand the variation of the pressure distribution within the turbine for changes in the steam flow. In order to develop the desired relationship we may refer to Napier's formula, Eq. (9-4), for the flow through a nozzle and observe that the flow is approximately proportional to the inlet pressure. Since the flow through a stage of a turbine is essentially the same as the flow through a nozzle regardless of the type of blading, the increase or decrease of the steam flow will cause a corresponding increase or decrease in the pressure differential for each stage and therefore for the entire turbine. However, the exhaust pressure for most turbines remains nearly constant for changes in steam flow, so that the initial pressure for a group of stages will increase when the flow is increased. Although this straight-line relationship is not absolutely correct, it is so close to the more complicated but more accurate solutions that it is used extensively in power-plant design. Since no turbine operates with zero exhaust pressure, the stage-pressure curve will bend at very low flows, as indicated by the solid line in Fig. 9-30.

Governing of a turbine may be accomplished in two basic ways by: (1) reducing the flow by decreasing the inlet pressure and (2) reducing the steam flow with constant inlet pressure by reducing the area available to the steam jet.

The first method is used on small turbines by throttling the steam with a valve. Because of its simplicity, this method lends itself to the automatic control of mechanical-drive turbines; but since a throttling process is inefficient, it is used only on small units. The effect of throttle governing on the condition curve is shown in Fig. 9–31. Since a throttling process is one of constant enthalpy, the thermal energy of the steam entering the turbine is the same for all loads; but because of the increase in entropy, there is less available energy during the expansion, and because of the decreased initial pressure, the flow is also less for fractional loads.

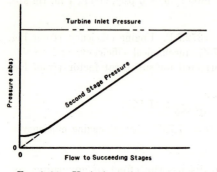

FIG. 9–30. Variation of stage pressure with flow.

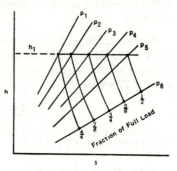

FIG. 9–31. Effect on the condition curve of throttle governing.

The second method of control involves the use of several valves instead of one. These governing valves are located in the steam chest of an impulse turbine or an impulse-reaction turbine, and each admits steam to a different portion of the nozzle ring, Fig. 9–32. The governing mechanism of the turbine moves the pilot-valve stem up to increase steam flow when the speed decreases and moves the stem down to decrease steam flow when the speed increases. Upward motion of the stem admits high-pressure governor oil to the bottom of the power piston, causing it to rise against spring pressure. Since the pilot-valve sleeve is attached to the power piston, the sleeve will rise with the piston and shut off the high-pressure oil. The bar lift is directly connected to the power piston. Note that loss of oil pressure for any reason will shut the governor valves, i.e., the governor mechanism *fails safe*.

The valves may open sequentially or in pairs, but in any event the throttling effect will be reduced because only a portion of the total steam flow will be throttled. For any given group of valves open at one time, the pressure drop across the first impulse row will be the difference between the constant inlet pressure and the pressure curve for the second stage, as determined by the method previously outlined, Fig. 9–30. Therefore, as the flow through the turbine decreases, the pressure drop and the thermal head for the first stage will increase, or the percentage of the turbine load that is developed by the first stage will increase.

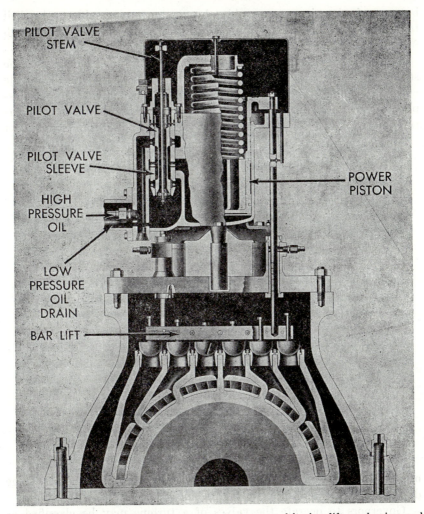

PILOT VALVE
STEM

PILOT VALVE

PILOT VALVE
SLEEVE

HIGH
PRESSURE
OIL

LOW
PRESSURE
OIL
DRAIN

BAR LIFT

POWER
PISTON

Fig. 9–32. Steam chest with governor valves operated by bar-lift mechanism and oil servomotor. (General Electric Co.)

Since the reverse of this is also true when increasing load, it can be seen that a point will be reached where the pressure differential across the first stage will become so small that it will be impossible to pass sufficient steam through the first stage to develop the required load by the remaining stages of the turbine. When this point is reached, the steam may be by-passed around the first stage. Under this condition, the remaining stages in the turbine will operate under unfavorable conditions with a resulting poor efficiency. For this reason the first stage is by-passed only when the turbine is operating under heavy load and efficiency becomes of minor importance. Also, by-pass governing subjects the shell to abnormally high temperatures and the seals to abnormally high pressures.

When by-passing any stages of a turbine, some small quantity of the steam must be allowed to pass through the by-passed stages to provide adequate cooling. The windage effect of the blades in the dense atmosphere of steam can easily be great enough to increase the blade temperature and to reduce the strength of the metal. Fig. 9–33 shows the effect

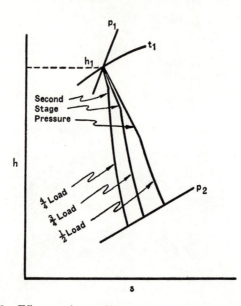

Fig. 9–33. Effect on the condition curve of multivalve governing.

of multivalve governing on the condition curve. Note that the first stage develops a larger percentage of the turbine load as the flow is reduced.

Governors to operate the valves of a turbine may be classified as mechanical or hydraulic. First consider the simple mechanical governor shown in Fig. 9–34. This type of governor is used on small turbines that are controlled by throttle governing.

The speed-sensitive element of the governor is direct-connected to the turbine shaft and therefore revolves with the shaft. *Two weights,* or *flyballs,* are pivoted in such a way that they move outward from the *governor stem* as the shaft speed increases. The motion of the weights and levers is restrained by the governor spring.

When the turbine speed is very low, the weights are close to the shaft, and the governor stem, the linkage, and the governor valve are in a position to admit the maximum steam to the turbine. As turbine speed in-

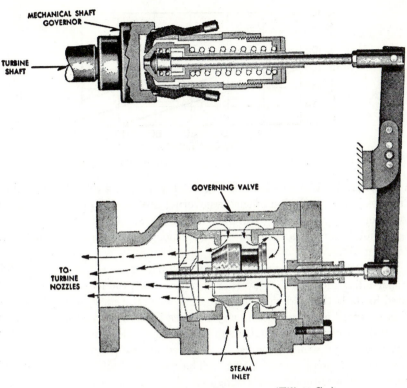

Fig. 9–34. Mechanical speed governor. (Elliott Co.)

creases, the flyballs move away from the stem by overcoming the spring pressure, and the linkage reduces the valve steam-flow area.

Each successive position of the governor valve, from open to nearly closed, requires a distinct position of the flyballs, which in turn means a different shaft speed. Thus, a small turbine load means that the governor valve will be nearly closed, the weights must be away from the stem, and the speed must be high. At a maximum turbine load, the valve must be wide open, the weights will be close to the stem; this is accomplished by a low turbine speed. Thus, turbine speed decreases as load increases, Fig. 9–35.

This change in turbine speed with load is the *steady-state regulation* (where steady state means for gradual speed changes) and is defined as

$$S = \frac{N_n - N_f}{N_f} \qquad (9\text{--}35)$$

where S = governor regulation, %
N_n = no-load speed, rpm
N_f = full-load speed, rpm

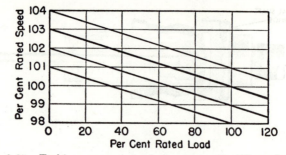

Fig. 9-35. Turbine governor regulation curves for 3% regulation.

A governor of this type, shown in Fig. 9-34, will have 6 to 10% regulation. Changes in the regulation of such a governor are obtained by changing the governor spring to produce different spring forces. A governor with a 0% regulation is termed an *isochronous* governor; it is not normally used for steam turbines.

If a turbine had a governor with 3% regulation, as shown in Fig. 9-35, proper adjustment of the governor spring would produce design speed at rated load. If the load were to decrease to zero, the speed would be 103% of rated speed without making any changes to the governor spring. From Fig. 9-35, it can be seen that a suitable change in the compression of the governor spring would produce rated speed at two-thirds load at a corresponding no-load speed of 102%. The value of 102% can be obtained analytically by proportional triangles.

Obviously, the force that can be exerted by a governor of this type is limited. A more elaborate governor that is designed to supply a much greater force, such as required for a multivalve governing system, is shown in Fig. 9-36. This governor is driven directly from the turbine shaft and has two weights as before. However, the force developed by the governor is multiplied by the *oil relay* of the *servomotor*. As the weights move outward with increased shaft speed, the stem positions the oil pilot valve. A downward movement of the governor stem admits high-pressure oil to the top of the governor servopiston. A downward motion of the servopiston closes the steam governing valves. However, the same oil pressure that operated the servopiston also acts on the restoring piston to return the pilot valve to its neutral position; otherwise, continued oil pressure on the servopiston would completely close the governing valves.

Since the further compression of the servopiston spring will require an increased oil pressure, and since this oil pressure also acts on the restoring piston to oppose the force from the governor weights, the governor will have a regulation curve, and the per cent of regulation will be determined by the spring scale for the servopiston spring.

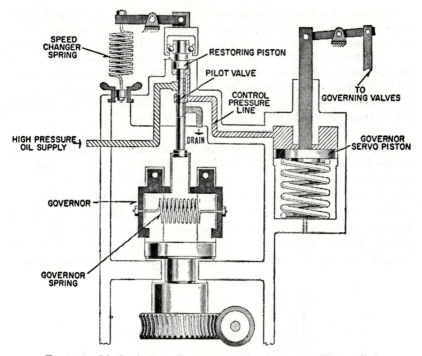

Fig. 9-36. Mechanical speed governor with servomotor. (Elliott Co.)

The family of regulation curves, as shown in Fig. 9-35, are obtained by
the *speed-changer spring* rather than by adjustments to the governor
spring. Hand adjustments are required by the governor of Fig. 9-36, but
for automatic speed adjustment of turbine generators, the hand device is
replaced by an electric motor called the *speed-changer motor*.

Hydraulic governors employ a pump, direct-driven from the turbine
shaft, in place of the weights. The pump may be either of the centrifugal
type or the gear type. The pump-discharge pressure is proportional to
the square of the pump, or turbine shaft, speed.

The regulation for most turbine generators is adjustable from 2 to 6%.
It is desirable to have all units of a system operate with the same regulation so that all units will have similar speed changes for a definite
percentage of load change.

EXAMPLE 9-8. A system contains two units: Unit A is 40,000 kw and has a
3% regulation, and Unit B is 50,000 kw and has a regulation of 4%. Both units
are electrically connected and are generating 50,000 kw total (30,000 kw for Unit
A) at 3600 rpm, 60 cycles. If the load reduces to 45,000 kw, what will be the
system frequency, and what will be the load on each unit if no adjustment is made
on the governor speed changer?

SOLUTION. Let a equal the decrease in load for Unit A, and b equal the decrease in load for Unit B. The percentage decrease in speed for each unit will be the same and will be $(a/40,000) \times 3$ for Unit A and $(b/50,000) \times 4$ for Unit B. Thus $3a/40,000 = 4b/50,000$ and $a + b = 5000$. Solving these two equations simultaneously produces $a = 2580$ kw and $b = 2420$ kw, or the load for Unit A will be 27,420 kw and for Unit B it will be 17,580 kw. The speed increase will be $(2580/40,000) \times 3 = 0.1935\%$, or the speed will be $3600 \times 1.001935 = 3607$ rpm and the frequency will be $60 \times 1.001935 = 60.116$ cycles.

9–14. Curves of Steam Rate and Willans Line. Reduction of the turbine load may be accomplished by either throttling the steam flow with one valve or by closing one of several control or governor valves. The effect of both these types of control on the condition curve has been noted, and it was observed that in both cases the enthalpy differential per pound of steam for the turbine was decreased as the load decreased. This would indicate that the steam rate increases as the load decreases, as shown by the solid line on Fig. 9–37, for a turbine that has an infinite number

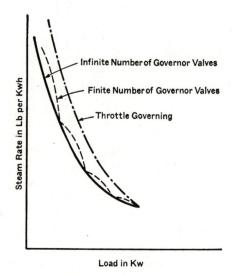

FIG. 9–37. Steam-rate curves.

of governor valves, and that the nozzle area decreases in infinitely small increments as the load decreases. Under such conditions, no wasteful throttling action would occur at any time, and the steam-rate curve would be at a minimum for all loads. The decrease in efficiency, as represented by the increase in steam rate for loads above and below ME (Most Efficient) load, would then be due to the decrease in blade efficiency and to leakage, windage, and friction. Since some steam would be required to operate the turbine generator at rated speed but with no load, the steam-rate curve would reach infinity at zero load.

With a finite number of governor valves, the steam-rate curve would have a number of *humps*, as shown in Fig. 9–37, and part of the steam would be throttled as each valve gradually closed. The envelope of this curve, i.e., a curve through the points representing complete closure of one or more valves, would be the curve for an infinite number of valves. As the number of governor valves is decreased, the steam rate at low load will be further removed from the optimum until the limit is reached when only one valve (throttle governing) is used.

The total steam required by the turbine, of course, can be found from the product of load and steam rate, and when plotted against the load, becomes a Willans line, Fig. 9–38. The three cases of governing that were shown for steam rates are used for this figure, and the same reasoning can

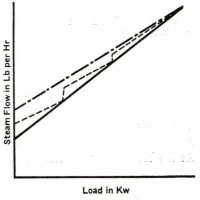

be applied to the Willans lines. Note that the Willans line for throttle governing and for an infinite number of governor valves is a straight line and will conform to the general equation

$$y = a + bx \qquad (9\text{–}36)$$

where $y =$ throttle steam flow, lb per hr

$a =$ no-load steam consumption, lb per hr

$b =$ slope of the curve, lb per kwhr

$x =$ load, kw

Load in Kw

Fig. 9–38. Willans lines.

The effects of a governor valve on Willans lines are illustrated in Fig. 9–39. These performance curves are for a turbine controlled by throttle governing but with two hand-operated governor valves. With these two valves closed, the maximum steam flow that can pass through the first section of the first-stage nozzle block is 13,850 lb per hr. This steam flow will develop 214 hp. When more load is required, the steam flow can be increased only by increasing the nozzle area, i.e., by opening the first hand valve. Note that as soon as this valve is opened, the throttle valve must reduce the steam pressure at the nozzle block. The resultant throttling of the steam is wasteful, and the steam flow increases to 14,700 lb per hr at the same 214 hp. Now the load and steam flow may be increased to 257 hp and 16,400 lb per hr. Opening the second hand valve permits a maximum load of 300 hp at 18,900 lb of steam per hr. Observe

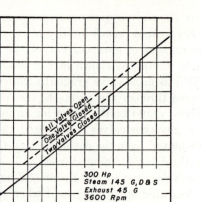

Fig. 9–39. Willans lines for a 300-hp turbine with two hand valves. (Murray Iron Works Co.)

that at low loads it is uneconomical to open more hand valves than are necessary to carry the load.

The Willans line is a convenient check on the test results of a turbine if it is tested at the valve points, but it is not indicative of the efficiency of a turbine if operated extracting. The turbine manufacturer has no control over the feed cycle and can only guarantee approximate bleed pressure. As the bleed pressure increases it will be found that the extraction steam

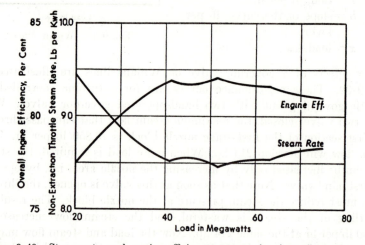

Fig. 9–40. Steam-rate and engine-efficiency curves showing effect of governor valves. (75,000-kw turbine generator, 1800 rpm, air cooled, 315 psig, 725 F, 1 in. Hg abs.)

will also increase, with an attendant increase in the throttle flow, even though the turbine efficiency remains constant. Thus, the Willans line would increase in slope due to no fault of the turbine. Large extracting turbines are best tested by determining the heat rate rather than the steam flow or throttle steam rate. Fig. 9–40 shows the expected performance of a steam turbine.

9–15. Extraction Factor. The *extraction factor,* or replacement factor, is sometimes useful to the power-plant designer when he is considering an extraction type of turbine. Referring to Fig. 9–41, the enthalpy differ-

ence from the inlet to the outlet of an extraction type of turbine is Δh, as determined from the condition curve. This may be divided into two parts: (1) the drop from throttle to the extraction point, and (2) the drop from the extraction point to the exhaust, designated by Δh_1 and Δh_2, respectively. If 1 lb of steam enters the turbine at the throttle and leaves at the extraction point, it will do 235.6 Btu of work, or 242.1 Btu of work less than another pound of steam that enters at the throttle and leaves at the exhaust.

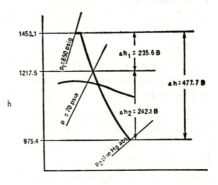

FIG. 9–41. Development of extraction factor.

In order to keep the load on the turbine constant, sufficient additional steam must be added at the throttle to replace the 242.1 Btu lost by extracting 1 lb of steam from the turbine. The replacement factor or extraction factor for these conditions will be

$$\text{Extraction factor} = \frac{\Delta h_2}{\Delta h}$$

$$= \frac{242.1}{477.7}$$

$$= 0.5068$$

(9–37)

Thus, 0.5068 lb of steam must be added to the turbine throttle for each pound of steam extracted in order to maintain a constant load on the turbine. If the turbine is to deliver 65,000 kw with 3000-kw losses and 101,950 lb per hr extracted, then the total throttle flow will be

Nonextracting throttle flow	$= 68,000 \times 3413 \div 477.7$	$= 485,800$ lb per hr
Replacement flow	$= 101,950 \times 0.5068$	$= \underline{51,670}$ lb per hr
Throttle flow		$= 537,470$ lb per hr

The values of extraction factors change with the bleed pressure, being larger for high pressures and smaller for low-extraction pressures.

9–16. Performance of Controlled-Extraction Turbines.   The extraction from turbines may be uncontrolled, and therefore the extraction pressure will be a direct function of the flow to the succeeding stage, or the extraction may be controlled to maintain at a predetermined extraction pressure. Central-station turbines using extraction are predominantly of the uncontrolled type, since there would be no advantage in a fixed extraction pressure, and there might be disadvantages from the standpoint of more complicated equipment and poorer efficiency. Many cases have arisen in industrial installations where a fixed steam pressure is advisable and sometimes mandatory for process requirements.

Automatic control of the extraction pressure may be obtained by using a *grid valve*, a steam chest with *governor valves*, or a steam chest with *balanced poppet valves*. A *double-automatic extraction* turbine is shown in Fig. 9–42. Steam enters the turbine through the main steam chest and the governor valves, the same as in any other large turbine. After expansion through the first stage, the steam enters a bleed belt where some of it may be extracted through the bleed nozzle at the bottom of the turbine. The remaining steam flows into a second steam chest which is essentially identical to the first. The governor valves in the second steam chest are controlled to maintain a given pressure at the outlet of the first stage and thus a predetermined extraction pressure. After leaving the governor valves of the second steam chest, the steam passes through more stages to another bleed belt. The pressure at this point in the turbine is controlled by a balanced poppet valve. A grid valve to control extraction pressure, Fig. 9–43, may be used for extraction pressures up to about 50 psig. Warpage and binding of the plates make this valve unsuitable for higher pressures. The valve consists of two discs, one stationary and the other movable. Both discs contain ports that act as valves. The movable disc is operated by the governor to open or close the ports. Poppet valves are used for extraction pressures of 20 to 150 psig, while conventional steam chests are used for higher pressures.

If the flow to the low-pressure section of the turbine were completely shut off, sufficient heat would be generated from the friction of the moving blades in the dense atmosphere of steam to cause the blades to overheat. This is prevented by allowing a small portion of the steam to flow to the low-pressure section of the turbine at all times to keep the blades cool. This minimum quantity of steam is called *cooling steam,* and it may or may not do sufficient work in the low-pressure section of the turbine to overcome the friction of the moving blades. In the event that the cooling

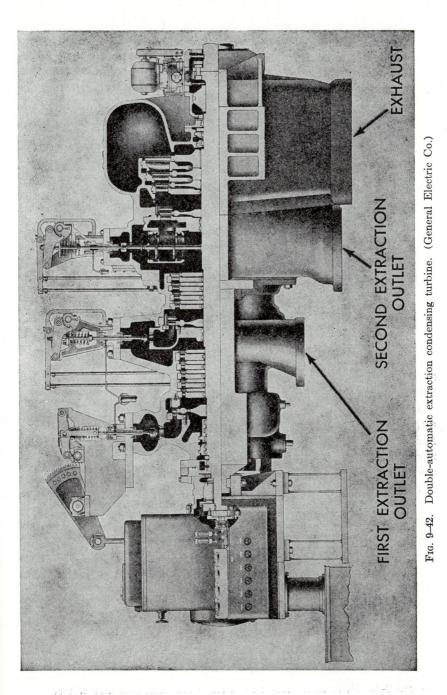

FIRST EXTRACTION OUTLET

SECOND EXTRACTION OUTLET

EXHAUST

FIG. 9-42. Double-automatic extraction condensing turbine. (General Electric Co.)

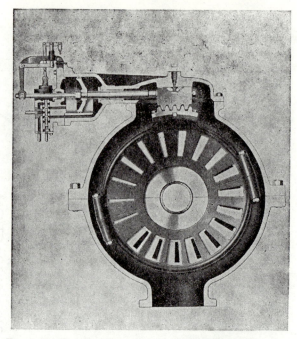

FIG. 9–43.   Details of a grid valve. (General Electric Co.)

steam does not provide sufficient work, some power to overcome the frictional loss must be supplied by the other sections of the turbine.

Performance curves for turbines with a single point of automatic extraction are similar to the sketch of Fig. 9–44. The parameter for this family of curves is the quantity of steam extracted, which ranges from zero to the maximum available. The extremes of the curves on the right are determined by the maximum load available from the turbine. A line of maximum extraction for any given turbine load is shown on the left.

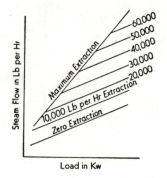

FIG. 9–44.   Performance curves for a single automatic-extraction turbine.

To attempt to extract more steam than indicated by this line would mean an attempted extraction of more steam than was being inducted into the turbine at its throttle, which is an obvious impossibility.

When determining the performance of a single-extraction turbine, the unit may be considered as two turbines: the first one above the extraction point and the second one below the extraction point. Performance curves can be calculated by estimating the Willans lines for the high-pressure and low-pressure sections and estimating the cooling steam required. The procedure can be explained best by the following example:

EXAMPLE 9–9.   Draw the performance curves for a 3000-kw, geared, single automatic-extraction, condensing turbine generator with steam conditions of 400 psig, 150 deg superheat, and 28 in. Hg vacuum with extraction at 45 psia. Cooling steam at 2000 lb per hr will be required. The high-pressure section will need 69,700 lb per hr at 3000 internal kw and 17,000 lb per hr at 500 internal kw. The low-pressure section will require 37,300 lb per hr at 2000 internal kw, 11,500 lb per hr at 500 internal kw, 5500 lb per hr at 100 internal kw, 2900 lb per hr at zero internal kw, and −50 internal kw for zero flow. The mechanical and electrical losses are as follows:

Generator load, kw	750	1500	2250	3000
Gear and mechanical losses, kw	120	120	120	120
Generator losses, kw	122	128	164	208

SOLUTION.   In order to plot the curves shown in Fig. 9–44 it is necessary to assume the throttle flow. For the line of 40,000 lb per hr extraction, assume a throttle flow of 55,000 lb per hr. From the Willans lines for each element of the turbine it will be seen that the high-pressure section will develop 2325 internal kw and the low-pressure section will develop 730 internal kw with 15,000 lb per hr, or a total of 3055 internal kw for the turbine. By plotting the losses, both mechanical and electrical, versus internal kw, the total losses for this load will be 316 kw. One point on the curve for 40,000 lb per hr extraction will then be at a throttle flow of 55,000 lb per hr for 2739-kw output by the generator. The end point on the left for this curve, i.e., the point of minimum kilowatt output for 40,000 lb per hr extraction, will be for a throttle flow of 40,000 lb per hr plus 2000 lb per hr cooling steam. At this condition the output will be 1700 internal kw from the high-pressure element, minus 25 kw for the low-pressure element, and minus 245 kw for mechanical and electrical losses, or 1430 kw net. The −25 kw for the low-pressure element represents energy that is lost by the blades of the low-pressure section as they move through the dense steam atmosphere, over and above the energy developed by the cooling steam. After determining several more points at other extraction flows in this manner, the performance curves of Fig. 9–44 may be drawn.

9–17. Constructional Details.   A steam turbine may be divided into three important parts: the rotor consisting of blades mounted on a shaft or drum, the top half and the bottom half of the casing including steam connections and bearings, and the governing system. In order to discuss these parts with their accessories, let us start with a small single-stage turbine employing a two-row Curtis stage, Fig. 9–45.

Turbine shafts are forged of steel and are accurately balanced to ensure smooth operation. The blades of this turbine are mounted on a

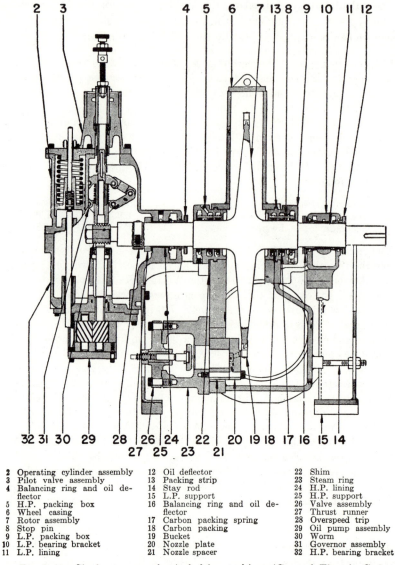

2	Operating cylinder assembly	12	Oil deflector	22	Shim
3	Pilot valve assembly	13	Packing strip	23	Steam ring
4	Balancing ring and oil deflector	14	Stay rod	24	H.P. lining
		15	L.P. support	25	H.P. support
5	H.P. packing box	16	Balancing ring and oil deflector	26	Valve assembly
6	Wheel casing			27	Thrust runner
7	Rotor assembly	17	Carbon packing spring	28	Overspeed trip
8	Stop pin	18	Carbon packing	29	Oil pump assembly
9	L.P. packing box	19	Bucket	30	Worm
10	L.P. bearing bracket	20	Nozzle plate	31	Governor assembly
11	L.P. lining	21	Nozzle spacer	32	H.P. bearing bracket

FIG. 9–45. Single-stage mechanical drive turbine. (General Electric Co.)

disc or wheel that is made, along with the shaft, from a solid forging. The whole rotating element, including the blades, is dynamically balanced. The blades shown in Fig. 9–45 have dovetail roots, but many other shapes are used, depending on operating conditions and manufacturer's preferences [see Figs. (9–10), (9–12), and (9–21)]. The blades are inserted into machined grooves through a slot in the periphery, and each blade is held in place by a soft-iron calking strip. The last blade fills this slot and must be fastened to the wheel by a pin, Fig. 9–12.

Steam leakage along the shaft must be controlled by a leak-off gland at each end of the shaft. This gland consists of several carbon rings composed of interlocking segments pressed against the shaft by spring clips. Each gland of the turbine shown in Fig. 9–45 has three rings with a space between the second and third rings. The pressure differential across the last ring is maintained constant by connecting the space to the turbine exhaust nozzle or some region with a pressure just slightly above atmospheric. In this way there is a small pressure drop across the last ring, and only a small quantity of steam may leak to atmosphere.

Steam enters the turbine through the governor valve at the bottom of Fig. 9–45 and then passes to a group of nozzles. Other nozzle groups may receive steam from the governor valve if the hand valves are opened. The effect of hand valves was discussed in Art. 9–14, with Willans lines shown in Fig. 9–39.

Nearly all modern turbines have at least two governors—one for an overspeed emergency trip and the other to maintain constant speed. The emergency governor for a turbine is shown in Fig. 9–45. When the shaft speed reaches a predetermined safe maximum value (usually about 10 to 20% over the rated speed), the centrifugal force of the eccentric pin overcomes the resisting spring force and the pin disengages the trip mechanism, closing the throttle valve. The unit also may be tripped by a hand button.

Casings of turbines are made of either cast iron, when the operating temperatures are low, or cast steel with or without an alloying metal. The upper half of the casing is kept free of piping connections wherever possible to make removal easier. Nearly all modern turbines have a horizontal flange dividing the upper and lower halves of the casing. The turbine casing is anchored to the bedplate at the exhaust end and is permitted to expand toward the outboard end, Fig. 9–42. Note that the outboard end is attached to the bedplate through flexible channels. Expansion in the axial direction will bend the channels.

An interesting cross-compound, impulse marine turbine with articulated double-reduction gear is shown in Fig. 9–46. For capacities up to 25,000 hp, the low-pressure element is a single-flow design, as shown; for

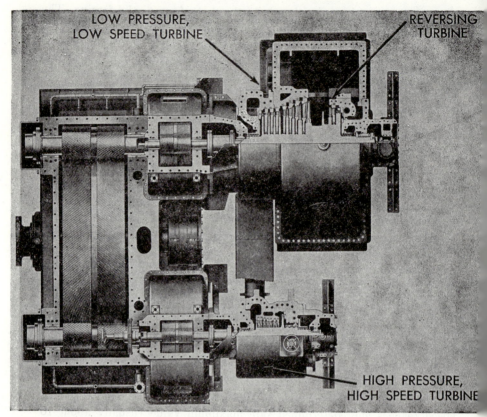

Fɪɢ. 9–46. Cross-compound marine turbine with articulated double-reduction gear. (General Electric Co.)

larger units the low-pressure element is of the double-flow design. Steam enters the high-pressure element at the forward end and exhausts at the afterend through the cross-over pipe, shown in the illustration, to the inlet of the low-pressure unit.

The full-load speed of the high-pressure element is about 6500 rpm, while for the low-pressure element it is about 4500 rpm. The propeller shaft connects to the coupling on the afterend of the bull (large) gear shaft.

Articulation is provided in each shaft connecting the first reduction gear and second reduction pinion. Quill shafts and flexible couplings provide the articulation and permit each member to align itself at all loads. Articulation also cushions shock loads that are transmitted to and from the propeller. The entire thrust of the ship is transmitted to the hull through the Kingsbury type of thrust bearing that is shown mounted on the forward stub of the bull gear shaft.

Since this unit is connected through reduction gears to the propeller shaft, the direction of the ship can be reversed only by reversing the turbine, which is accomplished by the astern blading located at the outboard end of low-pressure element. The astern section consists of a Curtis stage having two rotating rows.

A turbine generator with its condenser is shown in Fig. 9–47. Rated steam conditions for this impulse-reaction unit are 1450 psig, 1000 F, 1000 F; there are 23-in. exhaust blades. Steam enters the turbine through a stop valve (not shown in the illustration), flows into the steam chest at the extreme right, and flows through a single, rotating row of impulse blades and the reaction stages. The steam then leaves the high-pressure section at 400 psig to be reheated. Note the dummy piston at the right end of the shaft and the diaphragm between the high-pressure section and the reheat or intermediate-pressure section. Reheated steam enters the reheat section through the intercept valve and leaves by two cross-over pipes (only one is shown) beneath the floor. Blowout patches may be seen in the top of the low-pressure section casing.

A turning gear and motor is located between the low-pressure casing and the generator. Shafts of medium- and large-sized turbines (15,000 kw and above) will warp if allowed to remain at rest for several hours, even when absolutely cold. The turning gear and motor rotate the shaft at 3 to 10 rpm when the unit is being cooled or heated. When the unit is shut down for several days, the shaft is rotated a fraction of a turn every few hours after it has completely cooled.

Observe that the shafts of the large turbines are sealed by labyrinths (Fig. 9–52) rather than by the carbon rings used on the smaller units.

A straight reaction unit, Fig. 9–48, has certain unique governing features which are not found on either impulse or impulse-reaction turbines. Each row of reaction blading must flow full of steam, i.e., there is no such thing as partial peripheral admission for a reaction-blade row. Therefore, a small volume of steam will require a small blade area, or rather an annular area. During low loads, steam is admitted to the first group of reaction stages. As the load increases, a point will be reached where the pressure drop required to force all the steam through the small blades will be so large that the efficiency of the unit will be reduced. Then the first group of reaction stages will be by-passed, except for a small amount of cooling steam, and the steam will enter the second group of reaction stages. As more load is added, more area is needed and more stages are by-passed.

The tandem-compound turbine of Fig. 9–49 was designed for 2350 psig, 1100 F, 1050 F steam conditions. It has several interesting features. Steam enters each section of the two-row Curtis stage by individual pipes. This extremely high-temperature section is surrounded by steam that has

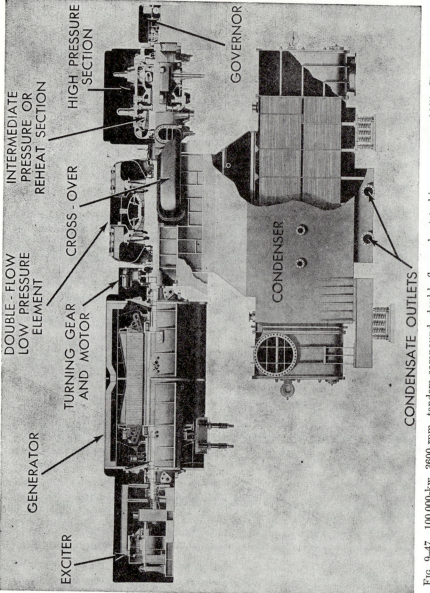

Fig. 9–47. 100,000-kw, 3600-rpm, tandem-compound, double flow, reheat turbine generator. (Allis-Chalmers Manufacturing Co.)

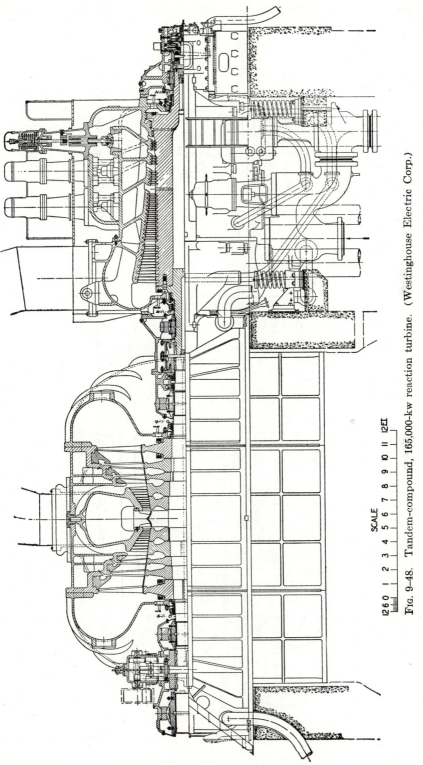

SCALE

1260 1 2 3 4 5 6 7 8 9 10 11 12FT

Fɪɢ. 9-48. Tandem-compound, 165,000-kw reaction turbine. (Westinghouse Electric Corp.)

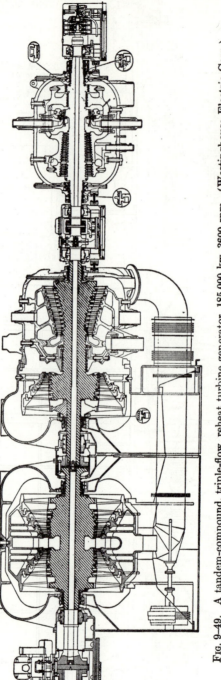

Fig. 9-49. A tandem-compound, triple-flow, reheat turbine generator, 185,000 kw, 3600 rpm. (Westinghouse Electric Corp.)

already expanded through the first stage. This reduces the temperature differential and pressure differential across this portion. Observe the double-shell construction around the high-pressure reaction stages. Extraction points are clearly shown for the intermediate-pressure or reheat section. Approximately one-third of the steam leaving the reheat section passes to the single-flow, low-pressure section via the double casing. The other two-thirds (approximate) of the flow passes to the double-flow, low-pressure section by way of the cross-over pipe that is located below the floor.

Application of all impulse stages to a large triple-flow, 3600-rpm, tandem-compound, reheat turbine is shown in Fig. 9–50. Because of the impulse stages, no dummy piston is necessary for this unit. Double-shell construction is used, indicating that the unit is for high steam pressures.

The limiting factors on the maximum size of a turbine generator are the

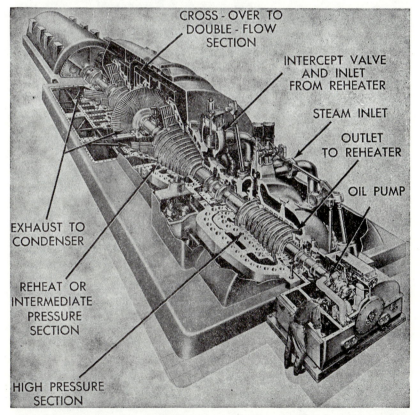

Fig. 9–50. Tandem-compound, 3600-rpm, triple-flow turbine generator. (General Electric Co.)

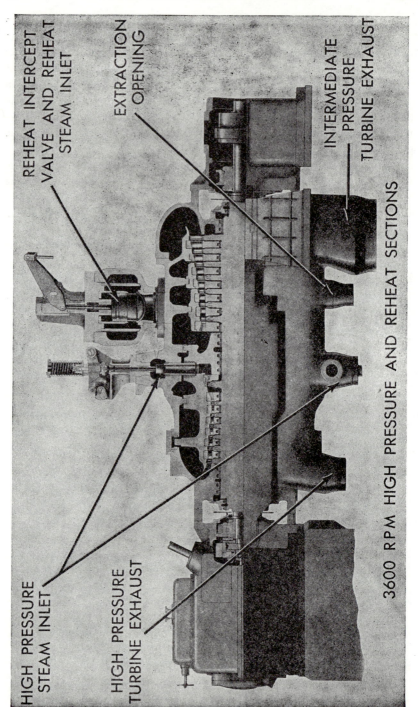

REHEAT INTERCEPT VALVE AND REHEAT STEAM INLET

EXTRACTION OPENING

INTERMEDIATE PRESSURE TURBINE EXHAUST

HIGH PRESSURE STEAM INLET

HIGH PRESSURE TURBINE EXHAUST

3600 RPM HIGH PRESSURE AND REHEAT SECTIONS

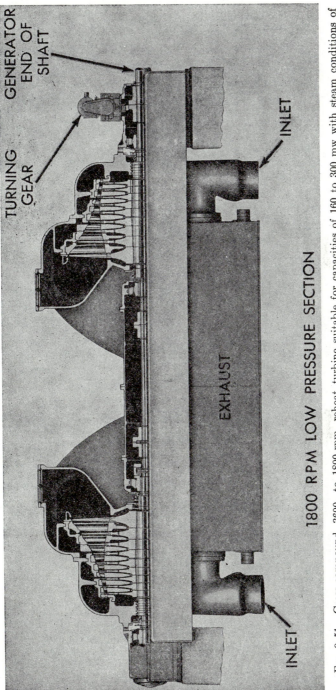

Fig. 9-51. Cross-compound, 3600- to 1800-rpm, reheat turbine suitable for capacities of 160 to 300 mw with steam conditions of 1800 psig, 1000 F, 1000 F to 2400 psig, 1050 F, 1050 F. (General Electric Co.)

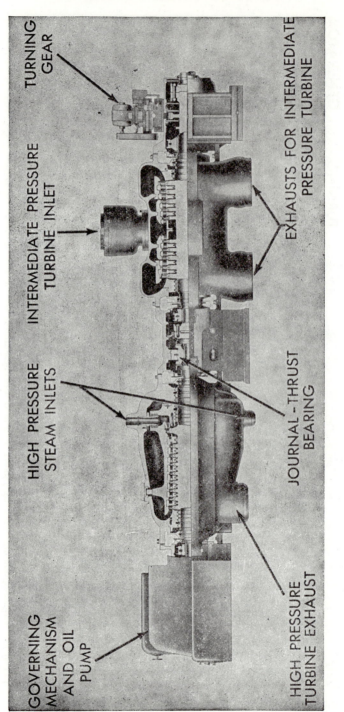

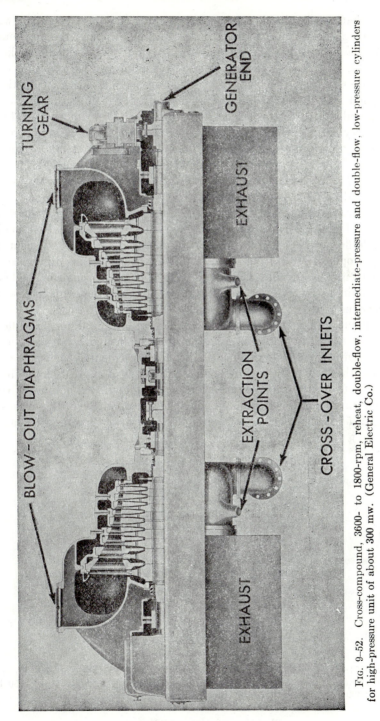

Fig. 9-52. Cross-compound, 3600- to 1800-rpm, reheat, double-flow, intermediate-pressure and double-flow, low-pressure cylinders for high-pressure unit of about 300 mw. (General Electric Co.)

energy that can be developed in one generator and the steam that will pass through the exhaust annulus of the turbine. In turn, the maximum area of the exhaust annulus is determined by the stresses in the buckets and wheel. With the same stresses, the buckets of an 1800-rpm unit can be twice as long as the buckets of a 3600-rpm unit. From a practical standpoint, one 1800-rpm stage with buckets that are 43 in. long has the same annulus area as three 3600-rpm stages with 26-in. buckets. Thus, where very large turbines are to be designed for high vacuum, cross-compound units can provide reduced exhaust loss and therefore improved efficiency. This high-pressure element operates at 3600 rpm (Fig. 9–51) while the low-pressure element operates at 1800 rpm. This arrangement can be used for ratings up to 300 mw.

Another arrangement of a large impulse-reheat unit is shown in quarter-section in Fig. 9–52. Note that the intermediate-pressure reheat section is a double-flow type and that the double-flow, low-pressure element is arranged opposite to that of Fig. 9–51. The reason for this difference in arrangement is to permit a bearing to be located in the center of the long shaft required for very large units (over 300 mw) and to permit the use of two condensers where desired.

A unique design of a low-pressure element arranged for axial-flow exhaust is shown in Fig. 9–53 as a quarter-section and as an end view. This design permits raising the condenser to approximately the same elevation as the turbine, thereby reducing the amount of excavation required or height of foundation. The turbine heat rate is improved because of the reduced turbine-exhaust pressure due to the venturi shape of the exhaust.

Fig. 9–54 is a section of a main stop valve of a type used on large

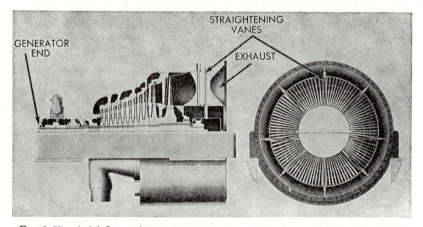

Fig. 9–53. Axial-flow exhaust, low-pressure section of a cross-compound turbine having a total capacity of 125 to 250 mw and steam conditions of 1250 psig, 950 F, 950 F to 2400 psig, 1050 F, 1050 F. (General Electric Co.)

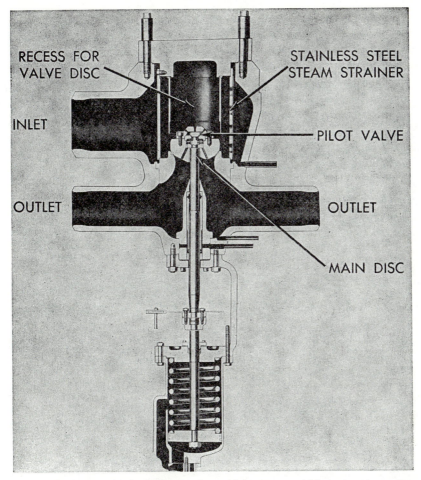

Fig. 9–54. Main stop valve for large high-pressure, high-temperature steam turbine. (General Electric Co.)

high-pressure, high-temperature turbines. Notice the parts that are labeled, and particularly notice the valve disc and the hydraulic operating mechanism. The first movement of the valve stem opens the inner or pilot valve to admit a small amount of steam through the ports shown. This steam reduces the pressure differential across the main disc to aid in opening the valve. Oil pressure applied to the bottom of the operating piston holds the valve open against spring pressure; loss of oil pressure closes the valve.

A quarter-section of the first supercritical-pressure turbine to go into commercial operation is shown in Fig. 9–55. The unit is rated at 125,000 kw, 4500 psig, 1150 F, 1050 F, 1000 F. The heat balance diagram for this

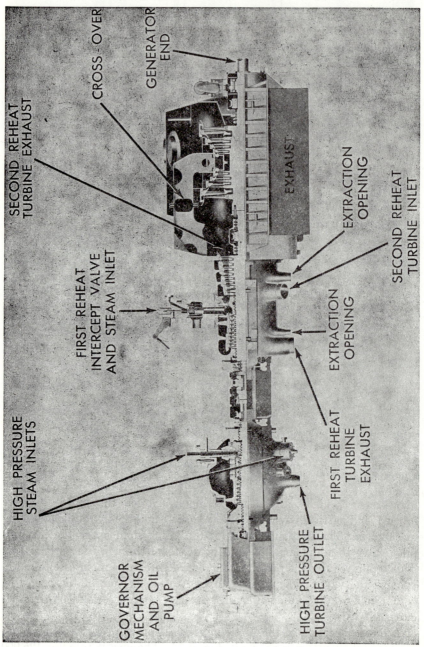

FIG. 9-55. Supercritical-pressure, tandem-compound, double-reheat turbine. (General Electric Co.)

unit is shown in Fig. 11–19. The turbine will operate on throttle governing. This reduces the pressure differential across the nozzle boxes at part load, thereby permitting thinner walls. The first stage buckets are ⅜ in. and the high-pressure casing is approximately spherical. The high-temperature parts of this unit and of all the other very high-temperature units are austenitic. In order to reduce the size of inlet valves and pipes, four steam leads are used, each lead having its own stop valve.

PROBLEMS

9–1. Plot a curve of steam flow, in pounds per square inch per hour, through an orifice with pressure ratio as the abscissa, varying from zero to unity, for steam conditions of (a) 1250 psig and 925 F, (b) 850 psig and 925 F, (c) 600 psig and 800 F, (d) 400 psig and 750 F, (e) 215 psig and 550 F, and (f) 125 psig and 450 F.

9–2. For the steam conditions of Prob. 9–1, plot the flow with critical pressure as ordinate and initial absolute pressure as abscissa.

9–3. Determine the throat area, mouth area, and length from throat to mouth for each nozzle listed (assume 100% efficiency).

	(a)	(b)	(c)	(d)	(e)	(f)
Steam flow, lb per hr	10,000	6,000	1,500	12,000	900	15,000
Inlet press, psia	500	100	230	4,500	75	160
Inlet temperature, F	600	...	550	1,000	...	...
Inlet quality, %	...	97	...	...	98	99
Outlet pressure, psia	3	16	40	1,000	8	25
Divergence angle β, deg	15	14	12	10	13	12

9–4. Repeat Prob. 9–3, using the following efficiencies based on enthalpy for the corresponding parts of Prob. 9–3: (a) 93%, (b) 90%, (c) 97%, (d) 85%, (e) 94%, (f) 89%. Efficiencies apply to the section of the nozzle between the throat and the mouth; use 100% efficiency from entrance to throat.

9–5. What would be the efficiency of a nozzle based on velocity if the efficiency based on thermal energy were 94%?

9–6. A throttling calorimeter uses an orifice of ⅛-in. diam for steam at 450 psig and 3% moisture. What would be the flow through the calorimeter if the efficiency of the orifice were 80%?

9–7. Determine the flow in pounds per hour through a nozzle of 2.5-sq in. throat area when steam enters at 75 psig and 400 F. Exhaust pressure is 40 psia.

9–8. Steam is supplied to a nozzle at 185 psig and 520 F. Find the area of the throat and mouth and the length of the nozzle from throat to mouth if $\beta = 12.5$ deg when the exhaust pressure is 16 psia. Use 95% as the efficiency and 5000 lb per hr of steam as the flow.

9–9. Calculate the total nozzle area required for the first stage of each turbine listed:

	(a)	(b)	(c)	(d)	(e)	(f)
Steam flow, lb per hr	1,100,000	780,000	530,000	110,000	700,000	1,000,000
Inlet pressure, psia	1,265	4,500	865	600	1,450	2,000
Inlet temperature, F	900	1,150	900	825	1,000	1,050
First stage pressure, psia	965	2,000	615	330	1,050	1,450
Nozzle efficiency, %	94	90	96	93	95	92

9–10. Using the data of Prob. 9–3, draw cross-sections to scale of each nozzle from throat to mouth when the nozzle is shaped for (a) uniform pressure drop per unit length, (b) uniform enthalpy drop per unit length, (c) uniform change in specific volume per unit length, and (d) straight sides.

9–11. Using the data given in Prob. 9–8, draw a cross-section to scale of the nozzle from throat to mouth when the nozzle is shaped for (a) uniform pressure drop per unit length, (b) uniform enthalpy drop per unit length, (c) uniform change in specific volume per unit length, and (d) straight sides.

9–12. A steam jet impinges on a turbine blade whose entrance (α_1) is zero degrees and $\beta_1 = \beta_2$. Calculate the power developed for a steam flow of 1 lb per min. Then calculate the power if the entrance angle (α_1) were increased to 15 deg.

	(a)	(b)	(c)	(d)	(e)	(f)
Steam velocity, fps	2000	1500	3000	1800	4000	2500
Blade velocity, fps	800	700	850	600	900	650

9–13. A steam jet impinges on a Rateau stage. For the conditions listed, determine: all angles and vectors, the power for a steam flow of 1 lb per sec, the blade efficiency, the leaving loss, and draw a vector diagram for the blade (assume $\beta_1 = \beta_2$):

	(a)	(b)	(c)	(d)	(e)	(f)
Nozzle angle, α_1, deg . .	14	30	16	15	14.5	20
Steam velocity, V_1, fps	1200	1500	4000	2000	3000	1700
Velocity ratio, ρ	0.47	0.50	0.30	0.41	0.35	0.39
Bucket coefficient	0.90	0.75	0.85	0.80	0.75	0.70

9–14. Construct four theoretical velocity diagrams for a single-stage Rateau turbine of 25 hp with 91% mechanical efficiency, 1100-fps mean blade speed, 20-deg nozzle angle, symmetrical blades, steam entering at 125 psig and 1% moisture and leaving at 2 psig. Determine all velocities and angles, stage efficiency, velocity ratio, leaving loss, throat and mouth areas, and the hourly quantity of steam required.

9–15. A Rateau blade receives steam at an absolute velocity of 1500 fps. The angles β_1 and β_2 are equal, and α_1 is 20 deg. The mean wheel diameter is 36 in., and ρ is selected for maximum efficiency. Neglecting friction, find (a) the wheel speed in revolutions per minute (b) all vectors, (c) all angles. Draw four types of vector diagrams.

9–16. Design a single-stage turbine with a Rateau wheel for each of the following conditions and find the steam flow, all vectors and angles, the diagram efficiency, the leaving loss. Construct the vector diagram. Take $\beta_1 = \beta_2$, and assume that leakage and rotational losses are a part of the mechanical efficiency.

	(a)	(b)	(c)	(d)	(e)
Initial steam pressure, psig......	95	150	75	120	50
Initial steam temperature, F....	. . .	450	. . .	325	. . .
Initial steam quality, %........	99	. . .	98	. . .	97
Exhaust pressure, psia..........	20	35	5	16	15
Nozzle angle, α_1, deg..........	17	20	15	16	17
Bucket speed, fps..............	500	900	600	1000	800
Nozzle efficiency, E_{nh}..........	0.90	0.88	0.91	0.92	0.89
Bucket coefficient..............	0.77	0.70	0.65	0.80	0.60
Horsepower output............	25	50	100	75	35
Mechanical efficiency, %.......	87	85	86	89	85

9–17. Estimate the number of Rateau stages in a steam turbine for the following conditions and optimum efficiency:

	(a)	(b)	(c)	(d)
Inlet steam pressure, psig...............	250	100	400	150
Inlet steam temperature, F..............	500	400	600	425
Exhaust pressure, psia..................	18	2	1	3
Blade speed, fps.......................	800	600	900	500

9–18. A 75-hp, single-stage, two-row Curtis turbine receives steam at 110 psia saturated and exhausts at 18 psia. The nozzle angle is 19 deg, nozzle efficiency 90%, 25-in. mean diam, 4000 rpm, $\beta_1 - \beta_2 = 4$ deg, $\alpha_3 = 29$ deg, $\beta_3 - \beta_4 = 4$ deg, $V_3 = 0.95V_2$, $C_2 = 0.80C_1$, $C_4 = 0.88C_3$, and mechanical losses are 6 hp. Determine all angles and velocities, draw three types of velocity diagrams, and find the work for each rotating row, the exhaust enthalpy, stage efficiency, steam flow, leaving loss, and the nozzle throat and mouth areas.

9–19. Find the work done per pound of steam per second for a 50% reaction stage, all vectors and angles, the diagram efficiency, and the isentropic enthalpy drop per stage, and draw a vector diagram for the stages given in the table on the next page.

	(a)	(b)	(c)	(d)	(e)	(f)	(g)
V_1, fps.........	1000	600	2000	1200	900	1350	800
U, fps...........	700	400	1500	900	600	1000	500
α_1, deg....	20	25	28	21	18	20	22
N_n, %...........	100	96	97	95	98	96	95
N_{co}, %........	100	80	70	85	60	65	75

9–20. All the rows in a group of reaction stages are to be made identical in shape at the mean diameter. Calculate the work done per pound of steam (Btu) by the group, the velocity diagram (all angles and vectors), and the diagram efficiency. Assume 50% reaction.

	(a)	(b)	(c)	(d)	(e)	(f)
Stages per group.....	6	4	8	5	7	3
α_1, deg........	20	30	18	22	25	18
Shaft speed, rpm.....	3600	3600	3000	1800	1500	6000
Mean blade diam, in..	33	27	33	58	60	20
N_n, %...............	96	94	97	93	95	94
N_{co}, %............	60	65	50	70	75	55
ρ.................	0.80	0.90	0.85	0.75	0.80	0.50

9–21. Estimate the number of stages in a reaction turbine for optimum efficiency and the conditions of Prob. 9–17.

9–22. One stage of a 50% reaction turbine has conditions as follows: $U = 600$ fps, $\rho = 0.80$, $\alpha_1 = 15$ deg, $e_{co} = 50\%$, $e_n = 96\%$. Find (a) all angles and velocities, (b) power developed for 1 lb of steam per sec, (c) enthalpy drop per stage, (d) stage efficiency. Draw velocity diagrams for the jet entering and leaving the moving row.

9–23. A turbine generator unit at 7000-kw load receives 88,100 lb per hr of steam at 900 F and 850 psig and exhausts at 5 psig with an enthalpy of 1166.4 Btu per lb. What are the combined mechanical and electrical losses?

9–24. The steam rate of a nonbleeding 3000-hp turbine is 17 lb per hp-hr when operating at 250 psig, 125-deg superheat, 5-psig back pressure. The mechanical losses are 15 hp. What is the exhaust enthalpy and RCR?

9–25. A 7000-kw turbine, 435 psig, 720 F, and 1¾ in. Hg abs, has the following steam rates: 7000 kw, 10.88 lb per kwhr; 6400 kw, 10.86 lb per kwhr; 5800 kw, 10.50 lb per kwhr; 1920 kw, 14.03 lb per kwhr. Estimate the no-load steam flow by drawing the Willans line and projecting to zero load. (Use large scales.)

9–26. Briefly explain the reason for by-passing the first stage of a five-stage Rateau turbine under certain conditions.

9–27. The engine efficiency of a 75-hp, mechanical-drive turbine is 37.8%. The ideal steam rate is 24.3 lb per kwhr, the initial enthalpy is 1256 Btu per lb, and the mechanical losses are 4 hp. Find (a) the RCR and (b) the exhaust enthalpy.

9–28. A 350-hp turbine receives steam at 175 psig and 75-deg superheat. The RCR is 59%, the mechanical losses are 15 hp, and the back pressure is 15 psig. What is (a) the engine efficiency, (b) exhaust enthalpy, (c) the steam flow, and (d) the steam rate?

9–29. It is required to find the initial temperature for a topping unit of 86% internal efficiency to be used with a condensing unit that operates at 300 psig and 1½ in. Hg abs with 77% internal efficiency. If there is a 15-psi pressure drop between the two units and the low-pressure unit must exhaust at 12% moisture, what will be the initial temperature for the topping unit? Sketch a Mollier diagram for this operation. Initial pressure for topping unit is to be 1800 psig.

9–30. A 4000-kw turbine generator unit receives 48,000 lb per hr of steam at 900 F and 1250 psig and exhausts at 5 psig with an enthalpy of 1137 Btu per lb. What are the combined mechanical and electrical losses? What is the RCR?

9–31. The steam rate of a nonbleeding 2000-hp turbine is 17 lb per hp-hr when operating at 300 psig, 125-deg superheat, 5-psig back pressure. The mechanical losses are 50 hp. What are the exhaust enthalpy and RCR?

9–32. A turbine operating at 600 psig, 875 F, and 1 in. Hg abs has an RCR of 81% and bleeds 15,000 lb per hr at 40 psia. The mechanical efficiency is 96.5%, and the generator is hydrogen-cooled with an efficiency of 98.5%. (a) If the generator output is 50,000 kw, what is the steam flow to the throttle of the turbine and to the condenser? (b) Same as (a) except with an additional bleed of 12,000 lb per hr at 8.5 psia. (c) Same as (b) except with a third bleed of 20,000 lb per hr at 215 psig. Use straight-line condition curves throughout.

9–33. A 1500-hp turbine operates at 350 psig, 600 F, and 25 in. vacuum. The engine efficiency is 67% and the losses are 4%. It is necessary to bleed enough steam at 60 psia to heat 150,000 lb per hr of water from 170 F to 280 F in a surface type of feedwater heater. There is a 2-psi pressure drop from the turbine to the heater. What are the throttle, bleed, and exhaust flows, and the throttle and exhaust steam rates? Use straight-line condition curves.

9–34. A turbine at a load of 2700 hp with 240 psig and 1290.0 Btu per lb of steam exhausts into a closed feedwater heater as shown in the accompanying illustration. The mechanical losses are 75 hp and the internal efficiency is 66%. Find the pressure of steam in the heater if there is a pressure drop of 5 psi between the turbine and heater. The 30,000 lb per hr quantity is drains from another heater.

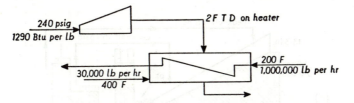

9–35. Estimate the engine efficiency and RCR for a 50,000-kw, 3600-rpm, hydrogen-cooled turbine generator operating at 1250 psig, 925 F, and 1½ in. Hg abs.

9–36. Estimate the engine efficiency and RCR you would expect for a 20,000-kw turbine generator operating at 3600 rpm, 600 psig, 850 F, and 1 in. Hg abs.

9–37. Determine an approximate RCR and engine efficiency for a geared, 4000-rpm, 600-hp mechanical-drive turbine.

9–38. What RCR and engine efficiency would you expect from a 40,000-kw topping unit operating at 3600 rpm, 1250 psig, 900 F, 230 psig?

9–39. Estimate and sketch on a Mollier diagram the condition curve for the unit of Prob. 9–35.

9–40. Calculate the extraction factors for Prob. 9–32.

9–41. Determine the extraction factor for Prob. 9–33.

9–42. Plot the performance curves for the single-extraction turbine used in the example of Art. 9–16.

9–43. Determine the force on the thrust bearing (and its direction) for each impulse-reaction turbine shown in the accompanying illustration. Assume 50% reaction.

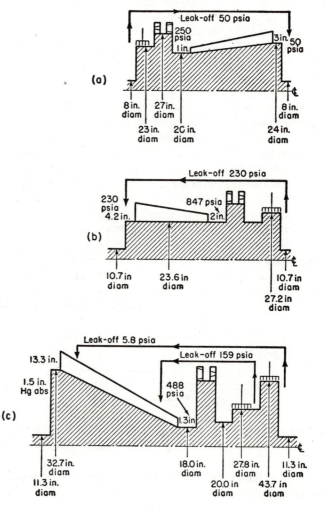

9–44. For each of the following systems, determine the frequency and the load on each unit after the load changes to the new valve when there has been no adjustment of the speed changer. All units are operating at rated speed before the load swing.

	(a)	(b)	(c)	(d)	(e)	(f)
New System						
Load, kw..............	65,000	24,000	20,000	50,000	130,000	90,000
Unit A, kw..............	60,000	20,000	10,000	10,000	40,000	22,000
Rated speed, rpm.......	3,600	3,600	1,800	1,200	1,500	1,200
Regulation, %..........	3	4	5	6	5	6
Load, kw..............	40,000	15,000	5,000	8,000	30,000	5,000
Unit B, kw..............	40,000	15,000	20,000	30,000	40,000	44,000
Rated speed, rpm.......	3,600	3,600	3,600	1,800	3,000	3,600
Regulation, %..........	5	6	2	4	4	5
Load, kw..............	15,000	6,000	12,000	24,000	30,000	20,000
Unit C, kw..............	. . .	. . .	. . .	40,000	80,000	66,000
Rated speed, rpm.......	. . .	. . .	. . .	3,600	3,000	3,600
Regulation, %..........	. . .	. . .	. . .	3	5	3
Load, kw..............	. . .	. . .	. . .	35,000	60,000	55,000

9–45. For each of the following systems, determine the speed-changer setting in terms of the no-load speed of each unit so that each unit will carry its assigned load at standard frequency.

	(a)	(b)	(c)	(d)	(e)	(f)
Unit A, kw..............	800	7,000	100,000	80,000	156,000	36,000
Rated speed, rpm.......	6,000	1,500	3,600	3,600	3,600	3,000
Regulation, %..........	5	4	3	3	4	7
Assigned load, kw.......	600	4,000	80,000	65,000	136,000	14,000
Unit B, kw..............	1,000	15,000	165,000	30,000	50,000	8,000
Rated speed...........	3,600	3,000	1,800	1,800	3,600	1,500
Regulation, %..........	4	2	6	5	6	9
Assigned load, kw.......	900	13,000	150,000	12,000	15,000	4,000
Standard frequency, cps....	60	50	60	60	60	50

BIBLIOGRAPHY

CHURCH, EDWIN F. *Steam Turbines*. New York: McGraw-Hill Book Co., Inc., 1950.

KEARTON, WILLIAM J. *Steam Turbine Theory and Practice*. New York: Pitman Publishing Corp., 1948.

KENT, R. T. *Mechanical Engineers' Handbook*. New York: John Wiley & Sons, Inc., 1950.

MARKS, LIONEL S. *Mechanical Engineers' Handbook*. New York: McGraw-Hill Book Co., Inc., 1951.

SALISBURY, J. KENNETH. *Steam Turbines and Their Cycles*. New York: John Wiley & Sons, Inc., 1950.

STODOLA, A., and LOEWENSTEIN, L. C. *Steam Turbines*. New York: D. Van Nostrand Co., Inc., 1927.

CHAPTER 10

STEAM ENGINES

10–1. General.   While the turbine converts the energy of the steam jet into kinetic energy and then into work, the steam engine makes use of the thermal energy of the steam by permitting it to expand against a piston during a nonflow process.

Fig. 10–1 shows a steam piston and cylinder of a double-acting engine with a pressure-volume history for the head end of the unit.   As the piston

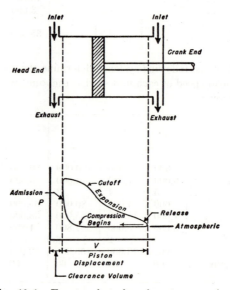

Fig. 10–1.   Events of stroke of a steam engine.

nears the extreme of its travel toward the cylinder head on the left of the diagram, steam is admitted to the cylinder by a valve not shown on the figure.   Pressure within the clearance space immediately rises to the full steam pressure, and the piston starts its stroke toward the crank end. Part way through the stroke, steam admission is stopped at the point known as the cut-off point. From then until the exhaust valve (not shown) is opened at the point of release, the steam is permitted to expand in a nonflow process.   The exhaust valve remains open until the exhaust stroke has been nearly completed, and from then until admission, the steam remaining in the cylinder is compressed.

The cushioning effect of compression during the latter part of the exhaust stroke promotes smooth operation by helping to retard the motion of the reciprocating parts of the engine. There is also a thermodynamic advantage in that the temperature of the steam is raised during compression, which in turn raises the temperature of the cylinder and piston higher than the exhaust temperature. Thus, there is some reduction in the chilling effect on the incoming steam as it comes in contact with the metal walls.

The other side of the cylinder would have a similar *PV* diagram but opposite in phase. The diagram shown in Fig. 10–1 is known as an *indicator card*.

Steam engines were developed in a crude form in 1698 for the purpose of pumping water. However, James Watt's refinement of the Newcomen engine in about 1763 was the real beginning of the steam-power prime-mover era. Since that time the steam engine has been largely replaced by the steam turbine except in small installations. One notable exception is the use of the steam engine as the prime mover for many of the merchant vessels built for the United States and England during World War II. The choice was due to the simplicity of the working parts of the steam engine compared to those of the steam turbine. It was much easier to train machinists and to build machine tools for the manufacture of large numbers of steam engines than it was for steam turbines.

This does not mean that the steam engine does not have advantages over the steam turbine for certain applications. While steam engines operate at low speed and have large space requirements per unit output, they are simple and reliable and have efficiencies comparable to small turbines. They maintain more nearly constant efficiency as load is decreased.

The best vacuum for maximum economy of a steam engine will depend on the type of engine and its design. Usually the best vacuum will be from 22 to 25 or 26 in. High vacuums require large exhaust ports and increased size of cylinders. The latter will increase the friction.

10–2. Steam Engine Nomenclature. Some of the important parts of the steam engine are shown in Fig. 10–2. Steam enters the unit through the *stop valve* and then the steam chest (not shown in Fig. 10–2). The stop valve is either in a wide-open position, as when the unit is in operation, or in a tight-shut position. The various types of governing valves that permit steam to enter and leave the *cylinder* are one of the major distinguishing features of different types of steam engines and as such will be the subject of another article.

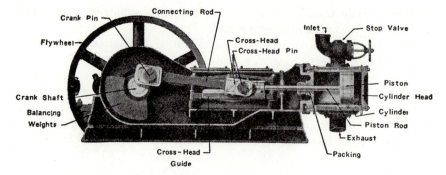

FIG. 10–2. Longitudinal section of a steam engine. (Skinner Engine Co.)

Nearly all steam engines are *double-acting*, Fig. 10–2, i.e., steam acts on both faces of the *piston*. After its expansion in the cylinder, steam leaves through the *exhaust* at a pressure either above or below atmospheric. The energy imparted to the piston is transmitted to the *crankshaft* by the *piston rod, crosshead, connecting rod,* and *crank* in that order. Since the piston shown in the picture has horizontal motion only and the crank has rotary motion, a single rod replacing the piston and connecting rods would be difficult to seal as it passes through the cylinder head. Therefore, the purpose of the crosshead is to permit the reciprocating motion of the piston rod to be transformed into the rotary motion of the crank.

The *head end* of the engine is the end farthest away from the flywheel, while the *crank end* is that nearest to the flywheel. When the piston is at its extreme position at either the head end or the crank end, it is in *dead center*. In order to start an engine, the piston must be in a position just after either dead-center position for its normal direction of rotation so that steam will be admitted to the cylinder and so that the steam force may start rotation of the engine.

Piston dsiplacement is the volume swept through by the piston during one stroke and is different for the head end and crank end of the engine shown in Fig. 10–2. For the head end, the piston displacement is the product of the piston area times the length of stroke, whereas for the crank end it is this product minus the cross-sectional area of the piston rod times the length of stroke.

Clearance volume is the volume between the piston and cylinder head when the piston is at dead center plus the volume in the ports leading from the piston to the valves. In general, clearance volume is kept as low as possible, since a large volume means that there are large areas of metal that will cool the incoming steam, causing condensation. Usually the valves of high-speed engines must be large to permit the flow of steam into the cylinders during the short time available. Therefore, the clearance volume of high-speed engines is usually larger than that for low-speed engines. Certain uniflow engines need large clearance volumes to prevent excessive pressures at the end of compression.

Piston speed may be based on the maximum speed attained by the piston during its stroke or may be taken as the average speed during the stroke. In the latter case the value is twice the stroke expressed in feet times the revolutions per minute of the engine.

The size of an engine is given as the *bore*, or diameter of the cylinder, and the *length of stroke*, all expressed in inches and always given in that order.

An engine is said to be running *over* when the crankpin is above the centerline of the cylinder during the time that the piston is moving from head-end dead center to crank-end dead center. If the crankpin is below the cylinder centerline, the engine is running *under*.

A steam engine, without a generator, has two bearings to support the crankshaft. When both these bearings are on one side of the connecting rod, the engine is referred to as a *side-crank* engine. The flywheel is located between the two bearings. The designation right-hand or left-hand indicates the side of the engine, when standing at the engine head end, on which the bearings and flywheel are located. If the connecting rod is between the bearings, the engine is called a *center-crank* engine.

10–3. Slide Valves. The purpose of the valves in the engine is to control the admission of steam into the cylinder in accordance with the position of the piston and to control the release of steam pressure in the cylinder as well as the beginning of compression, Fig. 10–1.

The simplest valve is known as the **D** slide valve because of its shape, Fig. 10–3. Motion of the valve is controlled by means of an eccentric on the crankshaft. Just before the piston reaches dead center, steam must be admitted to the clearance space, and therefore the slide valve must uncover the steam port. The amount by which the port is opened to steam flow when the piston is at dead center is the *lead*, Fig. 10–4(a).

When the valve is in mid-position of its travel, it will cover the steam port by an amount known as *steam lap*. Similarly, the release edge of the valve will lap the other edge of the port.

Lead and lap are the measurements by which the valves of the engine are set to give the necessary cut-off of steam admission, release of pressure, and compression. A valve without lead or lap would give admission of the steam throughout the entire stroke, as is done in the direct-acting pump.

The eccentric must have a position on the shaft ahead of the crank to overcome the steam lap of the valve. Advancing the eccentric 90 deg

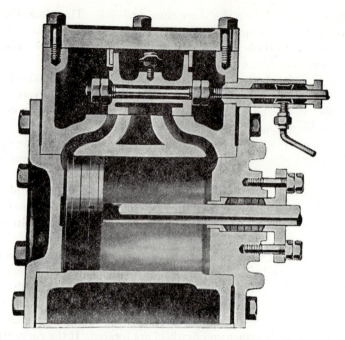

FIG. 10-3.   D slide valve. (Troy Engine & Machine Co.)

ahead of the steam piston or of the crank would put the valve at the mid-position when the piston is at dead center. However, the valve should be advanced farther by a sufficient angle to overcome the steam lap and the lead. Fig. 10–4(c) shows the *lap angle* and the *lead angle*. The sum of these two is the *angle of advance*. Thus, the eccentric must be ahead of the crank by 90 deg plus the angle of advance.

10–4. Other Types of Steam Engines. The simple slide-valve engine is easy to construct and is useful when low first cost of equipment is of importance, but it has some disadvantages. Other types of engines, such as the poppet-valve engine, Corliss engine, and uniflow engine, were developed to overcome the shortcomings of the slide-valve engine.

Steam surrounding the slide valve represented in Fig. 10–4 presses the valve against its seat with such force that lubrication may be difficult. Also, wear may be excessive, and distortion, with resulting leakage past the valve, is probable with high-temperature steam. Although it is possible to balance the slide valve, poppet valves are readily adaptable to the highest steam pressures and temperatures, Fig. 10–5. Some poppet valves are double-seated and are subject to leakage from unequal expansion due to either different metals or different temperatures on each disc. If the

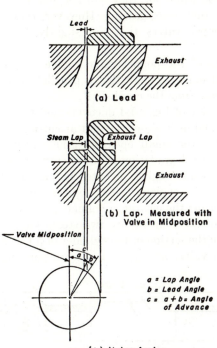

(a) Lead

(b) Lap. Measured with Valve in Midposition

a = Lap Angle
b = Lead Angle
c = a + b = Angle of Advance

(c) Valve Angles

FIG. 10-4. Lap and lead.

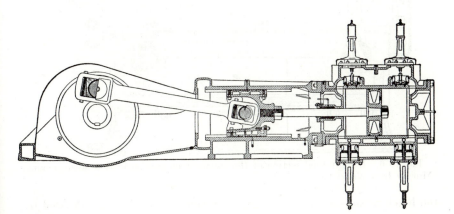

FIG. 10-5. Longitudinal section of a poppet-valve steam engine. (Erie City Iron Works.)

valves are ground-in while hot, they will not leak for the particular temperature at which they were ground. However, they will leak at other temperatures. Poppet valves that are slightly flexible will not leak normally.

The four valves of the poppet-valve engine shown in Fig. 10–5 are operated by cams from a lay shaft that is parallel to the piston rod and is driven by bevel gears from the crankshaft.

As the **D** slide valve of Fig. 10–4 uncovers the steam port at the beginning of admission and as it closes the port at cut-off, Fig. 10–1, the steam is throttled through the small opening. Since throttling is a wasteful process because of the degradation of the energy, there is a loss of available energy. The Corliss engine, invented by George H. Corliss in 1846, was designed to improve engine efficiency by the quick closing of the steam valves and by reducing the clearance volume and port areas.

Cross-sections of the four valves, two steam valves and two exhaust valves are shown in the section of the cylinder of a Corliss steam engine in Fig. 10–6. All four valves are cylindrical, with portions removed to form

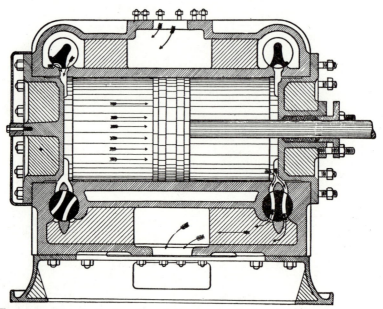

Fig. 10–6. Longitudinal section of the steam cylinder of a Corliss steam engine. (Murray Iron Works Co.)

the black areas shown on the figure. A complicated linkage, devised by Corliss, operates the valves from an eccentric on the crankshaft. Quick closing of the steam valves is obtained on the slow-speed engines by a releasing mechanism and a dash pot. As the valve is rotated into the open position by a lever, the dash pot is raised and a partial vacuum is created below the piston of the dash pot. When the valve is released, the partial vacuum below the dash-pot piston snaps the valve closed, thereby reducing the throttling effect.

All the engines that have been discussed may be classified under the general headings of *reverse-flow engines*. In this type of engine, steam enters the cylinder and flows toward the piston face and then reverses its flow during the exhaust stroke as the steam leaves the cylinder. When fresh steam enters the cylinder it comes in contact with metal that has been chilled by contact with the cooler exhaust steam from the previous cycle and is partially condensed with a resultant loss in efficiency. Several methods of reducing the loss due to condensation will be discussed in Art. 10-5, but one of the methods is so radical that it constitutes a definite type of engine: the *uniflow engine*, Fig. 10-7.

In a uniflow engine, steam enters the cylinder through the poppet valve and forces the piston to travel through its cycle. However, as the piston nears the end of its stroke, it uncovers slots or exhaust ports cut in the cylinder wall. The steam escapes through the exhaust ports at the center of the cylinder and thus does not travel back across the cylinder wall. The ends of the cylinder in the uniflow remain at a higher temperature than in the counterflow designs. The center of the uniflow cylinder, or the exhaust belt, is relatively cool and is at approximately exhaust temperature.

The exhaust ports must be long enough to permit reduction of the steam pressure within the cylinder to exhaust pressure. The length of the ports is usually about 10% of the piston stroke.

Two auxiliary exhaust valves, one for each end of the cylinder, are shown on the uniflow engine of Fig. 10-7. These valves are not needed when the engine is operating condensing but are advantageous for an engine with small clearance when operating noncondensing. The valves are operated by a valve gear that actuates the lower of the two cams below the valve. The idler cam, i.e., the upper of the two cams, can be moved in the horizontal plane by a spring-loaded diaphragm. When the engine is operating under vacuum, the diaphragm withdraws the idler cam and makes the lower cam inoperative even though it is still rotated through a small angle by the valve gear.

The necessity of the auxiliary exhaust valves when the back pressure is approximately atmospheric can be shown by means of Fig. 10-8. Fig. 10-8(a) shows the pressure-volume history of one cycle of the engine when the exhaust pressure is less than atmospheric. Compression is normal, and the steam at the end of compression is less than the initial steam pressure. However, in order to keep the pressure in the cylinder less than the initial pressure when the back pressure is atmospheric or greater,

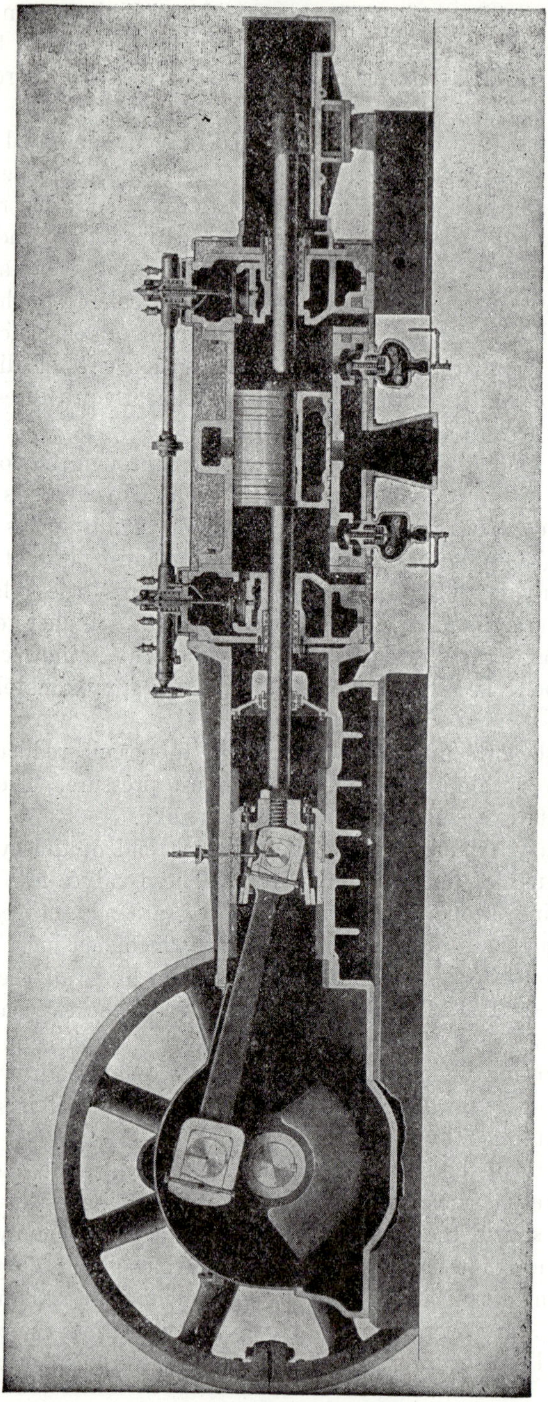

Fig. 10-7. Uniflow engine. (Skinner Engine Co.)

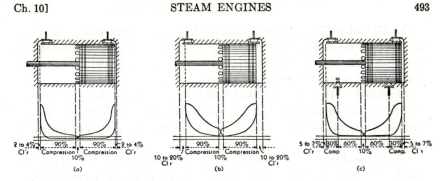

Fig. 10–8. Uniflow-engine indicator cards showing effect of clearance and auxiliary exhaust valves. (Skinner Engine Co.)

the clearance volume must be increased from 2 to 4% up to 10 or 20%, Fig. 10–8(b). Increasing the engine clearance reduces the economy. Also, it is impractical to change the clearance volume during operation if the exhaust pressure should change for any reason.

Inclusion of the auxiliary exhaust valves prevents the beginning of compression until 70% of the exhaust stroke has been completed and thus reduces the pressure at the end of compression with an engine having a relatively small clearance volume, Fig. 10–8(c).

10–5. Cylinder Condensation. When a fresh charge of live steam enters the clearance space of a counterflow engine, it strikes a cooled surface and some of the steam is condensed. The loss caused by cylinder condensation may vary from 10 to 35% of the dry steam entering the cylinder. The cool surface encountered in the clearance space and on the cylinder walls throughout the admission period and part of the expansion was cooled by contact with exhaust steam at the low exhaust pressure.

Since cylinder condensation is a matter of heat transfer, it is obviously affected by (1) surface area, (2) temperature difference between the live steam and the surface, and (3) the time that steam is exposed to the cold surface. Each of the following methods of reducing condensation employs one or more of the three fundamental factors listed above and should be analyzed in view of them.

(a) *Size of Clearance Space.* Reduction of the surface area exposed to the steam may be accomplished by making the clearance volume as small as possible and by locating the valves in such a way that the ports are short. Because the piston moves at a relatively slow speed during the beginning of its stroke, the steam is exposed to the clearance surfaces for a relatively long time. Minimum clearance volume cannot be determined on the basis of condensation only, since the pressure at the end of compression must be considered.

(b) *Cylinder Size.* Large cylinders have less surface area per unit volume than do small cylinders. In general, large engines are more efficient than small ones. This is particularly true when comparing extremely small engines with medium-sized engines.

(c) *Speed.* High-speed engines will reduce the time that the steam is exposed to the cool surfaces but may not increase the efficiency of the unit. Friction losses increase with speed, and the type and action of the valve gear may be restricted at high speeds.

(d) *Steam Jackets.* Live steam may be passed through a jacket surrounding the cylinder to keep the cylinder walls at a high temperature.

(e) *Superheated Steam.* Steam in the superheated state not only will increase the efficiency of the theoretical and actual cycle but also will reduce cylinder condensation in two ways: (1) Heat transfer from the superheated vapor is probably less than that for a moist vapor; (2) also, after expansion there is less moisture in the exhaust steam when it has been initially superheated. Probably a major part of the cooling of the surfaces during exhaust comes from the hot walls attempting to evaporate the moisture that clings to the surfaces.

(f) *Compression Period.* During part of the compression the temperature of the entrapped steam is higher than the temperature of the walls. The temperature of the steam and the time of exposure to the walls will affect the temperature of the wall in contact with the fresh steam charge.

(g) *Cut-Off.* The more steam that is in the cylinder during admission, the smaller will be the percentage of steam condensed by the cool surfaces.

(h) *Compounding.* Expanding the steam in two or more stages, and therefore in two or more cylinders, reduces the temperature difference between the live steam and the exhaust steam in each cylinder. The efficiency of uniflow engines, however, is seldom exceeded by double- or triple-expansion engines.

(i) *Reheating.* Reheating the steam as it passes from one cylinder to the next of a compound engine will have the same beneficial effect on the low-pressure cylinder as superheat has on the high-pressure cylinder.

(j) *Lagging.* All steam cylinders should be insulated whether or not they are steam-jacketed.

10–6. Governing. Some mechanism must be provided on the steam engine to regulate the amount of steam supplied to the cylinder. Since most steam engines operate at essentially constant speed, only two types of governing system will be considered. The two methods of governing steam engines are the same in principle as for the steam turbine: *throttle governing* and *variable cut-off governing.*

As the name implies, throttle governing controls the pressure of the steam entering the cylinder. Because of the degradation of energy during throttling, the process is wasteful and is applied only to small engines where efficiency is not of paramount importance. Effect of throttle governing on the indicator card is shown in Fig. 10–9; low steam pressure reduces the area of the indicator card and likewise reduces the power output of the engine.

Flyball governors used on current designs of steam engines are identical to those shown in Chap. 9 for small steam turbines except that those used on steam engines are driven by a belt or chain from the crankshaft. Observe that if the belt should break or slip off the pulley, the valve will go wide open. This will cause overspeeding of the engine, with consequent damage.

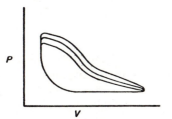

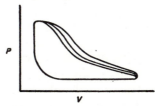

FIG. 10–9. Effect of throttle governing on indicator cards.

FIG. 10–10. Effect of variable cut-off governing on indicator cards.

A throttling governor will not control the actual admission of steam to the cylinder; it merely controls the pressure of the steam. A **D** slide valve or other valve system will be required to control admission, cut-off, exhaust, and compression.

Variable cut-off governing resembles the multivalve governing used for steam turbines. Fig. 10–10 shows the effect on the indicator card. Initial steam pressure at the beginning of admission is constant at all loads, but the duration of the admission process is much shorter at light loads than at heavy loads. With extremely small cut-off at very light load, condensation losses are high; with large cut-off, the expansion period is seriously reduced. If the cut-off were 100% of the stroke, there would be no time for expansion, and the engine would become nonexpanding, with the same rectangular indicator card produced by direct-acting pumps. Many engines operate at about 25% cut-off (cut-off occurring at about 25% of the stroke) at rated load. Performance data of various engines are given in Table 10–1.

Variation of cut-off with load changes are controlled either by *centrifugal* or *inertia* flywheel governors, sometimes called *shaft* governors. These governors regulate the engine valves to change the angle of advance. A centrifugal governor consists of two weights on the flywheel that are restrained by springs. Centrifugal force causes the weights to move outward against the spring force and move the eccentric to change the valve advance.

An inertia governor has a weight arm pivoted about a point on one of the flywheel spokes near the hub. Centrifugal forces, created as the flywheel revolves, causes a slight rotation of the weight arm about its pivot point and against the spring force. This rotation changes the angle of advance, since the valve rod is operated from the eccentric connected to the weight arm. During a sudden change in engine speed, inertia effect causes the weight arm to lag the flywheel in changing speed. This accentuates the change in angle of advance and reduces the speed fluctuations encountered during a load change.

The inertia force is more predominant in the inertia governor than in the centrifugal governor, although both effects are present in each type of governor.

10–7. Engine Performance. Eqs. (9–30), (9–31), (9–32), (9–33), and (9–34) apply to the steam engine as well as to the steam turbine and should need no further discussion. However, it is often advisable to analyze the pressure-volume history of the steam within the steam cylinder by means of the indicator and the indicator card. The engine indicator is an instrument that traces lines on a card in proportion to the pressure in the cylinder and the volume of steam in the cylinder. The result, for a steam engine, is an indicator card or *PV* diagram such as Fig. 10–1.

The mean ordinate of the indicator diagram represents the average pressure on the face of the piston. This mean ordinate, when measured in pounds per square inch, is known as the *indicated mean effective pressure* (imep) and may be visualized as follows: Imagine that the actual indicator card is replaced with a hypothetical card that is rectangular but has the same area and the same length as the actual card, Fig. 10–11.

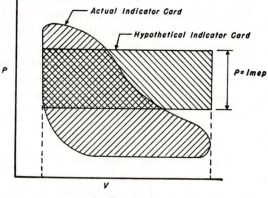

FIG. 10–11.

The hypothetical card represents an imaginary engine doing exactly the same work as the actual engine and with the same stroke but with 100% cut-off. The mean pressure acting on the face of the piston of the hypothetical engine is the same as that for the actual engine and is proportional to the height of the hypothetical indicator card. Because the mean height is the difference between the lower pressure on the hypothetical indicator card and the maximum pressure, imep has the units of psi and not psig or psia.

Work, for the hypothetical engine, is force times the distance through which the force acts. Force is the mean pressure times the piston area, and the distance is the stroke. This work is expended once during each revolution. Horsepower is the work developed per minute divided by 33,000 ft-lb per min. Therefore,

$$\text{ihp} = \frac{PLAN}{33{,}000} \qquad (10\text{--}1)$$

in which ihp = indicated or internal horsepower for one end
P = indicated mean effective pressure (imep), psi
L = engine stroke, ft
A = piston area, sq in.
N = rpm

Since by definition the area of the hypothetical indicator card is the same as the area of the actual indicator card, Eq. (10–1) is the indicated horsepower for one end of the actual engine. Both the areas of the piston faces and the mean effective pressures may be different for each end of the cylinder, so that the total indicated horsepower of the engine is the sum of those calculated for each end.

Of course, the indicated mean effective pressure is found without actually drawing the hypothetical card; it is obtained by dividing the area of the actual card in square inches by the card length in inches and multiplying the result by the spring scale for the indicator (the spring scale is the steam pressure required to move the indicator pointer 1 in.).

Once the engine has been built, the length of stroke and the area of each piston face are constants. The product of the stroke times the piston area, divided by 33,000, is the engine constant and may be different for each end of the cylinder. Thus

$$\text{Engine constant} = \frac{LA}{33{,}000} \qquad (10\text{--}2)$$

Only a part of the energy delivered to the faces of the piston is obtained from the engine shaft—the difference being the friction horsepower (fhp). The engine output may be determined in several ways, but the method most commonly used in college laboratories is the Prony brake. The use of this method has given rise to the name *brake horsepower* (bhp) for the shaft horsepower. Thus

$$\text{ihp} = \text{bhp} + \text{fhp} \tag{10-3}$$

and $$\text{Mechanical efficiency} = E_m = \frac{\text{bhp}}{\text{ihp}} \tag{10-4}$$

Sometimes it is convenient to express the mean effective pressure as the product of the imep and the mechanical efficiency, giving rise to the term *brake mean effective pressure* (bmep). Then

$$\text{bmep} = E_m \times \text{imep} \tag{10-5}$$

Eq. (10-1) can then be used to calculate brake horsepower.

Referring to Fig. 10-12, the work done by the engine per revolution is

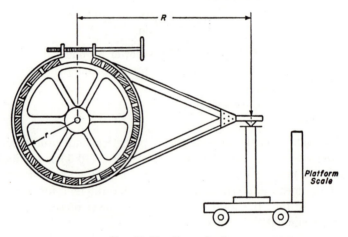

Fig. 10-12.   Prony brake.

the frictional force (F) between the flywheel and the wood blocks times the circumference of the flywheel $(2\pi r)$. Again, using power as work per minute,

$$\text{bhp} = \frac{2\pi r F N}{33,000} \tag{a}$$

It is not convenient to measure the frictional force, but the Prony brake will transmit the frictional moment (rF) to the scale such that the moment (RW) will exactly equal (rF). Making this substitution in Eq. (a), we get

$$\text{bhp} = \frac{2\pi R W N}{33,000} \tag{10-6}$$

and $$\text{Brake constant} = \frac{2\pi R}{33,000} \tag{10-7}$$

where R = radius of the brake arm, ft
$\quad W$ = net scale force, lb
$\quad N$ = engine speed, rpm

W is not the reading on the scales because of the weight of the brake pedestal and brake. These constant weights, called the *tare weight*, should be found before the engine is tested by loosening the brake band and inserting a piece of small pipe between the band and the flywheel directly above the center of the crankshaft to act as a knife edge. The tare weight will be the scale reading when the brake band is so supported and should be deducted from all readings of the scale taken during the test to get the correct value of W

EXAMPLE 10–1. A 12 × 14 in. steam engine was tested at 195 rpm. The following data were taken: piston rod diam, 2½ in.; spring scale, 100 psi per in.; brake arm, 5 ft; scale reading, 800 lb; tare, 150 lb; head-end card, 3.55 sq in.; crank-end card, 3.90 sq in.; length of cards, 3.30 in. Find the ihp, bhp, mechanical efficiency, and bmep.

SOLUTION.

$$\text{imep, head end} = \frac{3.55}{3.30} \times 100 = 107.7 \text{ psi}$$

$$\text{imep, crank end} = \frac{3.90}{3.30} \times 100 = 118.2 \text{ psi}$$

$$\text{Piston area, head end} = (12)^2 \frac{\pi}{4} = 113.1 \text{ sq in.}$$

$$\text{Piston area, crank end} = 113.1 - (2.5)^2 \frac{\pi}{4} = 108.2 \text{ sq. in.}$$

$$\text{ihp, head end (Eq. 10–2)} = \frac{107.7 \times 14 \times 113.1 \times 195}{12 \times 33,000} = 83.9$$

$$\text{ihp, crank end} = \frac{118.2 \times 14 \times 108.2 \times 195}{12 \times 33,000} = 88.1$$

Then
$$\text{Engine ihp} = 83.9 + 88.1 = 172.0$$

$$\text{bhp (Eq. 10–6)} = \frac{2\pi 5(800 - 150)195}{33,000} = 120.5$$

$$E_m \text{ (Eq. 10–4)} = \frac{120.5}{172.0} = 70.1\%$$

$$\text{bmep (Eq. 10–5), head end} = 70.1 \times 107.7 = 75.4 \text{ psi}$$

$$\text{bmep, crank end} = 70.1 \times 118.2 = 82.9 \text{ psi}$$

10–8. Diagram Factor. A prediction of the probable output of an engine may be made by comparing the imep from actual and theoretical indicator cards. So long as all comparisons are made with theoretical cards of the same shape, the processes assumed for the theoretical card are of no importance. Therefore, it is convenient to assume a theoretical card for an engine without clearance and to use a perfect gas expanding in a constant temperature process rather than with a reversible adiabatic expansion, Fig. 10–13.

Some authorities refer to curve B-C of Fig. 10–13 as a hyperbolic curve

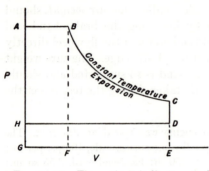

or a curve of $PV = $ constant. The name given to the curve is unimportant so long as all engines are compared to the same theoretical cycle. For ease in calculation, the actual expansion will be replaced by a perfect-gas isothermal expansion.

The work output for one cycle of the theoretical card would be the area $ABCDHA$, but

Fig. 10–13. Theoretical indicator card for determining diagram factor.

$$\text{Area } ABCDHA = \text{area } ABFGA + \text{area } BCEFB - \text{area } DEGHD \qquad \text{(a)}$$

$$\text{Area } ABFGA = P_B V_B \qquad \text{(b)}$$

$$\text{Area } BCEFB = P_B V_B \ln \frac{V_C}{V_B} \qquad \text{(c)}$$

$$\text{Area } DEGHD = P_D V_C \qquad \text{(d)}$$

Substituting Eqs. (b), (c), and (d) into Eq. (a),

$$\text{Work} = P_B V_B + P_B V_B \ln \frac{V_C}{V_B} - P_D V_C \qquad \text{(e)}$$

Divide Eq. (e) by the piston displacement V_C to get the mean effective pressure, and substitute the units of psia for P_B. Then,

$$\text{Theoretical mep} = p_B \frac{V_B}{V_C} + p_B \frac{V_B}{V_C} \ln \frac{V_C}{V_B} - p_D \qquad \text{(f)}$$

$$\text{Cut-off ratio} = r = \frac{V_B}{V_C} \qquad \text{(g)}$$

$$\text{Theoretical mep} = p_B r + p_B r \ln \frac{1}{r} - p_D \qquad \text{(h)}$$

$$= p_B r \left(1 + \ln \frac{1}{r} \right) - p_D \qquad \text{(10–8)}$$

By definition,

$$\text{Diagram factor} = f = \frac{\text{imep}}{\text{theoretical mep from Eq. (10–8)}} \qquad \text{(10–9)}$$

Most steam engines have a mechanical efficiency of about 90%, although some small engines may not do quite so well. Other data, representative of most engines, will be found in Table 10–1.

TABLE 10–1

Steam Engine Performance Factors at Rated Load

	Cut-Off, % of PD *	Diagram Factor, %	Clearance Vol, % of PD
Slide valve	25	73	8
Poppet valve	25	73	5
Corliss	15	85	5
Uniflow, with auxiliary exhaust valves	20	76	6

* PD = piston displacement.

EXAMPLE 10–2. Estimate the capacity of a 10×10-in. slide-valve steam engine operating at 300 rpm with 110-psig steam pressure and 16-psia exhaust pressure. The piston rod diam is $1\frac{3}{4}$ in.

SOLUTION. Assume 90% mechanical efficiency, 25% cut-off, and 73% diagram factor. From Eq. (10–8),

$$\text{Theoretical mep} = 124.7 \times 0.25 \left(1 + \ln \frac{1}{0.25} \right) - 16 = 58.5 \text{ psi}$$

From Eq. (10–9),

$$\text{imep} = 58.5 \times 0.73 = 42.7 \text{ psi}$$

From Eq. (10–1),

$$\text{Head-end ihp} = \frac{42.7 \times 10 \times (10)^2 \frac{\pi}{4} \, 300}{12 \times 33{,}000} = 25.4$$

$$\text{Crank-end ihp} = \frac{42.7 \times 10(10^2 - 1.75^2)\frac{\pi}{4} \, 300}{12 \times 33{,}000} = 24.7$$

Then

$$\text{bhp} = 0.90(25.4 + 24.7) = 45.1$$

PROBLEMS

10–1. What is the average piston speed of an 18×18-in. steam engine operating at 200 rpm and with piston rod diam of 3 in.?

10–2. Calculate the average piston speed of a 13×12-in. steam engine operating at 275 rpm.

10–3. The speed of an engine changes from 250 to 275 rpm with maximum change in load. Which speed is for full load and what is the governor-speed characteristic?

10–4. Determine the engine constants for the engine of Prob. 10–1.

10–5. Calculate the engine constants for the engine used in Example 10–1 .

10–6. What is the brake constant for the Prony brake of Example 10–1?

10–7. The following data were taken from the test of a 10 × 24-in. Corliss engine whose piston rod diam was $1^{11}\!/_{16}$ in.: brake tare, 37 lb; brake arm, 72.4 in.; spring scale, 60 psi per in.; length of indicator cards, 2.85 in.:

Scale Reading, Lb	Speed, Rpm	Area of C.E. Indicator Card, Sq In.	Area of H.E. Indicator Card, Sq In.
450	103	2.39	2.77
350	105	1.86	2.00
300	107	1.66	1.76
225	109	1.19	1.30
150	111	0.74	0.66
91	113	0.45	0.39

Calculate the imep, ihp, bmep, bhp, and mechanical efficiency for each load, and plot curves of speed, ihp, mechanical efficiency, and friction hp using bhp as the abscissa. Also calculate the speed characteristic of the governor.

10–8. Data taken from a test of a 6 × 8-in. **D** slide-valve steam engine are as follows: piston rod diam, 1⅛ in.; spring scale, 50 psi per in.; brake tare, 28 lb; length of brake arm, 36.5 in.; length of indicator cards, 2.65 in.:

Scale Reading, Lb	Speed, Rpm	Area of C.E. Indicator Card, Sq In.	Area of H.E. Indicator Card, Sq In.
100	244	2.10	2.01
83	251	1.54	1.65
66	259	1.15	1.18
49	266	0.71	0.78
35	272	0.30	0.39
0	281	0.23	0.25

Calculate imep, ihp, bhp, bmep, and mechanical efficiency for each load. Plot curves of speed, friction hp, ihp, and mechanical efficiency versus bhp as the abscissa. Determine the governor-speed characteristic.

10–9. Estimate the bhp of a 15 × 14-in. double-acting steam engine with a 2⅜-in. piston rod when operating at 250 rpm. Steam pressure is 90 psig and the exhaust pressure is 1.5 psig. The engine is of the Corliss type.

10–10. What would you expect the output of a 17 × 16 uniflow engine to be with 150-psig steam and 5-psig exhaust pressure? Piston-rod area on both head end and crank end is 3 in.; the speed is 185 rpm. The engine has auxiliary exhaust valves and is double-acting.

10–11. Assume 20% investment charges and 4000 hr per yr of operation for a steam plant at a pressure of 125 psig saturated. Which would be the most economical installation, a 200-kw turbine generator set or a 200-kw uniflow engine-generator set, for an average load of 100 kw? Other data are: 76% boiler efficiency, 55¢ per million Btu fuel cost, 4 in. Hg abs exhaust pressure for engine, $18,000 cost of engine-generator set with 57% engine efficiency for set at average load; turbine generator set costs $25,700 with 54.1% engine efficiency for set at average load, 2 in. Hg abs exhaust pressure for turbine. Condensate leaves condenser saturated and is pumped directly to boiler. Maintenance on the turbine plant is estimated as 4¢ per million and 5¢ per million for the engine plant. Assume all other costs to be the same.

10–12. Same as Prob. 10–11 except use 3000 hr per yr and 18¢ per million Btu as the fuel cost.

10–13. A decision must be made on the purchase of either a uniflow or a Corliss engine at 500-kw rating. The plant will operate at 175 psig noncondensing, and water will be supplied to the boiler at 210 F regardless of which engine is purchased. Fuel will cost 35¢ per billion Btu, boiler efficiency will be 71%, 25% investment charges, 2700 hr per yr, average load will be 50% of rating, maintenance on the plant with a uniflow engine will be 5¢ per million Btu and 5.1¢ per million Btu for the Corliss engine plant. The uniflow engine and generator will cost $46,500; the engine efficiency for the set will be 58%. Similar values for the Corliss engine-generator will be $40,500 and 56%. Assume the exhaust pressure will be 15 psig for both engines.

10–14. Same as Prob. 10–13 except use 63¢ per million Btu as the fuel cost and 20% investment charges.

BIBLIOGRAPHY

KENT, R. T. *Mechanical Engineers' Handbook*. New York: John Wiley & Sons, Inc., 1950.
MARKS, LIONEL S. *Mechanical Engineers' Handbook*. New York: McGraw-Hill Book Co., Inc., 1951.
MOYER, J. A. *Power Plant Testing*. New York: McGraw-Hill Book Co., Inc., 1934.
SHOOP, CHARLES F., and TUVE, GEORGE L. *Mechanical Engineering Practice*. New York: McGraw-Hill Book Co., Inc., 1956.

CHAPTER 11

HEAT BALANCES

11-1. General. If the Law of Conservation of Energy is applied to any piece of equipment, or to several pieces of equipment, in which a transfer of energy takes place because of a temperature difference, the result is known as a *heat balance*. However, for the purposes of this text, the term "heat balance," without any qualification, will be used to indicate the application of the Law of Conservation of Energy to the power-plant prime mover and its associated equipment—condenser, feedwater heaters, drain coolers, etc.

The regenerative cycle involving the extraction of steam from turbines for the purpose of heating the feedwater was invented by Ferranti in about 1905 but was not applied commercially in this country until the early 1920's. The advantages of this cycle are (1) the reduction of temperature stresses in the boiler by introducing hot feedwater rather than cold and (2) the increased economy. The improvement of the regenerative cycle over the Rankine cycle is due to the retention in the cycle of the energy for each pound of extracted steam that would have been otherwise rejected in the condenser.

The importance of the heat balance in the design of steam power plants cannot be overestimated. Once the capacity of the station is determined, the proper size of the boilers, heaters, condenser, pumps, piping—in fact practically all the mechanical equipment—can be determined solely by a heat balance. The weights and sizes of most of the mechanical equipment to be used for determining the sizes of structural members will depend on the heat-balance requirements. Likewise, the electrical system for many of the station auxiliaries, such as pumps, will depend on heat-balance determinations. Such economic factors as the advisability of including certain equipment in the cycle or the cost of the electrical energy generated can be decided only after sufficient heat balances have been made.

This chapter will deal with the methods used in heat-balance calculations and with a discussion of the heat balances used in modern stations.

11-2. Effect of Pressure, Temperature, and Vacuum on Cycle Efficiency. Fig. 11-1 shows the basic equipment required in the turbine hall of a power plant, i.e., the turbine, condenser, pump, and piping. This is

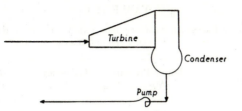

Fig. 11-1. Diagram of Rankine cycle.

the Rankine cycle, and any improvements in efficiency can be obtained for a given set of steam conditions only by the addition of more equipment. Although this cycle is not in use today except in the very smallest of plants, it may be used as a basis for determining (1) the effect of steam conditions on the efficiency of the turbine hall and (2) the effect of additional equipment on the cycle. It should be remembered that while the values determined in the first case will be applicable only to the Rankine cycle, their relative changes will apply to various arrangements of the regenerative cycle also.

Assume that the Rankine Cycle Ratio (RCR) and the mechanical and electrical efficiency of the turbine generator are both 100%, i.e., the steam expansion is a reversible adiabatic process. Then, for initial conditions of 100-psia saturated steam leaving the boiler (not shown in the heat-balance sketch, Fig. 11-1) and entering the turbine, the enthalpy will be $h_1 = 1187.2$ Btu per lb and $s_1 = 1.6026$ Btu per (lb) (F). If the condenser pressure is taken as 14.7 psia, then the enthalpy of the steam entering the condenser will be found on a Mollier diagram to be $h_2 = 1047.0$ Btu per lb at the same entropy as at the throttle. Since the steam will be condensed to saturated water in the condenser, the enthalpy of the water leaving the condenser and entering the boiler will be $h_3 = 180.1$ Btu per lb. The efficiency for this cycle will then be the *work done* by 1 lb of steam passing through the turbine, divided by the *heat added* in the boiler for 1 lb of steam, or

$$\text{Efficiency} = \frac{h_1 - h_2}{h_1 - h_3} = \frac{1187.2 - 1047.0}{1187.2 - 180.1} = 13.92\%$$

In a similar manner other values may be found for the efficiency as tabulated in Table 11-1.

Examination of these data will lead to several important conclusions. While an increase in throttle pressure results in an increase in the efficiency of the Rankine cycle, the rate of improvement of the efficiency decreases as the pressure rises. Observe in Table 11-1 that the efficiency increases $18.68 - 13.92 = 4.76\%$ for a doubling of the pressure from 100 to 200 psia, but that it increases $22.9 - 18.68 = 4.22\%$ for a doubling of the pressure

TABLE 11-1

EFFECT OF PRESSURE, TEMPERATURE, AND VACUUM ON RANKINE
CYCLE EFFICIENCIES

Initial Conditions	Condenser Pressure, Abs.	Efficiency, %	Ratio of Efficiencies Based on First Case
100 psia, saturated ..	14.7 psi	13.92	1.00
200 psia, saturated ..	14.7 psi	18.68	1.342
400 psia, saturated ..	14.7 psi	22.9	1.645
100 psia, 650 F	14.7 psi	15.88	1.141
100 psia, saturated ..	1 in. Hg	28.6	2.055

from 200 to 400 psia. Additional increases in pressure would yield still smaller increases in efficiency. Note that increasing the throttle pressure also increases the throttle temperature when saturated steam is considered. Doubling the throttle temperature will not increase the efficiency as much as a corresponding change in pressure. The greatest improvement in efficiency will be found by decreasing the condenser pressure.

Although this illustration has been based on the theoretical cycle, i.e., for a theoretically perfect turbine, the results are indicative of those that would be found for actual plants. In some cases improvements would be even greater for practical units. For example, it was noted in the chapter on turbines (Chap. 9) that increases in throttle temperature for a period just above the saturation temperature caused a marked improvement in the turbine efficiency. Thus, the addition of superheat in the fourth case would cause a more pronounced improvement in the cycle efficiency than indicated in Table 11-1.

The effect of improved vacuum in the condenser cannot be overemphasized. It has been shown that the condenser pressure is limited by the temperature of the available circulating water. However, any refinement in condenser design and performance that will increase the vacuum will have a marked effect on the cycle efficiency. An increase in vacuum of a small fraction of an inch is important in practice.

11-3. Basic Heater Arrangements. The first refinement in the heat balance to be considered is the addition of a surface type of feedwater heater which would receive steam from a bleed belt of the turbine and would cascade its drains to the condenser, Fig. 11-2. The extracted steam would be condensed in the heater by transferring its energy to the feedwater. A part of this energy would have been lost to the system in the condenser if the steam had not been bled, but the steam would have done more work. As previously discussed in the chapter on turbines (Chap. 9), an additional quantity of steam must be added to the throttle flow to

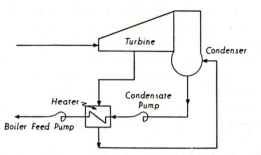

FIG. 11-2. Diagram of plant with surface heater and drains cascaded to condenser.

maintain the same turbine output after the steam has been extracted, as there was before the steam was extracted. However, the extraction factor must always be less than unity; therefore, the heat lost in the condenser by the replacement steam must be less, in Btu per hour, than the heat saved by the cycle from the extracted steam. With this net over-all saving of heat for the cycle, there will be an increase in the cycle efficiency.

Extracted steam that is condensed in the heater leaves as water, called *drains*. The flow of drains from the heater is regulated by a float controlled valve, called a *drainer*, to permit only water to leave the heater. A mechanical drainer is identical to the pump control shown in Fig. 3-31. Drainers may be air-operated by replacing the mechanical linkage with a pneumatic system, but the principle of operation is the same.

Drains from the heater still reject considerable energy to the circulating water, since they enter the steam space of the condenser with the enthalpy of saturated water for the pressure existing in the heater. The drains will be cooled to saturated water at condenser pressure. The higher the extraction pressure and the lower the condenser pressure, the greater is the energy lost in the condenser by the heater drains. This loss may be reduced by the inclusion of a drain cooler in the cycle, Fig. 11-3.

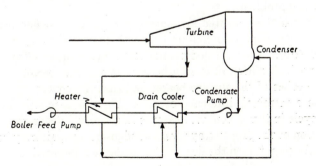

FIG. 11-3. Diagram for feed cycle with drains cascaded through a drain cooler.

Obviously, the lower the temperature of the drains leaving the drain cooler, the greater will be the efficiency of the cycle, but the surface requirements of the drain cooler, and therefore the cost, will be higher. The most economical size of drain cooler will depend on the local conditions, but many large central stations have found that a 10-deg terminal difference is best. The improvement in the turbine heat rate over the Rankine cycle for the addition of the heater only will be better than 4%, while the improvement for the heater and drain cooler will be nearly 7% over the Rankine cycle.

The next improvement that may be made is to pump the heater drains into the main feedwater system after they have left the heater, Fig. 11–4.

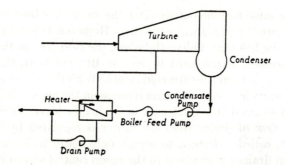

Fig. 11–4.   Diagram for feed cycle with drains pumped ahead.

Such a process is known as *pumping the drains ahead*. In this way all the energy of the extraction steam is saved for the cycle. As might be expected, the improvement here in heat rate over the Rankine cycle will be about 0.1% more than that for the cycle with the drains cascaded to the condenser through the drain cooler. This cycle arrangement has a distinct disadvantage in that an additional pump (a drain pump) is required with sufficient spare equipment and more auxiliary power. It is one more piece of equipment that must be operated and that may fail, thus causing maintenance problems.

The fourth basic heater arrangement that will further improve the cycle efficiency employs a contact type of heater, Fig. 11–5, rather than a surface heater. The thermodynamic advantages of this cycle are that there is no degradation of energy because of the transfer of heat through the metal tubes of the surface heater and that the contact type of heater will have a terminal difference of zero. There is a practical advantage for this cycle in that a contact heater is inherently a deaerator; thus, with the addition of the proper trays, vent condensers, etc., air may be removed more easily than it is in the deaerating condensers required in the previous cycles.

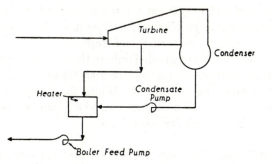

Fig. 11-5. Diagram for direct-contact heater in feed cycle.

In order to effect greater improvements in the cycle efficiency, it is necessary either to increase the number of heaters in the cycle or to add reheat to the cycle. Additional heaters may be added to the cycle either by the duplication of one of the heater arrangements or by a combination of several arrangements.

In the reheat cycle, the steam is removed from the turbine after completing only a part of its expansion. The steam is then returned to the boiler and is resuperheated, or reheated, to approximately its original temperature, Fig. 11-6. The most economical reheat pressure is usually about one-fourth of the initial steam pressure. The total improvement in cycle efficiency is about 1.5 percentage points. Although it would be theoretically possible to use the reheat cycle without the regenerative cycle, it would not be economical to do so. Therefore, the regenerative cycle is used without reheat, but the reheat cycle is never used without feedwater regeneration.

An intercept valve must be installed in the return line from the reheater. This valve is closed automatically whenever the main stop or

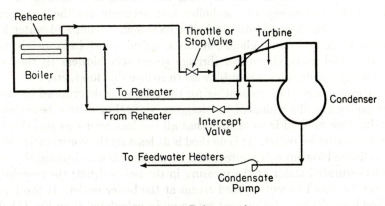

Fig. 11-6. Reheat cycle.

throttle valve is closed to prevent the steam trapped in the reheater and piping from entering the turbine and causing it to overspeed.

It is possible to have more than one stage of reheating. To date, two stages of reheating have been applied only to the supercritical plants. Most of the recent installations of turbines of 100,000-kw capability, or more, have employed one stage of reheat. While the reheat cycle has been employed on units as small as 40,000 kw, reheat can be justified only on such small units in the high fuel-cost areas.

11-4. Heat Rates.

In Chap. 1, gross station heat rate and net station heat rate were defined in terms of the fuel input to the plant. This concept is of use when testing a plant that is already installed. However, it is also necessary for the designer to predict the heat rates for the plant while it is still in the design stage. Therefore, we shall now consider the methods of predicting the heat rates.

In addition to the two heat rates mentioned above, the *turbine heat rate* is of importance to the turbine manufacturer because it comprises one of his guarantees. Although any of these heat rates may be based on the horsepower, the kilowatt is the more commonly used unit and therefore it will be used here. Then, for a non-reheat unit,

$$\text{Turbine heat rate} = \frac{W_1 h_1 - W_2 h_2}{\text{generator output in kw}} \tag{11-1}$$

where W_1 = steam entering the cycle, lb per hr

W_2 = feedwater leaving the cycle from the highest pressure heater, lb per hr

h_1 = enthalpy of steam entering the cycle, Btu per lb

h_2 = enthalpy of feedwater leaving cycle, Btu per lb

Gross station heat rate differs from the turbine heat rate in that it includes the inefficiency of the boiler and accounts for the water and steam lost from the cycle due to boiler blowdown, steam and water lost through valve and pump glands, steam and drains lost from air ejectors and aftercondensers, steam lost through steam-soot blowers (if they are used; many plants use air-soot blowers to reduce this loss), etc.

The exact method of accounting for these losses is determined by company policy and the whims of the designer, since they cannot be assigned exactly. One method is to assume that all the loss occurs as superheated steam at the boiler outlet. This method is at least on the conservative side. If the boiler blowdown is appreciable, it can be taken as leaving the system as saturated water at the pressure in the boiler drum; the remaining loss can be taken as superheated steam at the boiler outlet. If the latter method is used, gross station heat rate may be calculated from Eq. (11-2).

Gross station heat rate $= \dfrac{(W_1 + W_3)h_1 + W_4 h_4 - W_2 h_2}{E_b (\text{generator output, kw})}$ (11-2)

in which $W_3 =$ steam lost between boiler and turbine, lb per hr
$\quad W_4 =$ water lost as boiler blowdown, lb per hr
$\quad h_4 =$ enthalpy of blowdown, saturated liquid at boiler pressure, Btu per lb
$\quad E_b =$ boiler efficiency

and other symbols are as in Eq. (11-1).

The gross heat rate has been used to some extent in Europe, but it is not generally used in the United States because it does not give a fair comparison between stations that have steam-driven auxiliaries and stations that have electrically driven auxiliaries.

Net station heat rate

$$= \dfrac{(W_1 + W_3)h_1 + W_4 h_4 - W_2 h_2}{E_b (\text{generator output in kw} - \text{auxiliary power in kw})} \quad (11\text{-}3)$$

For a reheat unit, the numerators of Eqs. (11-1), (11-2), and (11-3) would each be increased by the steam flow through the reheater, times the enthalpy rise through the reheater. All three of these equations neglect small water quantities entering the system, such as the make-up water entering the evaporator or the condenser.

Equipment that requires auxiliary power includes fans, coal conveyors, crushers, pulverizers or stokers, air compressors for control and maintenance, shop equipment, electrical controls, ash removal, precipitators, elevators, station lighting, transformer losses for auxiliary equipment, condensate pumps, circulating pumps, drain pumps, raw-water pumps, boiler feed pumps, condensate-storage pumps, and screening equipment for circulating-water systems. Many of these devices may be either steam or electrically driven.

Consumption of steam-driven equipment should be considered in the heat balance if the equipment will be in operation for a major part of the day. The picture is often further complicated by having both steam and electrical drives for some of the major equipment, such as condensate and boiler feed pumps, circulating pumps, and fans. It is common practice to determine the auxiliary power requirements for the major equipment only, i.e., for the fans, pumps, ash system, pulverizers, and station lighting, together with the transformer losses for this auxiliary power.

In the event that the plant generates process steam, and electrical energy is a by-product as in Fig. 1-1, the turbine heat rate would reduce to

$$\text{Turbine heat rate} = \dfrac{3413}{E_{mg}} \quad (11\text{-}4)$$

where E_{mg} is the mechanical and generator efficiency.

EXAMPLE 11-1.   What is the unit fuel cost of a 44,000-kw unit when operating at an average load of 30,000 kw?   The unit operates at steam conditions of 1250 psig and 950 F with water entering the steam generator at 368 F.   The turbine requires 242,000 lb per hr of steam and the steam losses are expected to be 3000 lb per hr taken at the steam generator outlet.   Fuel costs 19¢ per 10⁶ Btu, the steam generator efficiency will be 86%, and auxiliaries will require 2000 kw.

SOLUTION.   From Eq. (11-3),

$$\text{Net station heat rate} = \frac{(242,000 + 3000)1468.0 - (242,000 + 3000)340.7}{0.86(30,000 - 2000)}$$
$$= 11,470 \text{ Btu per net kwhr}$$

From Eq. (1-4),

$$\text{Unit fuel cost} = \frac{11,470 \times 19}{10^6} = 2.18 \text{ mills per net kwhr}$$

11-5. Heat Rate Estimates.

Short-cut methods of estimating turbine heat rates find many applications in power-plant design, not only to provide a quick answer during the early stages of design, but also to show how changes in any one variable will affect the plant economy.   It should be emphasized that the only method of accurately determining the heat rate is by a complete heat-balance calculation, as will be demonstrated in a later article of this chapter.

Consideration of the cycles of Figs. 11-2, 11-3, 11-4, or 11-5 should reveal that there would be a wide range of extraction pressures that could be selected.   If the extraction pressure should be reduced to the exhaust pressure, then the saturated steam temperature in the heater would be the same as the saturated water temperature leaving the condenser and entering the heater.   Thus, since there would be no heat transfer, the heater would be useless.   The turbine heat rate would then be the same as for Fig. 11-1.

If the heater extraction were taken from the turbine throttle, the heater would be, from a thermodynamic standpoint, a part of the boiler, and the heat rate would also revert to that of Fig. 11-1.   Thus, there can be an extraction pressure that is too high and one that is too low.   Between these two extremes, then, there must be an extraction pressure that will give a maximum efficiency.

By a series of calculations for varying throttle pressures, throttle temperatures, exhaust pressures, and numbers of heaters, the curves of Figs. 11-7, 11-8, and 11-9 were developed.   These curves were based on a 5-deg terminal difference for closed heaters and a pressure loss in the extraction steam lines of 5% of the pressure at the turbine flanges.   The term *reduction in heat consumption* used on the ordinate means the reduction in turbine heat rate compared with the cycle arrangement without feedwater regeneration, Fig. 11-1.   Similarly, the term *total rise in feedwater temperature* used for the abscissa means the difference in temperature between the

1" Hg abs condenser based on 30.0 in Hg

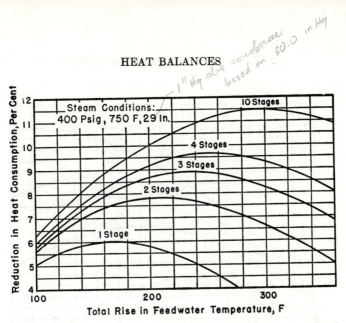

Fig. 11–7(a). Reduction in heat consumption obtained by regenerative feedwater heating. Correct for throttle temperature in accordance with Fig. 11–7(e). Correct for vacuum and load in accordance with Fig. 11–7(f). (Westinghouse Electric Corp.)

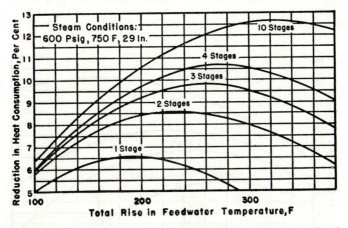

Fig. 11–7(b). Reduction in heat consumption obtained by regenerative feedwater heating. Correct for throttle temperature in accordance with Fig. 11–7(e). Correct for vacuum and load in accordance with Fig. 11–7(f). (Westinghouse Electric Corp.)

feedwater leaving the last heater and the feedwater leaving the condenser.

These curves warrant careful inspection. The following conclusions may be drawn from them:

1. The optimum extraction pressure increases as the number of stages increases. For an infinite number of stages of feed heating, the optimum feedwater temperature entering the boiler would be at saturation temperature corresponding to the throttle pressure.

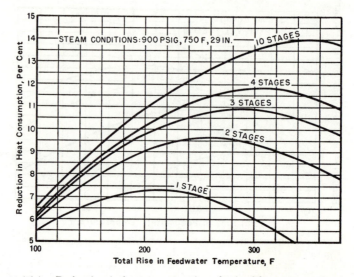

Fig. 11-7(c). Reduction in heat consumption obtained by regenerative feedwater heating. Correct for throttle temperature in accordance with Fig. 11-7(e). Correct for vacuum and load in accordance with Fig. 11-7(f). (Westinghouse Electric Corp.)

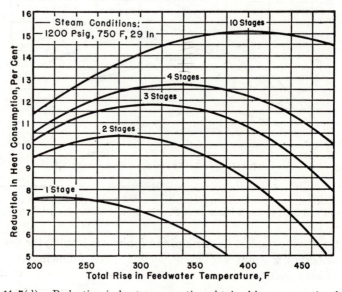

Fig. 11-7(d). Reduction in heat consumption obtained by regenerative feedwater heating. Correct for throttle temperature in accordance with Fig. 11-7(e). Correct for vacuum and load in accordance with Fig. 11-7(f). (Westinghouse Electric Corp.)

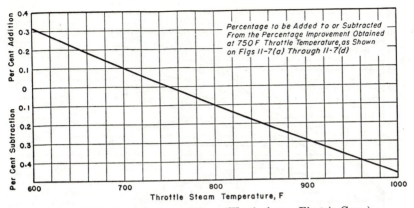

FIG. 11-7(e). Superheat correction. (Westinghouse Electric Corp.)

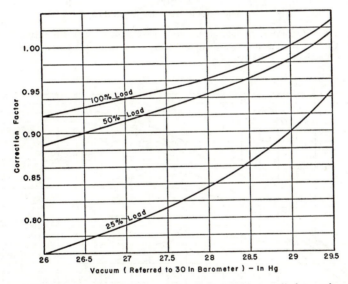

FIG. 11-7(f). Vacuum and load correction factors to be applied to values of gain due to feedwater heating obtained from curves on Figs. 11-7(a) through 11-7(d). (Westinghouse Electric Corp.)

2. Improvement in efficiency increases with increased stages of heating, but the improvement is small for more than six stages. For this reason very few plants have more than six or eight stages of feedwater heating.

3. The curves are very flat at the peaks, indicating that the extraction pressure may vary considerably from the optimum without seriously affecting the plant efficiency.

4. For a given number of stages, the optimum extraction pressure for the highest pressure heater increases with an increase in the throttle pressure.

5. The curves clearly show the diminishing improvements that may be ob-

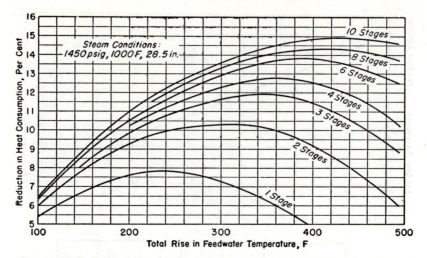

Fig. 11–8. Reduction in heat consumption obtained by regenerative feedwater heating. Correct for vacuum and load by using *ratio* of correction factors from Fig. 11–7(f). (Westinghouse Electric Corp.)

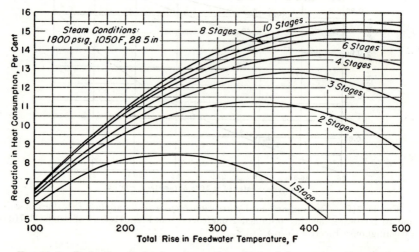

Fig. 11–9. Reduction in heat consumption obtained by regenerative feedwater heating. Correct for vacuum and load by using *ratio* of correction factors from Fig. 11–7(f). (Westinghouse Electric Corp.)

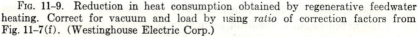

tained by increasing the throttle pressure or by increasing the stages of feedwater heating.

Values shown on these curves would change slightly for heater arrangements other than the one used, but these changes would be negligible. In selecting the proper extraction pressures for a particular installation, due consideration should be given to the average load that is anticipated. In general, it is preferable to select the best extraction pressure for the average load on the unit rather than for the maximum or rated load of the unit. Correction factors for throttle temperature, vacuum, and load should be applied to these curves as indicated by Figs. 11–7(e) and 11–7(f).

Corrected heat-rate improvement

$$= \frac{(\text{improvement} + \text{temp correction})}{\text{load and vacuum factor}} \quad (11\text{--}5)$$

EXAMPLE 11–2. Estimate the amount that can be economically paid for the use of five feedwater heaters rather than four in a cycle for a 75,000-kw turbine generator. Steam conditions are 850 psig, 900 F, 1 in. Hg abs. Other data are: 86% boiler efficiency, 5000 hr per yr operation at rated load, 12% investment charges, 975.4 Btu per lb turbine exhaust enthalpy including exhaust losses (refer to Fig. 11–10), mechanical and generator losses of 1850 kw, 377.3 F final feedwater temperature, 23¢ per 10^6 Btu fuel cost.

SOLUTION. If there were no extraction, the throttle steam flow would be the work done per pound of steam divided into the energy equivalent of the internal turbine output of 75,000 plus 1850 = 76,850 kw.

$$\text{Throttle flow} = \frac{3413 \times 76,850}{1453.1 - 975.4} = 549,300 \text{ lb per hr}$$

From Eq. (11–1) for the nonregenerative cycle,

$$\text{Turbine heat rate} = \frac{549,300(1453.1 - 47.1)}{75,000}$$

$$= 10,290 \text{ Btu per kwhr}$$

The feedwater total temperature rise is $377.3 - 79.0 = 298.3$. From Fig. 11–7(c), i.e., for the nearest pressure to the given conditions of 850 psig, and a rise of 298.3 F, the reduction in heat rate for four heaters would be 11.8% and about 12.2 for five heaters. There will be no vacuum or load-correction factors, and the superheat correction factor will be both added and subtracted to the difference in improvement. Then the reduction in turbine heat rate for five versus four heaters would be 10,290 $(0.122 - 0.118) = $ (approx) 41 Btu per kwhr, or $41 \div 86\% = 48$ Btu per kwhr reduction in gross station heat rate. The annual fuel savings would equal the (reduction in heat rate) × (turbine kilowatt output) × (hours of operation per year), or the annual monetary saving would be

$$\frac{48 \times 75,000 \times 5000 \times 0.23}{10^6} = \$4140$$

or the economical investment, from Eq. (1–5),

$$\frac{4140}{0.12} = \$34,500$$

Note that a practical problem of this type should be worked for at least three different loads.

Typical values of turbine heat rates for 1½-in. Hg condenser pressure and at rated load are:

Unit	Heat Rate, Btu per kwhr
12,650-kw; 600 psig, 825 F.............................	10,365
16,500-kw; 850 psig, 900 F.............................	9,740
44,000-kw; 1250 psig, 950 F...........................	9,350
100,000-kw; 1450 psig, 1000 F, 1000 F....................	8,150
150,000-kw; 2000 psig, 1050 F, 1000 F....................	7,700
125,000-kw; 4500 psig, 1150 F, 1050 F, 1000 F............	7,100

11–6. Estimation of Turbine-Throttle Steam Flow. The curves of Figs. 11–7, 11–8, and 11–9 are also useful in estimating the throttle flow as a preliminary step in making heat-balance calculations. Eq. (11–1) gives the relationship between turbine heat rate and the steam flow entering the turbine. Since the curves of Figs. 11–7, 11–8, and 11–9 were derived for a cycle without make-up or steam and water losses, W_1 will equal W_2 in Eq. (11–1).

EXAMPLE 11–3. Estimate the throttle flow for the 75,000-kw turbine generator of the previous example for four stages of feedwater heating.

SOLUTION. From the preceding example, the total temperature rise of the feedwater will be 298.3 F, the nonextracting steam flow will be 549,300 lb per hr, and the nonregenerative heat rate will be 10,290 Btu per kwhr. From Fig. 11–7(c) the heat-rate improvement will be 11.8, and from Fig. 11–7(e), the temperature correction will be −0.29%. The load and vacuum correction will be unity. Applying these corrections as indicated in Eq. (11–5),

$$\text{Turbine heat rate} = 10,290 \, [1 - (0.118 - 0.0029)] = 9108$$

From Eq. (11–1), for no losses or make-up,

$$W = \frac{9108 \times 75,000}{1453.1 - 350.6} = 620,000 \text{ lb per hr}$$

This figure compares very favorably with the value obtained by a heat-balance calculation of 630,000 lb per hr with make-up and losses as calculated in the next article.

11–7. Heat-Balance Calculations. The methods commonly employed for calculating heat balances will be illustrated by a cycle using a combination of heater arrangements.

Inspection of Fig. 11–10 will show that steam from the boiler goes to the turbine and to the air-ejector condenser by way of the air ejector proper. Steam is extracted from the turbine through four bleed belts. The highest pressure-bleed belt supplies steam to the fourth heater only, while the next bleed belt supplies steam to both the third heater and the evaporator. The evaporator vapor and the steam from the third extraction point supply the second heater. The lowest pressure heater receives

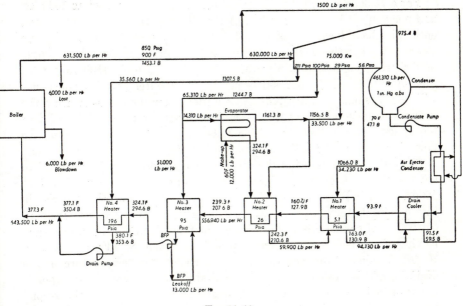

Fig. 11–10.

steam from the last bleed belt, and the remaining steam in the turbine flows to the condenser where it is condensed and mixed with the cascaded drains from the second and first heaters and the air-ejector condenser.

The condensate flows through the air-ejector condenser, drain cooler, first heater, second heater, and into the third heater by means of the pressure developed by the condensate pump. All these heaters except the last are surface-type heaters. Deaeration of the feedwater may take place in either the condenser or the third heater, or more likely in both these pieces of equipment. Feedwater from the contact heater is raised in pressure to something in excess of boiler pressure, after which it is further heated by the fourth heater and then mixed with the drains from the fourth heater. The leak-off from the boiler feed pumps, 13,000 lb per hr, returns to the third heater but has no thermodynamic effect since it is at essentially the same temperature as the water in the heater.

The extraction pressures and the pressure drops in the steam piping have been assumed from experience, and a knowledge of the station layout and the steam conditions at the extraction openings have been determined from a turbine condition curve. The temperatures at the heaters are determined on the assumption that there is no subcooling of the drains and that there is a 3-deg terminal difference for the surface heaters. Zero terminal difference is used for the contact heater. Terminal difference for the drain cooler has been taken as 10 deg.

For a 75,000-kw unit, the electrical losses may be assumed as 1050 kw and mechanical losses as 800 kw. After all temperatures, pressures, and enthalpies have been established, it is necessary to estimate the throttle steam flow. For this case we shall use 630,000 lb per hr (see the example of the preceding article). Since the air ejector requires 1500 lb per hr, the boiler flow will be 631,500 lb per hr. The water entering the boiler will be this same quantity plus 2% additional for the make-up from the evaporator to take care of the losses throughout the plant, such as boiler blowdown and soot blowing. Using 12,000 lb per hr as the make-up, the flow of water into the boiler will be 643,500 lb per hr.

Since the steam entering the air-ejector condenser can be assumed to have sufficient energy remaining after compressing the air to deliver 1000 * Btu per lb to the condensate, there would be an estimated rise in the condensate temperature of some 2.5 deg for an assumed flow of about 550,000 lb of condensate per hr. Thus, the temperature of the condensate leaving the air-ejector condenser would be about 81.5 F; and with a 10-deg terminal difference for the drain cooler, the temperature of the drains would be about 91.5 F. The enthalpy of the evaporator vapor is determined by assuming saturated steam at the pressure in the second heater. Usually, this assumption will be very nearly correct since the evaporator will normally be located near the heaters with a correspondingly small pressure drop in the vapor line. Furthermore, the vapor must be very nearly dry or the moisture will have entrained solids and will thus defeat the purpose of the evaporator.

The flow from the fourth heater will be $643,500 - S_4$, where S_4 is steam flow to the heater. Application of the Law of Conservation of Energy to this heater will produce the equation

$$S_4(1307.5 - 353.6) = (643,500 - S_4)(350.4 - 294.6)$$

$$S_4 = 35,560 \text{ lb per hr}$$

The temperature of the mixture of drains and feedwater entering the boiler may be found from the enthalpy of mixture, or

$$h = 350.4 + \frac{35,560(353.6 - 350.4)}{643,500} = 350.57 \text{ Btu per lb, or } 377.3 \text{ F}$$

Also, the water flow through the fourth heater will be $643,500 - 35,560 = 607,940$ lb per hr.

As previously mentioned, the boiler feed pump leak-off will have no effect on the third-heater calculations. Then, letting S_3 be the steam entering the heater,

$$S_3(1244.7 - 207.6) = 607,940(294.6 - 207.6)$$

$$S_3 = 51,000 \text{ lb per hr}$$

* Some designers use 1200 Btu per lb for this value.

For the evaporator

$$S_e(1244.7 - 294.6) = 12,000(1161.3 - 28.1)$$

$$S_e = 14,310 \text{ lb per hr}$$

The total quantity of steam bled from the third extraction opening is $51,000 + 14,310$ lb per hr. It is interesting to note that the evaporator requires $14,310 \div 12,000 = 1.193$ lb of steam for each pound of water evaporated. The water entering the second heater is $607,940 - 51,000 = 556,940$ lb per hr.

For the second heater,

$$1156.5S_2 + 12,000(1161.3) + 14,310(294.6) + 556,940(127.9)$$
$$= (S_2 + 12,000 + 14,310)(210.6) + 556,940(207.6)$$

and

$$S_2 = 33,590 \text{ lb per hr}$$

The total drains flowing from the second heater will be $33,590 + 12,000 + 14,310 = 59,900$ lb per hr.

The calculations for the first heater are more complicated than those for the other heaters, since the temperature of the feedwater entering cannot be found until the quantity of drains flowing through the drain cooler is known; but this depends on the temperature of the feedwater entering the heater. The easiest method is to consider the air-ejector condenser, drain cooler, and first heater all as one heat exchanger for the purpose of the calculations. Then, if S_1 is the extraction steam, application of the Law of Conservation of Energy will produce

$$S_1(1066.0) + 59,900(210.6) + 1500(1000.0) + 556,940(47.1)$$
$$= (S_1 + 59,900)(59.5) + 556,940(127.9)$$

$$S_1 = 34,230 \text{ lb per hr}$$

The drains leaving the first heater $= 34,230 + 59,900 = 94,130$ lb per hr. The temperature of the feedwater leaving the drain cooler may now be found:

$$h = 127.9 - \frac{34,230(1066.0) + 59,900(210.6) - 94,130(130.9)}{556,940}$$

$$= 61.8 \text{ Btu per lb, or } 93.9 \text{ F.}$$

Similarly for the air-ejector condenser,

$$h = 47.1 + \frac{1500(1000.0)}{556,940} = 49.8 \text{ Btu per lb, or } 81.8 \text{ F}$$

With these temperatures, the drain-cooler terminal difference is **9.8 deg** instead of the 10 deg assumed. However, this is close enough.

Before determining the turbine output for the throttle flow assumed, it is desirable to check our calculations so that we may be sure that there are no errors. This may be done by assuming that all the feedwater heaters, the evaporator, drain cooler, and air-ejector condenser are one heat exchanger. The energy entering must be equal to the energy leaving. Thus

ENTERING

$$
\begin{array}{rll}
556,940\ (47.1) & = & 26,232,000\ \text{Btu per hr} \\
34,230\ (1,066.0) & = & 36,489,200\quad\text{``}\quad\text{``}\quad\text{``} \\
33,590\ (1,156.5) & = & 38,846,800\quad\text{``}\quad\text{``}\quad\text{``} \\
65,310\ (1,244.7) & = & 81,291,500\quad\text{``}\quad\text{``}\quad\text{``} \\
35,560\ (1,307.5) & = & 46,494,700\quad\text{``}\quad\text{``}\quad\text{``} \\
12,000\ (28.1) & = & 337,200\quad\text{``}\quad\text{``}\quad\text{``} \\
[1,500]\ (1,000.0) & = & 1,500,000\quad\text{``}\quad\text{``}\quad\text{``} \\
\hline
737,630\ \text{lb per hr} & & 231,191,400\ \text{Btu per hr}
\end{array}
$$

LEAVING

$$
\begin{array}{rll}
94,130\ (59.5) & = & 5,600,700\ \text{Btu per hr} \\
643,500\ (350.57) & = & 225,591,800\quad\text{``}\quad\text{``}\quad\text{``} \\
\hline
737,630\ \text{lb per hr} & & 231,192,500\ \text{Btu per hr}
\end{array}
$$

Note that both the flow quantities and the energy quantities must check. The flow of 1500 lb per hr to the air ejector should not be included with the total flow entering the system since this quantity was not included in that leaving the system. The value of the enthalpy of the steam entering or leaving the air ejector is not known; only the difference between these two values is known.

The output of the turbine may now be determined in either of two ways: (1) application of the Law of Conservation of Energy or (2) the extraction factor method discussed in the chapter on turbines (Chap. 9). By the first method, all the energy entering must be equal to all the energy leaving, or

$$(\text{Internal kw})\,3413 + 35,560(1307.5) + 65,310(1244.7) + 33,590(1156.5)$$
$$+\,34,230(1066.0) + 461,310(975.4) = 630,000(1453.1)$$

This same equation may also be written as:

$$\text{Internal kw} = [\,630,000(1453.1 - 1307.5) + 594,440(1307.5 - 1244.7)$$
$$+\,529,130(1244.7 - 1156.5) + 495,540(1156.5 - 1066.0)$$
$$+\,461,310(1066.0 - 975.4)\,] \div 3413 = 76,878$$

Then the net output of the turbine will be

$$
\begin{array}{lr}
\text{Internal output} \dotfill & 76,878\ \text{kw} \\
\text{Electrical losses} \dotfill & 1,050\ \text{kw} \\
\text{Mechanical losses} \dotfill & 800\ \text{kw} \\
\hline
\text{Output} \dotfill & 75,028\ \text{kw}
\end{array}
$$

This is within 28 kw of the required load and would be close enough for most work. However, if it were necessary to obtain a more accurate answer, the reduction in throttle flow could be approximated by determining the steam rate based on the internal kilowatt. The correction in this case would be

$$28 \times \frac{630,000}{76,878} = 230 \text{ lb per hr}$$

In order to determine the output by the extraction factor method, first find all the extraction factors and the nonbleeding steam rate:

EXTRACTION FACTORS

$$\text{Fourth bleed } \frac{1307.5 - 975.4}{1453.1 - 975.4} = 0.69520$$

$$\text{Third bleed } \frac{1244.7 - 975.4}{1453.1 - 975.4} = 0.56374$$

$$\text{Second bleed } \frac{1156.5 - 975.4}{1453.1 - 975.4} = 0.37911$$

$$\text{First bleed } \frac{1066.0 - 975.4}{1453.1 - 975.4} = 0.18966$$

$$\text{Nonbleeding steam rate} = \frac{3413}{1453.1 - 975.4} = 7.1441 \text{ lb per internal kwhr}$$

The throttle flow required for an output of 75,000 kw with 1850-kw mechanical and electrical losses will be

```
Nonbleeding throttle flow = 76,850 × 7.1441 = 549,020 lb per hr
Replacement steam:
    Fourth bleed: 35,560 × 0.69520        = 24,720 "  "   "
    Third bleed: 65,310 × 0.56374         = 36,820 "  "   "
    Second bleed: 33,590 × 0.37911        = 12,730 "  "   "
    First bleed: 34,230 × 0.18966         =  6,490 "  "   "
    Throttle flow required when bleeding = 629,780 lb per hr
```

This checks the results of the previous method.

Effects of pump work and compressibility of the liquid have not been considered in this problem. Usually, these variations are not significant until the thottle pressure reaches 2000 psig.

For our problem we shall assume that the auxiliary power will be 5% of the turbine generator load and that the boiler efficiency is 86%. The plant net output will be 75,028 kw $\times$ 0.95 = 71,277 kw. Also, assume that the boiler pressure is 1300 psia.

From Eq. (11–1),

$$\text{Turbine heat rate} = \frac{631,500\,(1453.1) - 643,500\,(350.6)}{75,028}$$

$$= 9223 \text{ Btu per kwhr}$$

From Eq. (11–2),

Gross station heat rate

$$= \frac{(631{,}500 + 6000)(1453.1) + 6000(585.4) - 643{,}500(350.6)}{0.86(75{,}028)}$$

$$= 10{,}915 \text{ Btu per gross kwhr}$$

From Eq. (11–3),

Net station heat rate

$$= \frac{(631{,}500 + 6000)(1453.1) + 6000(585.4) - 643{,}500(350.6)}{0.86(71{,}277)}$$

$$= 11{,}489 \text{ Btu per net kwhr}$$

If the coal used in the boiler has a heating value of 14,000 Btu per lb, then the station will require $11{,}489/14{,}000 = 0.821$ lb per net kwhr of coal. The boiler would require $71{,}277 \times 0.821 \div 2{,}000 = 29.2$ tons per hr of coal at this generator output.

11–8. Partial-Load Heat Balances. Prediction of partial-load performance by means of heat-balance calculations is equally as important as that for full load. The general method to be followed is the same as that for full load with one exception—the extraction pressures have been established for full load and will vary as the flow to succeeding stages of the turbine, Fig. 9–30. In other words, the extraction pressures cannot be arbitrarily set for all loads but can be selected for one load only. At other loads they will follow the laws governing turbine performance.

Terminal differences for surface heaters will decrease when the load decreases. Very often, in practice, the terminal differences of the heaters are dropped one-third in value when the turbine load changes from full to three-quarters load, two-thirds when the turbine load changes to one-half load, and are considered as unity for all lesser loads. This rule of thumb is for heaters without desuperheating zones. Other designers maintain the terminal-difference constant at all loads on the assumption that their effect on the heat rate at partial loads is negligible. More accurate methods of determining heater performance are given in the chapter dealing with that type of equipment (Chap. 8).

Since reductions of the turbine load will decrease both the extraction pressure and the water flowing through the heater, the quantities of extracted steam will decrease sharply. The friction loss in pipes varies with the square of the flow and directly with the specific volume, Eq. (2–16). Since the extracted steam flow and the extraction pressure vary almost lineally with turbine load, the percentage of pressure drop in the extraction lines will remain almost constant.

For these reasons, partial-load heat balances involve more cut-and-try steps than full-load heat balances. For a first guess of extraction pressures, they may be assumed to vary as the turbine load. Obviously, this is not correct, since the Willans line for a turbine cannot go through the zero-zero axis. However, this will give a reasonably accurate first guess and is about the only basis that can be used at that stage of the calculations. Further assumptions for extraction pressures should be based on steam flows.

11–9. Selection of Extraction Pressures. The most efficient arrangement of feedwater heaters with several stages of extraction is that in which the quantities of steam to each heater are equal. That is, if there were four stages of heating, as in Art. 11–7, each heater would receive 25% of the total extraction steam. In this example, the total steam extracted was 168,690 lb per hr, so that the extraction for each bleed belt should have been about 42,170 lb per hr for most efficient operation. At three points the extraction was about 80% of this value, while the fourth was nearly 150%. These changes in extraction quantities from the optimum cause only small variations in the heat rate. For example, if one of the pressures were changed to cause 50% of the total extracted steam to flow to one heater, the heat rate would be changed only about 0.6% from the optimum value. For this reason, a common rule of thumb is to select extraction pressures that will give approximately equal temperature rise in each feedwater heater. This is not strictly the same as equal extraction quantities but is sufficiently accurate for most practical work.

However, the power-plant designer does not have full control over the heater pressures. The pressure for the last heater in the cycle will often be determined by the boiler design, since the temperature at which the feedwater enters the boiler will affect the design of the economizer and the temperature stresses in the boiler. It is usually considered best to use a standard boiler design, wherever possible, i.e., one which has been built and used in other installations, since redesigning a boiler will add to the expense.

If the turbine is of the compound type, as in the example of Art. 11–7, the ease of extracting steam from the cross-over pipe would dictate the pressure of the third heater. That particular turbine had a cross-over pressure of 100 psia at the rated load. Although this procedure would not be mandatory, a bleed belt placed a few stages either side of the cross-over would add considerably to the initial cost of the turbine.

Often the design of the turbine for best performance and life will dictate the pressure at which the first heater will operate. As discussed in connection with turbines, the efficiency of the low-pressure stages and their wearing qualities will depend on the moisture content of the steam. An extraction point at the proper pressure will aid in the elimination of moisture. If, for some reason, a heater cannot be placed at this pressure point in the cycle, turbine manufacturers will sometimes place an internal bleed in the turbine which will bleed some of the steam to the condenser. Since this by-passes the last few stages, the bled steam will not do any work in the turbine and its energy will be lost to the cycle. Some cases have existed where a turbine manufacturer furnished a low-pressure feedwater heater without extra charge in order to improve the efficiency of his turbine.

EXAMPLE 11-4. Select the extraction pressures for a 50,000-kw turbine generator unit having steam conditions of 1250 psig, 950 F, 2-in. Hg, with three stages of feedwater heating. Use 5 F TD on all surface heaters, 0 F TD on the contact heater (the middle heater), and 5% pressure drop in all extraction lines.

SOLUTION. Saturation temperature at 2-in. Hg condenser pressure is approximately 101 F. From Fig. 11-7(d) the optimum temperature rise is about 310 F. Thus, the rise in each heater will be 103 F, and the feedwater outlet temperature of each heater will be:

$$\text{First heater:} \qquad 101 + 103 = 204 \text{ F}$$
$$\text{Second heater:} \qquad 204 + 103 = 307 \text{ F}$$
$$\text{Third heater:} \qquad 101 + 310 = 411 \text{ F}$$

The saturation temperature in each heater will be the feedwater outlet temperature plus the terminal difference. The saturation temperatures and pressures will be:

$$\text{First heater:} \qquad 204 + 5 = 209 \text{ F}, \qquad 13.8 \text{ psia}$$
$$\text{Second heater:} \qquad 307 + 0 = 307 \text{ F}, \qquad 74.4 \text{ psia}$$
$$\text{Third heater:} \qquad 411 + 5 = 416 \text{ F}, \qquad 295.7 \text{ psia}$$

With a 5% pressure loss in the extraction piping, the optimum extraction pressures will be:

$$\text{First heater:} \qquad 13.8 \div 0.95 = 14.5 \text{ psia}$$
$$\text{Second heater:} \qquad 74.4 \div 0.95 = 78.4 \text{ psia}$$
$$\text{Third heater:} \qquad 295.7 \div 0.95 = 311 \text{ psia}$$

11-10. Preferred Standard Turbine Generators. In order to obtain the benefits derived from standardization, a joint committee of the ASME and the AIEE established a set of standard sizes and designs of steam turbine generators, Table 11-2. In addition to setting standard ratings, certain electrical standards are also established. Recommended extraction pressures are listed in terms of the saturated steam temperature; this does not mean that the steam is saturated but merely provides a method of establishing the pressure with a tolerance.

The ratings listed are the maximum ratings of the turbines, read as kilowatts at the generator terminals. Due to manufacturers' design tolerances, it is possible to obtain 5 to 10% higher output than listed, but these are not guaranteed.

TABLE 11-2

PREFERRED STANDARDS FOR LARGE 3600-RPM, 3-PHASE, 60-CYCLE CONDENSING STEAM TURBINE GENERATORS

	Air-cooled Generators				Hydrogen-cooled Generators Rated for 0.5-psig Hydrogen Pressure†				
Turbine rating,* kw	12,650	16,500	22,000	33,000	44,000	66,000	100,000†		150,000
Generator rating, kva	13,529	17,647	23,529	35,294	47,058	70,588	106,951	106,951	160,428
power factor	0.85	0.85	0.85	0.85	0.85	0.85	0.85	0.85	0.85
short-circuit ratio	0.80	0.80	0.80	0.80	0.80	0.80	0.8	0.8	0.8
Throttle pressure, psig	600	850	850	850	1250	1250	1450	1450‡	1800‡
Throttle temperature, F	825	900	900	900	950	950	1000	1000	1000
Reheat temperature, F								1000	1000
Number of extraction openings	4	4	4	5	5	5	5	5	7
Saturation temperatures at openings at "turbine rating," with all extraction openings in service, F—1st	180	180	180	180	180	180	185	180	150
2nd	240	240	240	240	240	240	250	245	185
3rd	290	290	290	290	290	290	310	305	235
4th	360	360	360	360	360	360	385	375	280
5th	..	..	..	420	420	420	450	450	325
6th	..	..	..	..	..	..	..	..	375
7th	..	..	..	..	..	..	..	..	455
Exhaust pressure, in. Hg. abs.	1.5	1.5	1.5	1.5	1.5	1.5	1.5	1.5	1.5
Generator capability at 0.85 power factor and 15-psig hydrogen pressure, kva		20,294	27,058	40,588	54,117	81,176	122,994	122,994	184,492
Generator capability at 0.85 power factor and 30-psig hydrogen pressure, kva		22,059	29,411	44,118	58,822	88,235	133,689	133,689	200,535

A tolerance of 10 F shall apply to above saturation temperatures. Tolerances shall be unilateral, if possible, so as not to reduce the spread in temperature between adjacent extraction openings.

* The "turbine rating" is guaranteed continuous output at generator terminals when the turbine is clean and operating under specified throttle steam pressure and temperature and 2.5-in. Hg abs exhaust pressure, with full extraction from all extraction openings.

‡ A 10% pressure drop is assumed between high-pressure-turbine exhaust and low-pressure-turbine inlet for the reheat machines.

† There are two different turbines; the first for regenerative-cycle operation, and the second for reheat-regenerative-cycle operation.

Turbine generators at these and many other ratings—kilowatt, steam conditions, and extraction points—are obtainable from the manufacturers.

11–11. Actual Station-Heat Balances. In order to interpret a heat-balance diagram properly, the student should be able to visualize the equipment represented by the symbols on the diagram. To aid the student in this process and to help him become familiar with the more common cycles, Figs. 11–11 to 11–19 are included. The cycle shown on

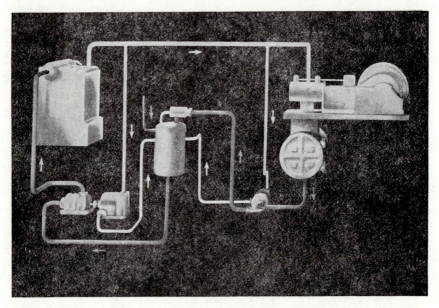

Fɪɢ. 11–11. A small condensing power plant with a steam engine as the prime mover. (*Power* magazine.)

Fig. 11–11 is the most common arrangement used for small plants. In some cases the steam engine is replaced by a turbine. Steam used in the prime mover is condensed and fed to a contact heater. Steam for the heater is obtained from the exhaust of the pump drives which receive steam at full boiler pressure. Note that the condensate enters the deaerator by way of a vent condenser and that the system make-up is introduced into the contact heater.

The cycle of Fig. 11–12 is identical with that of Fig. 11–4 except for the location of the pumps. The surge tank shown here is a very important feature of any plant although it does not enter into the heat-balance calculations. As the load changes and as the boiler pressure and level change, the quantity of fluid in the system will fluctuate. A surge tank or its equivalent must be in the system to take care of such fluctuations.

In the previous cycle the surge effects were taken care of by the storage section of the deaerator. One big difference between these two cycles, however, is the manner of deaeration. The cycle of Fig. 11–11 is known as a *closed cycle,* since air cannot enter the system except through leaks in the condenser or prime mover. In this arrangement air is removed by the air ejector on the condenser. In the *open system* of Fig. 11–12, a considerable amount of air comes into the system with the make-up as it enters the surge tank. If the boiler pressure and temperature were very

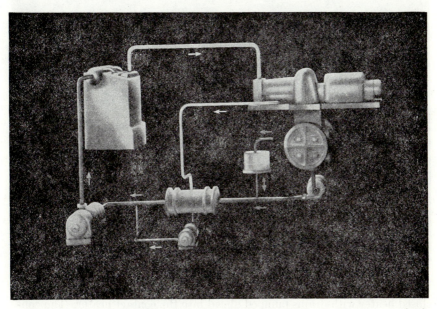

Fig. 11–12. A small turbine plant with drains pumped ahead. (*Power* magazine.)

low, this might not have a deleterious effect. One way to avoid these destructive effects would be to induct the make-up into the condenser where the air would be removed by the air system on the condenser and to install large closed tanks under the condenser to accommodate the system surges.

Surge-tank level is usually controlled by float valves so that, when the water level drops to a predetermined value, more water can be brought in until the level has reached a safe maximum. Water must be pumped into the surge tank shown in Fig. 11–12. If the surge tanks discharged into the condenser, pumps would not be necessary, since the difference between atmospheric pressure and that in the condenser would be sufficient. There must also be provision on the surge tank to take care of any overflow. A condensate storage tank located at some convenient place in the plant may be used to receive the overflow from the surge tank and to supply it with make-up. If there are two boilers in the station, then the condensate storage tank should be at least large enough to store enough water to fill one of the boilers after it has been emptied for inspection, repairs, etc., plus some capacity for surge-tank make-up.

The next stage in the improvement of a plant is that of Fig. 11–13. Two stages of feed heating have been employed: a deaerating heater and a

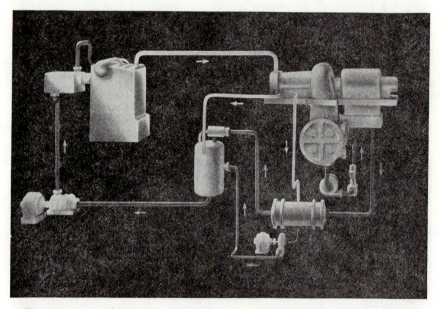

Fig. 11–13. A turbine plant with two stages of feed heating. (*Power* magazine.)

closed heater with drains pumped ahead into the contact heater. The condensate condenses the steam used by the air ejector and also provides cooling water for the generator. There has been much discussion about the advisability of using condensate for generator cooling. One advantage is that part of the losses in the generator are saved for the cycle, but the disadvantages are: (1) the condensate piping is more complicated and (2) the generator cooling water is at a rather high pressure. Therefore, any leaks in the cooler tubes would admit water into the generator with possible serious damage. The cooling water used for most generators is under a partial vacuum so that defective tubes do not inject water into the windings. This is accomplished by arranging the cooler at the top of a hydraulic loop where the pressure is less than atmospheric. Note that an economizer has been included with the boiler.

The power plants depicted in Figs. 11–11, 11–12, and 11–13 are typical of the basic designs used for small industrial plants and for small central stations. One very common addition to the industrial plant is a means of providing low-pressure steam for heating and processes. This steam often comes from the exhaust of the turbines; thus, the heating system or industrial process acts as the condenser for the system. Extreme caution must be exercised in such a plant to prevent contamination of condensate or drains returned from the process to the boiler feed cycle. Oil or chemicals in the boiler feed may be disastrous. Special equipment should be installed to purify contaminated condensate, or in extreme cases the condensate should be discharged to waste and 100% make-up used for the boiler.

Basically there is little, if any, difference between the industrial plant and the central station. Many industrial, or isolated, plants are larger than some central stations. Then, too, many central stations supply process steam to nearby industrial plants or to district steam-heating mains. Of course, any small plant cannot justify several feedwater heaters regardless of whether it is an industrial plant or a central station. *The proper number of heaters for any cycle can be determined only by an economic balance between savings in operating costs and increased investment.*

The following figures show power plants that are normally thought of as being central stations. However, many industrial plants have superposed turbines and several stages of feedwater heating. Therefore, these pictures should not be considered as portraying central stations only.

Fig. 11–14 shows a method that was used during the 1930's to improve the economy of an old plant consisting of a low-pressure turbine, con-

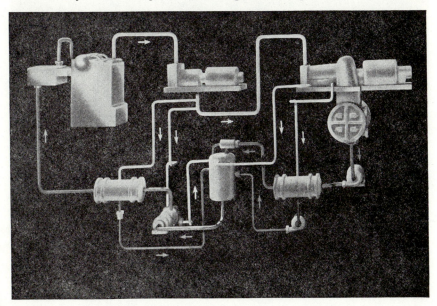

FIG. 11–14. Superposed plant. (*Power* magazine.)

denser, condensate pump, surface heater, and deaerator. When the cycle was rejuvenated, a topping turbine, new high-pressure boilers, high-pressure boiler feed pumps, and a high-pressure surface heater were added. The superposed turbine then received its steam directly from the high-pressure boiler and exhausted into the old low-pressure unit, the boiler feed pump turbine, and the cross-over heater (so named because it receives steam from the cross-over pipe between the turbines). The cross-over heater had its drains cascaded back to the deaerator, while the low-pressure surface heater had its drains pumped ahead.

Fig. 11–14 could also represent a cross-compound unit.

Every effort has been made by the designers of the plant of Fig. 11–15 to produce a highly efficient station. The cycle employs reheat and five

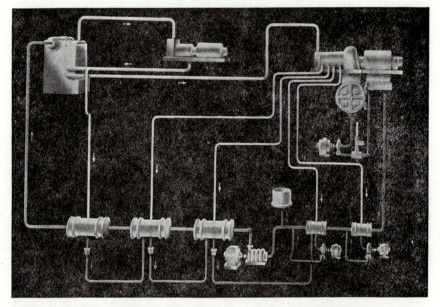

Fig. 11–15. Reheat and regenerative cycles. (*Power* magazine.)

stages of feedwater heating. Advantage is taken of the losses in the generator by using condensate in the coolers, and the drains are pumped ahead from the first two stages of the cycle. The last stages of heating are on the discharge of boiler feed pump, which necessitates the use of high-pressure heaters. Steam for the final heater is taken from the cross-over pipe before the steam enters the boiler reheater. The turbine is of the cross-compound type.

The reheat cycle portrayed in Fig. 11–15 has been used extensively since World War II to combat increased fuel costs. Some of the variations of this cycle are: incorporation of the surge tank with the condenser by having a large condenser hotwell, use of desuperheating zones and integral drain coolers with all surface heaters, use of six to eight stages of feedwater heating, use of a deaerator with large feedwater storage capacity for either the second or third feedwater heater.

The heat-balance diagram for one of the two 80,000-kw units at the Oswego plant is shown in Fig. 11–16. Six stages of feed heating are used, three with surface heaters with integral drain coolers and three with contact heaters. The condensate is used for oil and hydrogen coolers in order to reduce maintenance by eliminating the cleaning problem for these coolers and to keep the heat transfer to a maximum. Boiler feed pumps at this plant are variable-speed controlled by means of reduction gears and eddy-current electrical couplings. Note that the air ejector has three condensers, with the aftercondenser located between the heaters for the twelfth and fourteenth stages. The heat-flow diagram for this plant is shown in Fig. 11–17. Although this type of diagram has little use in design work, it is convenient for use in portraying the distribution of heat throughout the system. The cross-section of Oswego Station is shown in Fig. 11–18.

A heat-balance diagram for the first supercritical-pressure power plant is shown in Fig. 11–19. Pressure and temperature conditions are given on the diagram. Note that there are seven stages of feedwater heating and two stages of steam reheating for this 125,000-kw unit. Because of the extremely high feedwater pressures involved, some 5450 psia, two boiler feed pumps and one condensate pump and one condensate booster pump are used in series to reduce the design pressures for the feedwater heaters. Two sets of these pumps are required to carry full load.

11–12. Binary Cycles. In general, the efficiency of a cycle depends on the highest and lowest temperatures in the cycle. For this reason there has been a steady increase in steam pressures and temperatures over the years. The lowest temperature, the condenser temperature, is, of course, fixed by the temperature of the available circulating water. However, the difficulty in increasing steam temperature economically has been the inherent increase in steam pressure in order to provide an expansion in the turbine that will produce a reasonable amount of moisture in the turbine exhaust steam (say 10 to 13% moisture). High steam pressure means thicker walls for the piping and pressure vessels. Therefore, there has been a search for fluids that vaporize at high temperatures with correspondingly low pressure. Such a fluid, used in conjunction with water, would increase the cycle area on the temperature-entropy diagram over that of water only, Fig. 11–20. The lower part of the cycle is for steam and the upper part is for the top fluid. If the steam part of the cycle were regenerative, the customary change would take place in the diagram.

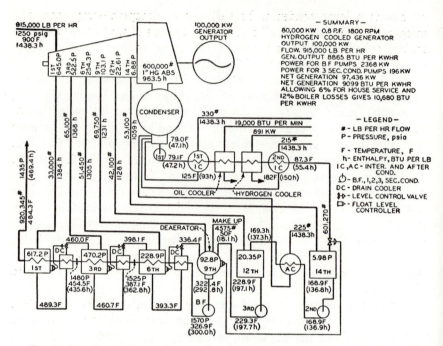

FIG. 11–16. Heat-balance diagram for 100,000-kw output from Oswego turbine generator. (*Trans. ASME.*)

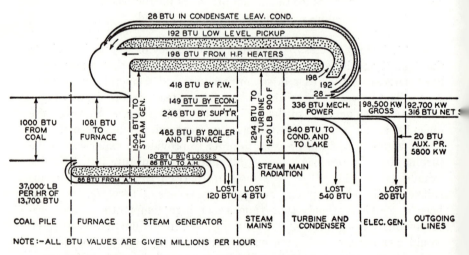

FIG. 11–17. Heat-flow diagram for 900,000 lb per hr output from Oswego steam generator. (*Trans. ASME.*)

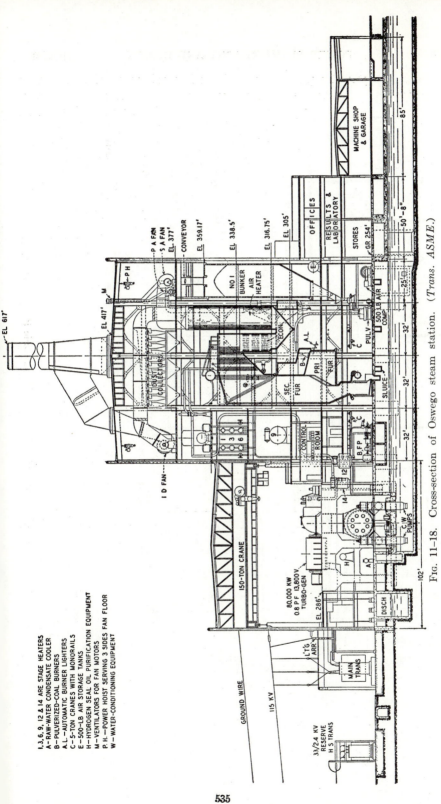

I,3,6, 9, 12 & 14 ARE STAGE HEATERS
A—RAW-WATER CONDENSATE COOLER
B—PULVERIZED-COAL BURNERS
A.L.—AUTOMATIC BURNER LIGHTERS
C—5-TON CRANES WITH MONORAILS
E—500-LB AIR STORAGE TANKS
H—HYDROGEN SEAL OIL PURIFICATION EQUIPMENT
M—VENTILATORS FOR FAN MOTORS
P. H.—POWER HOIST SERVING 3 SIDES FAN FLOOR
W—WATER-CONDITIONING EQUIPMENT

FIG. 11-18.   Cross-section of Oswego steam station.   (*Trans. ASME.*)

535

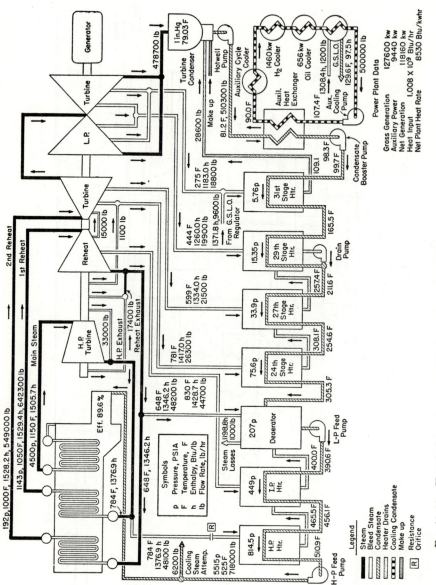

Fig. 11-19.   Heat balance for supercritical-pressure plant of Ohio Power Company. (*Trans. ASME.*)

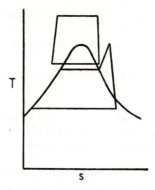

FIG. 11–20. Temperature-entropy diagram for a binary cycle.

The basic equipment for a binary cycle is shown in Fig. 11–21, assuming mercury to be the additional fluid. Note that the mercury leaves the boiler and does work in the mercury turbine. Then it passes to its condenser and, in giving up its remaining latent heat, evaporates the water to nearly saturated steam. This steam is superheated in a section of the mercury boiler and is then used in a conventional steam turbine.

Mercury has been found to be the most suitable medium for a binary cycle, but several others have been investigated, such as ammonia, methyl chloride, ethyl chloride, sodium dioxide, and ethyl bromide. None of these has had any commercial success to date. The ideal fluid for use with water in a binary cycle should have the following nine properties.

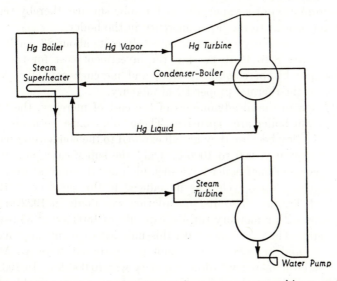

FIG. 11–21. Basic cycle arrangement for a mercury-steam binary cycle.

1. High vapor temperatures at reasonably low pressure
2. High critical temperature
3. High heat of vaporization to keep the weight of fluid in the cycle to a minimum
4. Temperature of solidification below room temperature
5. Chemical stability
6. Nontoxic
7. Noncorrosive to the metals normally used in power plants
8. Available in large quantities at reasonable cost
9. Ability to wet tube walls to promote heat transfer

Although mercury does not have all these attributes, it is more favorable than any of the other fluids investigated. At a pressure of 100 psia the saturation temperature is 910 F, and at 300 psia, 1090 F. For the condenser conditions, mercury at 6 psia will evaporate 1200 psi water with a reasonable terminal difference for the condenser-boiler. Since this pressure is below atmospheric, a vacuum system is required. The critical temperature of mercury is about 2240 F, well above operating temperature, and its freezing temperature is −38 F, so that there is no danger of solidification in the tubes and piping.

The heat of vaporization of mercury is low (about 125 Btu per lb), but the other fluids that have been proposed are also in this general range. Since mercury is an element, it is stable under all operating conditions. Although mercury does not cause any corrosion, it is extremely dangerous to human life, and elaborate precautions must be taken to prevent its escape. A wetting agent, magnesium and titanium, must be added to prevent the formation of a vapor film at the tube surface, thereby retarding the transfer of heat to the liquid mercury in the boiler.

The price of mercury varies but currently costs about $3 per lb. Since a plant requires 8 to 10 lb per kw, the investment cost for the fluid is about $30 per kw of combined steam and mercury capacity. See the Appendix for the thermal properties of mercury.

In addition to the disadvantages of the cost of mercury, the tubes of mercury-steam boiler are expensive. These tubes must be made of high temperature alloy because they are all exposed to the condensing temperature of the mercury, 900 to 1000 F. Only the superheater and reheater tubes of a conventional boiler have such high metal temperatures.

The mercury-vapor cycle was introduced in this country in 1913 by Dr. W. LeR. Emmet. The first installation was made in 1922 at Dutch Point Station with a mercury turbine capacity of 1800 kw at 35 psig mercury pressure. The experience with this unit led to many improvements in the cycle, and in 1928 the next unit was installed at South Meadow Station where the mercury turbine capacity was 10,000 kw. In 1932 units were built at Schenectady and at Kearny Station, each for double the rat-

ing of South Meadow unit and at 140 psig mercury pressure. The Schenectady plant is an outdoor station. The latest installation was at the Public
Service Company's Schiller Station at Portsmouth in 1950. There are two
7500-kw mercury turbines operating at 105 psig and 934 F and one
25,000-kw steam-turbine set operating at 600 psig and 825 F. The net heat
rate for the plant is about 9400 Btu per kwhr at a total output of
43,000 kw. Exact details of the plants have varied, but it has been found
practical to obtain the mercury boiler pressure by locating the mercury
condenser over the boiler and taking advantage of the static pressure of
the mercury rather than using a boiler feed pump for the mercury.

The specific heat of mercury is very low—about one-thirtieth of that of
water at 80 F and one sixty-sixth of that of water at 650 F. This produces

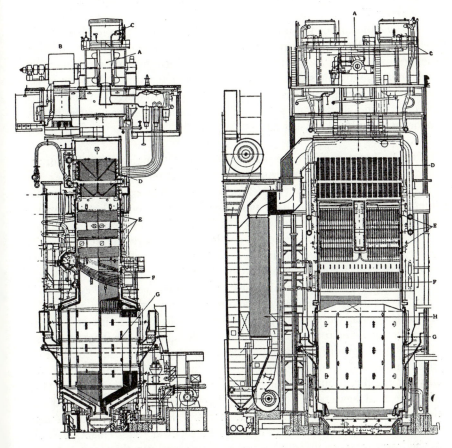

FIG. 11–22. General Electric mercury boiler installed at the Kearny Station of
the Public Service Electric and Gas Company of New Jersey, 1940. (*Trans. ASME,*
Vol. 64, No. 7.)

a steep saturated liquid line on the temperature-entropy diagram and eliminates the advantages of a regenerative cycle for the mercury phase.

The new boiler installed at Kearny Station in 1940 to replace the original 1932 boiler is shown in Fig. 11–22. The porcupine tubes used for many years were eliminated because their small passages were potential sources of plugging with foreign materials and because they were hard to clean and manufacture. The walls are covered with mercury tubes in the same way that waterwalls are used in any large steam boiler. The new boiler has a single drum 54-in. ID with 4½ in. thick walls and is 30 ft long. The drum is baffled to prevent liquid carry-over. The pressure drop from the drum to the turbine is sufficient to produce 30 F superheat.

In the mercury-steam cycle, thermal efficiencies of more than 37% have been realized on tests. The engine efficiency of mercury turbines appears to be about 72.5%.

The calculation of the mercury cycle is no different from that for the steam cycle. The steam part of the cycle may be calculated first and then the requisite mercury capacity may be determined or, for a given mercury turbine, the available steam capacity may be determined. The use of the regenerative steam cycle is just as advantageous when combined with mercury as it is with pure steam. The terminal difference of the boiler-condenser should be at least 30 F.

11–13. Location of Pumps in the Cycle.   Although the location of the condensate and boiler feed pumps is of no importance insofar as the thermodynamic heat-balance calculations are concerned, it may have considerable effect on the auxiliary power requirements, on the ease of operations, and on the initial investment. In the Rankine cycle, Fig. 11–1, only one pump is necessary to raise the condensate pressure from condenser to boiler level. Plants that operate on the Rankine cycle are usually very small or old, so that the pump is of the reciprocating type.

When one or more feedwater heaters are included in the cycle, as in Figs. 11–2 and 11–4, the flows and pressures involved make the use of centrifugal pumps advantageous. It is usually desirable to use two pumps in series to develop the required head. Because of the characteristics of condensate pumps, as explained in Chap. 3, they are inherently slow speed and therefore larger in proportion than the main boiler feeder, not only for the pump but also for the driver. The arrangement of Fig. 11–3, with the feed heaters in between the two pumps, has the advantage of cheap, low-pressure heaters and drain coolers. However, the arrangement of Fig. 11–4 has the advantage of lower horsepower requirements because of the lower temperature of the water in the pump. In this case the drain pump would have to develop more head than if the boiler feed pump were located after the junction of the two water lines. These differences would be small for plants that would have the cycles of the types shown in Figs. 11-3 and 11-4 but would be important for the larger plants with several stages of feedwater heating.

In the event that a contact heater is used, a pump must be located at the outlet of each contact heater. Since the water in a contact heater is at saturation temperature corresponding to the pressure in the heater, it could hardly flow into a higher-pressure heater without the aid of a pump. One of the disadvantages of a cycle that employs only contact heaters is the number of pumps involved and the complicated control that is required for them.

Control of boiler feed pumps was mentioned in Arts. 3-12 and 3-13. Water level in the boiler drum acts to control the pump flow either by regulating the speed of the pump or by a control valve throttling the pump discharge, Fig. 11-23.

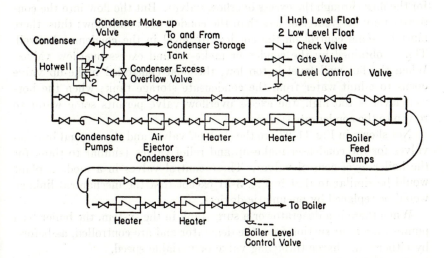

Fig. 11-23. Schematic diagram of condensate and feedwater piping for surface type of heater.

11-14. Condensate Pump Control.

Several methods have been used successfully to regulate the flow from condensate pumps. All the schemes regulate the flow at constant speed because there would not be enough saving in pump power to justify variable speed.

When a system contains only closed heaters, as in Fig. 11-15, but without the surge tank, the schematic diagram of the piping may be as shown in Fig. 11-23. This sketch shows two condensate and two boiler feed pumps. Since it is standard practice to provide spare pump capacity, the use of two of each of these pumps signifies that each pump can carry the full cycle flow by itself; or there is 100% spare capacity. If there had been three of each type of pump, then there would be 50% spare capacity, and it would require any two of the condensate pumps and any two of the boiler feed pumps to carry full system flow. Note that any heater may be by-passed and that the boiler feed pump is controlled by the throttling action of the boiler-level control valve. If the pumps were to be operated at variable speed, the boiler-level control valve could be omitted and the pump speed controlled from boiler water level.

Because of the pressures involved, and because of the desire for remote operation of the high-pressure water valves, the boiler feed pump discharge-gate valve and possibly other valves in the discharge system would be motor-operated.

In Fig. 11–23, the boiler feed pump acts as the control device for the condensate pump, since there can be no more or less water flow through the condensate pump than there is through the boiler feed pump (except for the flow through the excess overflow valve). But the flow into the condenser may be greater or less than the condensate pump flow; thus, there must be some control on the condensate level in the condenser hotwell. This is obtained by the condenser make-up and excess overflow valves. When the hot-well level is too low, the float-controlled make-up valve opens to admit water from the condensate storage tank; when the hot-well level is too high, the excess overflow valve permits some water to return to the condensate storage tank.

Not shown in Fig. 11–23 are the shut-off valves and the manual by-pass valves for the condenser make-up and relief valves (similar to those for the boiler-level control valve). These control valves in a modern plant would be similar to that in Fig. 3–31 except that the mechanical linkage would be replaced by pneumatic devices.

When there is a deaerator or a surge tank in the system, the boiler feed pumps take their suction from the deaerator and are controlled, as before, by either a discharge throttling valve or variable speed.

Fig. 11–24 shows a method of condensate pump control wherein the pump flow is controlled by the level in the hotwell. Make-up and excess overflow valves are then controlled from low- and high-level floats attached to the deaerator.

A variation of this scheme would be to control the condensate-pump throttling valve from deaerator level and to control the condenser make-up and excess overflow from hot-well level.

There is still a third variation of Fig. 11–24 wherein the condensate-pump control valve is omitted. This may be referred to as cavitation control, since the condensate pump will immediately pump the hotwell nearly dry. This will reduce the NPSH for the condensate pump, and the pump will operate on a break-off curve and at a point such as point G,

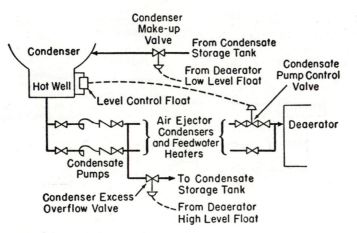

Fɪɢ. 11–24. Schematic piping diagram for condensate pump control by throttling.

Fig. 3–17. Make-up and excess overflow valves are operated from deaerator level. If the inflow to the condenser decreases in accordance with system demands, the NPSH will decrease and the condensate pump will operate at a point such as H.

Cavitation under these conditions may not be too severe, since the density in the vapor bubbles is very low at normal condenser pressures. Of course there will be practically no water storage in the condenser hot-well, and all the surge capacity of the system must be in the deaerator.

One additional method of condensate pump control is by continual recirculation, as in Fig. 11–25. Condenser make-up and excess overflow valves are controlled from deaerator level. The recirculation valve is so arranged that as the hot-well level drops, the recirculation valve opens to permit more of the condensate pump discharge to return to the hot well; thus, there will be less flow to the deaerator. Part (b) of Fig. 11–25 shows the condensate-pump characteristic curve and a break-off curve for 3-ft NPSH; assume that the hot-well level control is set to provide 3-ft NPSH for the pump.

If the system requires a flow corresponding to point A on the system head curve of Fig. 11–25(b), the pump will operate at point B on the 3-ft NPSH curve. The flow to the deaerator will be gpm A, the pump flow will be gpm B, and the flow through the recirculation valve will be gpm B minus gpm A. Note that the head at point B is slightly higher than at point A because of the increased flow through the air ejector condensers and the piping up to point T.

There are two serious disadvantages to condensate pump control by continuous recirculation, Fig. 11–25. The first disadvantage is that the horsepower required by the pump at gpm B is much greater than at

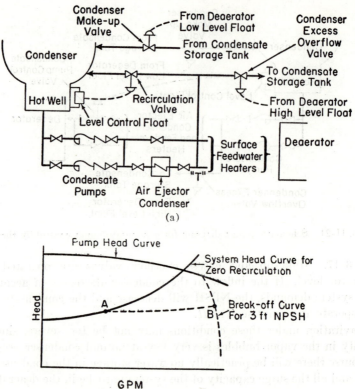

FIG. 11-25. Schematic piping diagram and pump characteristic curve for condensate pump control by recirculation.

gpm A. The second, and less important, disadvantage is that some of the energy transferred to the condensate in the air ejector condenser via the recirculated water is wasted.

Note that in all the above schemes of condensate pump control, except control by continuous recirculation, low flow protection should be provided for the air ejector condensers, as shown in Fig. 8–15 and the associated discussion.

11–15. Selection of Equipment Capacities. The first step in the design of a power plant is to establish the size of the turbine generator. This is usually accomplished by a study of the load growth of the system. The next step is to establish the pressure and temperature at the turbine throttle and the number of feedwater heaters. Heat balances and economic studies of costs are the bases for establishing pressures, temperatures, and the number of heaters and their terminal differences (see the example of Art. 11–5). Selection of the most economic condenser with its auxiliaries, selection of constant-speed versus variable-speed boiler feed pumps and fans have also been considered.

However, there are several other pieces of equipment that must be considered. No specific rules can be given for the selection of design capacities and pressures of pumps, fans, boilers, etc. Each company has its own factors and procedures that best fit its needs and beliefs. The following values and factors can be considered only as typical of those used by some designers for large units.

Steam Generator. Maximum continuous capacity of a steam generator is established from the maximum guaranteed turbine flow. Usually, the boiler is specified to be capable of a 2- or 4-hr peak of 10% greater than the maximum continuous load. This should provide sufficient steam to operate the turbine at more than maximum guaranteed load. Most turbines will develop about 10% extra load due to design tolerances used by the manufacturers.

Boiler-drum design pressure, or the steam working pressure (SWP), equals the turbine throttle pressure, plus the pressure drops through the steam piping and superheater, plus an allowance of 3 to 8% (the larger values for the lower pressure units) to permit pressure fluctuations during operation. The drum safety valves are set at the design pressure.

Superheater-outlet design pressure is established by the turbine throttle pressure, plus the steam pipe friction, plus an allowance of 3 to 8% to permit pressure fluctuations during operation. The superheater safety valves are set at the design pressure, but the design pressure is selected so that the superheater safety valves operate before the drum valves (at rated load) to ensure steam flow through the superheater tubes to keep them cool.

The economizer design pressure is slightly higher than drum pressure because of the pressure drop through the economizer.

Boiler Feed Pump. Pump capacity, without use of spare equipment, should be at least 10% greater than the maximum boiler output to provide for water surges. The pump head should be about 3% greater than the head required to put water into the boiler at the boiler SWP.

Condensate Pump. Pump capacity, without use of spare equipment, should be at least 20% greater than the maximum boiler load. The head should be sufficient to pump water into the deaerator at the maximum conceivable deaerator pressure or should provide ample (150 psi or more) NPSH at the boiler feed pump suction if there is no deaerator.

Heater Design Pressure. Usually, surface-heater shell and water-space design pressures are 150, 300, 450, 600 psig, etc. The shell design pressure should be at least 10% greater than the maximum turbine extraction pressure, and the water-space design pressure should be sufficient to withstand approximately boiler feed pump shut-off head with cold water.

Fans. Probably there is more variation in selecting the capacities of forced- and induced-draft fans than for any of the other major power-plant auxiliaries. The minimum allowance is for 10% extra of developed pressure and 10% extra of quantity based on about 5% more excess air for the forced-draft fan and 10% more excess air for the induced-draft fan than given in the boiler manufacturer's guarantee as the absolute maximum boiler load. The use of 10% allowance on both quantity and head is inconsistent with the system head curve for forced- and induced-draft fans. Use of these percentages can cause the fans to limit the maximum output of the boiler. If the limitation should be caused by the induced-draft fan, the situation can be hazardous because the forced-draft fan might supply more air to the boiler than the induced-draft fan can remove.

Some designers base the fan capacity on an allowance of 25% on capacity and 55% on developed pressure at 14% CO_2 in the flue gas (for coal), all for the peak load. For an average coal, 14% CO_2 would correspond to about 30% excess air (see the example of Art. 5–10).

In setting the capacity and pressure allowances it is well to consider the leakages of air and gas that can occur throughout the steam-generator unit. Infiltration can occur through the boiler casing and will be greater for an old style of brick casing than for a modern bolted or welded steel casing. Also, there is some leakage in the duct work (1 or 2%), the economizer (less than 4%), and the air preheater. The latter leakage is small (not over 2%) for a tubular heater but may be 7 to 10% for a regenerative heater. The boiler manufacturer should take these leakages into account in making his guarantees of excess air at the air heater outlet.

Conservative allowances for both forced- and induced-draft fans would be not less than 15% allowance on capacity and 32% allowance on developed static pressure at the boiler peak load (provided the design pressures have been conservatively calculated).

EXAMPLE 11–5. Use the heat balance of Fig. 11–10 and select the pressures and capacities of (a) the steam generator, (b) boiler feed pumps, (c) condensate pumps, (d) heaters, and (e) fans. Assume that the heat balance is for the maximum guaranteed capability of the unit. Other data are as follows for the heat balance flows:

High-pressure feedwater heater pressure drop, 15 psi
Low-pressure heater pressure drop, 10 psi each; drain cooler, 7 psi
Condensate-piping pressure drop, 25 psi; air ejector condenser, 3 psi
Boiler feed piping pressure drop, 50 psi
Steam-piping pressure drop, 50 psi
Static elevation from condenser hot-well level to deaerator inlet, 90 ft

Static elevation from minimum deaerator water level to boiler-drum inlet, 30 ft
Flue-gas pressure drop through mechanical dust collector, 3-in. WG
Steam generator:
  Excess air in furnace, 20%
  Excess air at air heater outlet, 25%
  Total draft loss (flue-gas side) of the boiler, superheater, economizer, air heater,
    and boiler manufacturer's ducts, 8.9-in. WG
  Loss (air side) through air heater, windbox, and burners, 8.6-in. WG
  Air temperature to tubular air heater, 100 F
  Flue-gas temperature from air heater, 315 F
  Excess air in furnace, 18%
  Excess air at air heater outlet, 23%
  Water-pressure drop through economizer, 15 psi
  Steam-pressure drop through superheater, 45 psi
  Fuel, 65,600 lb per hr of 12,500 Btu per lb coal containing 8% ash
Boiler feed pump and condensate-pump characteristic curve rise, 18%
Minimum design barometric pressure, 28.5 in. Hg.

SOLUTION. Since 75,000 kw was taken as the maximum guarantee capability
of the turbine generator, it can be assumed that the unit will have an actual output of 10% greater (including effects of changing vacuum). Therefore all flows
will be assumed to increase 10% for peak load and will be rounded off to give convenient figures.

(a) Steam generator capacity will be 640,000 lb per hr of steam continuously,
the circulation shall be satisfactory for an output of 710,000 lb per hr, and the
unit shall be capable of supplying 710,000 lb per hr of steam for a period of 2 hr.

The drum pressure will be the throttle pressure plus the steam-piping and
superheater pressure drop with a 8% allowance; or the SWP = 1.08(850 + 45
+ 50) = say, 1025 psig.

(b) If two boiler feed pumps are to be used, giving 100% spare, the capacity
of each pump should be the flow from the deaerator increased by 10% for extra
load and increased 10% as an allowance; or 1.10 × 1.10 × 607,940 = 736,000 lb
per hr. Converting into volume flow at 324.1 F,

$$\frac{736,000}{500 \times 0.905} = \text{say, 1650 gpm}$$

For a large boiler, the feed pumps should be able to supply water to the boiler
at the maximum drum pressure. Thus, the pump-developed pressure would be,
after correcting the pressure drops for the 10% increased flow,

Drum pressure	= 1025 psig
Piping and heater, $(50 + 15)\,(1.10)^2 =$	79 psi
Economizer, $15\,(1.10)^2$	= 18 psi
Static elevation at 377.3 F, 30 ÷ 2.64 =	11 psi
Total	1133 psig
Less deaerator pressure	80 psig
Total	1053 psi
Allowance of 3%	32 psi
Pump-design developed pressure	1085 psi

or, at 324.1 F, 1085 × 2.55 = 2760-ft developed head at 1650 gpm.

(c) If there are two boiler feed pumps, then it would be good practice to have
two condensate pumps. The flow for each, with 20% allowance, would be 1.10

$\times 1.20 \times 556{,}940 = 735{,}000$ lb per hr, or $735{,}000 \div 500 = 1470$ gpm. The developed head would be, for a negligible condenser pressure and for a 10% increase in deaerator pressure due to load increase (theoretically, the increased deaerator pressure should have been used in determining the boiler feed pump developed head),

Deaerator pressure, 95×1.10	$= 105$ psia
Piping pressure loss, $25(1.10)^2$	$= 30$
Pressure loss in heaters, etc., $(10 + 10 + 7 + 3)(1.10)^2$	$= 36$
Static elevation at 239.3 F, $90 \div 2.44$	$= 37$
Total	208 psia
Less condenser pressure	0
Total	208 psi

It is entirely possible for the actual turbine extraction pressure to deviate from the intended value due to design tolerances. Also, if the turbine blades become dirty due to deposits of boiler carry-over impurities, the extraction pressure may increase appreciably. Therefore, allow an additional 25 psi to give a design developed head of $208 + 25 = 233$ psi, or $233 \times 2.31 =$ say 540 ft.

(d) Low-pressure heater design pressures for the shell side would be 150 psig, as there would be little, if any, price reduction for a lower pressure. At overload, the high-pressure heater could conceivably receive steam at $1.10 \times 181 = 199$ psig. Therefore, take a 300-psig design pressure as the next higher pressure classification and as providing ample margin.

For the low-pressure heaters, the water side could be subjected to a maximum pressure of the condensate pump shut-off head. For cold water and an 18% rise in the characteristic curve, shut-off head would be $1.18 \times (540/2.31) - 14.7 = 260$ psig. Therefore, use 300-psig design pressure on the water side.

At very low load, the deaerator pressure would be near atmospheric. Therefore, using cold water, the boiler feed pump shut-off head would be $1.18 \times (2760/2.31) = 1410$ psig.

The nearest standard design pressure would be 1500 psig.

(e) From Eq. (5–8),

$$W_{ta} = \frac{7.65 \times 12{,}500}{10{,}000} = 9.56 \text{ lb air per lb fuel}$$

Specific volume for the air entering the air heater,

$$v_a = \frac{RT}{P} = \frac{53.3 \times 560}{144 \times 28.5 \times 0.491} = 14.8 \text{ cu ft per lb}$$

and for the flue gas leaving the air heater,

$$v_g = \frac{RT}{P} = \frac{53.3 \times 775}{144 \times 28.5 \times 0.491} = 20.5 \text{ cu ft per lb}$$

At peak load,

$$\text{Fuel} = 65{,}600 \times 1.10 = 72{,}200 \text{ lb per hr}$$

The forced-draft fan capacity at 20% excess air in the furnace and 20% allowance would be

$$\text{Capacity} = \frac{1.2 \times 1.2 \times 72{,}200 \times 9.56 \times 14.8}{60} = 245{,}000 \text{ cfm}$$

After correcting for peak boiler load, the loss in the air circuit for the boiler would be 8.6 $(1.10)^2 = 10.4$-in. WG. To this must be added the loss in the secondary air duct, which may be assumed as 1.5 in., giving a total of $10.4 + 1.5 = 11.9$ in. For a 20% allowance for capacity, there should be a $(1.20)^2 - 1 = 44\%$ allowance on developed pressure. Therefore the design developed should be $11.9 \times 1.44 =$ say, 17.2-in. WG.

When calculating the induced-draft fan capacity, the flue-gas flow per pound of fuel may be taken as the actual air per pound of fuel plus 1 lb of fuel and minus the ash in 1 lb of fuel [see Eq. (5–15)]. Thus, at 25% excess air and a 20% allowance,

$$\text{Capacity} = \frac{1.2 \times 1.25 \times 72.200(9.56 + 1 - 0.08)20.5}{60}$$

$$= 388,000 \text{ (say, 390,000) cfm}$$

Correct for the peak boiler load, take the furnace draft as 0.2-in. WG below atmospheric, and assume breeching losses to be 1.2-in. WG to get 8.9 $(1.10)^2 + 0.2 + 1.2 = 12.2$-in. WG. Then, with a 44% allowance, the design developed pressure would be $1.44 \times 12.2 = 17.6$-in. WG.

PROBLEMS

11–1. Make up a table of Rankine cycle efficiencies, like Table 11–1, for the following steam conditions: (a) 100 psia, 460 F, 14.7 psia; (b) 200 psia, 460 F, 14.7 psia; (c) 400 psia, 460 F, 14.7 psia; (d) 100 psia, 920 F, 14.7 psia; (e) 100 psia, 460 F, 1.5 in. Hg abs; (f) 100 psia, 460 F, 1 in. Hg abs; and (g) 100 psia, 460 F, ¾ in. Hg abs. Comment on the results and compare them with Table 11–1.

11–2 to 11–6. For each of the following turbines, calculate the turbine heat rate, the gross-station heat rate, the net-station heat rate, and the unit fuel cost (neglect the pressure drop through the reheater) in mills per kilowatt-hour.

Problem No.	11-2	11-3	11-4	11-5	11-6
Steam pressure, psig..........	1,450	2,000	1,250	850	600
Steam temperature, F.........	1,000	1,050	950	950	825
Feedwater temperature to boiler, F................	449	501	439	414	274
Steam flow to turbine, lb per hr.	650,000	1,023,000	842,000	578,000	113,000
Steam losses, lb per hr........	3,000	5,000	4,200	3,000	1,000
Boiler blowdown, lb per hr.....	0	1,000	2,000	2,000	500
Reheater flow, lb per hr.......	606,000	866,000	...	...	...
Reheater pressure, psig........	380	380	...	...	..
Temperature entering reheater, F.....................	700	670	...	...	...
Temperature leaving reheater, F	1,000	1,000	...	...	...
Generator output, kw.........	100,000	156,250	100,000	66,000	12,650
Power to auxiliaries, in percentage of generator output....	4.9	4.7	5.6	5.2	6.1
Steam generator efficiency, %...	89.2	90.1	88.9	88.0	86.2
Fuel cost, cents per 10^6 Btu....	23.0	28.5	19.8	27.3	17.8

11-7 to 11-11. Estimate the increased present worth of steam turbine generators for each pair of conditions listed. Assume the hours of operation per year to be all at rated load and that the feedwater temperature rise is optimum.

Problem No.	11-7	11-8	11-9	11-10	11-11
Capability, kw...................	12,650	66,000	80,000	44,000	33,000
Boiler efficiency, %..............	87	88.3	89.6	86	87.4
Exhaust pressure, in. Hg abs........	1	2.0	1.5	1.5	1.0
Electrical and mechanical losses, kw..	440	1,140	1,710	865	650
Fuel cost, cents per 10^6 B...........	18	28.2	24.0	36	31
Operation, hr per yr...............	4,000	3,500	6,500	5,000	4,500
Investment charges, %.............	10	14	12.5	15	5
Alternate "A"					
Pressure, psig...................	400	850	1,450	1,250	850
Temperature, F..................	750	900	1,000	950	900
Number of heaters...............	3	5	5	4	4
Exhaust enthalpy, B..............	986	1,017	995	998	1,002
Alternate "B"					
Pressure, psia...................	600	1,250	1,800	1,250	850
Temperature, F..................	825	950	1,050	950	900
Number of heaters...............	3	5	5	5	5
Exhaust enthalpy, B..............	990	1,010	990	998	1,002

11-12. Estimate the throttle flow for the turbines of Prob. 11-7.

11-13. Estimate the throttle flow for the turbines of Prob. 11-8.

11-14. Estimate the throttle flow for the turbines of Prob. 11-9.

11-15. Estimate the throttle flow for the turbines of Prob. 11-10.

11-16. Estimate the throttle flow for the turbines of Prob. 11-11.

Data for Probs. 11-17 to 11-28 inclusive: 65,000-kw unit operating at 850 psig, 900 F, and 1 in. Hg abs. Mechanical losses at full load are 1000 kw, and electrical losses are 2000 kw. Pressure drop through throttle valve and steam chest is 4% of initial pressure. Turbine is impulse reaction. Plot the condition curve for the turbine from the data of Fig. 11-10.

0 F terminal difference (TD) on open heater, 4 F TD on closed heater, 5% pressure drop on bleed lines, and 10 F TD on drain coolers. Find all steam and water flows, temperatures, pressures, enthalpies, and heat rate in Btu per kilowatt-hour for a load of 65,000 kw on the generator. Also find percentage of improvement over Rankine cycle.

11-17. Operation on Rankine cycle (Fig. 11-1).

11-18. One closed heater. Bleed point at 70 psia at turbine flange (Fig. 11-2).

11-19. One closed heater and drain cooler at same bleed pressure (Fig. 11-3).

11-20. One closed heater with drains pumped ahead and at same bleed pressure (Fig. 11-4).

11-21. One open heater with same bleed pressure (Fig. 11-5).

11-22. Two closed heaters with the drains cascaded to the condenser through a drain cooler. Bleed pressures at flanges are 120 psia and 140 psia.

11-23. Three closed heaters with the drains cascaded to the condenser through a drain cooler. Bleed pressures at flanges are 200 psia, 50 psia, and 8.0 psia. Assume that the auxiliaries use 5% of the generator output and that the boiler is 82% efficient. Also find the heat rate in Btu per net kilowatt-hour.

11-24. One closed heater and a drain cooler with bleed pressure at flange of (*a*) 6.0 psia and (*b*) 200 psia (Fig. 11-3). Use these data, together with Prob. 11-19, to draw a curve for one heater similar to Fig. 11-7(c).

11-25. One contact heater with flange pressure at 70 psia. Work by extraction factor method (Fig. 11-5).

11-26. Two closed heaters and a drain cooler. Flange bleed pressures are 120 psia and 14.0 psia. Work by extraction-factor method.

11-27. Same as Prob. 11-23 except with an evaporator receiving motivating steam from the highest-pressure extraction line and discharging vapor and drains to the middle heater. Evaporator receives 8000 lb per hr of 60 F water. Losses are from the steam line.

11-28. Same as Prob. 11-8 except at one-half load. Assume 7 F TD for drain cooler, 2 F TD for closed heaters, 3% pressure drop for bleed lines, 1000-kw mechanical losses, and 1700-kw electrical losses.

Data for condition curve:

145 psia, 1290.8 Btu per lb	3.4 psia, 1051.3 Btu per lb
60 psia, 1219.5 Btu per lb	1 in. Hg abs, 976.0 Btu per lb

Data for Probs. 11-29 to 11-38, inclusive: 50,000-kw unit operating at steam conditions of 1250 psig, 950 F, 1½ in. Hg abs with mechanical losses of 700 kw and electrical losses of 800 kw. The unit is assumed to be of the impulse-reaction type, hydrogen cooled, 3600 rpm; pressure drop through the throttle valve and steam chest is 4% of the initial pressure.

Data for the condition curve (including exhaust loss):

1250 psig, 950 F, 1468.0 Btu per lb	30 psia, 1159 Btu per lb
700 psia, 1406 Btu per lb	7 psia, 1084 Btu per lb
300 psia, 1322 Btu per lb	1½ in. Hg abs, 993 Btu per lb
110 psia, 1240 Btu per lb	

0 F TD for the open heater, 3 F TD for the surface heaters, 5% pressure drop for the bleed lines, and 10 F TD for the drain coolers. Find all steam and water flows, temperatures, pressures, enthalpies, and turbine heat rates. Also find the percentage of improvement in heat rate over that of the Rankine cycle.

11-29. Operation on the Rankine cycle (Fig. 11-1).

11-30. One closed heater. Bleed point at 90 psia at the turbine flange (Fig. 11-2).

11-31. One closed heater with drains cascaded to the condenser through a drain cooler for the same bleed pressure as Prob. 11-30 (Fig. 11-3).

11-32. One closed heater with drains pumped ahead for same bleed pressure as Prob. 11-30 (Fig. 11-4).

11-33. One contact heater for the same bleed pressure as Prob. 11-30 (Fig. 11-5).

11-34. Two closed heaters and a drain cooler with drains cascaded to the condenser. Extraction pressures at the turbine flanges are 220 psia and 26 psia.

11-35. Same as Prob. 11-34 except with an evaporator receiving motivating steam from the high-pressure extraction line, and the drains and the vapor from the evaporator cascading to the low-pressure heater. Evaporator receives 10,000 lb per hr of 60 F raw water. Losses consist of 6000 lb per hr saturated liquid, 1400 psia from the boiler blowdown connection, and 4000 lb per hr from the high-pressure steam line. Auxiliary power amounts to 5% of the generator output; the boiler efficiency is 83%. Find the plant net-heat rate for a throttle flow of 425,000 lb per hr.

11-36. Three closed heaters and a drain cooler with drains cascaded to the condenser. Extraction pressures at the flanges are 300 psia, 80 psia, and 12 psia. Assume that the auxiliary power is 6% of the generator output and that the boiler efficiency is 85%. Find the station net-heat rate.

11-37. One surface heater and a drain cooler with the drains cascaded to the condenser. Extraction pressure is 650 psia at the turbine flange.

11-38. Same as Prob. 11-37 except 7 psia at the turbine flange. Draw a curve using the percentage reduction in turbine heat rate as the ordinate and temperature rise of the feedwater as the abscissa for one feedwater heater in the cycle by using the data from heat balances of Probs. 11-31, 11-37, 11-38, and such other points as can be determined by deduction.

Data for Probs. 11-39 to 11-50, inclusive: 100,000-kw unit operating at 1450 psig, 1000 F, and 1.5 in. Hg abs. Mechanical and electrical losses are a total of 1710 kw at full load. The turbine is an impulse unit with a 4% pressure drop through the throttle valve and steam chest.

Data for the condition curve (including exhaust loss):

810 psia, 1439 Btu per lb	40 psia, 1166 Btu per lb
425 psia, 1367 Btu per lb	14 psia, 1101 Btu per lb
220 psia, 1302 Btu per lb	5 psia, 1044 Btu per lb
100 psia, 1234 Btu per lb	1.5 in. Hg abs, 993 Btu per lb

0 F TD for open heaters, 5 F TD for all closed heaters, 5% pressure drop on bleed lines, and 10 F TD on drain coolers. Find all steam and water flows, temperatures, enthalpies, and turbine heat rates in Btu per kilowatt-hour for a load of 100,000 kwhr on the generator. Also find the percentage improvement over the Rankine cycle (Fig. 11-1).

11-39. Operation on the Rankine cycle (Fig. 11-1).

11-40. One surface heater with an extraction point at 120 psia (Fig. 11-2).

11-41. One surface heater and drain cooler at same extraction pressure (Fig. 11-3).

11-42. One surface heater with drains pumped head and at same extraction pressure (Fig. 11-4).

11-43. One contact heater at same extraction pressure (Fig. 11-5).

11–44. Two surface heaters with the drains cascaded to the condenser through a drain cooler. Extraction pressures are 280 psia and 33 psia.

11–45. Three surface heaters with the drains cascaded to the condenser through a drain cooler. Extraction pressures are 425 psia, 103 psia, and 14.5 psia. Calculate the gross- and net-station heat rates by using a 90% boiler efficiency and an auxiliary power consumption of 4.8% of the generator output.

11–46. Same as Prob. 11–45 except that there is an evaporator producing 7000 lb per hr of vapor from 60 F water when receiving motivating steam from the highest-pressure extraction point and discharging drains and vapor to the middle heater. Take all the losses from the steam line from the boiler to the turbine.

11–47. Same as Prob. 11–45 except for a reheat cycle. Steam leaves the turbine at 425 psia and returns to the turbine at 385 psia.

Data for the condition curve (including exhaust loss):

1080 psia, 1464 Btu per lb	100 psia, 1353 Btu per lb
700 psia, 1414 Btu per lb	65 psia, 1320 Btu per lb
425 psia, 1360 Btu per lb	30 psia, 1250 Btu per lb
385 psia, 100 F	10 psia, 1168 Btu per lb
220 psia, 1450 Btu per lb	1.5 in. Hg abs, 1050 Btu per lb
140 psia, 1399 Btu per lb	

11–48. Same as Prob. 11–45 except at one-half load. Assume that the terminal differences and percentage pressure drops do not change. Combined mechanical and electrical losses are 1375 kw.

Data for the condition curve:

350 psia, 1392 Btu per lb	28 psia, 1171 Btu per lb
200 psia, 1334 Btu per lb	10 psia, 1107 Btu per lb
100 psia, 1268 Btu per lb	1.5 in. Hg abs, 988 Btu per lb

11–49. Same as Prob. 11–41 except at an extraction pressure of (*a*) 400 psia and (*b*) 10 psia. Using these data and that of Prob. 11–41, draw a curve for one heater similar to Fig. 11–8.

11–50. Work Prob. 11–43 by the extraction-factor method.

11–51 to 11–54. For each of the following problems, determine the extraction pressures if there is 5% pressure drop in the piping (S = surface heater, C = contact heater):

Problem No.	11–51	11–52	11–53	11–54
Steam press, psig.............	600	900	1450	1800
Condenser press, in. Hg abs....	2	1.0	1.5	1.5
Heater arrangement from high to low pressure.............	C,S,S	S,S,C,S	S,S,S,C,S,S	S,S,S,S,C,S,S
Terminal difference, F........	0,10,10	2,5,0,10	−3,0,3,0,5,5	−3,0,0,3,0,5,5

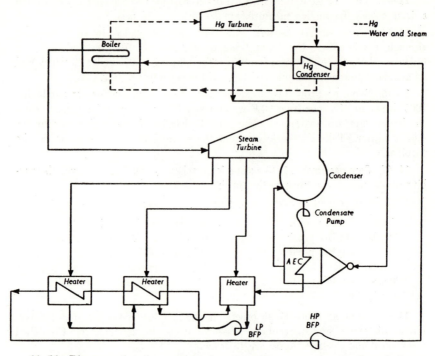

11–55. Binary cycle using water and mercury (see accompanying heat-balance diagram). Mercury turbine operates at 125 psig saturated and 3-in. back pressure. The steam turbine operates at 385 psig, 900 F and 1 in. Hg abs.

	Mercury Turbine	Steam Turbine
Internal efficiency	76%	85%
Over-all efficiency	72%	80%
Boiler efficiency = 86%		

Use straight-line condition curves with extraction points at 120 psia, 40 psia, and 7 psia. Assume 3 F TD on closed heaters, 0 F TD on contact heater. Neglect all pressure drops. If the mercury turbine generator output is 20,000 kw, find the output of the steam turbine generator and the plant net-heat rate with 4% of the combined outputs used by the auxiliaries. Use 700 lb per hr of steam for the air ejector and 1000 Btu per lb available for feed heating in the air-ejector condenser.

NOTE: Fig. 11–11 to Fig. 11–15 may be used to advantage as a basis for heat-balance problems. Sufficient data are presented throughout the text for the student to determine the performance of equipment for these problems.

BIBLIOGRAPHY

AIEE-ASME COMMITTEE ON STEAM TURBINE–GENERATORS. *Preferred Standards for Large 3600-RPM 3-Phase 60-Cycle Condensing Steam Turbine-Generators.* New York, 1953.

GIBSON, N. R., and CUSHING, H. M. "Advanced Design—Original Features Embodied in New 160,000 kw Oswego Steam Station," *Trans. ASME*, Vol. 64, No. 6, p. 541.

HACKETT, H. N. "Mercury for the Generation of Light, Heat, and Power," *Trans. ASME*, Vol. 64, No. 7, p. 647.

HACKETT, HAROLD N. "Mercury-Steam Power Plants," *Mechanical Engineering*, Vol. 73, No. 7 (1951), p. 559.

Power (December, 1938). McGraw-Hill Publishing Co., Inc., New York.

SELVEY, A. M., and KNOWLTON, P. H. "Theoretical Regenerative-Steam Cycle Heat Rates," *Trans. ASME*, Vol. 66, No. 6, p. 489.

SMITH, A. R., and THOMPSON, E. S. "The Mercury-Vapor Process," *Trans. ASME*, Vol. 64, No. 7, p. 625.

CHAPTER 12

GAS TURBINES

12–1. General. Commercial feasibility of extensive use of the gas turbine for electric power generation has yet to be proved. At the early part of 1956, 224 gas turbines were in service throughout the world; 101 of these were for power generation and 123 were industrial applications as mechanical drives. Only 21 were for power generation in the United States. Obviously, aircraft applications are not included in the above figures.

Gas-turbine ratings have ranged from 40 hp to 27,000 kw; present maximum size limitation is about 30,000 kw. Probably the most extensive industrial application of the gas turbine has been for pipeline pumping units. Other applications include supply of energy for chemical heating processes, supply of compressed air for processes, supply of process steam, and application to modes of transportation such a locomotives and ships.

When the gas turbine is used as a prime mover, applications may be divided into three classes: (1) as a means of increasing the capacity and decreasing the heat rate of steam generating plant, (2) as an independent source of electrical energy in direct competition with other prime movers, and (3) as a peak-load or back-up unit. Combinations of gas-turbine cycles with steam cycles depend upon the performance of the gas turbine and will be discussed later in this chapter.

As an independent source of electrical energy, the primary attribute of the gas turbine is its short delivery and installation time compared with that for steam equipment. The heat rate of a gas turbine generally is higher than the heat rate of either an equivalent steam turbine or diesel engine. However, the diesel usually has higher maintenance and lube oil cost than the gas turbine. Also of importance where cooling water is at a premium, is the low cooling water requirements of the gas turbine. A 5000-kw unit will require about 400 gpm of cooling water for both the oil cooler and generator air cooler.

All electrical systems have short periods of very high loads, as shown by the typical curves of Fig. 15–1, which are known as *peak loads* (see page 625). Because of the short duration of the peak loads, the fuel cost is unimportant. Therefore, most electrical systems use their old, inefficient, low-pressure units as peaking units.

The gas turbine has many attributes that make it ideal for peaking service because fuel cost is unimportant. A unit can be brought from dead cold to full load in 25 min or less. This is important in an emergency. Installation cost of a gas turbine is very low, $80 to $100 per kw for a peaking unit, due to the very small building-space requirements, simple foundations, and lack of piping and auxiliaries.

In addition, gas turbines may be operated by remote control and need little or no attendance while operating and no attendance when shut down. Remote control equipment costs about $2.00 per kw. They have relatively low maintenance costs.

A unit installed for back-up service is intended to provide electric power when some other piece of equipment is out of service. Thus, a gas turbine may be installed to provide power during an emergency such as the outage of a long transmission line to a remote location or when a hydro unit is inoperative during low-water periods. Units for back-up service must have low investment costs, quick starting, low operating labor cost, and low standby costs, the same characteristics as those of a peaking unit. Most of the current gas-turbine electric-generating units are for peaking or standby service; many of the schemes for combination steam-gas turbine installations will require development for future use.

The two important disadvantages of the gas turbine are its inability to use coal as a fuel or to use most heavy residual petroleums, such as Bunker C, without special processing to prevent high-temperature sodium-vanadium attack on the turbine blades and the combustion chamber. Natural gas is a suitable gas-turbine fuel when it is available in a reliable supply. Petroleum distillate fuels may also be used but are usually too expensive except as a back-up supply to a normally gas-burning unit. The life of a combustion chamber is about 2 yr when burning Bunker C and about 10 yr when burning natural gas. Much of the future of the gas turbine as an important utility prime mover depends on its adaptability to coal firing.

Nuclear reactors have been proposed as the heat addition element of the gas turbine. The reactor would replace the combustion chamber.

12-2. Gas-Turbine Cycles. The theoretical cycle for the gas turbine is the Brayton cycle * which is composed of isentropic compression, constant-pressure heat addition, isentropic expansion, and constant-pressure heat rejection. This is known as the *simple cycle* gas turbine.

As indicated in Fig. 12-1, air enters the compressor and is compressed, theoretically, at constant entropy to point *2'*. In the actual compressor the process is irreversible, producing an increase in entropy. Thus, the air leaves an actual compressor at *2* rather than at *2'*.

* See any standard text on thermodynamics.

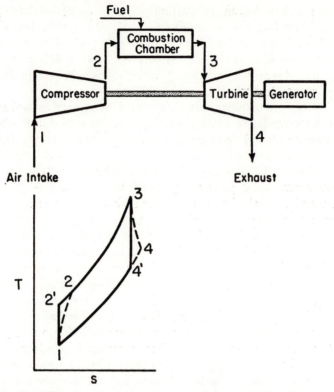

Fig. 12–1. Simple-cycle gas-turbine flow diagram and T-s diagram.

Heat addition takes place in the combustion chamber; theoretically from $2'$ to 3, but actually from 2 to 3. There is a small pressure drop in the combustion chamber that is not considered here.

Theoretical expansion through the turbine is isentropic from 3 to $4'$. As in a steam turbine, the actual expansion is accomplished at an increase in entropy. This is indicated as the process 3 to 4.

At this point it is well to distinguish between the *open* and the *closed* cycle. The flow diagram of Fig. 12–1 shows gases discharged to atmosphere from the turbine, and air entering the compressor from atmosphere. This is the open cycle. All commercial gas turbines in the United States are constructed on this cycle.

In the T-s diagram of Fig. 12–1, the point of turbine exhaust, point 4 (or $4'$) is connected to the point of air entrance, point 1, by a constant-pressure process, indicating that the gases leaving the turbine are cooled and then returned to the system. This is the closed cycle and requires an expensive heat exchanger between the turbine exhaust and the compressor intake and large amounts of air or water as the cooling medium.

Closed cycles have been proposed as a means of adopting the gas turbine to coal firing. In such an arrangement, energy would be added to the cycle in a heat exchanger between *2* and *3* and the erosive products of combustion from the coal would not enter the turbine. It appears more practical to rid the products of combustion of the erosive ash particles and use an open rather than a closed cycle.

Another advantage of the closed cycle, which can use gases other than air as the working medium, is that smaller passages are required for a given weight of gas flow, thus reducing machine size. Conversely, increasing gas density permits increasing gas flow, which in turn increases power output. In this way, it is possible to regulate the power output by regulating the pressure or density of the working fluid.

Both closed and open cycles have been proposed for nuclear gas-turbine plants.

Calculations of cycle performance will be based on the combined Mollier (*h-s*) and temperature-entropy diagram as reproduced in Appendix, Plate 8. More exact calculations should be based on the gas tables.* Using the nomenclature of Fig. 12–1, the following equations are obvious for the actual simple cycle for an air flow of 1 lb.

$$\text{Compressor work} = \frac{(h_2 - h_1)}{\eta_{mc}} = \frac{(h_{2'} - h_1)}{\eta_{mc}\eta_c} \qquad (12\text{--}1)$$

Note that for each pound of air flowing through the compressor, there is $(1 + W_f)$ pounds of gas flowing through the turbine.

$$\begin{aligned} \text{Turbine output} &= \eta_{mt}(h_3 - h_4)(1 + W_f) \\ &= \eta_t\eta_{mt}(h_3 - h_{4'})(1 + W_f) \end{aligned} \qquad (12\text{--}2)$$

$$\text{Cycle net output} = \eta_{mt}(h_3 - h_4)(1 + W_f) - \frac{(h_2 - h_1)}{\eta_{mc}} \qquad (12\text{--}3)$$

$$\text{Fuel burned} = W_f = \frac{h_3 - h_2}{\text{LHV}(\eta_b)} \qquad (12\text{--}4)$$

The weight of air required to produce a given output will be

$$W = \frac{\text{kw} \times 3413}{\eta_g\left[\eta_{mt}(h_3 - h_4)(1 + W_f) - \dfrac{(h_2 - h_1)}{\eta_{mc}}\right]} \qquad (12\text{--}5)$$

Cycle efficiency is of only academic interest, as the heat rate is more indicative of the unit performance. It is a custom of the industry to use the higher heating value of the fuel when determining the heat rate (HR). Thus,

$$\text{Heat rate} = \frac{W \times W_f \times \text{HHV}}{\text{kw}} \qquad (12\text{--}6)$$

* J. H. Keenan and J. Kaye, *Gas Tables* (New York: John Wiley & Sons, Inc., 1948).

where h = enthalpy of air or gas from Plate 8, Btu per lb

η_{mc} = compressor mechanical efficiency

η_c = compressor adiabatic efficiency

η_{mt} = turbine mechanical efficiency

η_t = turbine adiabatic efficiency, which is the same as RCR for a steam turbine

W_f = fuel/air ratio, lb fuel per lb air

η_b = burner efficiency

LHV = lower heating value of the fuel, Btu per lb

HHV = higher heating value of the fuel, Btu per lb

W = air flow, lb per hr

kw = generator output in kw

η_g = generator electrical and mechanical efficiency

and the subscripts are as defined in Fig. 12–1.

Lower heating values of heavy fuel oils (i.e., No. 6 Grade or Bunker C) range from 17,100 to 17,700 Btu per lb, while the lighter oils, such as the distillates, have lower heating values in the order of 18,500 Btu per lb.

The mechanical efficiencies of the compressor and turbine are about 99%, the burner efficiency is about 95% and includes heat losses and incomplete combustion, and the generator efficiency is about 98%. Turbine and compressor adiabatic efficiencies are in the order of 85%.

In the steam-power plant the consumption of electrical energy by the many auxiliaries represented an important part of the electrical output of the generators. The compressor of the gas-turbine plant consumes a large part of the turbine output; roughly about two-thirds of the turbine output. However, this energy is deducted from the turbine output, rather than from the generator output, and has been accounted for in Eq. (12–3) for the cycle output. The other auxiliaries in a gas-turbine plant (such as water pumps, electric lights, cooling-tower fans and pumps) consume a negligible amount of power. Therefore Eq. (12–6) may be taken as the net station heat rate.

EXAMPLE 12–1. Determine the air and fuel requirements and the heat rate for a 5000-kw, simple-cycle, gas turbine receiving 80 F air at 14.7 psia and having a compressor discharge and turbine inlet pressure of 85 psia. Turbine inlet temperature is 1400 F, the fuel LHV is 18,500 Btu per lb, and the HHV is 19,450 Btu per lb.

SOLUTION. From Plate 8, read the enthalpies as:

$$h_1 = 19.3 \text{ Btu per lb}$$
$$h_{2'} = 104.5 \text{ Btu per lb}$$
$$h_3 = 356.0 \text{ Btu per lb}$$
$$h_{4'} = 175.5 \text{ Btu per lb}$$

From Eq. (12–1),

$$h_2 = h_1 + \frac{h_{2'} - h_1}{\eta_c} = 19.3 + \frac{104.5 - 19.3}{0.85} = 119.5 \text{ Btu per lb}$$

which corresponds to a temperature t_2 of 494 F leaving the compressor. From Eq. (12–4),

$$W_f = \frac{356.0 - 119.5}{18,500(0.95)} = 0.01346 \text{ lb fuel per lb air}$$

Use Eq. (12–2) to obtain

$$h_4 = h_3 - \eta_t(h_3 - h_{4'}) = 356.0 - 0.85(356.0 - 175.5)$$
$$= 202.6 \text{ Btu per lb}$$

which corresponds to an exhaust temperature, t_4, of 822 F. Then, from Eq. (12–5),

$$W = \frac{5000 \times 3413}{0.98[0.99(356.0 - 202.6)(1 + 0.01346) - (119.5 - 19.3) \div 0.99]}$$
$$= 330,400 \text{ lb per hr}$$

The fuel input will be $W \times W_f = 330,400 \times 0.01346 = 4447$ lb per hr. Then the specific fuel consumption will be $4447 \div 5000 = 0.889$ lb per kwhr. Observe that the power input to the compressor is about 65% of the turbine output. From Eq. (12–6),

$$\text{HR} = \frac{4447 \times 19,450}{5000} = 17,290 \text{ Btu per kwhr}$$

The pound of air per pound of fuel would be the reciprocal of W_f, or 74.3 lb of air per lb of fuel. The theoretical air for this fuel would be in the order of 14.9 lb of air per lb of fuel, and the actual air would be about 498% of theoretical air, by using Eq. (5–8).

The addition of a regenerator to the gas-turbine cycle will improve the efficiency. A flow diagram and a T–s diagram for the regenerative gas-turbine cycle are shown in Fig. 12–2.

Exhaust gas passes through the regenerator, and heat is transferred from the exhaust gas to the compressed air. This increases the air temperature entering the combustion chamber from T_2 to T_6 and reduces the amount of fuel that must be added to the cycle. The temperature of the exhaust gases leaving the cycle is reduced from T_4 to T_5.

If perfect regeneration were possible, the exhaust gas would be cooled to the incoming air temperature or the effectiveness of the regenerator would be 100%, where

$$\epsilon = \frac{T_6 - T_2}{T_4 - T_2} \tag{12–7}$$

where ϵ = effectiveness
$\quad T$ = temperature, R or F
and the subscripts are as given in Fig. 12–2.

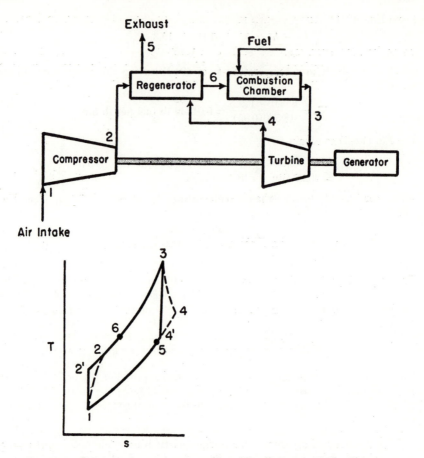

Fig. 12-2. Regenerative-cycle, gas-turbine flow diagram and T-s diagram.

In order to be economically justified, a regenerator must have an effectiveness of at least 50%, depending on the fuel cost and the hours of operation per year. Typical values of effectiveness would be from 75 to 80%. In some applications, such as for peaking service, the regenerator cannot be justified, regardless of the effectiveness value.

The regenerators are very similar to shell and tube feedwater heaters except for the metal thickness involved. The data of Chap. 4 can be used to calculate regenerator surface. The units usually have counterflow of the fluids.

When calculating a cycle that includes a regenerator, Eq. (12-4) should be changed by substituting h_6 for h_2, so that

$$W_f = \frac{h_3 - h_6}{\text{LHV}(\eta_b)} \tag{12-8}$$

where the symbols are as in Eq. (12-4) and Fig. 12-2.

EXAMPLE 12–2. Recalculate the preceding example to include a regenerator of 75% effectiveness.

SOLUTION. Using the data of the preceding example and Eq. (12–7),

$$T_6 = T_2 + \epsilon(T_4 - T_2) = 494 + 0.75(822 - 494)$$
$$= 740 \text{ F}$$
$$h_6 = 181.5 \text{ Btu per lb}$$

From Eq. (12–8),

$$W_f = \frac{356.0 - 181.5}{18,500(0.95)}$$
$$= 0.00993 \text{ lb per lb air}$$

The temperature of the exhaust gas leaving the regenerator, t_5, can be determined by an energy balance around the regenerator. From Eq. (12–5),

$$W = \frac{5000 \times 3413}{0.98[0.99(356.0 - 202.6)(1 + 0.00993) - (119.5 - 19.3) \div 0.99]}$$
$$= 333,800 \text{ lb per hr}$$

From Eq. (12–6),

$$\text{Heat rate} = \frac{333,800 \times 0.00993 \times 19,450}{5000} = 12,894 \text{ Btu per kwhr}$$

It is possible to add intercooling to the compressor by extracting the air part way through the compressor and passing the air through a cooler. Also, reheating the gases after partial expansion in the turbine is possible. Designs involving intercooling and reheating are being completed and probably will prove economical within a very few years.

Another variation of the gas-turbine cycle is known as the evaporative-regenerative cycle. Water is evaporated by the warm air leaving the compressor and before entering the regenerator, Fig. 12–3. The air leaving the drip-pan evaporator is saturated. The advantage of this cycle is that the mass flow through the regenerator and the turbine is increased without

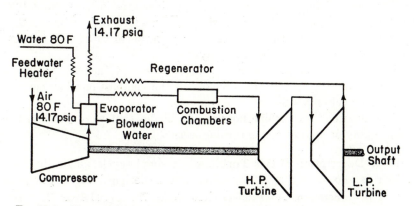

FIG. 12–3. Evaporative-regenerative-cycle gas turbine. (General Electric Co.)

increasing the compressor load. This increases the turbine output and also improves the heat transfer in the regenerator.

Although the combustion and regeneration processes have been taken as constant-pressure processes, it should be obvious that there is some loss of pressure as the air and gases flow through this equipment. The loss in pressure for the gases passing through a well-designed regenerator amounts to about 1 psi, while there is a pressure drop of about 3% of the absolute pressure for the air flowing through either a combustion chamber or a regenerator.

12–3. Performance. In order to show the relative effect of some variables on gas-turbine efficiency, it is convenient to assume the simple cycle and to assume that the perfect gas laws apply with constant specific heat. Then $\Delta h = c_p \, \Delta T$, and for a reversible adiabatic process,

$$\frac{T_A}{T_B} = \left(\frac{P_A}{P_B}\right)^{(k-1)/k}$$

The cycle thermal efficiency, using the nomenclature of Fig. 12–1, is

$$\eta = \frac{W(h_3 - h_{4'})\eta_T - W(h_{2'} - h_1)(1/\eta_c)}{W(h_3 - h_2)}$$

$$\eta = \frac{c_p(T_3 - T_{4'})\eta_T - c_p(T_{2'} - T_1)(1/\eta_c)}{c_p(T_3 - T_2)}$$

$$\eta^* = \frac{\eta_T T_3 [1 - (P_4/P_3)^{(k-1)/k}] - (T_1/\eta_c)[(P_2/P_1)^{(k-1)/k} - 1]}{T_3 - T_1\left[1 + \dfrac{(P_2/P_1)^{(k-1)/k} - 1}{\eta_c}\right]} \qquad (12\text{-}9)$$

From this equation the curves of Fig. 12–4 may be calculated. Although they are for a simple cycle, the shape, but not the magnitudes, of these curves will apply to the regenerative cycle.

Observe that for each turbine inlet temperature, there is an optimum pressure ratio. Thus, as new metals permit increases in the turbine inlet temperature, it may be expected that the pressure ratio will increase. Current designs of gas turbines use pressure ratios of 5½:1 and 6:1, but ratios as high as 9:1 have been employed.

Reducing the compressor inlet temperature will increase the cycle efficiency. Increasing the pressure loss in the combustion chamber will reduce the cycle efficiency.

The equations for turbine nozzles and blades that were developed in Chap. 9 may be applied to the turbine section.

12–4. Description of Gas Turbines. A simple-cycle, single-shaft gas turbine is shown in Fig. 12–5. This unit is designed for a generator

* See H. A. Sorensen, *Gas Turbines* (New York: The Ronald Press Co., 1951).

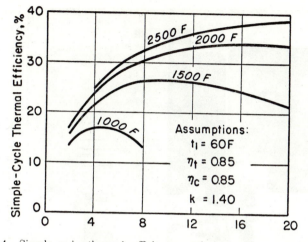

Fig. 12-4. Simple-cycle thermal efficiency variation with pressure ratio and turbine inlet pressure.

capacity of 21,800 kw. Although not shown, an electric motor is connected to the shaft at the left of the picture. Since a gas turbine must have an appreciable flow of gas under pressure entering the turbine impulse blades to provide sufficient power to drive the compressor without providing any excess power to the electric generator, the unit must have an independent drive to bring the compressor up to self-sustaining speed (approximately 60% of full-load speed). This independent drive is an electric motor, for power-plant applications, that ranges in size from 200 to 800 hp for single-shaft gas turbines of 5000 kw to 21,000 kw.

Air enters the 15-stage compressor at the left and is discharged into the annular space containing the combustors at the center of the unit at a pressure of about 90 psia and a temperature of about 500 F. The air enters the combustors through holes (not shown) in the left half of the combustor cylinder. Only about 20% of the oxygen in the air unites with the fuel.

Hot gases leave the combustors and enter the two-stage turbine at maximum full-load temperatures of 1450 F for oil fuel and 1500 F for natural gas fuel, the temperature for oil being lower because of the abrasive action due to the small amount of ash in the fuel.

The venturi-shaped turbine discharge permits an economic conversion of the kinetic energy of the exhaust gases into pressure energy. One manufacturer uses turbine speeds of 6900 rpm for 5000-kw units to 3290 rpm for 21,800-kw units. All units are connected to the generator through speed-reducing or speed-increasing gears except units rated at 13,200 kw and 16,500 kw; these units are direct-connected to the generator and operate at 3600 rpm.

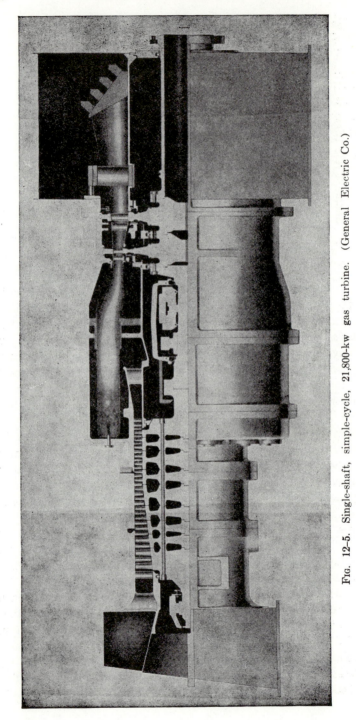

Fig. 12-5. Single-shaft, simple-cycle, 21,800-kw gas turbine. (General Electric Co.)

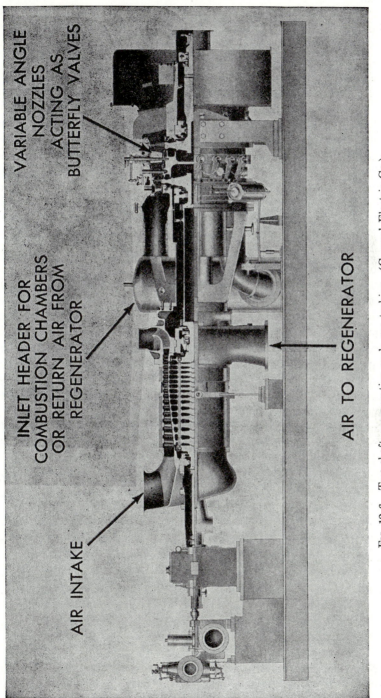

VARIABLE ANGLE NOZZLES ACTING AS BUTTERFLY VALVES

INLET HEADER FOR COMBUSTION CHAMBERS OR RETURN AIR FROM REGENERATOR

AIR INTAKE

AIR TO REGENERATOR

Fig. 12-6.   Two-shaft, regenerative-cycle gas turbine.   (General Electric Co.)

Air is extracted from the compressor to be used for cooling the mid-section bearing area and the base of the turbine wheels. This cooling air is discharged into the turbine exhaust to ensure a continuous flow of air.

Fig. 12–6 shows a gas turbine with two important features, either of which may be applied to a unit separately. First, note that the air leaves the compressor and does not go directly into the combustion chamber. After leaving the compressor, the air passes through a regenerator (which is not shown in the illustration). To conserve space, the regenerator may be located outside the building.

The other feature of this unit is that it has two shafts. The compressor and the first stage of the turbine are on one shaft, while the second stage of the turbine is on the other shaft which is connected to the load. The turbine blades are of the impulse design.

Observe that there are a series of butterfly valves, which constitute the second-stage nozzle diaphragm and which are operated from the external linkage and power piston and cylinder shown, to regulate the flow of gases through the unit.

With the single-shaft arrangement, only about 20% speed variation can be permitted because at lower speeds the compressor will not supply sufficient air to drive the unit. With the two-shaft arrangement, the compressor and the high-pressure stage can operate at a high enough speed to provide the necessary air, while the low-pressure turbine stage can operate at 35 to 100% of rated speed, as required by the driven equipment.

Since electric generators normally must operate at synchronous speed (variable-speed marine generators are an example of the exception), the more expensive two-shaft design is not justified for generator drive.

Two-shaft units have the added advantage of requiring a smaller starting motor than the single-shaft unit because the motor need not drive the second turbine stage.

Gas turbines have been proposed as drives for the boiler feed pumps of high-pressure steam plants. For this application, the two-shaft unit may be justified.

Notice in Fig. 12–6 that the unit is mounted much like a steam turbine in that the inboard end is supported by a heavy pedestal while the outboard end is supported by a flexible web. Also observe the equipment at the outboard end of the unit. This is a mechanical-drive expansion gas or steam turbine and is an alternate to the electric starting motor. The gas inlet and exhaust connections and the governing system are shown. Between the expansion turbine and the compressor is a reduction gear, oil pump, and control mechanism.

12–5. Gas Turbine Prime Mover Applications. Applications of the gas turbine to utility electric generation may be divided into four categories: (1) peaking power, (2) mechanical drives for auxiliaries, (3) supercharged boilers, and (4) exhaust heat recovery. To date, the first and fourth of these categories are the only ones that have been used commercially in the United States.

It has already been shown that the reduced first cost of the gas-turbine installation provides an important attribute for peaking applications and that the higher fuel costs are not prohibitive. Since the compressor power represents about two-thirds of the turbine output, anything that can be done to increase the flow through the turbine without increasing compressor power will be an advantage.

By injecting water into the hot gases before they enter the turbine (as distinguished from the evaporative-regenerative cycle), the turbine output can be increased about 40%, but the heat rate will be increased only about 23%. Since some of the fuel evaporates the water into steam and the steam leaves the turbine exhaust at a high enthalpy, the plant efficiency is reduced. The increased heat rate is not detrimental because of its short duration. Injection water must be demineralized and contain not more than 1 ppm of impurities to avoid turbine deposits.

Steam may be injected in place of water with similar results.

Peaking and standby units with water injection cost $80 to $100 per kw to install. However, operation at peak load with water injection is limited to about 1000 hr per yr because the exhaust temperature of the turbine is increased.

A 16,500-kw gas-turbine unit with regenerator would cost $130 to $150 per kw installed.

As mentioned previously, variable-speed, two-shaft gas turbines have been proposed for boiler feed pump drive. None has been installed to date.

Application of the gas turbine to supercharged boilers has been proposed for some time and has been referred to as the Velox cycle. The word "supercharged" is used rather than "pressurized" because in the latter case the gas in the furnace is at 1 to 3 psig, while in the former case it is at about 70 psig.

A flow diagram for a supercharged cycle is shown in Fig. 12–7 as applied to a 40,000-kw steam unit. The steam portion of the cycle is conventional except for the stack gas cooler which acts as a feedwater heater. Its purpose is to permit reduction of the stack gases to an economical value of 300 F. The conventional steam cycle would do this in the air preheater.

Air leaving the compressor is supplied to the boiler, where it unites with the fuel. After leaving the boiler at about 1450 F, the gases pass through the gas turbine and then through the economizer and stack gas cooler.

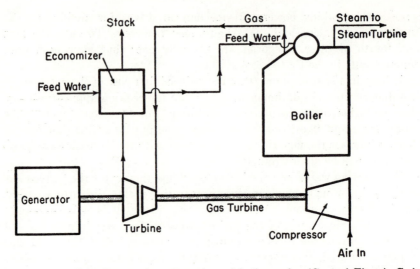

Fɪɢ. 12-7. Flow diagram for a supercharged boiler cycle. (General Electric Co.)

Because of the high gas density, heat-transfer rates in the boiler are increased, and it is claimed that the boiler weight is reduced by as much as 50%. Improvement in heat rate amounts to 7 to 8%. Also, station capacity is increased with only a minor increase in cooling water requirements.

A regenerator may be added to the cycle.

Use of a gas turbine in a combination steam-gas plant employing exhaust heat recovery is shown in Fig. 12-8. This plant differs from that of Fig. 12-7 in that the gas turbine has the conventional combustors and

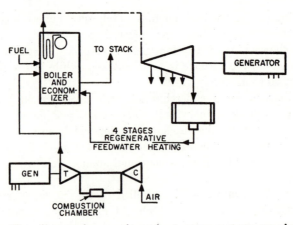

Fɪɢ. 12-8. Flow diagram for an exhaust heat-recovery steam-gas plant. (General Electric Co.)

that the gas passes through the gas turbine before entering the boiler, which is then pressurized rather than supercharged. Therefore, no forced- or induced-draft fans are needed. The heat rate is about $4\frac{1}{2}\%$ lower than for steam only. About 20 to 25% of the total fuel is fired in the gas tur- bine, and coal or any other fuel may be fired in the boiler without being detrimental to the gas turbine. As in the supercharged plant, the plant capacity is increased with only a small increase in cooling water.

Two-shaft gas turbines may be applied to both the exhaust heat recov- ery and the supercharged cycles to permit greater flexibility at partial loads.

PROBLEMS

12–1. Rework Example 12–1 to include a 3% pressure loss in the combustion chamber.

12–2. Rework Example 12–2 to include pressure drops in the combustion chamber and the regenerator.

12–3. Calculate the exhaust gas temperature leaving the regenerator of Ex- ample 12–2.

12–4. If $U = 5$ Btu per (hr)(sq ft)(deg F) and surface costs $3.00 per sq ft for a regenerator, what is the return on the investment for adding the regenerator to the turbine in Examples 12–1 and 12–2 if the turbine operates 5000 hr per yr at full load? Take the fuel cost as 45¢ per 10^6 Btu.

12–5. Sketch a T-s diagram to show intercooling and reheat for a gas turbine.

12–6. Neglect pressure losses and determine the economical investment in a regenerator for a gas turbine that is to operate 1000 hr per yr at a full load of 16,500 kw. Take the combustion efficiency as 94%, turbine and compressor in- ternal efficiency as 86%, pressure ratio as 5.5, fuel cost of 40¢ per 10^6 Btu, turbine inlet temperature as 1400 F, other efficiencies given in the text, and $12\frac{1}{2}\%$ in- vestment cost.

12–7. Calculate the thermal efficiency of a simple gas-turbine cycle from Eq. (12–8) for a pressure ratio of 8 and a turbine inlet temperature of 1600 F.

12–8. A steam cycle has a turbine heat rate of 10,500 Btu per kwhr at a steam turbine generator output of 60,000 kw when burning a Morgan County, Tennessee, coal with 25% excess air. A simple gas turbine is combined with the steam cycle, as shown in Fig. 12–7. Using 85% internal efficiency for the com- pressor and turbine, and other efficiencies as given in the text, calculate the out- put of the gas turbine generator and the combined turbine heat rate. The boiler efficiency is taken as 81%. Gas-turbine inlet temperature is 1500 F; the pressure ratio is 7.

12–9. Calculate a regenerative cycle using 87% internal turbine and com- pressor efficiency, pressure ratio of 6.5, 1400 F turbine inlet temperature, 75% regenerator effectiveness. Neglect pressure losses.

12-10. Calculate the reduction in thermal efficiency for a simple cycle at a pressure ratio of 6 and a turbine inlet temperature of 1700 F when air entering the compressor increases from 60 F to 100 F.

BIBLIOGRAPHY

Current Trends. No. 7. Milwaukee: Allis-Chalmers Manufacturing Co., 1942.

JENNINGS, BURGESS H., and ROGERS, WILLARD L. *Gas Turbine Analysis and Practice.* New York: McGraw-Hill Book Co., Inc., 1953.

MANN, JOHN W., JR. *Thermodynamic Performance and Design of Steam-Gas Turbine Power Plants.* The American Power Conference, 1956.

SORENSEN, HARRY A. *Gas Turbines.* New York: The Ronald Press Co., 1951.

VINCENT, E. T. *The Theory and Design of Gas Turbines and Jet Engines.* New York: McGraw-Hill Book Co., Inc., 1950.

WHITE, A. O. "The Place of the Gas Turbine in Electric Power Generation," *Trans. ASME,* 1956.

CHAPTER 13

DIESEL PLANTS

13–1. General. Although diesel engine-driven generators account for some 2% of the electrical energy generated in the United States, there are certain applications where their attributes make them superior to other prime movers. Some spark-ignition engines burning gasoline are used primarily for emergency service and usually in sizes not exceeding a few hundred horsepower. Spark-fired gas engines are available in large sizes, but the supply of natural gas for industrial purposes is seasonal in many sections of the United States. Therefore, the spark-ignition engines will not be given further consideration in this chapter.

Installation of a diesel engine plant can be accomplished much more quickly than a steam plant installation. Buildings and foundations are relatively simple, all the piping is for low-pressure service, and auxiliaries are not extensive. Since diesel engines are available in certain pre-established sizes for which the designs have been prepared, delivery time for the engines and generators is only a few months. A diesel engine can be started in a matter of minutes, and if the plant has been properly designed, it can start without any outside power or interconnections with other plants. This makes the unit ideal for standby emergency service.

Compared with the same size of steam unit, diesel engines have a lower installed cost per kilowatt-hour. In addition, they maintain their efficiency at fractional load, whereas the efficiency of a steam unit decreases rapidly below about three-quarter load. A diesel plant may require less operating personnel than a steam plant.

Among the disadvantages of the diesel unit, when compared with steam turbine plants, are the following: (1) inability to burn coal, (2) sizes limited to 7000 to 12,000 kw, (3) in many sections of the United States diesel fuel oil is more expensive per 10^6 Btu than other fuels, (4) lubricating oil costs are much higher for a diesel unit, (5) maintenance costs are higher for a diesel.

13–2. Engine Descriptions. Commercial diesel engines for electric-power generation may be classified in several ways: strokes per cycle, arrangement of cylinders, arrangement of pistons, type of fuel, and supercharging.

Reciprocating engines may operate on a two-stroke or a four-stroke cycle.* In the two-stroke cycle, there is a combination *expansion* and *exhaust* stroke and a combination *intake* and *compression* stroke to complete one revolution of the crankshaft. Many diesel engines operate on the two-stroke principle, as do many spark-ignition gas engines. By having an expansion stroke in each revolution, the power output of a given cylinder is increased over the output on a four-stroke cycle.

For each revolution of the crankshaft of a four-stroke-cycle engine, the piston moves through four strokes comprising intake, compression, expansion, and exhaust processes.

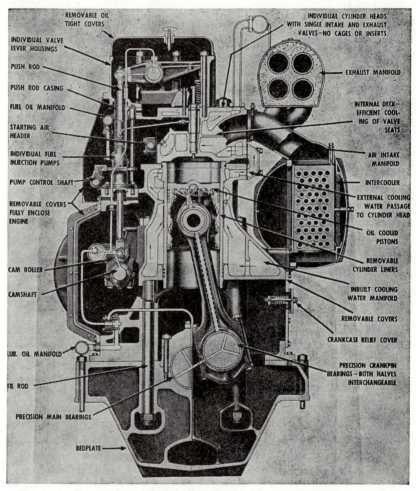

FIG. 13-1. Four-cycle, vertical, in-line diesel. (Nordberg Manufacturing Co.)

* See any standard text on thermodynamics.

Cylinder arrangement is of considerable importance to the plant designers because it has a bearing on the unit foundations, the building space, and maintenance problems. Perhaps the most common arrangement is vertical, in-line cylinders. Many diesel manufacturers have one series of engines in this category. A cross-section of a vertical, in-line engine is shown in Fig. 13–1.

When arranged in a V, Fig. 13–2, the engine is somewhat more compact. This particular unit operates on the four-cycle principle. Intake air and exhaust are controlled by valves (there are actually two of each) at the

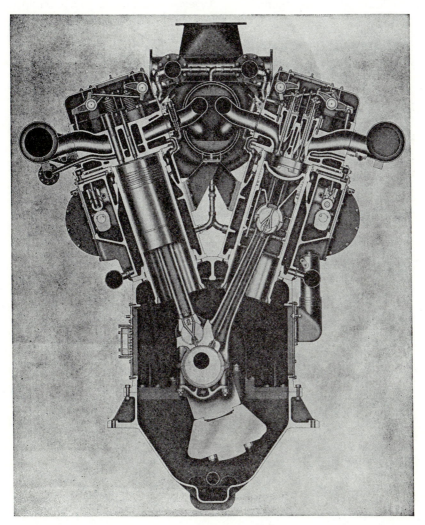

Fig. 13–2.　Four-cycle, V type of diesel engine. (The Cooper-Bessemer Corp.)

top of the cylinder. Although shown using a supercharger, the unit may also use atmospheric air.

Note that the renewable cylinder liners are of the wet type. That is, they are surrounded by the cooling water and form the cylinders; they are not liners within metal cylinders. The connecting rods are articulated, and the unit is of the *trunk type*. This indicates that there is no crosshead and that the connecting rod is attached directly to the piston wristpin.

Fig. 13–3 shows a cross-section of a two-cycle, radial, diesel engine. The advantage of this arrangement is that it requires a minimum of building space and foundation. An electric generator (or a pump or

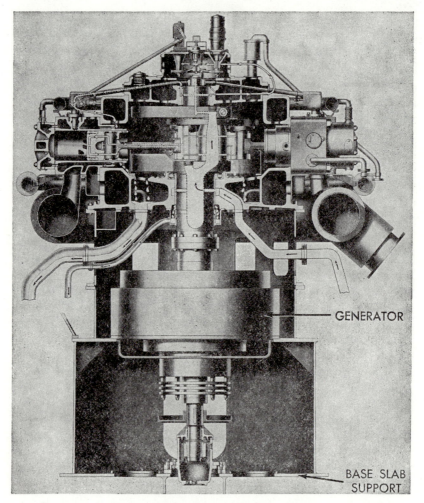

Fig. 13–3. Two-stroke radial diesel. (Nordberg Manufacturing Co.)

compressor) may be located below the engine, as shown. Of particular interest in this picture are the ports shown in the cylinder. As the piston nears the end of its expansion stroke it uncovers the exhaust ports. Further movement of the piston uncovers the intake ports and allows scavenging air to enter and sweep out the spent gases and provide air for the next cycle.

All two-cycle engines require scavenging air, which may also be used for a four-cycle engine. Scavenging air may be supplied from a separately driven compressor or, more commonly, by a centrifugal or piston type of compressor driven from the engine shaft. Most diesels for stationary plants employ a centrifugal compressor or supercharger to provide scavenging air. The Sun-Doxford marine engine is an excellent example of a unit with a piston compressor. One cylinder of the engine acts as an air compressor.

Most pistons for diesel engines are of the single-acting type wherein the working medium acts on only one face of the piston. Another arrangement that has proven to be successful is the *opposed piston* design, Fig. 13–4, in which the bottom ports are the exhaust ports, while the upper ports admit scavenging air. Because of the unidirectional flow of scavenging air, this design is sometimes known as a *uniflow* engine. The holes shown at the center of the cylinder are for fuel injection.

The lower pistons are connected to the main crankshaft, while those in the upper part of the cylinder are connected to the upper crankshaft. The blower is driven from the upper shaft, and the excess power from the upper shaft is transmitted to the main shaft by the bevel gears shown at the left end of the engine.

In this unit the crankshafts are out of phase so that the lower shaft leads the upper shaft by 12 deg. This causes the exhaust ports to open and close before the intake ports and permits a supercharging pressure of about 3.5 psig at the start of compression.

Opposed-piston marine engines of the Sun-Doxford design have 32-in. diam pistons with 55-in. and 40-in. strokes for a combined stroke of 95 in. One of these cylinders will develop 1500 hp at a speed of about 95 rpm.

Although the diesel engine is normally thought of as using oil as a fuel, oil is a high-cost fuel, per million Btu, in many sections of the United States. Where available, natural gas is usually a cheaper fuel. However, natural gas has become so popular as a fuel for home heating that nearly all the pipeline capacity is used for this purpose in the colder winter days. Thus, natural gas is not a fuel for power generation.

All the engines illustrated may be used for fuel oil only, fuel oil and natural gas in combination, or spark plugs may be added so that the engine can burn natural gas only. Other gases such as sewer gas, butane, etc., may be used.

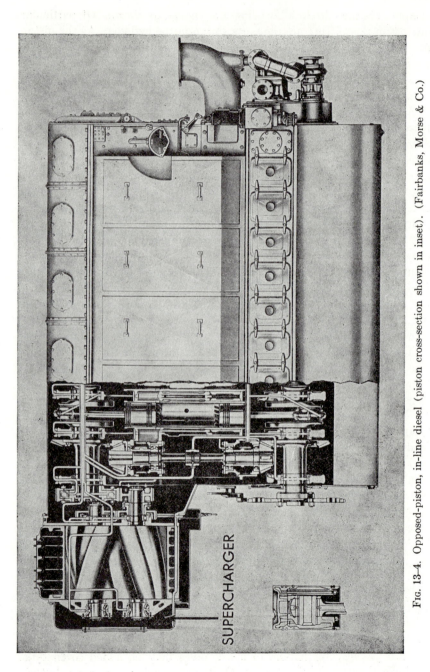

Fig. 13-4. Opposed-piston, in-line diesel (piston cross-section shown in inset). (Fairbanks, Morse & Co.)

SUPERCHARGER

The use of both oil and natural gas in the diesel is important economically and classifies the engine as a dual-fuel unit that may be either two-cycle or four-cycle. Fuel oil must be used in the diesel engine to cause ignition of natural gas. This oil is known as *pilot oil* and amounts to 4 to 6% of the full-load fuel-oil requirement. The governor systems are so designed that any proportion (except for pilot oil) of the total fuel may be gas. If the gas supply should fail, the system will automatically throw over to 100% oil supply.

Using Fig. 13–3 as an example, the gas is admitted to the cylinder through the valve shown in the top center of the cylinder wall. Gas admission takes place just after compression is started. Pilot oil is in-

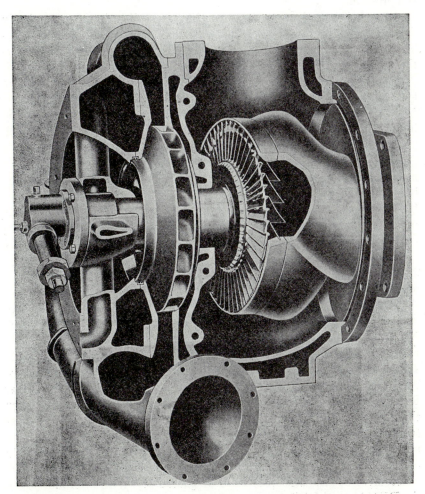

FIG. 13–5. Turbine-driven supercharger. (Nordberg Manufacturing Co.)

jected at the end of the compression stroke and is ignited by the high temperature of the air in the cylinder. The pilot oil ignites the gas.

As noted previously, scavenging air is required by all two-cycle engines. This scavenging air may or may not supercharge the cylinder. Supercharging is defined as the practice of supplying air to the cylinder under pressure; the air must be under pressure greater than atmospheric at the start of the compression stroke. Increased air pressure provides a greater weight of air in the same cylinder volume and therefore the ability to burn more fuel and produce more power.

Thus, for scavenging air to supercharge an engine, the air pressure in the cylinder must be maintained above atmospheric until the start of compression.

Supercharging of an engine may be accomplished by mechanically driven blowers or by turbine-driven blowers. In Fig. 13–4 is shown a lobular or Roots type of supercharger direct-driven from the engine crankshaft. The spiral lobes form a positive displacement blower.

A turbine-driven centrifugal blower is shown in Fig. 13–5. This is the type of supercharger that is used on the diesel of Fig. 13–1. Exhaust gases from the engine enter at the right and pass through the gas turbine, which in turn drives the blower. Notice that the impeller of the blower is much the same as the impeller of a centrifugal pump.

Since compression of any gas increases its temperature, a water-cooled intercooler may be added to reduce the air temperature leaving the blower discharge. This permits a further increase in the air capacity of the engine cylinders. An intercooler is shown in Fig. 13–1. Discharge pressures from the compressors may amount to 2 or 3 psig for scavenging purposes only, or 5 to 15 psig for supercharging.

13–3. Engine Performance. Diesel-engine performance guarantees are made on the basis of a fuel-oil higher heating value of 19,350 Btu per lb, 90 F air intake, and barometric pressure surrounding the engine of 28.25 in. Hg, which corresponds to an elevation of 1500 ft above sea level. No corrections are made for lower air temperature or higher barometer, but corrections are employed for higher air temperatures or elevations. It should be noted that the barometric pressure is that surrounding the engine and not the pressure corrected to sea level as furnished in Weather Bureau reports. There is a large variation in the difference between the higher and lower heating value of various gaseous fuels; it is customary to use their lower heating value. Performance guarantees are made for one-half, three-quarters, and full load, but the engine should be capable of developing a 10% overload continuously for 2 hr of each consecutive 24 hr. This procedure of guarantees is in accordance with the DEMA standards.*

Performance guarantees for oil-burning diesel engines are based on specific fuel consumption, but for gas-burning and dual-fuel diesel engines, they are based on the heat rate.

$$HR = \frac{W_0 HHV_0 + W_g LHV_g}{kw} \qquad (13\text{--}1)$$

where HR = heat rate, Btu per kwhr
   W_0 = weight of oil, lb per hr
   HHV_0 = higher heating value of the oil, Btu per lb
   W_g = weight of gas, lb per hr
   LHV_g = lower heating value of the gas, Btu per lb
   kw = generator net output

Fuel consumption guarantees for diesels are not made below one-half load because small changes in friction and in cooling-water temperatures may give erratic results at one-quarter load or less. Also, it has been found that the fuel injection valves of many diesels do not properly atomize the oil at low load. Of course this will have an important effect on combustion and on fuel consumption. Some operators avoid loads of less than one-third to one-quarter of rating wherever possible because of the poor combustion conditions.

Observe that diesel engines are not guaranteed to supply a continuous overload, whereas many steam turbine units will carry five-fourths load continuously. Because of the diesel's lack of overload ability, electric generators for diesels should be rated at their maximum continuous load.

An important advantage of the diesel engine is its ability to maintain its high efficiency at partial load. An example of the performance of a diesel under standard guarantee conditions is shown in Fig. 13–6. Note that there is only a slight increase in specific fuel consumption from full to three-quarter load. A Willans line may be plotted for the fuel consumption in pounds per hour versus load (see Art. 9–14). For many diesels this will be a straight line from about one-third to full load.

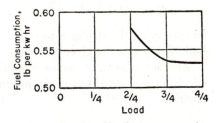

Fig. 13–6. Specific fuel consumption of a 600-rpm, 8-cylinder, 12 × 15-in., supercharged, 4-cycle, 1200-hp diesel engine with a 900-kw generator.

For approximate mental calculations it may be considered that the energy input to the engine by the fuel is divided equally between the useful power output, the losses in the cooling system, and the energy in the

* *Standard Practices for Low and Medium Speed Stationary Diesel Engines,* published by the Diesel Engine Manufacturers Association, 1951.

exhaust. This rule implies a thermal efficiency of 33⅓%, which is within the normal range for a diesel engine at full load.

Care must be taken in specifying the output of internal combustion engines. The purchaser is interested in the output of the engine after deducting for the power consumption of all connected pumps and blowers and with all necessary equipment such as silencers, manifolds, mufflers, and air filters in operation. Manufacturers occasionally rate their engines without these accessories attached, i.e., for a "stripped" engine.

Large diesel engines intended for the continuous service and long life of a stationary plant are usually slow-speed units. Thus, units ranging from 2000 or 3000 hp up to the largest, operate at 189 or 240 rpm and have piston speeds of about 1250 fpm. In the range of 1000 up to 3500 hp, speeds of 400 to 600 rpm are common, and piston speeds range from 1200 to 1600 fpm. Smaller units have speeds up to 900 rpm and possibly 1200 rpm. Diesels for transportation service are characteristically in the higher speed brackets because weight is an important factor and because they seldom operate for long periods at full load. Piston speed is defined as

$$PS = \frac{2L'N}{12} \qquad (13\text{-}2)$$

where PS = piston speed, fpm
$\quad L'$ = stroke, in.
$\quad N$ = speed, rpm

In specifying the size of a piston, it is always written *bore × stroke* and both dimensions are given in inches.

Another important factor in engine performance is the *brake mean effective pressure* (bmep). As indicated in Chap. 10, for a single-acting engine,

$$bhp = \frac{PLANn}{33,000} \qquad (13\text{-}3)$$

and

$$\text{Engine constant} = \frac{LAn}{33,000} \qquad (13\text{-}4)$$

where bhp = coupling, or brake horsepower
$\quad P$ = brake mean effective pressure, psi
$\quad L$ = stroke, ft
$\quad A$ = piston area, sq in.
$\quad N$ = power strokes per min for each cylinder
$\quad n$ = number of cylinders

Conservative values of bmep for nonsupercharged heavy-duty engines range up to about 85 psi, although higher values may be obtained. There has been a general tendency to increase bmep, since that would provide a means of reducing the unit cost of the engine. Low-pressure supercharging, using air pressures of 3 to 5 psig, was the first step in increasing bmep. Intercooling permits an additional increase in bmep, and high-pressure supercharging (about 15-psig supercharger discharge pressure) with intercooling permits a maximum bmep of some 160 psi. As an example, one manufacturer quotes an 8-cylinder, 360-rpm, 4-cycle, 13 × 20 engine for 80.5-psi bmep nonsupercharged, 121-psi bmep supercharged, 133-psi bmep supercharged with intercooling, and 161-psi bmep for high-pressure supercharging with intercooling. Of course, as the bmep is increased for a given bore and stroke, all bearings and other parts must be built to withstand the increased forces.

Diesel engines operate on compression ratios of 12:1 to about 18:1. Compression ratio is the ratio of the maximum cylinder volume divided by the minimum cylinder volume.* As the compression is increased for any internal-combustion engine (spark ignition or compression ignition), the efficiency of the cycle increases and the temperature of the gaseous mixture in the cylinder increases. Of course, for any given compression ratio, the mixture temperature at the end of compression is also affected by the temperature of the cylinder walls and by the temperature of the mixture at the beginning of compression. At the lower compression ratios, the temperature of the cylinder contents at the end of compression may not be sufficiently high to ignite the oil when the engine is cold. Generally, "hot starting" engines have compression ratios of 12:1 to 14:1 and require some accessory to assist the starting process. Engines above 14:1 compression ratio are "cold starting" engines and will not need auxiliary means to fire the first few charges when the engine starts at normal room temperatures.

Spark-ignition gas engines operate at compression ratios up to 12:1. At that value, their efficiency may exceed many compression ignition (diesel) engines of higher compression ratios because of the inherent better efficiency of the spark-ignition cycle.

If an engine is in a cold location and must start automatically, it is well to provide some means of ensuring immediate firing of the fuel. Several means are used successfully, such as an electric heater in the jacket water system (immersion heater), an electric heater in the air intake manifold, or a gas or gasoline heater in the intake manifold.

13–4. Auxiliary Equipment. (a) *Governing.* Speed regulation for a diesel is defined the same as for any other prime mover; Eq. (9–35) applies. While the curves of speed versus load are not straight lines for any prime mover, they will be so considered for the purposes of problem solutions.

---

* See any standard thermodynamics text.

Modern diesel engines of all but the very small sizes can be equipped with either nonisochronous or isochronous governors. An isochronous governor is one that has, or can be adjusted to have, a zero speed droop or zero speed regulation. A nonisochronous governor may be either a mechanical governor in which the power to regulate the fuel flow is obtained from flyweights, or it may be a relay type employing a hydraulic or electric (including electronic) system. Isochronous governors are of the relay type, and either mechanical or relay governors may be supplied with a means for changing either the engine speed or the characteristic.

When applied to engines that are intended for variable-speed service, governors with speed adjustment down to one-half speed or less are available. Where voltage or frequency is not important, such as for machinery drives or single generator industrial applications, 6% regulation is acceptable.

Isochronous governors are usually supplied for diesels that will be paralleled or that are prime movers for an electric distribution system. In that event, one of two diesels may be operated with an extremely small regulation (1% or less) and the other at a higher regulation (say, 3%). The unit with the small regulation will take the load swings, while the one with the higher regulation will be "base loaded."

All diesel engines should be supplied with emergency overspeed governors to stop the unit when it reaches an excessive speed, say, 10% overspeed.

(b) *Air Intake System.* Small engines located in a relatively large room may take their combustion air from within the room. Larger engines may also have their air intakes within the building when the engine is used at extremely infrequent intervals and for short periods; the windows are opened when the engine is operating so that the room rapidly cools to outside temperature.

Since diesels take from 2 to 6 cfm of air per rated horsepower (depending on the amount of scavenging air, etc.), it is customary to draw combustion air from outside the building. The location of the intake should be selected to provide as short and straight an inlet pipe to the engine as possible, yet it should not be in a location where it will receive ground dust, snow, or hot air from an exhaust or cooling system (cooling tower, etc.). The intake system should include combination intake muffler and air filter if located in the building, or an air filter, pipe, flexible tubing, and possibly a muffler if outside air is used.

(c) *Exhaust System.* Masonry exhaust systems should not be used because they cannot withstand thermal expansion. The exhaust pipe from the engine should have at least one, and frequently two, flexible-tubing sections to provide for expansion and to isolate the system from engine vibration. Obviously, the exhaust pipe should be as short and have as few bends as possible. A separate exhaust system should be used for each engine.

Most exhaust systems include an exhaust silencer, frequently of the spark-arresting type. In some instances, mufflers (from the standpoint of noise) are not necessary for turbo-supercharged engines, since the turbine will provide sufficient muffling effect.

Many schemes have been tried to conserve some of the sensible energy in the exhaust gases. In merchant marine applications, it is common practice to provide *waste-heat boilers* which utilize the engine exhaust to provide steam for many uses on the ship. No attempt should be made to lower the exhaust gases below the sulfur trioxide dewpoint, probably 250 F to 300 F, as mentioned for steam boilers. Thus, only about two-thirds the energy of the exhaust gas is actually available for waste-heat purposes.

Two common applications of exhaust heat recovery in electric generating plants are to provide warm water for various purposes within the plant and to heat ventilating air for the plant. In the first instance, a water coil may be placed in the exhaust muffler. To obtain warm air, the exhaust pipe may be jacketed for warming the air, or the air may be passed over the muffler.

(d) *Starting Systems.* Small diesels may be started by electric starters driven by batteries such as used for an automobile engine. Electric motors, air motors, gasoline engines, and even blank gun shells have been used on the smaller units. Air motors may be used on medium-sized engines.

Compressed air injected into one or more cylinders of the engine is the most common method of starting low- and medium-speed diesels of several hundred or more horsepower rating. Air at 250 psig, or up to 350 psig, is used. Air is supplied to the storage tanks by a motor-driven compressor that automatically maintains the proper pressure in the air tanks. Frequently, the compressor is arranged for dual drive by either a gasoline engine or a motor so that starting of the diesel is not dependent on electricity from an outside source. Storage tanks may be typical compressed-air tanks, but for large units a battery of welders' oxygen bottles has proved to be most satisfactory. The storage capacity of the tanks or bottles may be sufficient to start the largest diesel in the plant six times without recharging the tanks. As an example of the amount of starting air required, a 16-cu ft tank at 250 psig should start a 1200-hp diesel seven times without recharging the tank.

(e) *Cooling-Water Systems.* It has previously been indicated that roughly one-third the energy of the fuel input is rejected in the cooling-water system. This represents, for the medium- and large-sized engines, a sizeable amount of energy to be dissipated by means of cylinder-jacket water cooling, exhaust-manifold-jacket water cooling, and lubricating-oil cooling. Care must be taken to assure that the water will not be so cool in winter that the water will freeze or the lube oil will congeal. Water circulated through engine and manifold jackets should be treated to reduce corrosion and scale formations.

Cooling water systems may be divided into two types, *open system* and *closed system*. In the open system, cooling water from the jackets and lube-oil cooler is discharged to waste or to a cooling tower or spray pond. If wasted, the make-up jacket water must be continually treated to reduce corrosion and scale. If a cooling tower or spray pond is used, continual evaporation of some of the water will increase the concentration of impurities. For these reasons, the open system is not recommended.

Jacket and lube-oil cooling water is continually recirculated through the system and is treated to reduce scale formation and corrosion. The treating system may provide for zeolite softeners. This treated water is passed through a heat exchanger in one of three arrangements: (1) an air-cooled radiator, (2) an evaporative cooler, (3) or a cooling tower. In any event the closed, treated-water system must contain a surge tank to provide for water expansion, venting, and a place to introduce make-up water.

Radiators may represent a high first cost; the fans may require appreciable power. The radiators should be located so that air will not recircu-

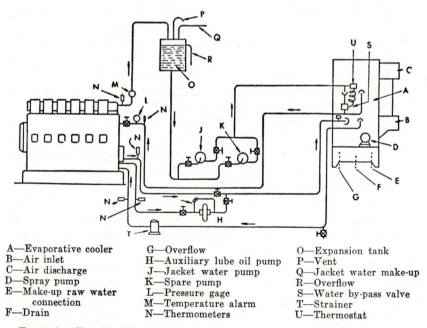

A—Evaporative cooler	G—Overflow	O—Expansion tank
B—Air inlet	H—Auxiliary lube oil pump	P—Vent
C—Air discharge	J—Jacket water pump	Q—Jacket water make-up
D—Spray pump	K—Spare pump	R—Overflow
E—Make-up raw water connection	L—Pressure gage	S—Water by-pass valve
	M—Temperature alarm	T—Strainer
F—Drain	N—Thermometers	U—Thermostat

Fɪɢ. 13–7. Closed cooling system with evaporative cooler. (Diesel Engine Manufacturers Assoc., *Standard Practices.*)

late through them or into the engine intake and so that the natural wind velocity will not retard air flow through the radiators. Hot air from the radiators can be used in the winter for building heating. Radiator performance is vitally affected by the air temperature.

Evaporative coolers, Fig. 13–7, are compact and therefore may be located inside the building. A pump sprays raw water over the cooling coils (containing jacket water), and an induced-draft fan in the top of the cooler pulls air through the water spray. Heat is transferred from the jacket water inside the coils to the raw water on the outside, which in turn is cooled by partial evaporation into the air. Extreme care should be taken to assure that moist air from the evaporative cooler does not enter the engine intake where moisture or ice accumulation would be injurious.

A cooling tower to remove heat from the heat-exchanger raw water is shown in Fig. 13–8, in which the heat exchangers are shown located at the cooling tower. In order to facilitate winter repair and to prevent freezing, the heat exchangers may be located inside the building in cold climates. Again, moist air from the tower should not enter the engine intake.

For medium and large engines, water pumps should be motor-driven, not driven by the engine shaft.

(f) *Lubricating-Oil Systems.* Lubricating-oil consumption is one of the important elements in diesel power costs. Medium- and large-sized engines will consume 0.0005 to 0.0009 gal of lube oil per kwhr. Therefore, the lube-oil system is worthy of the designer's careful attention.

Lube-oil storage capacity should be sufficient to store reclaimed oil and to store the plant consumption of new oil between deliveries.

The main lube-oil pumps may be driven from the engine shaft; they are usually gear-type pumps, Fig. 13–9. An auxiliary motor-driven oil pump should be supplied for emergency purposes. This pump may also be used as a *before-and-after* pump, i.e., to supply oil to the engine before starting and after shut-down.

An oil system should always include a strainer to remove the larger impurities and to accommodate the full oil flow of the engine. Also, a filter or purifier for further purifying the lube oil should be provided. The filter will remove smaller particles than will the strainer; it usually accommodates only a part of the oil pump flow, i.e., the part which is by-passed back to the sump. These filters are made of unbleached cotton waste, cellulose, clay, fuller's earth, etc. Fuller's earth may react adversely with some of the additives in the oil.

For medium and large units that are not just emergency units, reclaiming the lube oil by distillation and filtration through fuller's earth or similar synthetic materials, or by centrifuging, may be economical. The reclaiming systems may operate continually or on a "batch" basis.

A lube-oil meter should be included in systems for the larger units.

(g) *Fuel-Oil Systems.* When delivered by railroad tank car, there

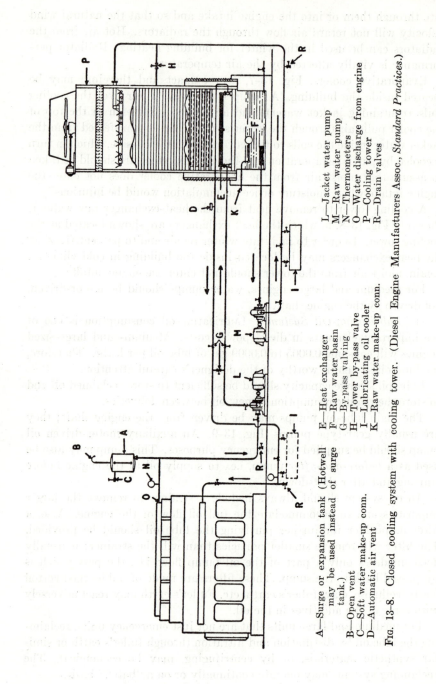

A—Surge or expansion tank (Hotwell may be used instead of surge tank.)
B—Open vent
C—Soft water make-up conn.
D—Automatic air vent
E—Heat exchanger
F—Raw water basin
G—By-pass valving
H—Tower by-pass valve
I—Lubricating oil cooler
K—Raw water make-up conn.
L—Jacket water pump
M—Raw water pump
N—Thermometers
O—Water discharge from engine
P—Cooling tower
R—Drain valves

Fig. 13–8. Closed cooling system with cooling tower. (Diesel Engine Manufacturers Assoc., *Standard Practices*.)

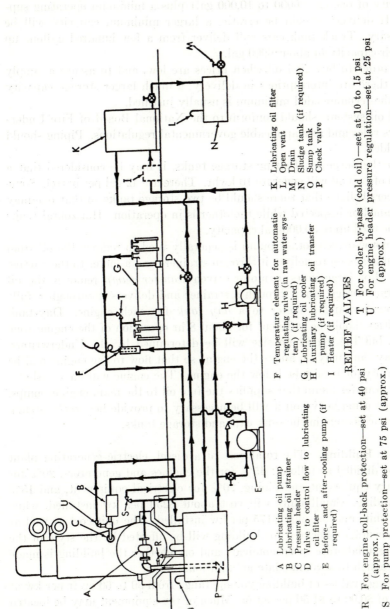

A Lubricating oil pump
B Lubricating oil strainer
C Pressure header
D Valve to control flow to lubricating
    oil filter
E Before- and after-cooling pump (if
    required)

F Temperature element for automatic
    regulating valve (in raw water sys-
    tem) (if required)
G Lubricating oil cooler
H Auxiliary lubricating oil transfer
    pump (if required)
I Heater (if required)

K Lubricating oil filter
L Open vent
M Drain
N Sludge tank (if required)
O Sump tank
P Check valve

RELIEF VALVES

R For engine roll-back protection—set at 40 psi
    (approx.)
S For pump protection—set at 75 psi (approx.)

T For cooler by-pass (cold oil)—set at 10 to 15 psi
U For engine header pressure regulation—set at 25 psi
    (approx.)

FIG. 13-9. Lubricating oil system. (Diesel Engine Manufacturers Assoc., *Standard Practices*.)

must be on hand at least sufficient storage capacity to accommodate the capacity of one car (6000 to 10,000 gal) plus a minimum operating supply. If delivery should be erratic, a larger minimum capacity will be necessary. Truck tank cars will deliver from a few hundred gallons up to their capacity of about 6000 gal.

In order to buy fuel oil when prices are low, and to assure a supply when there are interruptions in delivery, a much larger storage capacity than the recommended minimum is usually justified.

All oil systems should conform to the National Board of Fire Underwriters rules and any applicable governmental regulations. Piping should be welded.

For the purposes of sizing storage tanks, it may be considered that a gallon of fuel oil will generate 10 kwhr. There are 42 gal per barrel. Some engineers believe that there should be two storage tanks so that one may be cleaned or inspected while the other is in operation. Horizontal tanks may be used up to 30,000-gal capacity.

A motor-driven rotary pump is generally used to remove the oil from the tank car and to deliver it through strainers and a meter to the storage tanks. For large units, a motor-driven *transfer pump* removes the oil from the storage tanks through a strainer and delivers it through a full-flow purifier and a meter to a *day tank* for each engine. Day-tank capacities should be sufficient for about 8 hr operation of the engine (one shift), but the maximum size will be determined by the Underwriters. The day tank may be above the engine so that flow to the engine will be by gravity, or it may be below the engine. The engine will have a shaft-driven transfer pump that supplies the fuel oil to the main engine pumps.

When using heavy oil it will be necessary to provide hot water, steam, or electric heaters in the tank cars and storage tanks.

13–5. Building. The total cost of a diesel, electric-generating plant may be divided as follows: 45% for the engines and generators, 20% for auxiliary equipment and piping, 20% for electrical equipment, and 15% for building. Small units will cost about $250 per kw installed, while larger units may be only $175 per kw installed. The unit cost as well as the percentage allocated for building will be affected by the style of the building (elaborate or economical) and on whether the building is made large enough to accommodate an additional unit.

Average values of building volume range from 30 to 60 cu ft per kw at a cost of $0.90 to $1.50 per cu ft. Much of the equipment may be located either inside or outside the building. This, together with allowances for shop, electrical equipment, and office space, will have an important bearing on the size of the building.

Basements may not be necessary for stations with units of less than 1000 kw. Fig. 13–10, but it will be found advantageous to use basements

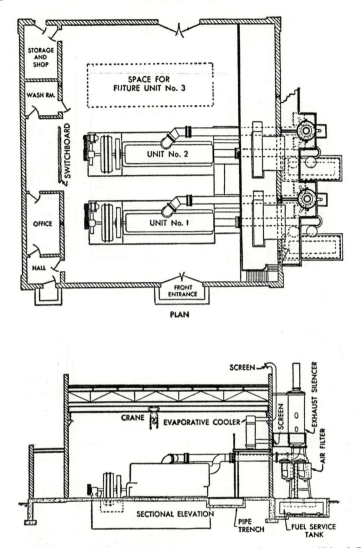

FIG. 13-10. Diesel power-plant arrangement without basement. (Diesel Engine Manufacturers Assoc., *Standard Practices*.)

for larger units, Fig. 13–11. Conventional arrangement of diesel engines is with parallel centerlines. Space should be provided for plant growth, either by allowing space for additional units or by installing a temporary wall at one end of the engine room. Metal and transite are common materials for temporary walls. Sufficient areas should be provided for working space or "lay down" space around the engine during maintenance periods. Heat exchangers within the building must be provided with space to pull the tubes.

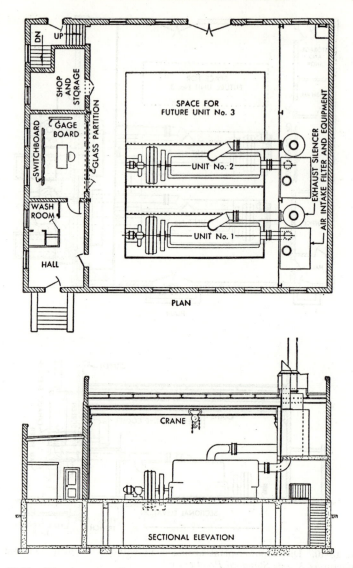

PLAN

SECTIONAL ELEVATION

Fig. 13–11. Diesel power-plant arrangement with basement. (Diesel Engine Manufacturers Assoc., *Standard Practices*.)

Note the locations of the intake and exhaust equipment in the illustrations. Some engineers prefer to locate the silencers and filters externally to the building to reduce vibration of the main structure.

Plants using natural gas may require a separate building for gas surge tanks and metering equipment. If gas pressure is inadequate, gas compressors may be necessary.

Main entrances to the building should be located so as to require visitors to pass the office before going into the engine room.

Ventilation within the plant should be adequate to remove odors emanating from the engine and auxiliaries. One complete air change every 2 min is not excessive.

PROBLEMS

13–1. A diesel plant contains three 2500-kw units that generate the equivalent of full capacity for 2000 hr per yr each. Fuel oil will be delivered monthly, but the storage capacity of the tanks should be sufficient to provide for the plant if one delivery is missed. How many tanks and of what size would you recommend?

13–2. Same as Prob. 13–1 except for two 500-kw units at 1500 hr per yr.

13–3. One diesel unit (A) in a plant has a rating of 3500 kw, and the other (B) has a rating of 1000 kw. The governor of unit A is set at $2\frac{1}{2}\%$ regulation at 3000 kw and 60 cycles. Unit B is set at $\frac{3}{4}\%$ at 600 kw and 60 cycles. What will be the frequency if the plant load increases to 3800 kw?

13–4. A diesel plant pays 10¢ per gal for fuel oil (19,350 Btu per lb) and uses an average of 0.583 lb per kwhr at an average load of 900 kw per engine for 3100 hr per yr. There are three 1200-kw units in the plant; seven employes are employed at an average salary of $4000 per yr. The plant cost $206 per kw to install; investment charges are at 9%. Lube oil amounts to 0.0008 gal per kwhr and costs 25¢ per gal. Maintenance and supplies each average 3.3 mills per kwhr. Calculate the cost of energy per kilowatt-hour and per year.

13–5. Calculate the bmep and piston speed for the 5-cylinder, 13×20 engine listed in the Appendix at 400 rpm (a) nonsupercharged, (b) supercharged, (c) supercharged and intercooled, and (d) high-pressure supercharged.

13–6. Calculate the bmep and piston speed for the 7-cylinder, $9 \times 11\frac{1}{2}$ engine listed in the Appendix at 600 rpm for (a) nonsupercharged and (b) for the three types of supercharging.

13–7. Calculate the volume of fuel input per cylinder per injection for the engine of Fig. 13–6. Assume the fuel to be 19,350 Btu per lb and 7.39 lb per gal of diesel oil.

13–8. For a 13×16 cylinder, estimate the weight of air drawn into the cylinder per suction stroke if the volumetric efficiency is 70% compared with atmospheric air at 14.7 psia and 100 F. If the engine received the same volume of air, but the air was at 6 psig and 100 F because of supercharging and intercooling, what would be the weight of air in the cylinder?

13–9. An 18×27 in. engine has a compression ratio of 12:1. If the air and gases (assume same as air) in the cylinder are at 14.0 psia and 190 F at the start of compression, what is the temperature at the end of compression if $n = 1.28$ (see any standard thermodynamics text for the equations of a polytropic process)?

13–10. If the engine of Prob. 13–9 had a 16:1 compression ratio, what would be the temperature at the end of compression?

13–11. If the engine of Prob. 13–9 had a 17:1 compression ratio, what would be the clearance volume?

13–12. Estimate the amount of waste heat recovery that could be expected from a 2500-kw diesel engine operating at three-quarter load if two-thirds of the energy in the exhaust could be used for heating the building.

BIBLIOGRAPHY

BOYER, GLENN C. *Diesel and Gas Engine Power Plants.* New York: McGraw-Hill Book Co., Inc., 1943.

FOSHOLT, SANFORD K. "Design of Diesel Power Plants," *Proceedings of the American Power Conference,* Vol. 15, 1953.

LICHTY, LISTER C. *Internal Combustion Engines.* New York: McGraw-Hill Book Co., Inc., 1951.

Standard Practices for Low and Medium Speed Stationary Diesel Engines. Chicago: Diesel Engine Manufacturers Assoc., 1951.

CHAPTER 14

NUCLEAR PLANTS

14-1. General. Estimates of the probable installed capacity of generating units operating on nuclear- and the fossil-fuel cycles were given in Chap. 1. It was also indicated that at some time in the future, economical sources of fossil fuels for power generation in the United States will become depleted. Since fuel costs vary throughout the world, nuclear power plants may be economical in some countries before they are economical in other countries.

Even in the United States, there is a wide variation in fossil-fuel costs; in the sections where there are natural gas and oil fields, the fuel costs less than 10¢ per 10^6 Btu; however, in upper New England, the interior of Florida, and in the sparsely settled sections of the upper Midwest and West, the cost is from 40 to 50¢ per 10^6 Btu. Many well-populated sections have fuel costs of 30 to 35¢ per 10^6 Btu. Based on 30¢ per 10^6 Btu, the cost of steam supplied to a very large, modern, high-pressure turbine would be from 50 to 60¢ per 10^6 Btu when fuel, investment costs, depreciation, taxes, and maintenance for the steam-generating part of the plant are included. The reactor portion of a present-day nuclear plant would have to be able to produce steam at this cost in order to compete with the fossil-fuel plant of today. Nuclear plants of the future must be even more economical because fossil-fuel plants of the future will be more efficient than the current installations.

If the fuel cost of an average modern coal-burning power plant were reduced to zero, the cost of electrical energy to the customer could be reduced only about 3 mills per kwhr. Or, conversely, at the same price of electrical energy to the customer, reducing the cost of fuel for a modern plant from 30¢ per 10^6 Btu to zero would permit an increased investment cost of about $150 per kw; or a total station cost that is slightly less than double the present station investment cost.

Present indications are that nuclear plants will differ from fossil-fuel plants only in replacement of the steam boiler by the nuclear reactor.

In addition to the advantage of a vast supply of nuclear fuel on the earth's surface (uranium deposits are found in Belgian Congo, the Colorado Plateau of the United States, and in northern Canada), nuclear plants require less space because of elimination of a huge coal-storage pile, a very low fuel transportation cost, greater mobility, and less difficulties with plant sites in remote areas.

14–2. The Atom.   All matter is composed of one or more of the 102 known elements. A periodic chart of 100 of the elements is shown in Plate 20 of the Appendix. When two or more of the elements are chemically combined, they become a *compound,* the smallest particle of which is the *molecule.* If the molecule of a compound is subdivided, it will lose its chemical characteristics and will be broken down into *atoms.*

At one time it was believed that the atom was the smallest possible unit and was indestructible. The word atom comes from the Greek language and means "an uncuttable thing."

Modern physicists conceive the atom to be a miniature solar system, in which there is a *nucleus* surrounded by one or more small particles revolving in an approximately circular orbit, much as the planets revolve about the sun. The dense nucleus contains one or more *protons* and *neutrons* (except for hydrogen which has no neutrons), and the particles revolving about the nucleus are *electrons.* The small proton particles each have a single, positive electric charge numerically equal to the negative charge of one electron. Neutrons have no electric charge but are tightly bound to the protons to form the nucleus.

To give some idea of the size of the particles being discussed, consider the hydrogen atom, which consists of a single proton with a single electron in the orbit around the proton nucleus. The diameter of the atom would be the diameter of the electron orbit; that is about 10^{-8} in. But the diameter of the proton is $1/10,000$ of the orbit diameter.

One convenient way of portraying the three-dimensional atom on a two-dimensional surface of a page is shown in Fig. 14–1. Although the shells or orbits of electrons are shown as concentric circles, the actual orbits are not all in the same plane. Obviously, these sketches are not to scale.

Observe that there are the same number of protons in the nucleus as there are electrons in the shells of the atoms. Thus, the atom is electrically neutral, since the neutron has no charge. If one or more electrons are removed from an atom, the atom has a *positive valence,* while an atom of *negative valence* has an excess of electrons. The *atomic number* of an atom is the number of protons in the nucleus. For an electrically neutral atom, this is also the number of electrons in the orbits.

In addition to a definite number of protons in the nucleus of the atom, each atom has a definite mass. Based on *atomic mass units (amu),* a mass of 16 is arbitrarily assigned to the oxygen atom. Then the mass of a proton or a neutron is approximately one each, and the mass of an electron is essentially zero. These values are only approximate; more exact values

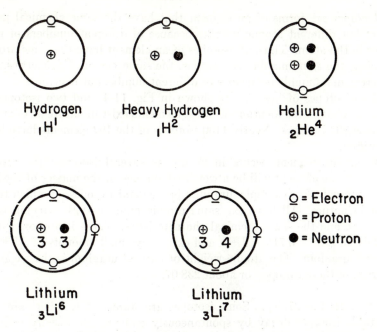

Fig. 14–1. Schematic representation of a few atoms (not to scale.)

are 1.00759 amu for protons, 1.00898 for neutrons, and 0.00055 for electrons. Therefore, the weight of an atom is the sum of the protons and the neutrons in the atom on the amu scale.

It is customary to identify an atom by its atomic number Z and its approximate atomic weight A. The atomic number Z is the number of protons in the nucleus, which is the same as the number of electrons. Letting N represent the number of neutrons in the nucleus,

$$A - Z = N \tag{14–1}$$

and the chemical symbol is written with Z as a preceding subscript and A as a superscript. Thus, in Fig. 14–1, the symbol for helium is written $_2\text{He}^4$ to indicate that the atomic number and the number of protons is two but that the atomic weight is approximately 4, since there are two protons and two neutrons. Of course there are two electrons.

Note that there are two forms of hydrogen and lithium shown in Fig. 14–1. Each form of hydrogen has one electron and one proton, producing an atomic number of 1. However, there are no neutrons in normal hydrogen, but there is one neutron in heavy hydrogen. Heavy hydrogen (deuterium) has the same atomic number as ordinary hydrogen but has an atomic weight of 2, while tritium is a third hydrogen isotope with a mass of 3 amu.

Isotopes are forms of an element that have the same chemical properties but different atomic weights because of different numbers of neutrons in the atom. Thus, all isotopes of an element have the same atomic number, the same number of protons, the same chemical properties, but different mass numbers because of different numbers of neutrons.

Two isotopes of lithium are shown in Fig. 14–1, and two isotopes of hydrogen have been mentioned. Most elements exist in nature as mixtures of some 300 isotopes. Nearly 1100 isotopes of the 102 elements have been identified.

When an element occurs in nature as several isotopes, the atomic weight of the element will be a weighted average of the masses of the isotopes present. For example, uranium has several isotopes, of which three exist in nature. All have the same atomic number (92), varying mass numbers, and the same chemical characteristics. There is approximately 99.28% of $_{92}U^{238}$, 0.71% $_{92}U^{235}$, and extremely small amounts of $_{92}U^{234}$ in natural uranium. The atomic weight of natural uranium is the weighted average of these isotopes, or about 238.07.

14–3. Radioactivity. Some isotopes are stable while others are not. Unstable isotopes decay by spontaneously emitting electrically charged particles or electromagnetic waves. All isotopes of elements having atomic numbers of 80 (mercury) or less are stable as found in nature, but unstable isotopes of all elements can be produced artificially. Elements having atomic numbers of 81 (thallium) or higher have unstable isotopes occurring in nature.

Unstable isotopes are *radioactive* and decay by emitting *alpha particles, beta particles,* and *gamma rays.*

Alpha particles carry a positive charge and have a mass of 4. They are composed of two protons and two neutrons; thus, they are the nucleus of the helium atom. When emitted from an atom, the alpha particle can travel only a short distance in air. A sheet of paper is ample shielding for an alpha particle. Therefore, discharge of an alpha particle from uranium would decrease the atomic number by 2 and the atomic weight by 4 to produce thorium,

$$_{92}U^{238} \rightarrow {}_2He^4 + {}_{90}Th^{234} \tag{14–2}$$

Beta particles are electrons emitted from the nucleus of an atom. Since the nucleus is supposed to contain only protons and neutrons, but no electrons, it is believed that the electron is ejected from the nucleus by the disintegration of a neutron into a proton. This disintegration has no effect on the mass number of the atom but does increase the atomic number. Thus, if an electron or beta particle has an atomic number of 1 and a mass number of 0, and Pa is protactinium,

$$_{90}\mathrm{Th}^{234} \rightarrow _{-1}e^0 + _{91}\mathrm{Pa}^{234} \qquad (14\text{-}3)$$

A thin sheet of metal or a piece of wood will provide shielding against beta particles.

Gamma rays are similar to X-rays in that they are electromagnetic. They are emitted from a nucleus after an alpha particle is expelled. When an alpha particle leaves the parent atom, there is an unbalance of energy in the daughter nucleus which causes the emission of the gamma rays. Satisfactory shielding from gamma rays requires thick lead or concrete.

Note that these three types of radiation are all from the nucleus of the atom; thus, they are nuclear reactions.

Decay of radioactive elements is measured in terms of the time required for one-half of the atoms of the material to decay. This is known as the *half-life* of the material. The half-life of a material is a constant and is independent of the amount of material considered and of the pressure or temperature. If one-half of a material decays in 1 min, then at the end of 2 min only one-fourth of the original material will remain; and at the end of 3 min, only one-eighth will remain. Radioactive decay is therefore exponential with time.

There is a wide variation in the half-life of various radioactive materials. The half-life of uranium U-238 (an abbreviation for $_{92}\mathrm{U}^{238}$) is about 4.5×10^9 yr, while for polonium, $_{84}\mathrm{Po}^{210}$, it is about 140 days; and for $_{84}\mathrm{Po}^{214}$ it is less than 2×10^{-6} sec.

The decay of U-238 to lead, $_{82}\mathrm{Pb}^{206}$, takes place by the emission of 11 alpha particles and 9 beta particles.

14-4. Fusion and Fission. Since an atom is composed of protons, neutrons, and electrons, it might appear that the mass of an atom could be determined by a summation of the masses of these constituents. Thus, the mass of a helium atom containing two protons, two neutrons, and two electrons would give you a summation of $2 \times 1.00759 + 2 \times 1.00898 + 2 \times 0.00055 = 4.03424$ amu. However, accurate measurement of the mass of the helium atom indicates a value of 4.0038 amu, or a discrepancy of about 0.030 amu.

Einstein proposed the principle of equivalence of mass and energy by the relationship that

$$E = c^2 m \qquad (14\text{-}4)$$

where E = energy, erg
   c = velocity of light, 2.9979×10^{10} cm per sec
   m = mass, gr

Therefore, in the formation of the helium atom mentioned previously, an amount of energy was released, and this energy was equivalent to the mass defect. Conversely, to break down the atom into its particles would require that this amount of energy be supplied to the atom to restore the mass defect.

The combination of light elements into heavier elements is the *fusion* process; the mass defect may be considered as the *binding energy* for the nucleus. Arranging the calculations for mass defect into an equation and combining with Einstein's equation produces

$$E_b = 1.412 \times 10^{-13}[Zm_p + (A - Z)m_n - m] \text{ *} \qquad (14\text{--}5)$$

where E_b = binding energy, Btu

Z = number of protons, or the atomic number

A = mass number

m_p = mass of the proton, amu

m_n = mass of the neutron, amu

m = mass of the atom, amu

For the mass defect calculated for helium, the binding energy would be $1.412 \times 10^{-13} \times 0.030 = 4.24 \times 10^{-15}$ Btu for one helium atom, or 1.06×10^{-15} Btu for each nuclear particle (sometimes called *nucleon*) of helium.

Although this amount of energy may seem small, it should be remembered that the amu is also small. For a pound of helium, the energy released when creating an atom from the atomic particles would be $1.06 \times 10^{-15} \times 0.2735 \times 10^{27} = 0.29 \times 10^{12}$ Btu, or about 85,000,000 kwhr. Conversely, it would require this same amount of energy to decompose a pound of helium into its atomic particles.

Substitution into Eq. (14–4) shows that when coal having a heating value of 13,000 Btu per lb is completely burned, there is a mass defect of about 3×10^{-10} lb.

Calculations of binding energy per nuclear particle for other elements produce the curve of Fig. 14–2. Some of the lighter elements do not have a binding energy per nuclear particle that falls on a smooth curve. However, there is a definite rising trend from helium to a peak value of binding energy at $A = 60$. This value of mass number is approximately the value for cobalt (58.94). Thereafter, the binding energy per nucleon decreases.

---

* erg = $1.058 \times 10^{10} \times$ Btu; cm = $2.54 \times$ in.; gr = $453.6 \times$ lb; amu = 0.6024×10^{24} $\times$ gr; amu = $0.2735 \times 10^{27} \times$ lb. The conversion factor from amu to gr or lb is based on the number of molecules per mol of a substance, called Avogadro's number (determined experimentally to be 6.024×10^{23} atoms per gr mol), and the molecular weight of the substance. Thus, using hydrogen as an example, $2 \div 6.024 \times 10^{23} = 3.32 \times 10^{-24}$ gr per molecule. Since hydrogen has an amu value of 2, there are 1.66×10^{-24} gr per amu.

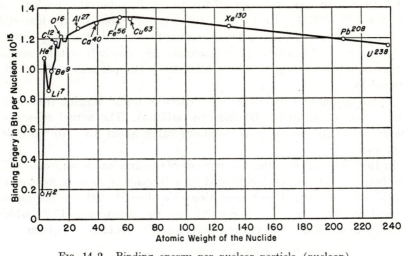

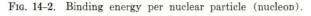

FIG. 14–2. Binding energy per nuclear particle (nucleon).

Elements at the peak portion of the curve, say, mass numbers of 40 to 100, are the most stable, since their binding energy per nucleon is the greatest.

Remember that binding energy is released directly in proportion to the mass defect occurring during the formation of a nuclide. Combining light nuclides into heavier nuclides is fusion and can occur to nuclides shown on the left portion of the curve. There is a release in energy for the fusion process. Splitting of very heavy atoms into lighter atoms will produce more stable nuclides, will provide a release of energy, and can occur to elements on the right of the curve. In both fusion and fission there is a mass defect, an increase in stability of the nuclide, and a consequent release in energy. However, the release in energy for fusion is greater than for fission, as indicated by the slope of the curve on each side of the peak of the curve or point of maximum stability. Nuclear reactions progress in the direction of greater stability. Energy released in the process first develops kinetic energy of the nuclei; the kinetic energy is converted into thermal energy.

In order to produce fusion, two nuclei must approach each other at very high velocity to overcome their natural electrostatic repulsion. Since temperature is proportional to kinetic energy,* high temperatures are necessary to support a fusion process.

As an example of the energy released during fusion, suppose that it were possible to combine two beryllium nuclei, $_4Be^9$, into oxygen, $_8O^{18}$. The difference in binding energies for these nuclei, from Fig. 14–2, is

* See the kinetic theory of gases in any standard text on thermodynamics.

2.1 $\times$ 10^{-6} Btu per amu, or 37.8 $\times$ 10^{-16} Btu for the nucleus, or (2.1 $\times$ 10^{-16}) (0.2735 $\times$ 10^{27}) = 5.8 $\times$ 10^{10} Btu for a pound of the material.

For the portion of Fig. 14–2 representing fission, note that if U^{235} were to be fissioned into two nuclei of approximately one-half of that weight, there would be an energy release of 1.2 $\times$ 10^{-16} Btu per amu, or 0.3 $\times$ 10^{11} Btu per lb.

Actual fission of U^{235} produces a variety of nuclei ranging from mass numbers of about 70 to 160 (zinc to terbium). The actual release of energy is slightly greater than the above value, about 3.5 $\times$ 10^{10} Btu per lb of material.

In 1939, it was shown that the nucleus of the uranium atom would split into two fragments when bombarded by a *thermal neutron* (a neutron of low kinetic energy). Experimentation also showed that the portion of natural uranium that was responsible for the fission was U^{235}; which amounts to about 0.7% of natural uranium. The reaction is written

$$_0n^1 + {}_{92}U^{235} \rightarrow {}_{92}U^{236}$$
$$_{92}U^{236} \rightarrow {}_{z_1}F^{41} + {}_{z_2}F^{42} + 2.05_0n^1$$

(14–6)

These equations state that when U^{235} is bombarded by a neutron it first absorbs the neutron to become U^{236}. This isotope is unstable and decays into two fission fragments (F) and two or three (an average of 2.5) neutrons. The sum of A_1 and A_2 will be 234 if two neutrons are produced and 233 if three neutrons are produced. Also A_1 and A_2 are the variety of mass numbers previously mentioned as ranging from 70 to 160; the largest percentages of nuclei being at about 95 and 139 mass numbers. Also $Z_1 + Z_2$ must equal 92.

Each of the fission fragments give off radiation and decay, with many different half-lives, into stable elements. The danger from fission is due to the beta and gamma rays given off during decay. The half-lives range from seconds to over 2 $\times$ 10^6 yr. The 2.5 neutrons that remain from the reaction may be used to cause other nuclei to fission and produce a chain reaction, Fig. 14–3. If these free neutrons are absorbed by nonfissionable material or are lost, then the process stops.

14–5. Nuclear Reactor Materials. Materials used for a reactor may be classified as (a) the fissionable material or fuel, (b) *fertile* material, (c) *coolant*, (d) *moderator*, and (e) structure (including reflectors, container, and shielding material).

Fissionable materials are uranium $_{92}U^{235}$, plutonium $_{94}Pu^{239}$, and uranium $_{92}U^{233}$. These three materials will fission when bombarded by neutrons, but U^{235} is the only one occurring in nature. Since it represents such a small percentage of natural uranium, it is possible to *enrich* the

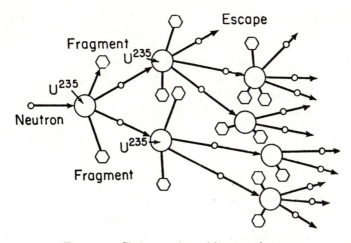

Fig. 14-3. Chain reaction without moderator.

uranium by increasing the U^{235} concentration. Several processes are possible. In the gas diffusion process, a compound of uranium is exposed to a porous membrane. The lighter compound of U^{235} passes through the membrane more readily than the heavier compound of U^{238}. Another method is electromagnetic separation.

U^{235} will fission when it absorbs a high-velocity neutron but will be more likely to absorb a low-velocity or *thermal* neutron. It appears that there is some advantage of U^{235} over the other materials due to its higher fission percentage.

Absorption of high-velocity neutrons in U^{238} is very low. However, thermal neutrons cause a radioactive disintegration

$$_0n^1 + {}_{92}U^{238} \longrightarrow {}_{92}U^{239} \xrightarrow[23 \text{ min}]{} {}_{93}Np^{239} + {}_{-1}e^0$$
$$_{93}Np^{239} \xrightarrow[2.3 \text{ days}]{} {}_{94}Pu^{239} + {}_{-1}e^0 \tag{14-7}$$

This equation states that the absorption of a neutron by U^{238} produces U^{239}, which has a half-life of 23 min and decays into neptunium with a beta emission. Neptunium has a half-life of 2.3 days and decays into the fissionable material plutonium by another beta emission. Pu decays at a slow rate (half-life of 2.4×10^4 yr) by alpha emission.

The other fertile material, thorium, $_{90}Th^{232}$, occurs in nature as 100% of this isotope. By a similar process, thorium produces U^{233}, which is fissionable.

$$_0n^1 + {}_{90}Th^{232} \longrightarrow {}_{90}Th^{233} \xrightarrow[23 \text{ min}]{} {}_{91}Pa^{233} + {}_{-1}e^0$$
$$_{91}Pa^{233} \xrightarrow[27 \text{ days}]{} {}_{92}U^{233} + {}_{-1}e^0 \tag{14-8}$$

where Pa is protoactinium.

The processes described above provide a fissionable atom for each atom destroyed; this is called *breeding*. Because of neutron losses from the reactor and absorption of neutrons by nonfertile atoms, there must be more than two neutrons produced by fission to make breeding possible. Fissioning of U-235 to breed U-238 into Pu-239, as described by Eq. (14-7), is portrayed in Fig. 14-4. Since there is an average of 2.5 neutrons produced by fissioning U-235, breeding is possible.

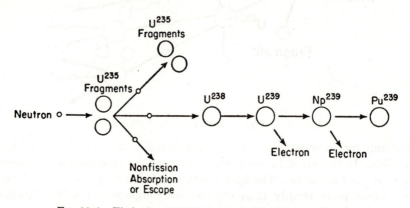

Fig. 14-4. Fissioning of U-235 to breed Pu-239 from U-238.

The collision of a neutron with a fissionable material is more likely to occur if the neutron is a thermal or low-energy neutron. Since the neutrons produced by fission, Fig. 14-4, are high-energy or fast neutrons, the reactors may include a material, called a *moderator*, to slow down the neutrons. The most effective moderators are materials that have a mass similar to that of the neutron. Hydrogen, deuterium (heavy hydrogen), water, heavy water, beryllium, and carbon have been found to be successful moderators.

It has already been indicated that large amounts of energy are liberated by fission. About 80% of the energy is represented by the kinetic energy of the fission fragments and about 20% by neutron kinetic energy and absorption, gamma rays, and beta decay. Collision of these particles with the materials in the reactor reduces the kinetic energy of the particles and converts the energy to thermal energy that must be removed from the reactor by a *coolant*. Air, helium, water, mercury, and liquid sodium are possible coolants.

In addition to a structural shell (frequently of stainless steel) to contain the materials and the pressure of the coolant, there must be a *reflector* inside the shell to reduce the escape of neutrons. When reflected back into the core by the reflector, neutrons may produce fission and reduce the amount of fissionable material needed for the reactor. While a reflector cannot contain any fissionable material, it may contain thorium or U-238 for breeding purposes. Lead, carbon, and beryllium are other reflectors.

The whole reactor must be enclosed by a biological shield to protect living organisms from the fast and slow neutrons, beta particles, and gamma rays. Dense concrete, several feet thick, has been satisfactory.

14–6. Classification of Reactors. It is possible to classify reactors according to (a) the fuel or fissionable material used, (b) the neutron energy, (c) the physical arrangement of the fissionable material, and (d) breeder or nonbreeder reactors.

Natural uranium with its low percentage of the fissionable U-235, or enriched uranium with a higher percentage of U-235, may be used in the reactor. The other possibilities are the artificial fissionable materials plutonium and U-233. Natural uranium is cheaper than enriched uranium, but because of the lower percentage of U-235, it will require a larger reactor. Both artificial materials are more expensive than enriched uranium.

A reactor containing no moderator would employ fast or high-energy neutrons and would be a *fast reactor*. Presumably, the atomic bomb would be of this type. Without a bulky moderator, a fast reactor would have a smaller size than a *thermal reactor* wherein the neutrons have been slowed down. Fast reactors may lose more neutrons by leakage than the thermal reactor and will make a reflector more necessary. Also, the fast reactor should require a more concentrated or enriched fuel because fewer of the fast neutrons will cause fission. *Intermediate reactors* employ neutrons having an energy somewhere between fast and thermal reactors.

Fissionable material for a reactor may be in the form of a lump such as a cylinder or plate, in which case it is known as a *heterogeneous reactor*; or the fuel may be in a liquid form, when the reactor is *homogeneous*. In the latter case, the fuel is a salt, such as uranyl sulfate, and is mixed with the moderator, which is water.

14–7. Nuclear-Power Reactor Steam Plants. Some 150 reactor designs have been proposed for steam-power generation. Five of these are being given intense study and are being designed or constructed.

The first large unit to be put into operation in the United States (Shippingport Power Plant of the Duquesne Light Company) has a *pressurized water reactor* (PWR) of the heterogeneous type. A sketch showing a longitudinal section of the reactor vessel is given in Fig. 14–5. The vessel is 33 ft high, 9-ft ID, 8.5 in. wall thickness of carbon steel with a stainless-steel cladding 0.25 in. thick, and has a dry weight of 250 tons. The vessel is insulated with 4 in. of glass wool. The core assemblies are located in the internal cage. Some of the core assemblies are rods containing en-

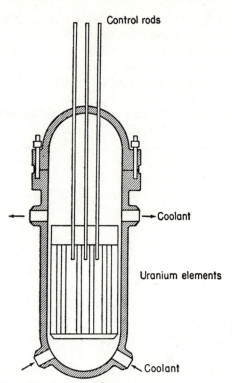

FIG. 14-5. Longitudinal section of PWR vessel.

riched uranium, and some are of natural uranium oxide pellets. Since the uranium oxidizes readily, the elements must be clad in aluminum, zircaloy, or stainless steel. This also reduces the possibility of particles of uranium or radioactive material and radioactive gases resulting from the fissioning process from entering the coolant system. Some PWR's use flat strips of clad fuel assembled into sandwiches or rods or tubes of the fuel.

Since fission of the uranium produces an average of 2.5 neutrons per fission, and one neutron is required to produce the next fission, there is an excess of 1.5 neutrons per fission. Waste and leakage of neutrons cannot exceed this value of 1.5 neutrons per fission if the reaction is to continue at a constant rate. If less than 1.5 neutrons per fission is wasted or absorbed by nonfissioning material, then there will be an increase in the chain reaction and an increase in the heat output of the reactor. Thus, the absorption of neutrons by a nonfissioning material forms a means of reactor control.

Control rods, of which only three are shown in the schematic diagram of Fig. 14-5, may be made of hafnium, boron steel, or stainless-steel-clad cadmium-silver alloy. Crystal-bar hafnium control rods are used in the Shippingport unit.

Light water serves as both the coolant to remove the energy generated in the reactor and the moderator to slow down the fast neutrons. There are four coolant loops for the Shippingport unit, of which one loop and its boiler are shown in the heat-balance diagram of Fig. 14–6. Only three

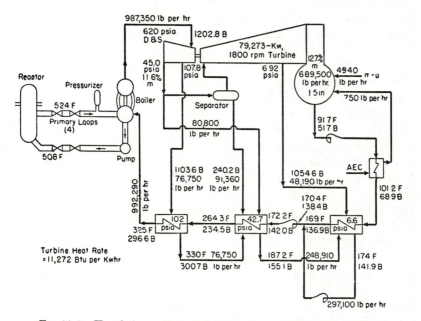

Fig. 14–6. Heat-balance diagram for Shippingport PWR steam plant.

loops are required for full load. There are two isolating values at the reactor inlet and two at the outlet. Each loop has a pump that displaces 15,000 gpm with a developed head of 105 psi and a 1200-kw motor.

Coolant pressure of 2000 psig is maintained in the coolant system by the 300-cu ft pressurizer tank. This tank contains 200 electric heaters having a rating of 500 kw. The tank is approximately half-filled with water; the remaining space is occupied by steam generated by the electric heaters. Control of the heating elements maintains the desired system pressure.

A small amount of primary system coolant is continuously withdrawn from the circuit and purified in a by-pass demineralizer. Extremely pure water is required for the reactor.

The 2000-psig operating pressure in the primary system would have a vapor temperature of 636 F. The coolant temperature leaving the reactor must be appreciably lower than this value to prevent vaporizing the coolant. Thus the maximum steam pressure for the secondary or turbine cycle is limited by the coolant pressure. For the data shown in the heat-balance diagram, the saturation temperature of steam leaving the boiler is 490 F, or there is a 34 F temperature difference between the incoming hot primary fluid and the saturation temperature in the boiler.

Note that a PWR vessel must be designed for a high pressure, that there is a degradation of energy in the boiler, and that the steam pressure to the turbine is limited to value in the order of 600 psig at full load in order to keep the primary circuit-design pressure within reasonable limits.

Each boiler consists of two shells connected by tubes. Primary fluid circulates through tubes in the lower shell. Hot boiler water containing steam bubbles circulates to the upper drum where the steam is withdrawn through a steam separator.

In order to prevent the escape of radioactive gases or of fission products in the vapor (this might be caused by a casualty resulting from loss of primary coolant in the reactor and consequent melting of the reactor or possible failure of the core elements), the nuclear components of the plant are housed in four steel containers. In turn, these containers are enclosed in concrete. An artist's sketch of the plant is shown in Fig. 14–7. The concrete also acts as a biological shield.

In the center is the spherical vessel with a dome that houses the reactor. Each of the two large cylindrical tanks at each side of the sphere house two of the primary coolant circuits and their boilers. Each of these tanks is 50 ft in diam by 97 ft long. The tank between the reactor and the turbine generator is 50 ft in diam by 147 ft long and contains the coolant pressurizing tank. These containers are air conditioned to maintain not more than 122 F.

Returning to the heat-balance diagram of Fig. 14–6, note that the only unusual feature of the steam cycle is that there is a steam separator in the cross-over line between the high- and low-pressure turbines. This separator is necessary to reduce the moisture in the exhaust of the turbine. The moisture content of the steam entering the separator is 11.6%.

Steam pressure entering the turbine will vary with load from 600 to 885 psig. Auxiliary power is expected to be 6000 kw for the nuclear portion of the plant and 3500 kw for the turbine generator portion.

An artist's sketch of the nuclear portion of another PWR plant is shown in Fig. 14–8. Note the differences in arrangement of this plant. Steam will leave the boiler at 405 psia saturated but will enter the turbine at 370 psia and 1000 F. The high steam temperature will be obtained by a separate, oil-fired superheater. There will be four stages of feedwater heating, and the generator output will be 236,000 kw; the heat rate has been calculated at 10,700 Btu net kwhr exclusive of reactor efficiency

Fig. 14-7. Artist's sketch of the Shippingport PWR Power Plant. (Duquesne Light Co.)

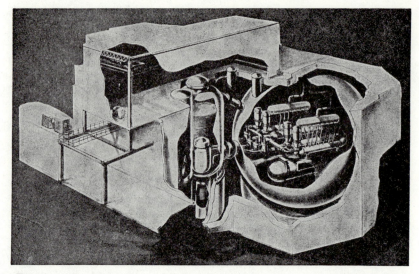

Fig. 14–8. Reactor portion of Indian Point PWR Plant. (Consolidated Edison Co. of N. Y., Inc.)

The submarine "Nautilus" employs a pressurized water reactor in its propulsion system.

Schematically, the reactor shown in Fig. 14–5 may be used to illustrate a second type of unit that is known as a *boiling-water reactor*. The difference between this reactor and the PWR is that steam is produced in the reactor vessel of the boiling-water reactor and that there is no boiler external to the reactor. The light water acts as the moderator. Steam generated in the reactor goes through an external separator and then directly into the turbine. Thus, the degradation of energy due to the temperature difference between the primary and secondary circuits is eliminated and the steam pressure at the turbine thereby can be increased. Since the steam leaving the reactor is radioactive, the turbine and other steam equipment must be biologically shielded. The radiation from steam and water is short lived and does not present a problem during maintenance. However, impurities from the steam are deposited on the turbine blades, and since these deposits may be radioactive, it may be necessary to flush the turbine to remove them.

A plant being designed for the Commonwealth Edison Company, for a 180,000-kw turbine generator to be located near Chicago, will have a *dual-cycle boiling-water reactor*. This unit will be a combination of the PWR and the boiling-water reactor in that saturated steam will be generated at 1000 psia within the reactor vessel and also some of the saturated water in the reactor vessel will be withdrawn as a liquid and supplied to an external boiler. Saturated steam at 500 psia will be generated in the external boiler. Steam will be admitted to the dual-pressure turbine at both steam pressures. All the water supplied to the reactor will be demineralized.

Enriched uranium in the form of UO_2 will be used in the reactor. The heat rate for the cycle is expected to be 11,925 Btu per net kwhr. The reactor portion of the plant may be housed in a pressure-tight sphere strong enough to confine all the dangerous products resulting from a failure of the reactor.

An important disadvantage of both the PWR and the boiling-water reactor is the moderator-coolant fluid pressure within the reactor vessel that requires a heavy vessel shell and that can release considerable energy in the event of a failure. Another fluid that could permit the use of high temperatures at moderate pressures would be advantageous. The *sodium-graphite reactor* (SGR) is such a unit. Graphite is the moderator, while liquid sodium acts as the coolant.

To illustrate the properties of sodium, note that it melts at 208 F and boils at 1620 F. It has good heat-transfer characteristics because of an extremely high thermal conductivity. There are no corrosion problems with common metals, but it reacts violently with water and oxygen. Also, it is readily absorbed by the porous graphite, necessitating a metal cladding or container for the graphite; zirconium is suitable for this purpose.

A proposed sodium-graphite reactor plant is shown in Fig. 14-9. The reactor and the primary circuit or intermediate heat exchangers, together

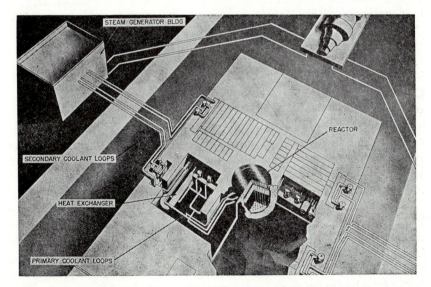

Fig. 14-9. A sodium-graphite reactor plant. (Atomics International, North American Aviation, Inc., and Consumers Public Power District of Nebraska.)

with pumps, surge tanks for the primary circuit, and a fuel storage area (not shown), are located in the reactor plant.

Liquid sodium is circulated through the reactor as the coolant. It enters the reactor at 500 F and leaves at 925 F to enter the heat exchanger. The heat exchanger is necessary because the primary circuit sodium becomes highly radioactive. There are four primary circuits, with a heat exchanger for each, and four secondary sodium-coolant circuits. Each pair of secondary loops conveys liquid sodium to a steam-generator building located at a distance from the reactor building. A failure resulting from a sodium-water reaction would not damage either the reactor building or the steam turbine and its associated equipment.

Secondary-circuit sodium from each pair of heat exchangers enters a superheater at 895 F and leaves to enter the evaporator section of the steam generator at 560 F. Next, the sodium enters an economizer and then returns to the intermediate heat exchangers.

Boiler feedwater enters the economizers at 300 F and leaves the super-heater at 800 psig and 825 F. Thus, the use of sodium as a coolant permits higher steam pressures and the use of superheat to reduce the plant heat rate. Estimates indicate a net-plant heat rate of 10,800 Btu per net kwhr at a load of 80,800 kw on the turbine generator for 2 in. Hg condenser pressure and with an assumed 5% auxiliary power consumption. Three stages of feedwater heating are used.

Construction of the reactor can be visualized from Fig. 14–9. Either slightly enriched uranium (1.8% U-235) or a thorium-U^{235} alloy may be used as the fuel, which is in the shape of a rod and is clad with stainless steel. These rods are inserted inside holes in the zirconium-canned graphite blocks. Primary circuit sodium can circulate around the rods; the sodium flows through a uranium reactor at the rate of 6.5×10^6 lb per hr and at the same rate in the secondary circuits for the 80,000-kw load. Control rods, inserted in some of the graphite block holes, are boron steel covered with stainless steel.

The reactor vessel is 17.5-ft ID with 2-in. thick stainless-steel walls and is 34 ft high, including a thick, removable, concrete upper shield on top of the reactor. Shielding consists of a 6-in. mild-steel thermal shield (a thermal shield absorbs fast neutrons and gamma radiation), thermal insulation, and dense concrete as a biological shield.

Regardless of whether uranium or uranium-thorium alloy is used, there is some conversion of the fuel. In the uranium unit, there is a conversion of U^{238} to U^{235}, and in the thorium-uranium unit the conversion is from U^{235} to U^{233}. Therefore, these units can be called *converters* because they produce less than one atom of new fissionable material for each atom destroyed. A breeder produces one or more new fissionable atoms for each atom destroyed.

Also, the reactors previously described are of the thermal heterogeneous type. A *fast breeder reactor* of the heterogeneous type uses fast rather than thermal neutrons and, therefore, has no moderator. However, fast neutrons are not so effective in creating fission so that there must be more costly fuel present in the reactor.

Fast breeder reactors are not too unlike the sodium-graphite unit described except that there are two layers of fertile blanket material around the fuel rods. The fuel may be U^{235}, U^{233}, or Pu^{239}. If the fertile material in the blankets is U^{238}, then Pu^{239} will be produced, but if the fertile blanket is Th^{232}, then the product will be U^{233}, regardless of which fuel is used.

Because a fast breeder reactor has no moderator, liquid-metal such as sodium or sodium-potassium alloy may be used as the coolant. This permits taking advantage of high coolant temperatures and low coolant pressures. With a liquid-metal coolant, an intermediate heat exchanger must be employed in a manner similar to that described for the sodium-graphite unit.

An example of the application of a fast breeder reactor to the generation of electric power is the plant that will be located south of Detroit and on the western shore of Lake Erie. Enriched uranium-235 alloyed with molybdenum and clad with zirconium will be used for the fuel elements. There will be about one atom of uranium-235 in the alloy for each of six total atoms present.

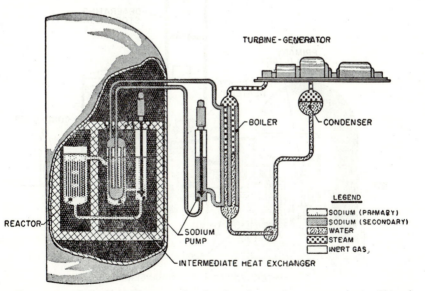

FIG. 14-10. Schematic diagram of a fast-breeder nuclear-power plant. (Atomic Power Development Associates, Inc.)

A schematic diagram of the plant is shown in Fig. 14–10. There are two liquid sodium systems. The primary system provides coolant for the reactor. Stainless steel is used to convey the liquid sodium which enters the reactor at 550 F, leaves at 800 F, and flows at the rate of 13,200,000 lb per hr. Energy from the primary system is transferred to the secondary sodium system in the intermediate heat exchanger. The secondary system flow rate is the same as for the primary system, but the secondary system temperatures are 515 F entering and 765 F leaving the intermediate heat exchanger. Use of the secondary system prevents radioactive contamination of the water or steam because the primary sodium does not emit neutrons.

Secondary sodium produces steam at 600 psia and 755 F in a counter-flow, once-through, shell-and-tube boiler. Water is transformed into steam at the rate of 1,000,000 lb per hr inside the tubes; sodium is on the shell side of the boiler. Water will enter the boiler at 400 F and leave as steam at 600 psia and 755 F.

Initial capacity of the plant is to be 100,000 kw, with a reactor power of 300,000 kw, providing a 33.3% thermal efficiency or a heat rate of only 10,240 Btu per gross kwhr. Uranium-235 will be consumed at the rate of

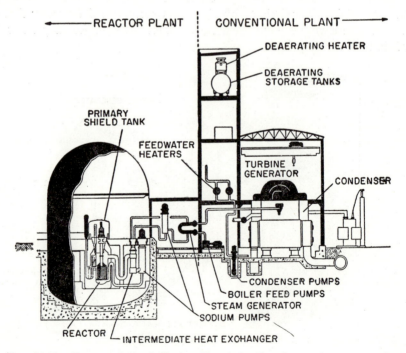

FIG. 14–11. Cross-section of Enrico Fermi fast-breeder nuclear-power plant. (Atomic Power Development Associates, Inc.)

87.5 kg per year, and plutonium will be produced at the rate of 106 kg per year, providing a breeding ratio of 1:2.

After preliminary operation of two years, steam flow will be increased to 1,470,000 lb per hr at 900 psia and 820 F to develop the full capacity of the 156,000-kw turbine. Fluid temperatures will increase for the increased load.

The schematic diagram of Fig. 14–10 does not show the complete steam cycle, which contains a live steam reheater, four surface heaters with a drain cooler, and a deaerator. A cross-section of the reactor plant and the steam plant is shown in Fig. 14–11, while an artist's conception of the completed plant is shown in Fig. 14–12.

Referring again to Fig. 14–11, note that the reactor vessel, containing the fuel elements and the liquid sodium, together with a transfer rotor and container to be used in installing new elements, the control rods, and handling mechanism, will be enclosed in a bottle-shaped primary shield tank (also called a secondary containment vessel) that has a maximum diameter of 24 ft and an over-all height of 53 ft. This secondary containment vessel will contain any liquid sodium resulting from a leak, even though unlikely, in the reactor vessel. The reactor vessel is 35 ft high, has a maximum diameter of 14.5 ft, and will be fabricated of 2-in. thick stainless steel.

Fig. 14–12. Artist's conception of Enrico Fermi atomic-power plant. (Atomic Power Development Associates, Inc.)

Enclosing the primary shield tank, the intermediate heat exchanger, and the primary circuit sodium pumps will be a cylindrical building constructed as an unfired pressure vessel of steel with a diameter of 72 ft. The over-all height will be 120 ft, the top will be a hemispherical head, and the bottom head will be hemiellipsoidal. Plates for the tank will vary from 1½ to 1 in. in thickness. The tank is to be designed to contain the products of a sodium-air reaction or the shock waves and missiles that would result from an explosion. The tank is to be incased in a concrete biological shield not less than 7 ft thick that will surround the bottom head and extend up to the concrete operating floor.

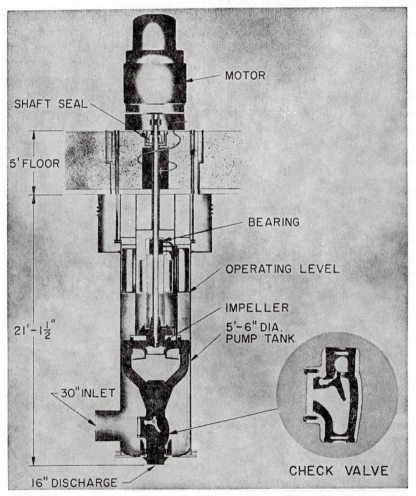

Fig. 14–13. Primary system sodium pump. (Atomic Power Development Associates, Inc.)

Construction of the sodium pumps is illustrated by the picture of a pump for the primary system shown in Fig. 14–13. This pump will deliver sodium to the reactor at 115 psig. The space above the sodium operating level in the pump will be filled with an inert gas. The pressure in the secondary sodium system will be greater than in the primary system. Thus, any leakage will be from the secondary to the primary system to prevent radioactive contamination of the secondary system.

Estimates of the cost for the plant are slightly over $40,000,000 for the reactor plant, including research, development, and business expenses during construction. Similar costs for the turbine plant are $14,000,000. Based on the ultimate anticipated capacity of 156 mw, the unit cost would be $350 per kw of installed capacity. A modern fossil-fuel plant at the same site should not cost over $175 per kw.

A homogeneous reactor would employ a fuel such as U^{235} or U^{233} in the form of uranyl sulfate in solution with light water, heavy water, or a liquid metal such as bismuth. If light water were used, it would act as a moderator. Graphite could also be used as a moderator. The operating pressure within the reactor would be high if either light or heavy water were used, but the bismuth would offer the advantage of low operating pressures together with high temperatures. Thorium, also in solution, may be used as a blanket to make the unit either a converter or a breeder.

An experimental homogeneous unit for the Atomic Energy Commission consists of three concentric and approximately spherical shells, Fig. 14–14. The inner shell, made of zircaloy, is the reactor core and contains the fuel solution. The space between the inner and middle shells contains either heavy water as a coolant or heavy water with a fertile material to form a breeder blanket. The outer shell is of stainless steel, is a blast shield, and contains cooling coils bonded to the outer surface.

A homogeneous reactor contains no control rods. Temperature control is accomplished by the concentration of uranium in the solution. In the AEC unit, the amount of uranium in the solution is established to produce a temperature of about 540 F. As the plant load decreases, the solution temperature tends to rise, but there is a negative temperature coefficient. This means that the reactivity of the solution decreases as the temperature rises. Thus, there is inherent control of the reactor energy output.

14–8. Economics of Nuclear Electric Power. Development of atomic electric-power plants in the United States has been conceived as progressing in three stages. The first stage consists of the first plants to be put into operation. These units need not compete economically with modern plants, since their purpose is to provide engineering and cost data that can be obtained only from experience. The second stage will consist of atomic plants designed on the experience gained from the first stage

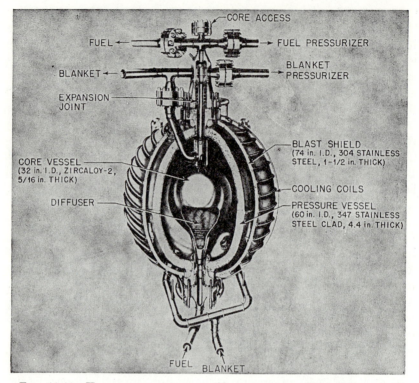

FIG. 14–14. Homogeneous reactor. (Newport News Shipbuilding & Dry Dock Co.)

and should be of the types previously proved to be the most economical and satisfactory. However, plants of the third stage must compete economically with new fossil fuel being installed in that period, not with the average of plants in operation at that time.

As an example of present-day large fossil-fuel plants, consider two of them; one located in the northeastern part of the United States and the other in the south central part. The northeastern plant would cost about $180 per kw to construct, would burn coal at about 35¢ per 10^6 Btu, and would have a heat rate of about 9500 Btu per net kwhr. At a capacity factor * of about 75%, electrical energy from this plant would be 4.5 mills per kwhr for fuel and operating costs and 3.5 mills for capital costs (at 12.5% investment charges).

The south central plant would be less expensive to build because it would burn natural gas at 10 to 12¢ per 10^6 Btu and would not require ash-handling facilities, bunkers, pulverizers, coal crushing and conveying equipment, and fly ash precipitators. Also, the building for a warm cli-

* See Article 15–1.

mate might be less expensive, and because of a lower fuel cost, an elaborate heat cycle could not be justified. This, together with higher circulating-water temperatures, would indicate a heat rate in the order of 12,000 Btu per kwhr and an investment cost of $120 to $130 per kw. Electricity from this plant might cost 4.5 mills per kwhr divided equally between capital costs and fuel and operating costs.

Plant heat rates have decreased from some 40,000 to 9300 Btu per net kwhr since just after the turn of the century until now. Plants currently under construction will lower this figure to nearly 8000 Btu per net kwhr with supercritical pressures, double reheat, and the cool circulating water of the northern states. Apparently, the supercritical-pressure plant of large size costs very little, if any, more than a large 2000-psig plant.

It should be noted that for plants of comparable size, the diesel plant costs about $50.00 less per kw of installed capacity than the coal plant. Fuel costs may also be lower for the diesel plant, but maintenance and lubricating-oil costs may be higher.

While atomic plants would not require the large coal-storage areas as do many steam plants, they must be located on large plots of ground to reduce hazard to the populace in the event of a failure. Likewise, fuel-transportation costs within and outside the plant are less for an atomic unit. However, the cost of the elaborate reactor, water-purifying equipment, instrumentation, shielding, etc., for the atomic plant are far greater than the cost of the steam generator and its accessories.

Another difficulty in estimating atomic plant costs is the cost of nuclear fuel. Since it is controlled by the government and not available on the free market, the fuel cost must be arbitrary and may not represent actual costs. This is also true of the resale of fissionable products developed in a breeder reactor.

Present indications are that atomic plants will vary between $250 and $600 per kw of installed capacity, depending on the size and type of plant. The total cost of the 68,000 kw Shippingport plant was $121,400,000. It has been indicated that the cost of a housing around the reactor to reduce hazard to the public would cost $1,000,000. The cost of natural uranium has been estimated at various amounts ranging from $10 to $50 per lb. The cost of enriched uranium or plutonium could be expected to be higher; values of highly enriched uranium have been taken as $7000 to $14,000 per lb in some reports.

Maintenance has been variously estimated as ranging from 1% of the initial cost to 1 mill per kwhr. Plant personnel, at least for the first few installations, will be about double that for a conventional plant, due to extra guards, chemists, test engineers, maintenance, and operation; say, about 130 employees versus 65.

EXAMPLE 14–1. A nuclear-power plant is to have a capacity of 150.000 kw but will operate at an average of 120,000 kw for 8760 hr per yr. The steam cycle will have a turbine heat rate of 11,300 Btu per kwhr at the average load; the net cost of the natural uranium used will be $350 per lb (the net cost is intended to mean the cost after allowing for the use of U^{238} converted to Pu^{239} and after allowing for the return of the spent fuel to the AEC); 15% of the neutrons released by fission are lost and are not effective for power purposes; investment charges are 15%, and the plant costs $600 per kw. Auxiliary power consumed within the plant will amount to 4% for the steam cycle and 7.5% for the nuclear plant. Average annual plant-labor cost will be $6000 each for 130 men, and maintenance will be 1 mill per kwhr. Calculate the cost of electrical energy from the plant for 50% depletion of the fissionable portion of the fuel. Note: The cost of natural uranium given above is a weighted average cost of natural uranium and the small amount of expensive highly enriched uranium needed to maintain the chain reaction.

SOLUTION. Annual generator output $= 120,000 \times 8760$
$$= 1.05 \times 10^9 \text{ kwhr per yr}$$
Annual reactor output $\quad = 1.05 \times 10^9 \times 11,300$
$$= 1.19 \times 10^{13} \text{ Btu per yr}$$

At an energy release of 3.5×10^{10} Btu per lb of fissionable uranium (Art. 14–4), the fuel consumed for no losses in the reactor is

$$\frac{1.19 \times 10^{13}}{3.5 \times 10^{10}} = 340 \text{ lb per yr of U-235}$$

With 15% losses in the reactor, the fuel consumption of fissionable U-235 is $340 \div 0.85 = 400$ lb per yr. But since natural uranium contains only 0.7% of fissionable uranium and it was stated that the fuel was to be depleted by only 50% (i.e., only half of the fissionable portion of the natural uranium would be consumed), the quantity of natural uranium required would be $400 \div (0.007 \times 0.50) = 114,300$ lb per yr of natural uranium. The annual cost of uranium would be $114,300 \times \$350 = \$40,000,000$. The annual plant investment cost is $150,000 \times \$600 \times 0.15 = \$13,500,000$. Annual labor cost is $130 \times \$6000 = \$780,000$ and the annual maintenance is $\$0.001 \times 1.05 \times 10^9 = \$1,050,000$. Since auxiliary power takes 11.5% of the generator output, the net output is $(1.0 - 0.115)1.05 \times 10^9 = 9.3 \times 10^8$ kwhr for an annual cost of $\$40,000,000 + \$13,500,000 + \$780,000 + \$1,050,000 = \$55,330,000$. The average cost of energy leaving the plant is 5.533×10^{10} mills $\div 9.3 \times 10^8$ kwhr $= 59.5$ mills per kwhr. Note that a modern fossil-fuel steam plant in a relatively high-cost fuel area would generate this energy at not more than 8 mills per kwhr.

PROBLEMS

14–1. How many kilowatt-hours of electrical energy could be generated at a thermal efficiency of 40% if the mass of 1 lb of coal could be converted into energy in accordance with the equivalence of mass and energy? How much difference if the fuel were wood?

14–2. A pound of fuel oil has a heating value of 19,000 Btu per lb. Calculate the mass defect for combustion of 1 lb of this oil.

14-3. For each of the following isotopes, calculate the mass defect and the binding energy; the mass of the isotope is the value given:

(a) $_3Li^7$ 7.0182
(b) $_8O^{16}$ 16.0000
(c) $_8O^{18}$ 18.0048
(d) $_{10}Ne^{20}$ 19.9989

14-4. Calculate (do not use Fig. 14-2) the energy released if fusion were possible for items a, b, c (masses are given in parentheses).

(a) $_6C^{12}$ (12.0038) to $_{12}Mg^{24}$ (23.9918)
(b) $_3Li^6$ (6.0170) to $_6C^{12}$ (12.0038)
(c) $_7N^{15}$ (15.0049) to $_{14}Si^{30}$ (29.9830)

14-5. Calculate the cost of power from the nuclear plant used in the example of Art. 14-8 if the plant were (a) subsidized by the government so that fuel cost was zero and (b) the initial cost of the plant was reduced to \$250 per kw.

14-6. Calculate the cost of electrical energy in mills per kilowatt-hour for each of the following fossil-fuel plants with a capacity of 200,000 kw if each operates for 8000 hr per yr at an average of 170,000 kw [plant (c) is a hydro plant].

	(a)	(b)	(c)
Fuel costs, cents per 10^6 Btu	32	10	....
Gross heat rate	9200	11,900	....
Auxiliary power, %	4.7	5.0	....
Cost of plant, dollars per kw	175	125	425
Investment charges, %	13	13	13
Maintenance, mills per gross kwhr	0.8	0.8	0.6
Avg annual operator's salary, dollars per yr	5500	5000	5000
Plant personnel	60	60	40

14-7. For each of the following nuclear plants calculate the cost of electrical energy in mills per kwhr at 50% fuel depletion:

	(a)	(b)	(c)
Plant capacity, kw	100,000	200,000	300,000
Avg load, kw	90,000	180,000	250,000
Operation, hr per yr	7,500	8,000	8,200
Fuel cost, dollar per lb	250	125	350
Turbine heat rate, Btu per kwhr	12,900	11,000	10,100
Auxiliary power (total), %	10	7	6.5
Cost of plant, dollars per kw	640	550	800
Investment charges	12.5	13	12
Maintenance, mills per gross kwhr	1	1	1
Avg annual plant-personnel salary, dollars per yr	6,100	5,800	6,800
Plant personnel	120	95	110
Neutron loss, %	20	10	12

BIBLIOGRAPHY

BALLINGER, R. MAXIL. *Atomic Energy Primer for Management.* New York: National Industrial Conference Board, Inc., 1955.

CISLER, WALKER L. "Electric Utility Progress in Atomic Power Development," *A Forum Report*. Atomic Industrial Forum, Inc. (June, 1955).

ELLIOTT, VORRAS A. *180,000-kw Dresden Nuclear Power Station*. Schenectady: General Electric Co., 1956.

HARLOW, J. H. "Projected Cost Outlook for Conventional Power," *Combustion* (January, 1956).

LANE, JAMES A. "Where Reactor Development Stands Today," *Nucleonics* (August, 1956).

MURRAY, RAYMOND D. *Introduction To Nuclear Engineering*. Englewood Cliffs, N. J.: Prentice-Hall, Inc., 1954.

STARR, CHAUNCEY. *A Sodium Graphite Reactor 75,000 Electrical Kilowatt Power Plant*. North American Aviation, Inc.

———. Articles on Nuclear Energy. *Power* (August, 1954, through July, 1956).

———. *Reports to the Atomic Energy Commission on Nuclear Power Reactors*. Vol. 1 and 2. Washington, D. C.: U. S. Government Printing Office, 1953 and 1955.

———. *Description of the Pressurized Water Reactor (PWR) Power Plant at Shippingport, Pa*. International Conference on the Peaceful Uses of Atomic Energy, 1955.

CHAPTER 15

ECONOMICS OF POWER PLANTS

15-1. Definitions. Fortunately for most power plants, all the electrical equipment connected to their lines is not in operation at any one time. Very few plants are large enough to supply sufficient current for such a demand. If a small unit, such as one comprising a city block, is considered, it will be noted that the habits of its customers vary. The time of rising in the morning and retiring at night, the evening's entertainment, the time of the meals, the appliances used in the housework—in fact nearly all details of the home life—will vary from dwelling to dwelling. If this is multiplied by several hundred thousand customers for the larger systems and consideration is taken of the different requirements of the industrial plants, it will be seen that there is very little likelihood of all the appliances requiring current at the same time.

Since there is no storage reservoir for large amounts of electricity, it must be generated as it is demanded by the customer. Before these effects of different demands on the power station can be considered, certain terms must be defined. The definitions given are those published by the American Institute of Electrical Engineers in their book, *American Standard Definitions of Electrical Terms,* as approved by the American Standard Association and the Canadian Engineering Standards Association.

Diversity factor is the ratio of the sum of the individual maximum demands of the various subdivisions of a system, or part of a system, to the maximum demand of the whole system, or part, under consideration. The particular subdivision under consideration may be one substation where there are several in a system or may be a group of residential customers supplied by one transformer. In the latter case it would be found that the sum of the maximum demands for all the residences would be more than the maximum demand on the transformer supplying these customers. As stated, the diversity factor will be a number greater than unity. At times the diversity factor is given as the reciprocal of that stated above.

Demand factor is the ratio of the maximum demand of a system, or part of a system, to the total connected load of the system, or part of the system, under consideration. Thus, in a residence the maximum demand would not be the sum of all lighting and appliance loads but would be some value less than that, since it would be extremely unlikely that all lights and appliances would be in operation at the same time.

Load factor is the ratio of the average load over a designated period of time to the peak load occurring in that period. The average load may be determined for any specified length of time such as a day, month, or year. The maximum demand may also be measured over various lengths of time such as a 15-min period, half-hour period, or an hour period. The first two are the most common. The load factor will have an important bearing on the cost of power. A low load factor indicates that the total capacity installed in the plant is used for short periods of time. Thus, a large investment is needed to generate a small load. The cost of the electricity will then be high to bring a return on an investment that is idle for a large percentage of the time.

Capacity factor is the ratio of the average load on a machine or equipment, for the period of time considered, to the rating of the machine or equipment. When applied to a plant, this factor is called *plant factor* or *plant-capacity factor*. If the capacity factor is applied to a plant for a period of a year, the plant is charged for energy that could have been generated during that period, as, for example, during the annual inspection or breakdowns. Therefore, some plant analysts prefer to use the output factor.

Output factor, or use factor, is the ratio of the actual energy output, in the period of time considered, to the energy output which would have occurred if the machine or equipment had been operating at its full rating throughout its actual hours of service during the period. This differs from the capacity factor only in that the load that could have been delivered by the plant is considered as the total capacity of the plant times the number of hours that the plant was in actual operation. However, it seems only right that the plant should be charged for imperfections, even though unavoidable, that cause shutdowns for any reason except a lack of load. The difference between the use factor and the capacity factor is indicative of the magnitude of the operation factor.

Load curve is a curve of power versus time, showing the value of a specific load for each unit of the period covered. The abscissa is usually time in hours, days, weeks, months, or years, and the ordinate is kilowatts generated. The most common period of time is hours. Figs. 15–1 and 15–2 are load curves.

Load duration curve is a curve showing the total time, within a specified period, during which the load equaled or exceeded the power values shown. Kilowatts are used as the ordinate, and normally, the 8760 hr of the year is the abscissa. The load duration curve of Fig. 15–3 indicates the number of hours during the year that the various loads are generated by the plant.

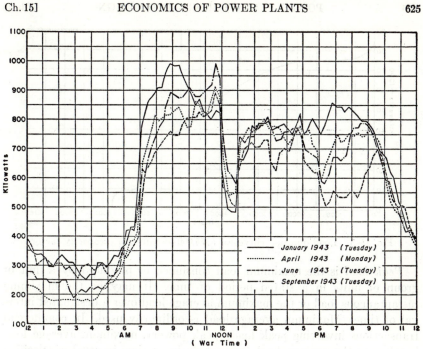
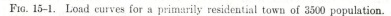

FIG. 15–1. Load curves for a primarily residential town of 3500 population.

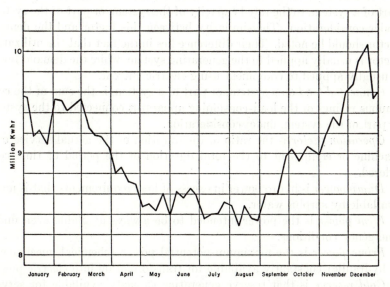

FIG. 15–2. Load curve for a large city and suburbs (industrial and residential for 1 yr).

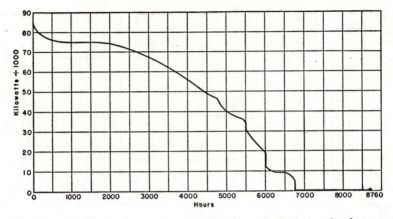

Fig. 15–3. Load-duration curve for a 75,000-kw unit, 50% capacity factor.

Peak load is the maximum load consumed or produced by a unit or group of units in a stated period of time. It may be the maximum instantaneous load or the maximum average load over a designated interval of time. Maximum average load is ordinarily used. In commercial transactions involving peak load (peak power), maximum average load is taken as the average load (power) during a time interval of specified duration occurring within a given period of time. The time interval is selected during the period in which the average power is greatest.

Utilization factor is the ratio of the maximum demand of a system, or part of a system, to the rated capacity of the system, or part of the system, under consideration. The similarity between this factor and the demand factor should be noted. Their difference lies in the fact that the utilization factor is usually applied to the generating system, while the demand factor is usually applied to equipment using electric current.

Connected load on a system, or part of a system, is the sum of the continuous ratings of the load-consuming apparatus connected to the system, or part of the system, under consideration.

Operation factor is the ratio of the duration of the actual service of a machine or equipment to the total duration of the period of time considered.

Dump power is hydro power in excess of load requirements that is made available by surplus water.

Firm power is the power intended to be always available even under emergency conditions.

Prime power is the maximum potential power (chemical, mechanical, or hydraulic) constantly available for transformation into electric power.

Cold reserve is that reserve generating capacity available for service but not in operation.

Hot reserve is that reserve generating capacity in operation but not in service.

Reserve equipment is the installed equipment in excess of that required to carry peak load. Reserve equipment not in operation is sometimes referred to as *standby equipment.*

Spinning reserve is that reserve generating capacity connected to the bus and ready to take load.

System reserve is the capacity, in equipment and conductors, installed on the system in excess of that required to carry the peak load.

Run-of-river station is a hydroelectric generating station which utilizes the stream flow without storage.

Spare equipment is equipment complete or in parts, on hand for repair or replacement.

Generating station auxiliary power is the power required for operation of the generating station auxiliaries.

House turbine is a turbine installed to provide a source of auxiliary power.

15–2. Actual Load Curves. The ideal load curve, from the standpoint of either the industrial plant or the central station, would be one of constant power generation for the entire 24 hr of the day. For such a condition, the units in the power plant could be operated at their most efficient point, the operating problems would be the simplest, and no banking losses would result during low-load periods. Another important factor would be that the first cost of equipment, per kilowatt hour generated, would be the lowest for any possible loading condition. Under wartime conditions, when industrial plants are operating on 24-hr schedules and when the machine tools are automatic and do not require time for inserting raw material, a constant load may be approached.

The load produced by street lights is practically constant in most cities, but the duration of the load varies from about 7 to 12 hr per day. However, street lighting represents only a small part of the total load in most communities.

The load curves from central stations serving both industrial and residential customers have certain common characteristics. From 2 AM to 7 AM the load is usually at a minimum. From 7 AM to 9 AM the load increases because of the commuter travel and industries beginning to operate. From 9 AM to noon the load remains nearly constant and is predominantly industrial. From noon to 1 PM the load drops sharply, representing the shutdown of industries for the noon hour. From 1 PM to 4 PM the load is usually at the same level as during the late morning hours, again due to industry. In winter the load will increase from 4 PM to 9 PM because of commuter travel, approaching darkness, and cooking

by electricity. The evening peak is often larger than that in the morning
or afternoon. From 9 PM to 1 AM the load will decrease. This decrease is
particularly sharp during the latter part of this period.

Load curves representing typical loads for the four seasons of the year
for a predominantly residential town of about 3500 population are shown
in Fig. 15–1.

These load conditions will be affected by many variables. Obviously,
the degree of industrialization of the community will have a most important bearing on the load curve. Sudden heavy storms, particularly around
4 PM, will cause severe loads to be put on the system because of the increase in lighting requirements. Winter will cause heavier night loads.
Usually, the heaviest residential load will occur at about 7 PM for a few
days before Christmas. Important athletic events, such as football, may
cause appreciable increases in load during a Saturday afternoon when
the load would otherwise be slack. Radio or television broadcasts of box-
ing matches or talks by prominent people have caused many central stations to use generators at night that normally would have been shut down.
Daylight saving time and the longer daylight hours in summer may reduce
the night peak until it is the same or less than the late morning and afternoon peaks.

An interesting display of the variation in load throughout the year is
often prepared by public utilities. The load curve is plotted on cardboard
for each day in the year. The cardboard is cut to leave the outline of the
load curve and these cards are stacked in order. The result shows the hills
and valleys of the load throughout the year.

The effects of the changing load versus the ideal load on the rates
charged by the utilities are to increase the first cost and to increase the
operating expenses. Sufficient generating capacity must be installed, along
with the necessary distribution facilities, to provide the maximum load
that is required at any time. However, this equipment is only working at
capacity for a very few hours per year. Also, during peak and low-load
periods it may be necessary to operate the equipment above and below its
most efficient point, thus increasing the operating cost. In the case of an
isolated station, as for example, an industrial plant, where the load is
practically zero at night, these conditions are even more important.
Utilities attempt to increase the off-peak loads by means of the rate structure as will be seen in subsequent sections on rates.

15–3. Rates, General. Public utilities operate under franchises issued
to them by government authority. In order to protect the large investments involved, these franchises are of necessity issued for long terms, and
in general, they permit the utility the sole use of the public right of ways
for the electric conductors. To a certain extent this gives the utility

monopolistic privileges in that no other concern may convey electric current along public streets. However, there are certain government agencies that have power to regulate the utility rates so that the public will be protected. By this means the utilities are allowed to make only certain limited profits.

This does not mean that the utility is free of competition. There are many small utilities that operate on private property such as in one city block. Their customers are the owners and tenants of buildings within the bounds of the block. In other cases small utilities will serve the customers located in small developments comprising several blocks where the streets are privately owned. Of much greater importance, however, are the private plants operated by industries to supply only the needs of that particular company. The public utility must keep its rates within reasonable bounds or many industries will build their own plants.

It is possible to divide private industrial power plants into two general classes—those that can economically justify their installation and those that are uneconomical but are retained because of other factors. In this latter group are those plants that are retained because of the high investment required to change over to utility service. Even though the return on the investment may be good, many concerns are not able to spare sufficient capital to make the necessary alterations. However, many times they are forced to make the change-over due to possible outages because of labor trouble or obsolescence of equipment.

In general, it may be said that with fair rates the small plant cannot compete with the large utility unless the small plant can sell or use its exhaust steam or its fuel is a surplus from the manufacturing process. Furthermore, the needs for exhaust steam must coincide reasonably well with the demand for electricity, so that the electricity really becomes a by-product. Paper mills require tremendous quantities of steam for water heating and for can dryers. Usually, the load cycles for steam and electricity coincide favorably. Other possible examples are hospitals, dye houses, tanneries, laundries, etc.

Diesel engines have been used, either with or without steam units, when the steam load is nonexistent or does not coincide with the electric load. A disadvantage to that source of power may be the time required to make repairs after a serious breakdown. If spare units are installed for 100% capacity for any type of prime mover, the initial cost is thereby doubled. One large public utility reports that during a recent 10-yr period only ten new diesel plants were installed within its territory, while eleven diesel plants were replaced with central-station service. A distinct advantage of diesel plants is their ability to shut down completely without any banking losses and their ability to start quickly from dead cold and to carry full load almost immediately.

The electric utility must also compete with equipment using other sources of energy, such as coal, oil, gas, and charcoal for melting furnaces, heat-treating furnaces and ovens, refrigeration, cooking, etc.

Sometimes manufacturing industries prefer to purchase their electrical energy rather than have their own plants because the cost of purchased power is an operating cost that is not subject to income tax. Also the manufacturing industries desire a high rate of return on their investment while a public utility is regulated by government agencies to permit only a low rate of return on the investment. Thus industries prefer to invest their capital in their manufacturing plant at a high rate of return and purchase their electric power.

The cost of electrical energy to the customer may be divided into several parts:

Investors' profit
Fixed costs (or investment charges)
Operating costs

Municipal plants do not come under the regulation of the government commissions since they are public property. Likewise, the rates charged by municipal plants would probably not include an item for investors' profit, but nearly all the customers will be taxpayers in the community who must share the expenses of the government. Therefore, there is an item in the rates to cover taxes in place of investors' profit. A distinct advantage in collecting taxes in this manner is that a delinquent customer will have his electric service disconnected. In addition, the tax payments are spread out into 12 installments per year so that there will be a minimum of opposition from the public. Large power users normally receive lower rates, and the tax portion of these rates will usually be reduced so that they will pay only their proportional part of the government expenses.

It is not unusual for utilities to allow a discount on the bills for prompt payment. Since all but a very small percentage of the customers take advantage of this discount, it is in reality added to the tariff so that it becomes a penalty on those who are delinquent in their payments.

15–4. Fixed Costs. These charges are defined as those expenses that do not vary with the output of the plant. For a public utility the exact items that should be included under this heading are determined by the Public Service Commission for the particular state in which the utility is located. The detailed list of items becomes long and highly involved; however, for any plant, private or public, the major items will be:

Cost of land, equipment, and transmission and
   distribution systems

   Interest on the debt
   Taxes and insurance
   Depreciation

There has been much discussion on the best method of determining the value to be carried on the company books for the cost of equipment. Two methods may be used—original cost or replacement cost new. Although the first method is usually preferred, there are advantages and disadvantages for both. The original cost method allows the value for the property to be established at the time of installation. This value will then remain on the books with proper allowances for depreciation for as long as the equipment is in operation. The values of the equipment will then depend on general prices at the time of installation. The latter method, replacement cost new, has the disadvantage of requiring a completely new estimate of the cost of equipment at periodic intervals, often 5-yr periods. This is a very costly procedure, but it indicates the replacement cost of the property at any particular time. Also, the cost on the books will change after each estimate and will depend on the general price levels existing at the time the estimate is made.

The costs of the building and land are usually unique for each plant. The type of district in which the plant is located is chiefly responsible for this variation. It is better to locate the plant as near as possible to the center of the load to reduce the transmission and distribution system cost as well as the losses involved with long lines. If the load center were in a congested industrial or possibly residential area, then, obviously, the cost of land would be high, the building would be attractive, and the smoke nuisance would be a factor.

The average cost of a large modern central station is about $175 per kw of installed capacity, not including the switch yard. This figure should be used with extreme care since there are so many variables. For example, if coal-handling facilities, bunkers, pulverizers, ash-handling facilities, dust collectors, etc., are eliminated by using oil or gas as the fuel, and outdoor or semi-outdoor construction is used, this cost may be reduced $25.00 to $60.00 per kw. Very small plants may cost $250 to $350 per kw.

Normally the cost of the plant is paid for by the money obtained from the sale of bonds and stocks. The interest that must be paid on the bonds will vary with business conditions. If the owner has the money and does not have to borrow it, an interest item should still be included to represent the income that would have been derived from the money if it had been invested or loaned on the market.

In recent years taxes on corporations have increased considerably. At the present time they constitute a major item in the cost of electricity to the customer. Taxes on private utilities have increased from 14.0% of the electric revenue in 1937 to 23.8% in 1956. Insurance is usually a much smaller item. Buildings and equipment must be insured and regularly inspected for possible accidents from fire and explosions.

The most important item in the fixed charge is that of depreciation. A part of the depreciation is due to the decrease in value of the property caused by wear and age, but the most important part for steam-power plants is due to obsolescence. The first part may be offset to a great extent by proper maintenance, but the latter part is difficult to predict. In order to obtain some uniformity among utilities, for accounting purposes the Public Service Commissions establish arbitrary limits on the life of equipment. It is customary to establish a reserve of sufficient amount to pay off the indebtedness at the end of the established life of the equipment.

The total depreciation reserve that must be accumulated during the life of the equipment will be the cost of the equipment minus the expected salvage value. The part of the fixed charges represented by depreciation will be the annual amount that must be set aside to build up the necessary depreciation reserve. Two common methods of determining this amount are *straight line* and *sinking fund*. In the first method, depreciation is assumed to occur in a straight line, or proportional to the years of service of the equipment. The same amount is put in the reserve fund each year throughout the life of the equipment. Any interest that is earned by the investment of the depreciation reserve is credited to the company income for that year. In the sinking-fund method a fixed sum is set aside each year, and interest on it is compounded periodically. The total amount placed in the sinking fund plus the interest that has been accumulated throughout the life of the equipment should equal the debt less the salvage value. Formulas for determining sinking-fund rates may be found in many engineering and accounting handbooks. Straight-line depreciation is a popular method for creating the reserve fund for power plants and will serve our purposes. (See Chap. 1.)

15–5. Operating Costs.   Those costs that are roughly proportional to the system load, as distinguished from those that have no relation to load, are referred to as operating costs. They include:

Labor
Fuel and supplies
Maintenance
Transmission and distribution costs
New business
Customers' expenses

The labor required averages about one man per 1000-kw capacity for both operation and maintenance for large central stations. This figure, which ranges from 0.5 to 2.5, will vary with the size of plant and the quality of labor employed. Labor will have an important bearing on the general economy of operation in the plant. Skilled labor will cost more per man-hour than unskilled labor; but by using skilled labor the plant will operate much more economically, emergencies will be handled better, and in general, maintenance expenses will be reduced.

The cost of fuel per unit of output will be affected by such factors as the efficiency of the plant, the load factor, cost of fuel delivered to the station, cost of storage of fuel to prevent shutdowns during strikes on the suppliers, premises, and strikes in the transportation system. The efficiency of large steam-power plants has been increased sharply during the last 20 yr. The efficiency is usually expressed in terms of the heat rate—the Btu input to the station required per kilowatt-hour transmitted from the station.

As an example of the increased economy due to improvements in design, consider the following case of a large central station. The heat rate of the plant as installed in 1905 with engine-driven units using steam at 175 psig and 400 F was 42,000 Btu per net kwhr. At that time this station was one of the most modern in existence. In 1909 vertical turbine generators were installed in this same plant using the same steam pressure but 25 F higher temperature, and the heat rate was reduced to 33,000 Btu per net kwhr. The plant was remodeled again in 1915 by the addition of horizontal turbine generators designed for steam at 215 psig and 550 F with a further reduction in heat rate to 21,000 Btu per net kwhr. In the late 1930's this station was remodeled for the third time by the addition of a topping unit with steam at 1250 psig and 900 F that exhausted into the old horizontal units. This time the heat rate was reduced to 12,400 Btu per net kwhr. While these figures are for only one particular station, they may be considered indicative of the times. The most modern plants employing throttle pressures of 2000 psig or higher, steam pressure of 1050 F, and single reheat to 1050 F have heat rates of about 9000 Btu per net kwhr. Supercritical-pressure plants with pressures up to 5000 psig and temperatures of 1200 F are expected to approach heat rates of 8000 Btu per net kwhr. The age of the plant, load factor, and method of operation will seriously affect the plant efficiency.

The pounds of coal required per kilowatt-hour may be determined by dividing the heat rate by the heating value of the coal. Thus, the coal rate for the plant just mentioned was reduced from 3 lb per net kwhr in 1905 to less than 0.89 lb per net kwhr based on coal having a heating value of 14,000 Btu per lb. The supercritical-pressure plant should have a fuel rate of less than 0.6 lb per net kwhr if the same fuel is used.

Plant heat rates will vary with the load from infinity at banking conditions to the best rate, which is often at about 80% of rated capacity. At night the load has been shown to be at the minimum. Even on large systems the night load may be so low that it can be carried by one large unit. Since the other units, however, will be required during the next day to carry the peak loads, they cannot be allowed to cool down completely because it may take several hours to bring them up to operating temperatures again. A boiler may take 10 hr or more to equalize the temperature stresses and a turbine 4 hr or more, depending on the size. Thus, the units that are not operating at night may have to be kept warm and near operating temperatures. This requires coal or other fuel in the boiler for which no electricity can be sold.

If one unit is carrying the entire system load, a fault on this unit or its auxiliaries would cause the entire load to be dropped. This can be very serious and it may take 3 or 4 hr to pick up the load again. Therefore, it is standard practice for some companies to carry spinning reserve, i.e., sufficient units on the line at all times, to take care of a failure of the one largest piece of equipment in operation at that time. All these requirements in the operating procedure tend to increase the heat rate for producing power, but the American public has expressed its willingness to pay for this service by its demands on the utilities. From an engineering standpoint the excess capacity that is maintained is reflected in the difference between the load factor and the capacity factor for the plant.

In the case of hydroelectric plants, the fixed charges completely overshadow the small operating costs, whereas for the steam plant the operating costs are a large part of the electric rate.

The customer's expense may be subdivided into such items as the cost of meter reading, billing, collection of revenue, publicity, promotion of electrical appliances, etc.

15–6. Rates. Much progress has been made in the past years in the methods of rate structures. The establishment of a satisfactory electric tariff that will fulfill all the many necessary requirements is exceedingly difficult. A tariff should have the following characteristics: simplicity, fairness, incentive, and uniformity.

The majority of customers for most utility companies will be residential users of electricity, and the next largest group will be general light and power consumers. These two groups may or may not use the greatest quantities of electricity, but from the standpoint of number of customers they are generally much larger than the large power users. It is essential that these customers should have a friendly attitude toward the utility and its policies. A simple domestic schedule and a small power-rate schedule that can be easily explained to these customers will do much to forward the company's public relations. The general public does not have the training required to understand thoroughly the complications involved in many rate structures. If the company cannot easily explain its rates to its customers and show them how to check their bill, many customers will feel dissatisfied.

Much trouble may be encountered by utilities serving large and widely separated communities if the rate structures are different in each section. Most utilities are gradually equalizing their city and suburban rates in spite of the fact that distribution and metering costs are higher for sparsely settled communities than for congested metropolitan areas.

Each customer should pay his proportionate share of the costs of service. A great many residential customers do not use more electricity than that covered by the minimum bill; but often the minimum bill does not cover the cost of current used plus the costs of the individual service, metering, and accounting, since public opinion will force low minimum charges. However, it is possible to proportion the various expenses between the large power users and the smaller consumers. United States courts have established that the property rights of a customer are violated by forms of rates that do not distinguish between regular and substantial users and those who use the utility's service occasionally and in slight quantities. The occasional user demands that the service be kept available at all times but pays only for the quantities used and pays nothing for the equipment required to keep service available to him on demand. Such customers decrease the capacity factor of a system.

One of the most important features of a tariff is its ability to induce larger consumption of electricity; as more customers take advantage of electrification, the cheaper it will become and the cheaper will be the appliances. Tariffs are sometimes referred to as *follow-on rate schedules* when there is an incentive feature included.

Electric rates may be divided into three classifications. The *flat rate* is based on the customer's installation, as a certain charge per lamp or per horsepower or per front foot of property. This type of rate was used to some extent in the early days of the industry to induce the installation of electric service. The charge was often based on the number of lamps installed and was not dependent on the hours of use of the lamps. A *customer's output rate* is a charge based on the customer's output, such as a certain charge per ton of ice manufactured by an ice plant. A *meter rate* is a charge based on the actual consumption of energy as determined by one or more meters. This type of rate may be subdivided into three variations: (1) a charge for each kilowatt hour of energy used, (2) a charge for the demand made on the lines as determined by assessment or by a demand meter, or (3) a combination energy and demand charge. Theoretically, the last method represents the fairest means of charge, since it combines both the cost of energy used and the cost of equipment needed to supply the customer's demand for energy. However, it will be realized that the expense of determining the customer's demand in the case of small residential users easily could be excessive.

The list of rates that follows was taken from the Report of the Proportional Rate Committee of the National Electric Light Association at the 52d Convention, with examples and additions taken from the published tariffs of utility companies.

(a) *Straight-Line Meter Rate.* The term *straight line* indicates that the price charged per unit is constant. This is the simplest of all meter rates. This rate is based on a flat charge per kilowatt-hour, such as 5¢ per kwhr. Under no circumstances could it be considered as encouraging the use of electricity, unless the rate were so low that it would be out of line with the usual charges in such rates.

(b) *Step Meter Rate.* The term *step* indicates that a certain specified price per unit is charged for the entire consumption; the rate depends on the particular step within which the total consumption falls each month. Disadvantages of the step rate are so outstanding that step rates should not be encouraged. It is irritating to a customer to realize that he can never predetermine his rate. In many instances this form of rate is practically equivalent to a straight-line meter rate because the top step is frequently so high that ordinary uses of electricity in the home would not bring the customer outside of the first step.

EXAMPLE 15–1.

 8¢ per kwhr for a monthly consumption of less than 200 kwhr
 6¢ per kwhr for a monthly consumption of more than 200 kwhr

(c) *Block Meter Rate.* The term *block* indicates that a certain specified price per unit is charged for all or any part of a block. Reduced prices per unit are charged for all or any part of suceeding blocks of units, each such reduced price per unit applying only to a particular block or portion thereof. This is one of the most popular types of rate schedules for residential customers. However, if the first block is so broad that only a few customers with electrical cooking and hot-water heating will use sufficient energy to come into the lower blocks, the rate becomes a straight-line meter rate.

EXAMPLE 15–2.

  75¢ covers the use of the first 12 kwhr per month
  5¢ per kwhr for any part of the next 28 kwhr per month
  3¢ per kwhr for any part of the next 35 kwhr per month
 1.7¢ per kwhr for excess over 75 kwhr per month
Minimum charge, 75¢ per month. For residential use only.

(d) *Wright Demand Rate.* The term *Wright demand* applies to that method of charge in which a maximum price per unit is charged for a certain amount of energy, depending on the number of rooms, floor area, maximum demand, connected load, etc., and one or more reduced prices per unit for the balance of the electricity used.

EXAMPLE 15-3.

10¢ per kwhr for electricity used equivalent to, or less than, the first 30 hr of use per month of the maximum demand

5¢ per kwhr for additional electricity used equivalent to, or less than, the next 30 hr of use of the maximum demand

3¢ per kwhr for electricity used per month in excess of the equivalent of 60 hr of use of the maximum demand

The obvious difficulty of this rate structure for residential use is the necessity of determining the demand. Meters for this purpose are too expensive. An assessed demand based on some percentage of connected load would be difficult to maintain with the constant addition of home appliances. Nevertheless, there is a distinct element of fairness in a rate that is based on both demand and consumption.

(e) *Block Hopkinson Demand Rate.* This term applies to a demand charge based on a maximum monthly demand plus a follow-on block rate for energy.

EXAMPLE 15-4.

$2.40 per month per kw for the first 50 kw of maximum demand in the month

$2.00 per month per kw for the excess of the maximum demand over 50 kw

0.5¢ per kwhr for the first 1000 kwhr used per month

0.3¢ per kwhr for the excess energy per month

This type of rate can be adapted to home use, but it has had very little, if any, acceptance. It is subject to the same limitations as the Wright demand rate for residential use.

(f) *Combinations of the Wright Demand and Block Hopkinson Demand Rates.* Combinations of the Wright demand and the block Hopkinson demand rates are popular when applied to small power, primary distribution, and high-tension customers. As in the block Hopkinson demand rate, the combination rate has a capacity charge, based on the demand, and an energy charge. However, in the combination rate the energy charge is based partly on the demand as in the Wright demand rate and partly on excess power as in the block Hopkinson demand rate. Three examples of this rate are given, one for each of the classes of service mentioned.

EXAMPLE 15-5.

(1) *General Light and Power (GLP) at 230 Volts.*

$1.00 per kw of demand, which includes the first 10-hr use of the demand

4.5¢ per kwhr for any part of the next 80-hr use of the demand

1.8¢ per kwhr for any part of the next 100-hr use of the demand

1.6¢ per kwhr for the excess use

Discount:

No discount on the first $10

10% discount on any part of the next $40

20% discount on the excess over $50

Demand: May be measured or assessed. If assessed, use

(a) Lighting, appliance, and cooking load

70% of the kw of the light load

40% of the kw of the appliance load

20% of the kw of the cooking load

(b) Power load

85% of the first 10 kw of the power load plus

55% of the excess over 10 kw of the power load

(c) Residence load

0.4 kw for each unit of residence

(Calculate demand to nearest 0.1 kw.)

(2) *Primary Distribution (PD)*. For use when the untransformed energy (2300 volts) is supplied to the customers' transforming and switching equipment.

Capacity Charge: Per kw of billing demand per month

$2.25 for any part of the first 50 kw

1.75 for any part of the next 50 kw

1.30 for any part of the next 100 kw

1.15 for the excess over 200 kw

Energy Charge: Per kwhr per month

1.0¢ for any part of the first 5000 kwhr

0.8¢ for any part of the next 100 kwhr per kw of billing demand

0.6¢ for any part of the next 200 kwhr per kw of billing demand

0.5¢ for the excess

Billing Demand: Determined each month by meter or not less than 25 kw

Minimum Charge: Capacity charge plus an energy charge for 100 hr of use of the demand

(3) *High-Tension Power (HT)*. For use when the untransformed energy (13,000 to 66,000 volts) is supplied to the customers' transforming and switching equipment.

Capacity Charge: Per kw of billing demand per month

$1.95 for the first 100 kw

1.25 for any part of the next 100 kw

1.10 for any part of the next 200 kw

1.00 for the excess over 400 kw

Energy Charge: Per kwhr per month

0.6¢ for any part of the first 300 kwhr per kw of billing demand

0.4¢ for the excess.

Billing Demand: Determined each month by meter or not less than 100 kw.

Minimum Charge: Capacity charge plus energy charge for 100 hr demand use of the demand.

The last two rates have a power-factor requirement of at least 80% for loads up to 50 kw, 90% for loads up to 1000 kw, and 95% for loads over 1000 kw. Since many plants, because of the nature of their loads, cannot reach the required power factor without installing correction of some type, provisions are made to adjust the billing in accordance with actual power factor of the plant. One method of doing this is to increase the measured demand by the ratio of the required power factor to the actual power factor.

One method of correcting the power factor is by means of static condensers. These may be either grouped at the service or installed on the various motors creating the low power factor. The latter method relieves circuit capacity and therefore is more desirable. Another method is to use synchronous motors operating at unity power factor. The investment for synchronous motors is so much greater than that for static condensers that they may not be economical. There is also the possibility that the cause of the poor power factor can be determined and corrected to some extent. Over-motoring is a very common cause. Arc welders and induction furnaces have extremely low power factor and nothing can be done with them except to install condensers.

Cost to the customer of installing switching and transforming equipment will be a very important factor in selecting the proper rate. Equipment may be located indoors in vaults or outside. Local laws and underwriters' requirements must be complied with. Metering equipment is owned and supplied by the power company.

15–7. Off-Peak Rates. As mentioned previously, the load on a central station will usually vary to a large degree during the 24-hr period. Any energy that can be sold during the low-load periods will not require additional generating capacity or other equipment. Therefore, large utilities have offered special rates, known as off-peak rates, to those who can use energy at these odd periods. The particular periods that are designated as off-peak hours will vary for different companies, depending on the load characteristics, and for one company may even vary with the type of service. In general, the off-peak hours will be from about 7 PM or 9 PM to 7 AM. These special rates are usually classed with many others into the general classification of riders.

EXAMPLE 15–6.

Off-Peak Water-Heating Rate.
>   This rate is intended primarily for residential use, at 115 to 230 volts, for use between 8 PM and 7 AM, between 9 AM and 11 AM, and between 2 PM and 4 PM. 0.9¢ per kwhr. Minimum Charge: $1.00 per month.

Primary Distribution Off-Peak Power.

Capacity Charge: 60¢ per kwhr of excess off-peak billing demand per month

Energy Charge: 0.5¢ per kwhr of night energy

Off-Peak Hours: 9 PM to 7 AM

Excess Off-Peak Billing Demand: Shall be the amount by which the greatest demand occurring during off-peak hours, as determined by measurement and adjusted for power factor, exceeds the billing demand for on-peak hours, whether the latter is minimum or actual demand.

Night Energy: Not less than 56% of the actual use during on-peak hours

Application: Rider may be applied to either or both the capacity charge and the energy charge.

High-Tension Off-Peak Power.

Capacity Charge: 50¢ per kw of excess off-peak billing demand per month

Energy Charge: 0.4¢ per kwhr of night energy

Off-Peak Hours: 7 PM to 7 AM

Excess Off-Peak Billing Demand: Shall be the amount by which the greatest demand occurring during off-peak hours, as determined by measurement and adjusted for power factor, exceeds the billing demand for on-peak hours, whether the latter is actual or minimum demand.

Night Energy: Not less than 67% of the actual use during on-peak hours

Application: Rider may be applied to capacity or energy charge, or both.

15–8. Riders. Riders are additions or amendments that are attached to the standard service contract to cover special conditions that frequently arise. They must be filed with the Public Service Commission as a part of the companies' tariffs. A few of the more common are listed together with a brief explanation.

Casualty Rider: Waiver of minimum requirements during riots, strikes, storms, floods, etc.

Construction Rider: Waiver of guaranteed minimum during construction.

Dual Service Rider: For service supplied from two sources to ensure continuous supply. This feature is used by many theaters, department stores, hospitals, etc. The throw-over may be accomplished manually or automatically with extra interlocking switches.

Emergency or Auxiliary Rider: Auxiliary source of power for customers that normally operate a private plant.

Extension Rider: Special charges where customers are located some distance from the power lines.

Farm Service Rider: Special rates for farms.

Meter Elimination Rider: For constant loads used during predetermined hours, such as bill boards, driveway lights, etc.

Seasonal Rider: For service supplied during certain months of the year only.

Temporary Service Rider: For service supplied for a short period.

Transformer Rental Rider: For supplying, operating, and maintaining transformer installations on the customer's property in connection with primary-distribution and high-tension rates.

PROBLEMS

15–1. A power station has a maximum demand of 55,000 kw when supplying three substations. For the following maximum daily demands on each substation, calculate the diversity factors (*a*) between substations and (*b*) for each substation:

	Substation A	Substation B	Substation C
Demand	22,300 kw	17,860 kw	26,940 kw
Distribution line 1.......	15,180 kw	9,570 kw	1,070 kw
Distribution line 2......	1,230 kw	4,800 kw	4,030 kw
Distribution line 3.......	5,370 kw	1,390 kw	7,380 kw
Distribution line 4.......	3,060 kw	1,680 kw	1,860 kw
Distribution line 5.......	6,760 kw	2,260 kw	9,570 kw
Distribution line 6......	....	1,930 kw	8,000 kw
Distribution line 7.......	....	2,720 kw	....
Distribution line 8.......	....	3,350 kw	....

15–2. Plot a load curve for the following hourly demands on a small plant. For what season of the year would you expect this load curve?

12 M	800 kw	6 AM	670 kw	12 N	940 kw	6 PM	940 kw
1 AM	680 kw	7 AM	720 kw	1 PM	890 kw	7 PM	770 kw
2 AM	630 kw	8 AM	750 kw	2 PM	1410 kw	8 PM	690 kw
3 AM	590 kw	9 AM	1280 kw	3 PM	1380 kw	9 PM	1070 kw
4 AM	590 kw	10 AM	1330 kw	4 PM	1430 kw	10 PM	1090 kw
5 AM	620 kw	11 AM	1360 kw	5 PM	1510 kw	11 PM	950 kw

15–3. If the plant in Prob. 15–2 had four 600-kw units, what would be the hours of operation of each unit? What would be the plant capacity factor? What would be the plant use factor? What would be the operation factor for each machine?

15–4. Assume that the load curve given in Prob. 15–2 is the average load during the month (30 days). Plot a load-duration curve using hours and percentage as the abscissa.

15–5. What size and number of turbine generators would you select to supply energy for the community represented by Fig. 15–1? Determine the hours of operation for each unit (show by replotting the load curve and adding block curves for the units), the plant capacity factor, plant use factor, plant operation factor, and the operation factor for each unit.

15-6. By using the load curves of Fig. 15–1 as average for their respective seasons, plot a load-duration curve.

15-7. Select the number and size of units that would best suit the community represented by Fig. 15–2.

15-8. A 35,000-kw plant has a utilization factor of 71% and a load factor of 39.6%. What is the average load on the plant?

15-9. A plant with an installed capacity of 300,000 kw has a utilization factor of 83% and a capacity factor 56%. Find the load factor.

15-10. After a 135,000-kw unit is installed, it is found that it can be overloaded. If the utilization factor on the machine is a maximum of 111%, what is the output of the machine?

15-11. If the operation factor of a unit is 78% and the output factor is 49%, what is the capacity factor?

15-12. A unit is rated at 80,000 kw. It operates for 6950 hr during the year and generates 350,000,000 kwhr. With a peak load of 75,500 kw, calculate: (a) load factor, (b) capacity factor, (c) output factor, (d) operation factor, (e) utilization factor.

15-13. A unit is rated at 135,000 kw, operates for 8237 hr during the year, generates 972,700,000 kwhr, and has a peak load of 156,250 kw. Calculate: (a) load factor, (b) capacity factor, (c) output factor, (d) operation factor, (e) utilization factor.

15-14. A 7500-kw plant cost $267.40 per kw to build. For an expected 20-yr life, a salvage value of 12%, and straight-line depreciation, what would be the annual depreciation reserve?

15-15. A 65,000-kw plant, costing $165.20 per kw to build, has a capacity factor of 45%, 20-yr expected life, 15% salvage value, average heat rate of 15,100 Btu per net kwhr, coal cost of $4.70 per ton in the bunkers, maintenance cost of 2.2 mills per gross kwhr, one employee per 1000-kw capacity at an average salary of $335.00 per month per man, 14% investment charges. Determine the cost of energy per kilowatt-hour at the station bus, using straight-line depreciation and neglecting the interest on the depreciation reserve. The HHV of the coal is 12,300 Btu per lb and auxiliary power is 5% of the generator output.

15-16. If the transmission, distribution, and billing costs are one and one-third times as great as the cost of energy at the station bus in Prob. 15–15, what would be the cost to the customer?

15-17. A plant contains two 150,000-kw units and costs $167 per kw to build. If fixed charges are 12.5%, calculate the cost of fixed charges in mills per kilowatt-hours for a capacity factor of (a) 93% and (b) 52%.

15-18. A certain plant contains three 20,000-kw units that cost $192 per kw to erect. If the capacity factor will be 47%, calculate the cost of fixed charges in mills per kilowatt-hours for (a) an REA plant that will have fixed charges of 5% and (b) a private utility that will have fixed charges of 13.5%.

15-19. Compare the fixed charges in mills per kilowatt-hours for a small REA plant containing three 20,000-kw units that cost $235 per kw to erect and a large private utility plant containing three 125,000-kw units that cost $158 per kw to erect. Take a capacity factor of 56% for each, REA fixed charges as 5%, and the private utility fixed charges as 14%.

15–20. A plant contains two 125,000-kw units that cost $182 per kw to build. The average heat rate for these units is expected to be 9100 Btu per gross kwhr with a capacity factor of 87.5%. Coal will cost 23¢ per 10^6 Btu and have a heating value of 12,200 Btu per lb. Fixed charges should be 15%, and there will be one employee per 2000 kw of capacity at an average salary of $4000 per yr. Auxiliary power amounts to 5.1%. Maintenance for the plant will be $1.00 per ton of coal burned. Calculate the cost of electrical energy in mills per kilowatt-hours.

15–21. A utility system has 1,000,000 kw of installed capacity. Because the plants were erected over a period of many years, the average cost is less than current erection prices and may be taken as $108 per kw. The plants burn coal that costs 33¢ per 10^6 Btu, and there is a maintenance cost of 4¢ per 10^6 Btu. The average heat rate is 16,200 Btu per gross kwhr at a capacity factor of 46%; the investment charges are 12.5%. Auxiliary power amounts to 6.1%, and there is one plant employee per 1800 kw of capacity with an average salary of $3900 per yr. Calculate the cost of electrical energy in mills per kilowatt-hours.

15–22. A customer used 500 kwhr of energy and has a connected load of: lighting, 2.5 kw; power, 1 kw; appliances, 2 kw; one residence unit, 1.9 kw. Calculate his bill for one month, using the GLP rate. What is his average cost of electricity?

15–23. A residence customer uses an average of 55 kwhr per month. What would be the increase in his average bill if he added an electric range that would consume 100 kwhr of energy per month? Use the block meter rate.

15–24. Calculate the bill for one month for a residence customer who uses 474 kwhr on the block meter rate.

15–25. What is the cost of electrical energy to a residential customer who uses 276 kwhr per month on the block meter rate?

15–26. A customer on the GLP rate uses 3438 kwhr of energy in one month and has a connected load of: 7.6 kw of lighting, 4.2 kw of power, 8.3 kw of appliance load, 3.2 kw of residence load, 6.6 kw of cooking load. Calculate the bill for one month.

15–27. The connected load for a customer on the GLP rate is as follows: 28.7 kw of power, 7 kw of appliance load, 5.3 kw of lighting load. What would be the bill for one month with 7482 kwhr of energy?

15–28. A customer must select either the PD or HT rate. He expects to have an average monthly use of 54,000 kwhr and a demand of 100 kw. It is estimated that he would have to spend $10,500 for transformers, switches, and room if he selects the HT rate and $2500 if the PD rate is selected. Which rate would be the best? What would be the load factor?

15–29. The cost of installation for a customer to receive energy would be about $200 under the GLP rate and $2600 under the PD rate. Which would you select if the average monthly demand were 100 kw and 12,000 kwhr of energy were used? What would be the load factor? Assume a transformer efficiency of 98%.

15–30. What would be the electric bill for the PD rate for one month with the conditions of Prob. 15–15 if the power factor were 75%?

15–31. A customer purchases service under the PD rate for a load of 80 kw and has a usage of 16,000 kwhr per month. His power factor is 68%. How much can he reduce his bill per month by increasing the power factor to the required value?

15–32. A prospective plant proposes to install approximately 60 hp in motor load and 10 kw of lighting. The plant will operate 40 hr a week at about 50% of full load. Peak demand will be about 60% of the full connected load. (*a*) What will be the load factor? (*b*) Will assessed or measured demand be better? (*c*) What rate will be the most beneficial? (*d*) What will service cost per month?

15–33. A large customer now purchasing service under the HT rate is considering installing a heat-treating furnace. It may be operated either by gas or by electricity. The demand of the furnace will be 500 kw. The furnace can operate entirely at night. This load, together with his existing load, will permit him to use economically the off-peak power rider. The increase in demand at night will be 200 kw over his day peak, and it is estimated that the furnace will require 80,000 kwhr a month. Gas will cost 40¢ per MCF, and 15 cu ft of gas is equivalent to 1 kwh. Which fuel will be most economical?

15–34. Consider the load curves of Fig. 15–1 to be average for their respective seasons of the year. If this energy is to be purchased by a local company and to be resold to the customers, what will be the cost per month (30 days) to the local company for each season of the year if (*a*) the PD rate is used, and (*b*) the HT rate is used? Would an off-peak rider be advisable? Why?

BIBLIOGRAPHY

AIEE. *American Standard Definitions of Electrical Terms.* New York, 1941.

FEDERAL POWER COMMISSION. *Statistics of Electric Utilities in the U. S.* New York: The Macmillan Co., 1943.

GRANT, E. L. *Principles of Engineering Economics.* New York: The Ronald Press Co., 1938.

KENT, W. *Mechanical Engineers' Handbook.* New York: John Wiley & Sons, Inc., 1950.

KNOWLTON, A. E. "Fourth Steam Station Cost Survey," *Electrical World* (December, 1939), p. 1585.

MARKS, L. S. *Mechanical Engineers' Handbook.* New York: McGraw-Hill Book Co., Inc., 1951.

CHAPTER 16

HYDROELECTRIC POWER PLANTS

16–1. General. Hydropower plant calculations may well be considered after the material discussed in Chap. 15. Definitions of load-duration curves, firm power, run-of-river stations, dump power, etc., should be reviewed because they play an important part in the design of a hydro plant.

Water wheels have been used for many years to produce power. They were of particular importance in the preparation of grain but have been used for saw mills, cotton mills, paper mills, blast furnaces, and even in chocolate factories. In 1607, the first application of water power on the North American Continent was for a grist mill at Port Royal, Nova Scotia.

It is common to classify water wheels into three types: *undershot* wheels, *breast* wheels, and *overshot* wheels. In the undershot wheel, the water enters the bottom of the wheel tangential to its periphery and impinges on the buckets or vanes. These wheels, used for heads up to 6 ft, have efficiencies in the order of 70%. The breast wheel is for heads up to about 16 ft, with efficiencies as high as 85%. Water enters between the bottom and top of the wheel at an angle and is prevented from leaving the wheel by a breast wall on the side of the wheel. Overshot wheels are used for the highest heads (the largest in the United States was for a head of slightly more than 60 ft) and may have efficiencies of up to 89%. Water enters the wheel at the top by being discharged from a flume.

Development of the hydraulic turbine began about 1820 in the United States. Francis and Boyden (the latter was the inventor of the hook gage for water-level measurement) and others in Europe developed many varieties of cast-iron turbines. The smaller size, metal construction, larger capacity, greater speed, and freedom from icing troubles proved to be the advantages of the turbine over the water wheel. Twenty-six days after the first operation of Edison's famous Pearl Street steam plant in New York, the first hydroelectric central station in the world was started in Appleton, Wisconsin, on September 30, 1882.

Since that time, the installed capacity of hydroelectric plants has increased to 26,000 mw, which supply 20.3% of the total electrical output of the United States. Much of this capacity is in plants operated and

645

owned by direct or indirect agencies and divisions of the federal government. Thus, these plants pay only meager taxes, whereas 23.8¢ of each dollar of electrical revenue for an average private utility is paid to local, state, and federal governments as taxes. Tax receipts lost to the government due to governmental plants must be paid in other ways by the public.

Many governmental hydroelectric plants supposedly combine recreational facilities, irrigitation, flood control, and/or navigation with the power-generation facilities. Part of the initial cost of the installation is allocated to these facilities. Since the major part of the electrical cost for a hydro plant is investment charges, allocation of part of the capital cost to other services reduces the major portion of the electrical cost. Private companies cannot make such allocations of the initial cost to other services even though their plants may also provide flood control, recreational facilities, pollution reduction, etc.

One disadvantage of most hydro plants is their dependence on the rate of water flow in a river. A prominent exception to this is the flow in the Niagara River. Except for most unusual conditions,* the Niagara River flow is relatively constant over many years. The extreme fluctuations in the water flow of most rivers between draught and flood conditions is well known. Nature's variations in river flows seldom coincide with man's use of electrical energy. To some extent, these fluctuations can be smoothed-out by water storage, but storage areas require vast amounts of flooded lands. For this reason, basically hydraulic systems, such as TVA, have found it necessary to supply *back-up* or *firming* generating capacity in the form of steam plants.

One method of storing water during low power-demand periods to supply energy during high-load periods is to use the *reversible pump-turbine*. The turbine may be of either the Francis, propeller, or Kaplan types, and when revolving in one direction, it acts as a turbine to develop power; when operating in the other direction, the unit acts as a pump. The pump-turbine is direct-connected to a generator-motor and may operate at different speeds when used as a pump than when used as a turbine.

Very large, modern steam plants require several hours to shut down and to start up but are not necessary during the weekends and at night to supply electrical energy. During these off-peak periods, electrical energy can be generated very cheaply, i.e., for the cost of the fuel only. Pump-turbine systems may employ off-peak power from steam plants to raise water from the reservoir at the hydroturbine outlet to the headwaters of the hydroturbine or to a higher reservoir. This water may be used during peak load periods to generate additional hydropower at a very low cost.

---

* On at least one occasion, flow in the Niagara River was completely stopped owing to extreme upstream winds.

Pump-turbines are usually physically larger than a turbine of the same capacity, and the motor generator is more costly than a standard generator.

In other cases, pumped-water storage in a special reservoir can be provided during high river-flow periods. During spring thaws or rainy seasons, the river flow may be able to develop more power than the electrical system can consume. Use of the surplus power to provide pumped storage can impound water for use during future peak load periods or dry seasons.

An interesting variation of this use of surplus hydraulic power is at Niagara Falls. During the tourist season the minimum flow over the falls during the daytime is maintained at 100,000 cfs by treaty with Canada. At night the flow is reduced to 50,000 cfs, providing an additional 50,000 cfs for power generation. However, this additional flow occurs at a time when the system's electrical load is low. Therefore, it can be used to drive a pump-turbine and transfer water from the river below the falls to special reservoirs 100 ft above the top of the falls, to be used for peaking power during the daytime.

It has been estimated that the undeveloped hydropower capacity of the United States is more than double the developed capacity. Some of this undeveloped capacity may not now be economical because suitable dam sites frequently are not near to load centers. However, changes in fuel costs or system characteristics may permit these sites to become economical in the future. Transmission costs may make the hydropower more costly than steam or other types of power; also, the cost of steam back-up, if firm power is required, increases the cost. In some instances it has been shown that steam power can be installed and operated at a lower total cost of electrical energy than for hydropower—when the costs are developed on an equal tax and interest structure and all hydro costs are included. It should be noted that steam plants require several hours to start up and carry load, whereas a hydro plant requires only minutes.

16–2. Hydraulic Equations. The horsepower developed by a hydraulic turbine may be calculated by deriving an equation based on the definition of a horsepower (33,000 ft-lb per min or 550 ft-lb per sec):

$$P = \frac{QH_n\rho E}{550} \qquad (16\text{-}1)$$

where P = turbine output, hp
Q = water flow, cfs
H_n = net effective head, ft
ρ = water density, lb per cu ft
E = turbine hydraulic efficiency

Taking the water density as 62.4 lb per cu ft,

$$P = \frac{QH_nE}{8.82} \qquad (16\text{--}2)$$

The term *net effective head* is worthy of further mention. It is defined [*] as the *difference in the total head for the water entering the turbine casing and the total head leaving the draft tube.* For the moment it will be sufficient to indicate that the draft tube is a conduit at the outlet of the turbine that conducts the water away from the turbine. Note that total head and not static head was used in the preceding definition.

Net effective head as defined here may be determined by deducting losses in the intake canal or intake conduit from the difference in elevation between the storage reservoir and the *tailrace*. The tailrace is the canal that is used to carry the water away from the plant.

For a turbine, *specific speed* may be defined as the speed of a hypothetical model turbine having the same configuration as the actual turbine, when the model would be of the proper size to develop 1 hp at a head of 1 ft. Note that the definition of specific speed for a turbine is in terms of horsepower and head, whereas for a pump it is in terms of flow and head.

If the turbine runner is considered as a form of orifice, then the flow, Q, will be proportional to the square root of the net head, H_n, times the runner diameter squared. The power, P, is the product of head and quantity,

$$P \propto H_nQ \propto D^2H_n^{3/2} \qquad (a)$$

or

$$D \propto \frac{P^{1/2}}{H_n^{3/4}} \qquad (b)$$

The ratio of peripheral velocity of the buckets to the velocity of the water jet is a dimensionless factor that may be called *speed factor*. In the case of steam turbines, this factor was called the velocity ratio and was designated as ρ. Then,

$$\rho \propto \frac{DN}{H_n^{1/2}} \qquad (c)$$

Since efficiency is a function of the speed or velocity ratio, and a hypothetical runner should have the same efficiency as the actual runner, then for constant efficiency,

$$D \propto \frac{H_n^{1/2}}{N} \qquad (d)$$

---

[*] Except for impulse turbines (See ASME Power Test Code for Hydraulic Prime Movers).

Equating Eqs. (b) and (d),

$$\frac{P^{1/2}}{H_n^{3/4}} \propto \frac{H^{1/2}}{N} \tag{e}$$

The constant of proportionality in this equation is the specific speed, N_s.

$$N_s = \frac{NP^{1/2}}{H_n^{5/4}} \tag{16-3}$$

The specific speed for a hydraulic turbine is usually taken either for fully open inlet gates and at design head or at the point of maximum efficiency. Since the actual head on the turbine may fluctuate with river flow, design head is usually selected as the weighted, average, net head when weighted on the basis of kilowatt-hours generated.

16-3. Types of Hydraulic Turbines.   Generally, hydraulic turbines are divided into three basic types; each type is suitable for a particular range of net heads.

Impulse turbines, sometimes known as *Pelton wheels* in honor of the inventor, Lester A. Pelton of California, may be used for heads above 200 ft but are most commonly applied to heads above 700 ft. However,

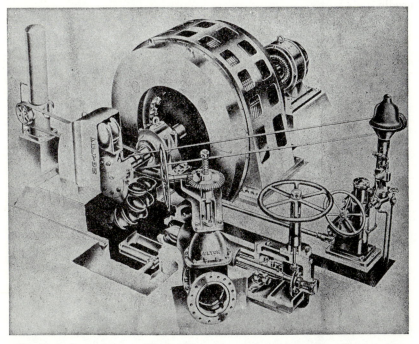

Fig. 16-1. Impulse type of hydraulic turbine with jet deflectors. (Baldwin-Lima-Hamilton Corp.)

the range of 200 to 700 ft is used only for small units so as to produce suitable generator speeds. Heads of 5000 ft have been used in European installations. Impulse turbine efficiencies vary from about 82 to 90%.

A view of an impulse turbine with part of the casing removed is shown in Fig. 16-1. This is a horizontal-shaft overhung unit. The generator, exciter, and belt-driven governors are also shown. Impulse turbines may have vertical shafts, although the horizontal shaft is probably more common. Large units have been designed with two runners (double over-hung), one on each side of the generator. It is possible to use multiple jets on a single runner. The characteristic shape of hydraulic impulse runners should be noted from this picture.

Turbine speed for the unit of Fig. 16-1 is controlled from the governor by the *jet deflector* shown. Full load is obtained when the deflector is rotated downward out of the path of the jet. Some jet deflectors cut in from the top rather than the bottom. The needle nozzle is operated by hand. With long intake pipes, slow operation of the manual nozzle mini-mizes water hammer produced in the pipes. Obviously, this procedure of governing is wasteful of water and should be used only when there is an ample supply. Where economy of water is essential, operation of the quick-acting deflector is followed by automatic, slow adjustment of the needle.

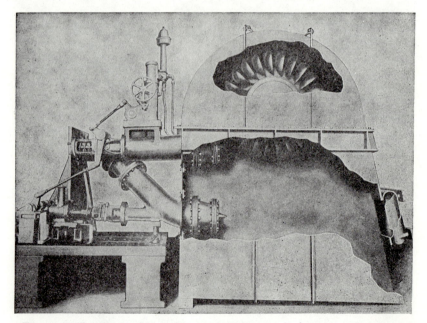

Fig. 16-2. Impulse turbine with synchronous nozzle by-pass and energy de-stroyer. (S. Morgan Smith Co.)

Another method of control to reduce water hammer in the intake pipes of a turbine having an automatically controlled needle valve is shown in Fig. 16–2. A by-pass nozzle, operated by the governor, is opened as the main nozzle needle is closed. A restoring spring gradually closes the by-pass nozzle.

In the case of impulse turbines, the *net effective head is measured at the bottom of the pitch circle of the runner with no consideration given to the water level at the tailrace* (because there is no draft tube). Note that there can be no suction head on an impulse turbine.

Reaction turbines, called *Francis type* after James B. Francis, the American who developed this type of runner, are medium head units. They are usually employed for heads of 70 to 900 ft, and at times, up to 1100 ft. The upper range of heads overlaps the values for impulse runners.

Suitable combinations of speed, horsepower, and head for various runner designs are related by the concept of specific speed. Fig. 16–3

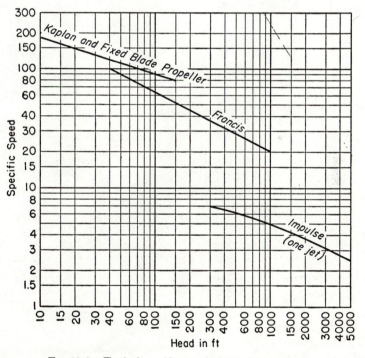

Fig. 16–3. Typical specific speeds for turbine runners.

shows typical values of specific speed for the three types of runners operating at sea level. These values are determined from experience and laboratory model tests. It can be seen from these curves that a Francis wheel for low horsepower and high head would produce a very high generator speed, making an impulse runner more adaptable in this range.

Francis turbines may have either a horizontal or vertical shaft. The latter is more common today, particularly on large sizes, and is shown in Fig. 16–4. Efficiencies at design point range from 90 to 94%. This illustration will be used to demonstrate several important terms. Water is conducted from the reservoir to the turbine by means of a *penstock* or *headrace*. When the powerhouse is some distance from the dam, steel penstocks may be used. Constructional features and economics will deter-

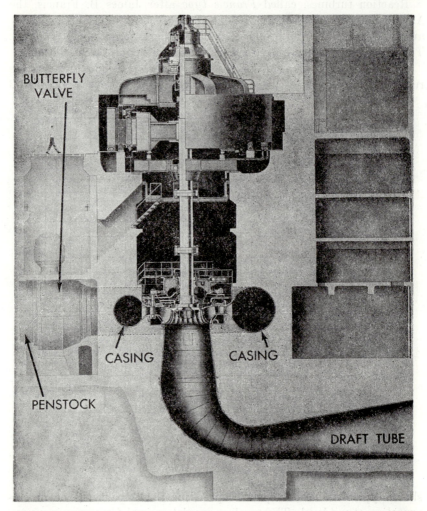

Fig. 16–4. Francis turbine for 115,000 hp at 480 ft head and 180 rpm. (Allis-Chalmers Mfg. Corp.)

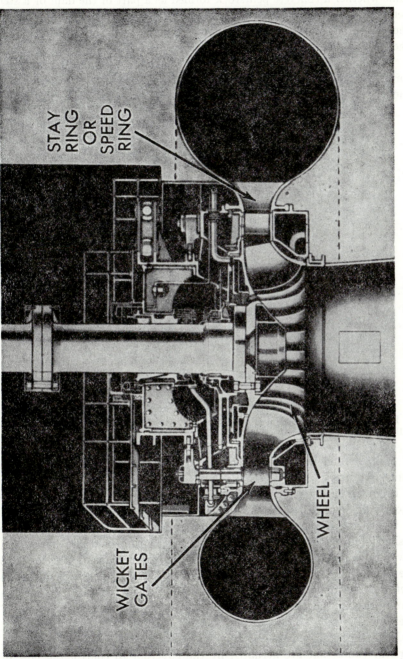

Fig. 16-4 (continued).

mine whether a concrete penstock formed as a part of the dam or a steel penstock is more suitable when the powerhouse is located at the base of the dam. A butterfly type of shut-off valve is shown at the entrance of the turbine casing.

The *spiral case* conducts the water around the turbine. It may be of formed concrete, cast iron, flanged or welded or riveted steel plate, as shown in Fig. 16–5. For low heads and small capacities, the casing may

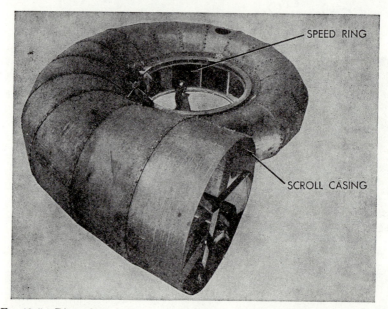

SPEED RING

SCROLL CASING

FIG. 16–5. Riveted steel-plate spiral or scroll case for a 26,000-hp Francis turbine. (Allis-Chalmers Mfg. Corp.)

be omitted, and the turbine may be located at the bottom of an open flume.

From the casing, the water flows through a *stay ring* or *speed ring*, frequently made of cast steel, shown in Fig. 16–5. The speed ring is fastened to the casing if the casing is steel, or grouted into the concrete if the casing is concrete. Vertical ribs in the speed ring hold the upper and lower rings together, act as a guide for the water, transmit the loads to the foundation, and hold the casing together (when the casing is made of metal). The upper and lower rings provide a smooth entrance for the water leaving the casing and entering the turbine and also accelerate the water.

Next, the water passes through the *wicket gates*. These are movable, vertical vanes that are actuated by the governor to control the flow of water and therefore the energy supplied to the runner. The combination of stay ring, wicket gates assembly, cover, and bottom ring is called the *distributor*.

After flowing from the wickets through the *runner*, the water leaves through the *draft* tube, which cannot be used on an impulse runner. The discharge from a Francis wheel fills the entire cross-section of tube. This makes the distance from the wheel outlet to the level of the tailrace a column of water and adds to the effective turbine head. In addition, there is an appreciable velocity of the water leaving the runner. Since the draft tube has a venturi shape, some of this kinetic energy is con-

Fig. 16-6. A 120,000-lb Francis runner for a 57,000-hp, 112-ft head, 85.7-rpm turbine. (Allis-Chalmers Mfg. Corp.)

verted to pressure energy. Thus, the draft tube performs two functions, i.e., it provides a column of water from the runner outlet to the tailrace and also converts kinetic energy to pressure energy. These two functions cause a vacuum condition to exist at the runner discharge.

Note in Fig. 16–4 that the cross-section of the runner bears a strong resemblance to a centrifugal pump impeller. This resemblance is also

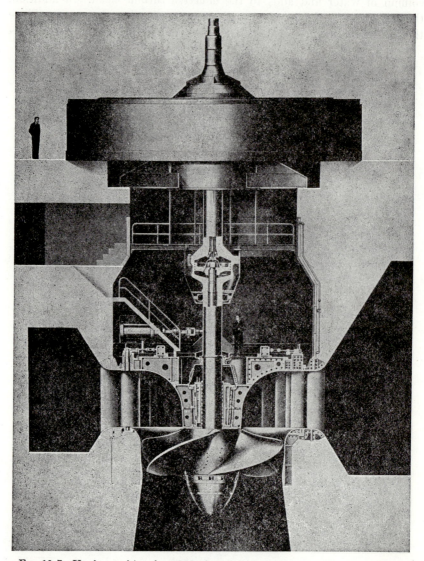

Fig. 16–7. Kaplan turbine for 42,000 hp at 52-ft head and 94.7 rpm. (Baldwin-Lima-Hamilton Corp.)

evident when comparing the small, cast, stainless-steel Francis runner in Fig. 16–6 with the Francis pump impeller of Fig. 3–19(a).

The propeller type of turbine wheel has a strong resemblance to a propeller-pump rotor, Figs. 3–29(c) and 16–7. There are two styles of these turbines, *fixed blade* and *adjustable blade* units. The latter are called *Kaplan turbines* after the man who perfected their design. Note that the scroll case shown in Fig. 16–7 is formed of concrete and is not circular in cross-section. Efficiencies at design point range up to 93%.

Although the range of heads for the propeller units shown in Fig. 16–3 indicates a maximum of 100 ft, it is possible to design these turbines for heads of 200 ft or even higher under certain special conditions.

As can be seen from Fig. 16–3, there is an overlap in the typical applications of Francis and propeller turbines. The lower range of heads shown for Francis wheels will produce very slow-speed generators except for small capacities. Because of the higher permissible specific speeds for propeller units, they can operate at higher generator speeds.

A Francis or a fixed-angle propeller runner will have a rather steep efficiency versus horsepower curve. Curves for six different blade angles of a Kaplan type of runner are shown in Fig. 16–8. Obviously, no one

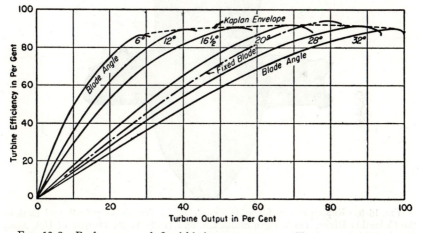

Fig. 16–8. Performance of fixed-blade runner versus Kaplan runner at same head. Individual Kaplan blade angles tested by the index method. (S. Morgan Smith Co.)

blade angle will be most suitable for all loads. The Kaplan unit provides a means of automatically changing the blade angle by the same governor mechanism that operates the wicket gates, and in coordination with the wicket gates by means of a mechanism inside the runner hub. Although more costly than a fixed-blade propeller unit, the Kaplan unit is much more economical when a limited supply of water is available or when the unit must operate at varying loads.

Note that the point of maximum efficiency for the fixed-blade unit has been selected at 80% of full-gate opening, i.e., 80% of full load; this would represent a typical comparison. Also observe that the maximum efficiency of the fixed-blade unit is slightly better than that of the Kaplan unit. This higher efficiency is due to smaller clearances at the blade periphery. Another factor is the smaller hub used on fixed-blade runners, since there is no internal mechanism to adjust the blade angles.

A Kaplan runner is shown in Fig. 16–9. The mechanism for varying

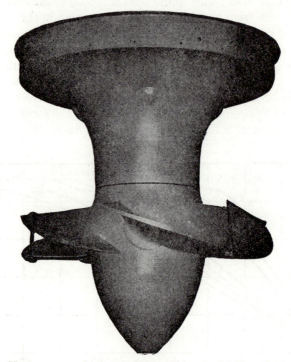

FIG. 16–9.   Kaplan runner for a 111,300-hp turbine for McNary Lock and Dam on the Columbia River. (S. Morgan Smith Co. and Walla Walla District of U.S. Army Corps of Engineers.)

the blade angle is inside the runner and is operated by a shaft inside the main shaft.

16–4. Turbine Speed. Nearly all modern hydro units in the United States are direct-connected to electrical generators that provide 25- or 60-cycle current (40-cycle frequency is used in some paper mills and

50-cycle frequency is employed in some foreign countries). These generators must have an even number of poles. For the speeds used for units of even a few hundred horsepower, it is preferable that the number of poles be divisible by 4 for greater flexibility in generator design. Turbine speeds are therefore limited by the following equation, in addition to the data of Fig. 16–3.

$$N = \frac{120f}{p} \qquad (16\text{-}4)$$

where N = turbine generator speed, rpm
f = electrical frequency, cps
p = number of generator poles

Large generators may have a speed of 60 rpm; speeds as low as 40 rpm have been considered. Most generators of more than a few thousand kilowatt capacity operate at speeds up to 300 rpm, although there are some exceptions.

EXAMPLE 16–1. A 30,000-hp turbine is to be installed for a 50-ft head. Select typical speeds for a Kaplan or Francis wheel for 60-cycle current.

SOLUTION. From Fig. 16–3, the specific speed for a Kaplan wheel should be about 111, while for a Francis wheel it should approximate 90. Rearranging Eq. (16–3), the speed for a Kaplan turbine would be

$$N = \frac{N_s H_n^{5/4}}{P^{1/2}} = \frac{111(50)^{5/4}}{(30,000)^{1/2}} = 85.2 \text{ rpm}$$

Next, it is necessary to determine the nearest generator speed from Eq. (16–4), wherein the number of poles is divisible by 4. Assume 84 poles. Then the actual speed would be

$$N = \frac{120 \times 60}{84} = 85.8 \text{ rpm}$$

or 90 rpm for 80 poles.

For the Francis wheel, from Eq. (16–3),

$$N = \frac{90(50)^{5/4}}{(30,000)^{1/2}} = 69 \text{ rpm}$$

Assuming 104 poles, the actual speed would be

$$N = \frac{120 \times 60}{104} = 69.3 \text{ rpm}$$

or 72 rpm for 100 poles. Therefore, either a Kaplan unit at 85.8 or 90 rpm or a Francis unit at 69.3 rpm could be used. However, 90 rpm would probably be more economical as a generator speed; in addition, the flat efficiency curve of the Kaplan unit might be more satisfactory if part-load efficiency is important. Since

Kaplan turbines usually have a higher than normal thrust and runaway speed, these must be investigated before a selection can be made.

16–5. Cavitation. The phenomenon of cavitation was defined in Art. 3–8 as the implosion of vapor bubbles in a liquid. These bubbles are formed by the flashing of some of the liquid into vapor caused by a reduction of the liquid pressure below the vapor pressure. When the liquid pressure is then increased above vapor pressure, the bubbles implode with a release of large amounts of energy. Some small amount of this energy is dissipated as sound. The remaining energy causes vibration of the equipment and also tears away part of the surface of the boundary metal. When cavitation occurs in pumps and turbines, the metal becomes pitted or honeycombed. The efficiency and maximum power of a unit may be badly impaired by severe cavitation.

Cavitation is most likely to occur on the outer edge of the back of Francis and propeller-runner blades and on the band of Francis runners. Because propeller runners operate at high specific speeds, they must be set lower to avoid cavitation. Cavitation on the runner can be controlled by the elevation of the runner above (or below) tailwater level. Other points at which cavitation can occur are on draft-tube walls, at sharp corners or restrictions, and on the needle and deflectors of impulse turbines. Impulse runners may encounter cavitation on the back edge of the bucket lip.

Tests of model runners are the most reliable means of predicting cavitation. From these tests the proper elevation of the bottom of a Francis wheel or the centerline of a propeller wheel can be determined. This elevation is one of the most important dimensions that must be determined for the plant and involves the use of the sigma function (σ).

However, preliminary calculations may be made by establishing the minimum permissible pressure at the wheel as being equal to the vapor pressure (H_{vap}); some manufacturers use an arbitrary value of 0.6 ft for H_{vap}. This will be equal to the barometric pressure (H_b), less the pressure due to the elevation of the wheel above tailwater level (H_e), less the velocity head at the wheel outlet (H_v), all expressed in feet of water; or,

$$H_{vap} = H_b - H_e - H_v \tag{16–5}$$

The velocity head at the runner outlet is proportional to the square of the velocity at this point. In turn, the velocity is proportional to the volume of water flowing, so that the velocity head is proportional to square of the flow. But the square of the flow is proportional to the turbine net head (H_n). Therefore, the velocity head is proportional to net head. Then, taking σ as the constant of proportionality,

$$H_v = \sigma H_n \tag{16–6}$$

Combining Eqs. (16–5) and (16–6),

$$\sigma = \frac{H_b - H_e - H_{\text{vap}}}{H_n} \qquad (16\text{–}7a)$$

and

$$H_e = H_b - H_{\text{vap}} - \sigma H_n \qquad (16\text{–}7b)$$

If the runner is above tailwater level, H_e is positive; if the runner is below tailwater level, H_e is negative. Values of the vapor pressure may be obtained from the steam tables and converted into feet of water for the summer water temperature. Barometric pressure is that existing at the plant elevation and not the barometric pressure corrected to sea level. For most purposes, it is satisfactory to assume the barometric pressure as 34 ft of water less 1.13 ft for each 1000-ft increase in elevation above sea level.

The constant of proportionality, σ, is called the cavitation factor and is assumed to be constant for all heads on a given runner and for all proportionally similar (*homologous*) runners. Actually, σ varies with gate and blade angle. The minimum σ necessary to prevent cavitation is the *critical σ*. The *operating σ* is the value at which the turbine actually operates and should exceed critical σ by an ample margin to prevent cavitation due to unforeseeable variations in equipment manufacture and in operating conditions. Approximate values of σ suitable for the solution of problems may be obtained from the following equations:

For propeller turbines,

$$\sigma = \frac{(N_s)^2}{15,000} - 0.2 \qquad (16\text{–}8)$$

For Francis turbines,

$$\sigma = \frac{(N_s)^2}{15,000} \qquad (16\text{–}9)$$

Some metals are more subject to the pitting effects of cavitation than others, porous materials being the most susceptible. Cast iron, which is used for small, low-head turbines of all three types, is the most susceptible to cavitation pitting. Some bronzes have about one-third the rate of pitting of cast iron, while cast carbon-steel exhibits only one-eighth the rate of pitting of cast iron. Cast stainless-steel (18% Cr, 8% Ni) aluminum bronze has about one-sixtieth the rate of cast iron. Most runners today are made of cast steel, and many are protected with a welded-on layer of stainless steel over the areas more likely to be subjected to pitting due to cavitation. Other runners may be provided with stainless-steel inserts or blade segments welded in place.

EXAMPLE 16–2. For the 30,000-hp, 50-ft head, 90 rpm Kaplan unit of the preceding example, determine the elevation of the propeller centerline for a 3000-ft elevation and 80 F water.

SOLUTION. For the actual speed of 90 rpm, the specific speed from Eq. (16–3) would be

$$N_s = \frac{90(30,000)^{1/2}}{(50)^{5/4}} = 117.2$$

From Eq. (16–8),

$$\sigma = \frac{(117.2)^2}{15,000} - 0.2 = 0.717$$

Barometric pressure equals $34 - 1.13 \times (3000/1000)$, say, 30.6 ft. At 80 F, from the steam tables, the vapor pressure is 1.0321 in. Hg or $(1.0321/12)$ $(0.01608/0.001183) = 1.17$ ft of water (0.01608 is the specific volume of water at 80 F from the steam tables, and 0.001183 is the specific volume of Hg).

From Eq. (16–7b),

$$H_e = 30.6 - 1.17 - 0.717 \times 50 = -6.4 \text{ ft}$$

Since this is a minus value, the centerline of the runner should be set at least 6.4 ft below the minimum tailwater level that can occur when the turbine is developing 30,000 hp.

16–6. Types of Hydro Plants. The dam provided for a hydro plant backs up a *reservoir*, of which the portion immediately upstream of the plant intakes is the *forebay*. This reservoir may be classified as either *storage* or *pondage*. Storage is intended to mean the impounding of excess river flow for use during dry seasons, while pondage indicates the accumulation of water to provide for the load changes throughout a period of one week. Storage provides more than sufficient water to accomplish the functions of pondage, but this is not true of pondage.

Hydro plants may be classified as (1) storage or reservoir plants or (2) run-of-river plants with or without pondage. Examples of storage plants are at Hoover Dam on the Colorado River and at Grand Coulee Dam on the Columbia. Reservoir plants may be for base load or peak load; the type of loading may change throughout the year, depending on river flow and other water demands such as for irrigation or flood control.

Most plants fall into the run-of-river category, of which there are several varieties. Those that are run-of-river plants without pondage may be base-load plants if their capability is equal to or less than the minimum river flow. The plants at Niagara Falls are in this class, as are some plants on the St. Lawrence River. Their capacity factor is very high, approaching 100%. Such installations are rare, since, for most rivers, the plant unit cost would be high because the plant capacity would be well under the river potential, i.e., for a slight increase in cost the plant capability could be greatly increased during most of the year. Most run-of-river plants without pondage are for peak load. Their deficiency in pondage may be due to lack of proper terrain, the necessity of maintaining certain water levels for navigation purposes, or the necessity of maintaining certain minimum flows because of prior riparian rights, preservation of fish life, sanitary conditions, etc.

Many hydro plants are in the category of run-of-river plants with pondage. The plants may provide base-load or peak-load power. During seasons of high-water flow, the plant may be base-loaded, while during dry seasons it may be peak-loaded. In either event, these plants are pondage plants if the pond is sufficient to take care of the load fluctuations during a week. During flood periods it is not uncommon for run-of-the-river plants, with or without pondage, to be made inoperative by excessively high tailwater levels.

16–7. Water-Flow Curves. Stream-flow data are collected for the important rivers of the United States by the U.S. Geological Survey (USGS). Some data for specific locations are also collected by other fed-

TABLE 16–1

MEAN WEEKLY DISCHARGE OF BETA RIVER FOR ONE YEAR, IN CFS

Week	Discharge	Week	Discharge
1	3,900	27	1,560
2	3,560	28	1,740
3	3,020	29	1,420
4	2,450	30	1,250
5	2,300	31	1,310
6	2,610	32	2,280
7	2,540	33	1,670
8	2,230	34	1,250
9	1,990	35	1,720
10	1,870	36	1,460
11	5,340	37	1,050
12	6,410	38	990
13	14,380	39	870
14	19,110	40	750
15	13,760	41	930
16	8,170	42	720
17	4,240	43	710
18	3,520	44	1,140
19	4,070	45	920
20	2,860	46	1,250
21	1,910	47	1,760
22	1,850	48	1,530
23	2,130	49	2,070
24	1,980	50	3,120
25	1,740	51	4,610
26	1,890	52	4,190

eral agencies such as the Bureau of Reclamation, the Corps of Engineers, etc., and by some private organizations.

To be of use for hydro-plant design purposes, the data should list the stream flow for at least 20 yr, preferably more. For complete analysis, the data should include the daily flow and should include periods of minimum flow and flood flows. Minimum flows are essential for establishing base load or pondage, while flood flows establish *spillway* or *overflow* capacity. USGS publications are usually on the basis of the average daily flow in second-feet (cfs) units. Tables of the average weekly stream discharge in second-feet may be used, but they do not provide such reliable data. In order to simplify the calculations, data on the basis of mean weekly discharge in second-feet are given in Table 16–1 for a hypothetical river, the Beta River, for only 1 yr of low-river flow.

There are various forms of duration curves, such as flow-duration curves and power-duration curves (either horsepower or kilowatt). Assuming a given head and efficiency, Eq. (16–2) gives the relationship between the flow- and power-duration curves.

A flow-duration curve for the Beta River could be developed by determining the percentage of time for which various flows are equaled or exceeded. For example, the maximum flow of 19,110 sec-ft is equaled for one week, or 1.92% of the time; 14,380 sec-ft is equaled or exceeded during two weeks, or 3.85% of the time; 13,760 sec-ft is equaled or exceeded

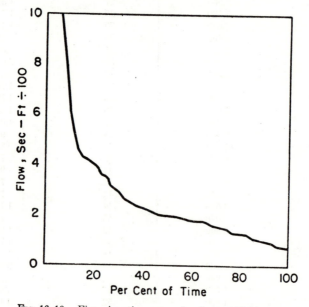

Fig. 16–10. Flow-duration curve for data of Table 16–1.

during three weeks, or 5.78% of the time; and 710 sec-ft is equaled or exceeded 100% of the time. The flow-duration curve is shown in Fig. 16–10.

If 710 sec-ft is taken as the lowest, average, weekly flow on record, a plant built for this maximum flow through the turbines could be base-loaded 100% of the time (neglecting time for maintenance) at full capacity. Since the flows given in Table 16–1 were average weekly discharge, it is assumed that there would be sufficient pondage at the plant to provide this average flow throughout the week. The power represented by this flow could be considered firm power.*

It is assumed that the plant would have a combined generator and turbine efficiency of 88% and a head of 100 ft. Substitution in Eq. (16–2) indicates a capacity of

$$P = \frac{710 \times 100 \times 0.88}{8.82} = 7080 \text{ hp}$$

or
$$0.746 \times 7080 = 5280 \text{ kw}$$

Then the generation during the entire 8760 hr of a year (assuming no outage) would be $8760 \times 5280 = 46{,}300{,}000$ kwhr.

Another method of graphing the data of Table 16–1 is by means of a *hydrograph,* Fig. 16–11 (the first week is taken as the week beginning January 1).

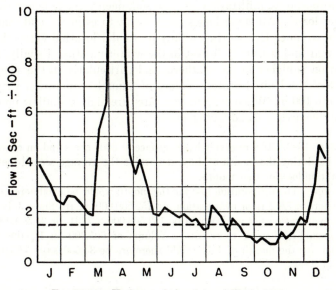

Fig. 16–11. Hydrograph for data of Table 16–1.

* Sometimes firm power for hydro plants is taken as power available for 95 to 97% of the time.

If the terrain and other factors are suitable, addition of storage to provide more water during the dry season would enable the generation of more energy with some increase in plant cost. A dashed line representing a flow of 1500 sec-ft has been drawn on Fig. 16–11. The reservoir would need to be large enough to store water during wet seasons so that the flow of 1500 sec-ft could be maintained during the dry periods of July through November. Assuming that the storage reservoir is full on the first day of January, the data of Table 16–1 or Fig. 16–11 show that during the dry periods, the storage reservoir must supply 5740 sec-ft weeks to maintain a discharge of 1500 sec-ft. A second-foot week is a flow of 1 cfs for a period of one week, or $3600 \times 24 \times 7 = 605,000$ cu ft. Since an acre has an area of 43,560 sq ft, an acre-foot represents a volume of 43,560 cu ft. Therefore, a second-foot week represents a quantity of $605,000 \div 43,560 =$ (about) 13.9 acre-ft. Then, the storage capacity would have to be $5740 \times 13.9 = 79,800$ acre-ft.

With this storage capacity, the plant could have a continuous output of 11,180 kw and generate 98,000,000 kwhr. Of course, with still larger turbines, more power could be developed during the early months of the year for which the data were tabulated and for other years of high river flow.

Observe that no corrections have been made from the storage reservoir for losses due to evaporation, irrigation, consumption for domestic or industrial purposes, infiltration into the soil, deep seepage below the dam, and other losses. All these losses are difficult to estimate and may vary from time to time. Evaporation at any particular location varies throughout the year and is usually lowest in the winter months. Usually, annual evaporation is higher in the southern, and particularly in the southwestern, part of the United States. Values of annual evaporation vary from about 15 to 100 in. When evaporation rates are high, a shallow pond with large surface area is disadvantageous.

EXAMPLE 16–3. If a reservoir has a surface of 6000 acres and an evaporation loss of 7 in. during the month of July, determine the average second-foot weeks of evaporation loss for that month.

SOLUTION. The evaporation loss for the month would be 6000 acres $\times$ 43,560 sq ft per acre $\times$ 7/12 ft of evaporation, or 152,500,000 cu ft. Since July has 31 days, the evaporation per week would be $152,500,000 \times 7/31$, or 34,500,000 cu ft. Since a second-foot week represents a volume of 605,000 cu ft, the average evaporation for the month of July would therefore be $34,500,000 \div 605,000 = 57$ second-foot weeks.

Another method of determining the amount of storage is by means of the *mass curve*, which is a plot of time in months, weeks, or days as the abscissa versus mass-flow quantity in second-foot months, second-foot weeks, or second-foot days as ordinate. The slope of a straight line on this graph gives the rate of flow in second-feet. At any period on this graph, where the slope of the mass curve is less than the slope of the demand curve, storage is required. An approximation of the amount of storage may also be determined from the mass curve.

16–8. Hydro Plant Economics. Extensive areas of land must be flooded to provide sufficient storage or pondage for a hydro plant. This requires the purchase of this land and the possible relocation of roads, railroad routes, residences, cemeteries, and even entire towns. In addition, the dam to provide the storage is usually a costly item. Typical cost distributions for the components of a hydro plant are: 55% for the dam, 20% for equipment, 15% for land, and 10% for structures.

Because of wide variations in conditions at the plant site, there is a wide variation in unit installation costs of hydro plants. Values of $400 to $700 per kw would probably be typical for the cost of current construction.

When government hydroelectric facilities are combined with navigation, irrigation, flood control, and other services, the cost of the installation is divided among the various facilities to be served. One popular method is on the basis of *proportionate-use-of-capacity*. In the case of turbines, penstocks, turbine house, etc., it is obvious that these costs are chargeable exclusively to the turbine. Also, there is little difficulty in assigning the costs of gates and flumes supplying irrigation water and in assigning the costs of locks and associated equipment for navigation.

However, if only irrigation and navigation facilities were provided, it would still be necessary to provide most of the dam structure. Conversely, if the generation of power were the sole purpose of the installation, there would be little reduction in the cost of the dam. Since investment costs are the major portion of electrical energy developed from hydropower, allocation of part of the cost of the dam, flooded land, etc., to other facilities permits a fictitiously low and unrealistic power cost.

Another item that may be of importance for a hydro plant is sometimes called *transmission liability*. This refers to the cost of electrical transmission systems, which convey the energy from a remote plant to the load center, over-and-above the cost for a steam plant. In a great many instances, it is possible to locate a steam plant, even with its large water and fuel requirements, much closer to the load center than is permitted by a hydro plant.

After construction of a hydro plant, its operating costs are much lower than for an equivalent steam plant. Although the water flowing through the turbines is normally free of cost, there are instances where payments must be made for use or storage of the water due to its effects upon other riparian rights. Operating costs for a hydro plant (which include station labor, maintenance, supervision, engineering, and supplies) probably vary from $2.00 per kw of installed capacity for large plants, to $7.00 for small plants, and to $12 or more for very small plants.

As mentioned previously, it is rarely possible to develop a practical electrical system in the United States solely by the use of hydropower because of the fluctuations in stream flow. When water is plentiful and the costs and difficulties of shutting-down and starting-up steam plants are not prohibitive, hydropower is used to supply a part of the system base load. In seasons when water is not plentiful, hydro plants are used to supply peak loads if water storage is available.

PROBLEMS

16–1 to 16–3. Calculate the horsepower and kilowatt output of the following hydro units:

Problem No.	16–1	16–2	16–3
Net head, ft	330	55	1000
Mechanical and electrical efficiency, %..................	87	90	83
Flow, cfs	4500	2000	1200

16–4. The average flow for the Mississippi River is approximately 700,000 cfs. If it is assumed that one-fourth of this flow would be available to develop power at a dam providing 35-ft net head, what would be the annual energy in kilowatt-hours that would be produced with an average of 88% efficiency?

16–5 to 16–7. Calculate specific speed for the following conditions:

Problem No.	16–5	16–6	16–7
Horsepower	115,000	34,000	20,000
Speed, rpm	180	69.2	360
Net head, ft	487	36	2,240

16–8. Determine the weighted average head (based on horsepower) for which the following unit should be designed; assume constant efficiency; calculate the plant and output factors:

Hr per yr.........................	760	1500	2000	3500	1000
Turbine flow, cfs....................	0	1000	1900	2600	3000
Gross head, ft.....................	55	50	60	52	47

The friction loss in the intake system will be 6 ft at a flow of 3000 cfs.

16–9 to 16–14. Select a suitable generator speed and type of runner for the following data:

Problem No.	16–9	16–10	16–11	16–12	16–13	16–14
Horsepower	66,000	111,300	21,000	55,000	9000	14,000
Net head, ft	50	80	482	465	465	800
Electrical frequency, cps..	60	50	60	25	50	60

16–15 to 16–18. Determine the maximum elevation of the runner above (or minimum below) tailwater level for the following data:

Problem No.	16–15	16–16	16–17	16–18
Type of runner	Francis	Francis	Propeller	Propeller
Elevation above sea level, ft....	1,260	2,050	4,350	600
Water temperature, F	70	75	65	80
Horsepower	150,000	20,000	7,000	34,000
Net head, ft	325	140	44	36
Actual speed, rpm.............	120	163.6	180	69.2

16–19. Plot to scale, the load-duration curve of Fig. 16–10.

16–20 to 16–22. For the data of Table 16–1, determine the storage capacity required to maintain a minimum turbine flow as listed; also prepare a storage and spill tabulation for each week of the year. In the tabulation, use units of second-foot week and column headings of week number, excess flow, deficiency of flow, spill, and line volume in reservoir.

Problem No.	16–20	16–21	16–22
Minimum turbine flow, sec-ft.	1400	1300	1200

16–23, 16–24. Compare the cost of electrical energy at a 60% plant factor for a steam versus a hydro plant.

Problem No.	16–23	16–24
Capacity, net kw	100,000	50,000
Average net heat rate, Btu per kwhr.....................	10,200	11,000
Cost of fuel, cents per 10^6 Btu...........................	20	28
Maintenance, mills per kwhr............................	0.4	0.6
Investment charges, steam plant, %.....................	14	14
Unit cost of steam plant, dollars per kw....................	170	190
Labor, supplies, and other operating expenses, mills per kwhr	0.7	0.7
Unit cost of hydro plant, dollars per kw.................	500	625
Transmission liability		
Length of line, miles	75	20
Initial cost, dollars per mile	25,000	15,000
Annual operating expense, dollars per mile.............	120	100
Annual operating labor and expenses, dollars per installed kw	2.00	2.60
Investment charges, hydro plant, %.....................	13	13

16–25. If the cost of the dam is 55% of the cost of the hydro plant, and 40% of the cost of the dam is charged to irrigation and recreation services, what will be the percentage of reduction in cost of electrical energy for the data of Prob. 16–23?

BIBLIOGRAPHY

CREAGER, WILLIAM P., and JUSTIN, JOEL D. *Hydroelectric Handbook.* New York: John Wiley & Sons, Inc., 1950.
DOLAND, JAMES J. *Hydro Power Engineering.* New York: The Ronald Press Co., 1954.
JASKI, FRANK E. "Pump-Turbines," *Allis-Chalmers Electrical Review*, 1954.

KENT, WILLIAM. *Mechanical Engineers' Handbook*. New York: John Wiley & Sons, Inc., 1950.

MARKS, LIONEL S. *Mechanical Engineers' Handbook*. New York: McGraw-Hill Book Co., Inc., 1951.

PFAU, ARNOLD. *Hydraulic Turbine Handbook*. Milwaukee: Allis-Chalmers Mfg. Corp., 1946.

UEHLING, EDWARD. "Water Over the Dam," *Allis-Chalmers Electrical Review,* 1954.

VOADEN, G. H. "Index Testing of Hydraulic Turbines," *Trans. ASME,* (July, 1951).

Power Test Codes for Hydraulic Prime Movers. New York: ASME, 1949.

APPENDIX

APPENDIX

PLATE	TITLE
1	Physical Properties of Pipe
2	Some Thermal Properties of Saturated Water
3	Mollier Diagram for Mercury
4	Enthalpy of Liquid Mercury
5	Enthalpy of Vaporization of Mercury
6	Saturation Temperature of Mercury
7	Surface Heater Dimensions
8	Temperature-Enthalpy-Entropy Diagram for Air at Low Pressures
9	Deaerator Dimensions
10	Fan Dimensions
11	Boiler Feed Pump Dimensions
12	Dimensions of Direct-acting Pumps
13	Boiler Dimensions
14	Condenser Dimensions
15	Turbine Generator Dimensions
16	Uniflow Engine Dimensions
17	Dimensions of Counterflow Steam Engines
18	Turbine Generator Engine Efficiencies
19	Diesel-Engine Generators
20	International Atomic Symbols and Weights

(NOTE: Space dimensions and performance of equipment are *approximate* and for problem use only. The manufacturer should be consulted for more exact data.)

PLATE 1

Physical Properties of Pipe

Nominal Pipe Size, In.	Outside Diam, In.	Wall Thickness, In.	Inside Diam, In.	Inside Diam, Fifth Power	Internal Cross-sectional Area		Weight of Pipe per Ft-Lb
	D	t	d	d^5	Sq In.	Sq Ft	

SCHEDULE 10

Nominal Pipe Size, In.	D	t	d	d^5	Sq In.	Sq Ft	Weight
14 OD	14.0	0.250	13.500	448,000	143.1	0.993	36.7
16 OD	16.0	0.250	15.500	895,000	188.7	1.310	42.1
18 OD	18.0	0.250	17.500	1,641,000	240.5	1.670	47.4
20 OD	20.0	0.250	19.500	2,820,000	298.6	2.073	52.7
22 OD	22.0	0.250	21.500	4,590,000	363.0	2.520	58.1
24 OD	24.0	0.250	23.500	7,170,000	434.0	3.013	63.4
30 OD	30.0	0.312	29.376	21,900,000	678.0	4.708	98.9

SCHEDULE 20

Nominal Pipe Size, In.	D	t	d	d^5	Sq In.	Sq Ft	Weight
8	8.625	0.250	8.125	35,400	51.8	0.359	22.37
10	10.750	0.250	10.250	113,000	82.5	0.572	28.0
12	12.750	0.250	12.250	276,000	117.9	0.818	33.4
14 OD	14.000	0.312	13.375	428,000	140.5	0.975	45.7
16 OD	16.000	0.312	15.375	859,000	185.7	1.289	52.4
18 OD	18.000	0.312	17.375	1,584,000	237.1	1.646	59.0
20 OD	20.000	*0.375*	19.250	2,640,000	291.0	2.020	78.6
24 OD	24.000	*0.375*	23.250	6,790,000	425	2.951	94.6
30 OD	30.000	0.500	29.000	20,500,000	661	4.590	157.6

SCHEDULE 30

Nominal Pipe Size, In.	D	t	d	d^5	Sq In.	Sq Ft	Weight
8	8.625	0.277	8.071	34,200	51.2	0.355	24.70
10	10.750	0.307	10.136	107,000	80.7	0.560	34.2
12	12.750	0.330	12.090	258,000	114.8	0.797	43.8
14 OD	14.000	*0.375*	13.250	408,000	137.9	0.957	54.6
16 OD	16.000	*0.375*	15.250	825,000	182.7	1.268	62.6
18 OD	18.000	0.438	17.124	1,472,000	230.3	1.599	82.2
20 OD	20.000	0.500*	19.000	2,480,000	283.5	1.968	104.1
24 OD	24.000	0.562	22.876	6,260,000	411.0	2.854	140.7
30 OD	30.000	0.625	28.750	19,600,000	649.0	4.506	196.1

NOTE: Wall thickness shown in italics is the same as for standard weight pipe.

* Wall thickness is the same as for extra strong pipe.

PLATE 1 (*Continued*)

PHYSICAL PROPERTIES OF PIPE

Nominal Pipe Size, In.	Outside Diam, In.	Wall Thickness, In.	Inside Diam, In.	Inside Diam, Fifth Power	Internal Cross-sectional Area		Weight of Pipe per Ft-Lb
	D	t	d	d^5	Sq In.	Sq Ft	

SCHEDULE 40

⅛	0.405	*0.068*	0.269	0.00141	0.057	0.0003	0.245
¼	0.540	*0.088*	0.364	0.00639	0.104	0.0007	0.425
⅜	0.675	*0.091*	0.493	0.02912	0.191	0.001	0.568
½	0.840	*0.109*	0.622	0.09310	0.304	0.002	0.851
¾	1.050	*0.113*	0.824	0.3799	0.533	0.003	1.131
1	1.315	*0.133*	1.049	1.270	0.864	0.006	1.679
1¼	1.660	*0.140*	1.380	5.005	1.496	0.010	2.273
1½	1.900	*0.145*	1.610	10.82	2.036	0.014	2.718
2	2.375	*0.154*	2.067	37.73	3.356	0.023	3.653
2½	2.875	*0.203*	2.469	91.8	4.79	0.033	5.794
3	3.500	*0.216*	3.068	271.8	7.39	0.051	7.58
3½	4.000	*0.226*	3.548	562.	9.89	0.068	9.11
4	4.500	*0.237*	4.026	1,058	12.73	0.088	10.79
5	5.563	*0.258*	5.047	3,275	20.01	0.138	14.62
6	6.625	*0.280*	6.065	8,210	28.9	0.200	18.98
8	8.625	*0.322*	7.981	32,400	50.0	0.347	28.56
10	10.750	*0.365*	10.020	101,000	78.9	0.547	40.5
12	12.750	0.406	11.938	242,000	111.9	0.777	53.5
14 OD	14.000	0.438	13.125	389,000	135.3	0.939	63.4
16 OD	16.000	0.500*	15.000	759,000	176.7	1.227	82.8
18 OD	18.000	0.562	16.876	1,369,000	223.7	1.553	104.7
20 OD	20.000	0.593	18.814	2,360,000	278.0	1.930	122.9
24 OD	24.000	0.687	22.626	5,930,000	402	2.791	171.1

NOTE: Wall thickness shown in italics is the same as for standard weight pipe.

SCHEDULE 60

8	8.625	0.406	7.813	29,100	47.9	0.332	35.6
10	10.750	0.500*	9.750	88,100	74.7	0.518	54.7
12	12.750	0.562	11.626	212,000	106.2	0.737	73.2
14 OD	14.000	0.593	12.814	345,000	129.0	0.895	84.9
16 OD	16.000	0.656	14.688	684,000	169.4	1.176	107.5
18 OD	18.000	0.750	16.500	1,223,000	213.8	1.484	138.2
20 OD	20.000	0.812	18.376	2,100,000	265.2	1.841	166.4
24 OD	24.000	0.968	22.064	5,230,000	382	2.652	238.1

* Wall thickness is the same as for extra strong pipe.

PLATE 1 (*Continued*)

PHYSICAL PROPERTIES OF PIPE

Nominal Pipe Size, In.	Outside Diam, In.	Wall Thickness, In.	Inside Diam, In.	Inside Diam, Fifth Power	Internal Cross-sectional Area		Weight of Pipe per Ft-Lb
	D	t	d	d^5	Sq In.	Sq Ft	

SCHEDULE 80

Nominal Pipe Size, In.	Outside Diam, In.	Wall Thickness, In.	Inside Diam, In.	Inside Diam, Fifth Power	Internal Cross-sectional Area		Weight of Pipe per Ft-Lb
⅛	0.405	0.095*	0.215	0.00046	0.036	0.0002	0.314
¼	0.540	0.119*	0.302	0.00251	0.072	0.0005	0.535
⅜	0.675	0.126*	0.423	0.01354	0.140	0.0009	0.739
½	0.840	0.147*	0.546	0.04852	0.234	0.001	1.088
¾	1.050	0.154*	0.742	0.2249	0.432	0.003	1.474
1	1.315	0.179*	0.957	0.803	0.719	0.004	2.172
1¼	1.660	0.191*	1.278	3.409	1.283	0.008	2.997
1½	1.900	0.200*	1.500	7.59	1.767	0.012	3.632
2	2.375	0.218*	1.939	27.41	2.953	0.020	5.022
2½	2.875	0.276*	2.323	67.6	4.24	0.029	7.662
3	3.500	0.300*	2.900	205	6.60	0.045	10.25
3½	4.000	0.318*	3.364	431	8.89	0.061	12.51
4	4.500	0.337*	3.826	820	11.50	0.079	14.99
5	5.563	0.375*	4.813	2,583	18.19	0.126	20.78
6	6.625	0.432*	5.761	6,350	26.1	0.181	28.58
8	8.625	0.500*	7.625	25,800	45.7	0.317	43.4
10	10.750	0.593	9.564	80,000	71.8	0.498	64.3
12	12.750	0.687	11.376	191,000	101.6	0.705	88.5
14 OD	14.000	0.750	12.500	305,000	122.7	0.852	106.1
16 OD	16.000	0.843	14.314	601,000	160.9	1.117	136.5
18 OD	18.000	0.937	16.126	1,090,000	204.2	1.418	170.8
20 OD	20.000	1.031	17.938	1,860,000	252.7	1.754	208.9
24 OD	24.000	1.218	21.564	4,660,000	365	2.534	296.4

SCHEDULE 100

Nominal Pipe Size, In.	Outside Diam, In.	Wall Thickness, In.	Inside Diam, In.	Inside Diam, Fifth Power	Internal Cross-sectional Area		Weight of Pipe per Ft-Lb
8	8.625	0.593	7.439	22,800	43.5	0.302	50.9
10	10.750	0.718	9.314	70,100	68.1	0.472	76.9
12	12.750	0.843	11.064	166,000	96.1	0.667	107.2
14 OD	14.000	0.937	12.125	262,000	115.5	0.802	130.8
16 OD	16.000	1.031	13.938	526,000	152.6	1.059	164.8
18 OD	18.000	1.156	15.688	950,000	193.3	1.342	208.0
20 OD	20.000	1.281	17.438	1,610,000	238.8	1.658	256.1
24 OD	24.000	1.531	20.938	4,020,000	344	2.388	367.4

SCHEDULE 120

Nominal Pipe Size, In.	Outside Diam, In.	Wall Thickness, In.	Inside Diam, In.	Inside Diam, Fifth Power	Internal Cross-sectional Area		Weight of Pipe per Ft-Lb
4	4.500	0.438	3.624	625	10.31	0.071	19.00
5	5.563	0.500	4.563	1,978	16.35	0.113	27.04

* Wall thickness is the same as for extra strong pipe.

PLATE 1 (*Continued*)
PHYSICAL PROPERTIES OF PIPE

Nominal Pipe Size, In.	Outside Diam, In.	Wall Thickness, In.	Inside Diam, In.	Inside Diam, Fifth Power	Internal Cross-sectional Area		Weight of Pipe per Ft-Lb
	D	t	d	d^5	Sq In.	Sq Ft	
SCHEDULE 120—Continued							
6	6.625	0.562	5.501	5,040	23.8	0.165	36.40
8	8.625	0.718	7.189	19,200	40.6	0.281	60.6
10	10.750	0.843	9.064	61,200	64.5	0.447	89.2
12	12.750	1.000	10.750	144,000	90.8	0.630	125.5
14 OD	14.000	1.093	11.814	230,000	109.6	0.761	150.7
16 OD	16.000	1.218	13.564	459,000	144.5	1.003	192.3
18 OD	18.000	1.375	15.250	825,000	182.7	1.268	244.2
20 OD	20.000	1.500	17.000	1,420,000	227.0	1.576	296.4
24 OD	24.000	1.812	20.376	3,510,000	326	2.263	429.4
SCHEDULE 140							
8	8.625	0.812	7.001	16,800	38.5	0.267	67.8
10	10.750	1.000	8.750	51,300	60.1	0.417	104.1
12	12.750	1.125	10.500	128,000	86.6	0.601	139.7
14 OD	14.000	1.250	11.500	201,000	103.9	0.721	170.2
16 OD	16.000	1.438	13.124	389,000	135.3	0.939	223.7
18 OD	18.000	1.562	14.876	728,000	173.8	1.206	274.3
20 OD	20.000	1.750	16.500	1,220,000	213.8	1.484	341.1
24 OD	24.000	2.062	19.876	3,100,000	310	2.152	483.2
SCHEDULE 160							
½	0.840	0.187	.466	0.02198	0.171	0.001	1.304
¾	1.050	0.218	.614	0.0873	0.296	0.002	1.937
1	1.315	0.250	.815	0.360	0.522	0.003	2.844
1¼	1.660	0.250	1.160	2.100	1.057	0.007	3.765
1½	1.900	0.281	1.337	4.27	1.404	0.009	4.866
2	2.375	0.343	1.689	13.74	2.240	0.015	7.445
2½	2.875	0.375	2.125	43.3	3.55	0.024	10.01
3	3.500	0.438	2.624	124.	5.41	0.037	14.33
4	4.500	0.531	3.438	480.	9.28	0.064	22.51
5	5.563	0.625	4.313	1,492	14.61	0.101	32.97
6	6.625	0.718	5.189	3,760	21.1	0.146	45.30
8	8.625	0.906	6.813	14,700	36.5	0.253	74.7
10	10.750	1.125	8.500	44,400	56.7	0.393	115.7
12	12.750	1.312	10.126	106,000	80.5	0.559	160.3
14 OD	14.000	1.406	11.188	175,000	98.3	0.682	189.1
16 OD	16.000	1.593	12.814	345,000	129.0	0.895	245.1
18 OD	18.000	1.781	14.438	627,000	163.7	1.136	308.5
20 OD	20.000	1.968	16.064	1,070,000	202.7	1.407	379.1
24 OD	24.000	2.343	19.314	2,690,000	293	2.034	542.0

PLATE 2

SOME THERMAL PROPERTIES OF SATURATED WATER

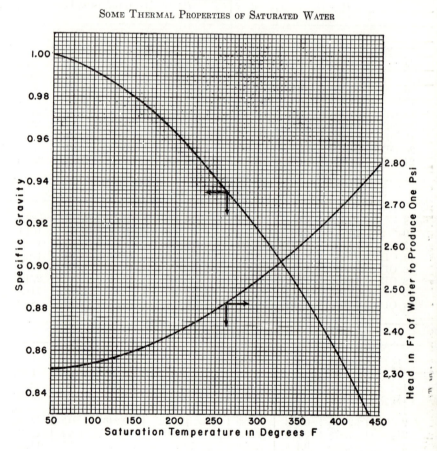

PLATE 4

ENTHALPY OF LIQUID MERCURY
(General Electric Co.)

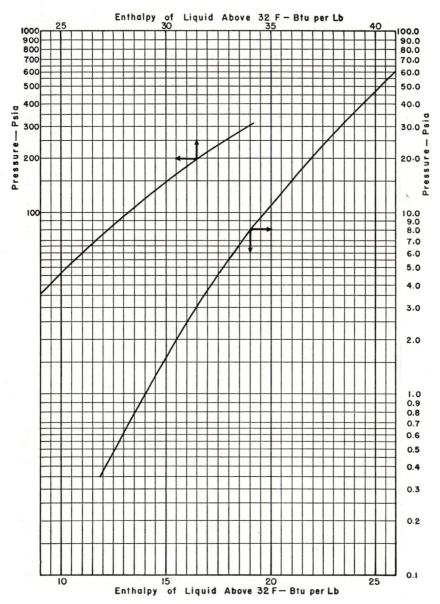

PLATE 5

ENTHALPY OF VAPORIZATION OF MERCURY
(General Electric Co.)

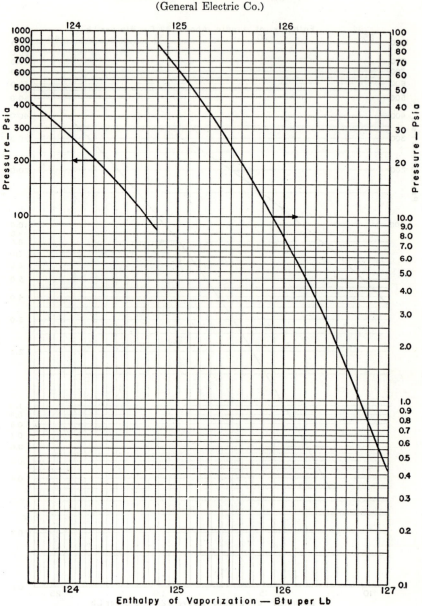

Enthalpy of Vaporization — Btu per Lb

PLATE 6

SATURATION TEMPERATURE OF MERCURY
(General Electric Co.)

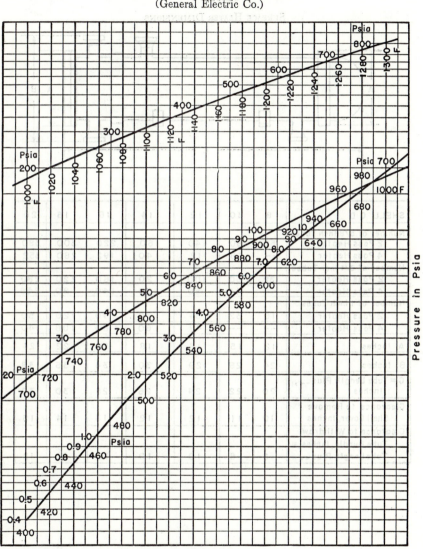

Temperature in Deg F

PLATE 7

Surface Heater Dimensions

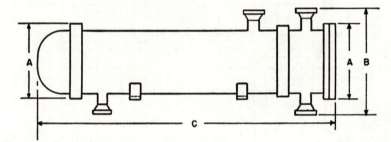

SHELL O D	8	10	12	14	16	18	20
Number of 5/8 Inch Tubes							
2 Pass Heater	50	90	140	160	220	300	380
4 Pass Heater	--	68	112	140	192	280	352
Number of 3/4 Inch Tubes							
2 Pass Heater	32	60	100	120	160	220	280
4 Pass Heater	--	44	80	100	140	180	248
Number of 1 Inch Tubes							
2 Pass Heater	16	32	54	70	90	120	150
4 Pass Heater	--	--	48	60	80	108	140
A Inches	12	15	16	18	20	22	24
B Inches	24	26	29	30	34	36	38
C = Tube Length Plus__, Inches	20	22	24	26	28	30	32
Clearance to Remove Shell Cover, Inches	10	11	12	12	13	14	15
Clearance to Withdraw Tube Bundle, Inches,= Tube Length Plus ___	8	8	10	10	10	12	12

PLATE 9

Deaerator Dimensions

(Vacuum to 50 psig)
(Cochrane Corp.)

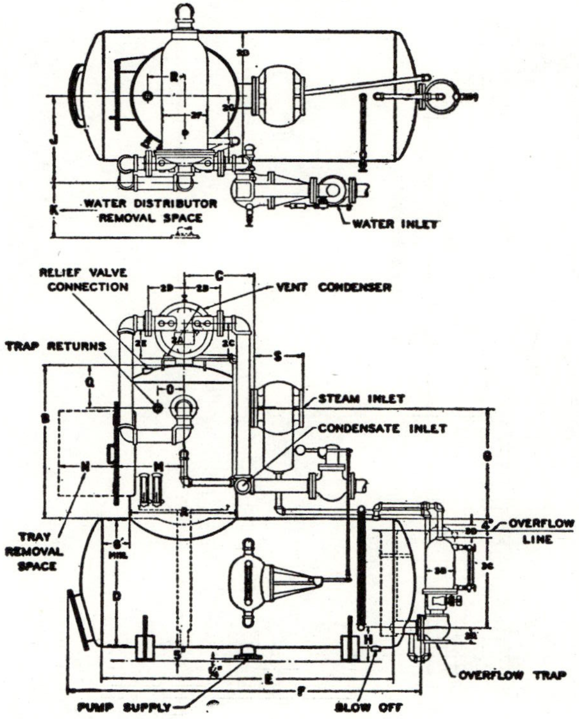

Capacity Lb Per Hr at Outlet	10,000 to 25,000	45,000	60,000	80,000	110,000
A	30"	36"	42"	48"	54"
B	57	60	66	69	72
C	20	23	26	29	32
D	as required	(Storage tank diameter should be equal to or larger than dia. heater top)			
E	"	"			
F	"	"			
G	45	45	48	51	51
H	24	24	15	18	18
J	21	24	27	30	33
K	30	36	39	45	51
M	20	23	26	29	32
N	18	24	30	30	36
O	7	7	9	12	10
Q	10	11	12	13	14
R	10	13	16	19	21
S	—	—	—	—	—
2A	14½	17¼	17¼	17¼	24
2B	32	50	40	59	48
2C	9¾	11½	11½	11½	14
2D	8	8½	10¾	10¾	15
2E	7¾	8½	8½	8½	10
2F	10	12	12	12	18
2G	—	—			
3A	Drainer	5½	5½	5½	6¼
3B		12	12	12	12
3C		32	32	32	33
3D		16	16	16	16
Blow Off	1½	2	2	2	2
Trap Returns	1½	2	2	2	2½
Relief Connection	2	2½	2½	2½	3
Regulating Valve	1½	2	2½	3	3
Pump Supply	5	5	6	6	8
Overflow	3	4	4	4	5
Separator	5	6	8	10	10

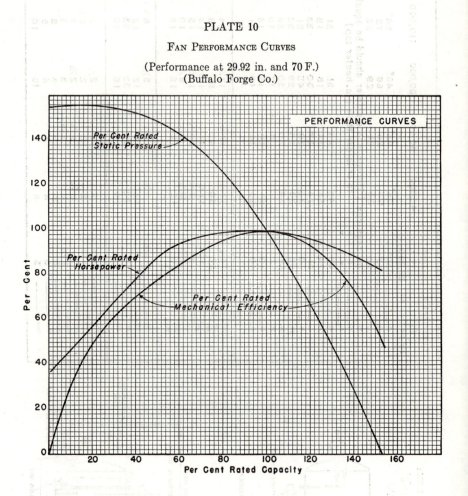

PLATE 10

FAN PERFORMANCE CURVES

(Performance at 29.92 in. and 70 F.)
(Buffalo Forge Co.)

PLATE 10 (*Continued*)

FAN DIMENSIONS

Fan Size Type SLD*	Motor Size	Speed	Limit, Hp	Average Operating Conditions	
				CFM**	Static Pressure, In.
3½	7½	1,760	7½	10,000	3
4	10	1,760	10	12,000	3½
4½	15	1,760	15.0	15,000	4
5	20	1,760	20.0	18,000	4½
5½-A	25	1,760	25	21,000	5
5½	30	1,760	30.0	23,000	5½
6-A	40	1,760	40.0	28,000	6
6	50	1,760	47.0	30,000	6½
6½-A	60	1,760	59.5	36,000	7
6½	20	1,160	20.0	25,000	3½
7	30	1,160	29.2	31,000	4
7½	40	1,160	40.0	38,000	4½
8	60	1,160	56.8	47,000	5
8½	75	1,160	75	58,000	5½
9	100	1,160	100.0	67,000	6½
10	75	870	73.5	72,000	4

* Inlet boxes will increase the bearing span of forced- and induced-draft fans.
** Quantity is for double-width fan; take one-half these values for single-width fan.

PLATE 10 (*Continued*)

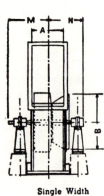

Single Width

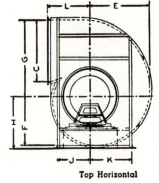

Top Horizontal

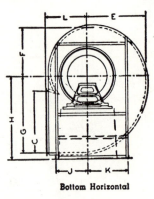

Bottom Horizontal

Fan Size	A	AA	B	C	E	F	G	H Top Hor	H Bot Hor	H Up Bl	K Dn Bl	J	K Hor Disch	K Up Bl	L	M	N	M¹	N¹
3½	12 5/8	23	18 13/16	21 5/8	21 1/2	17 1/32	26 13/16	19	31	23 1/2	15 1/2	12	14	17	16 3/4	23 3/4	19 3/8	28 3/4	24 3/8
4	13 1/2	24 1/2	20	23	22 7/8	18 3/32	28 1/2	20	32 1/2	25	16	13	15	18	17 5/8	24	19 5/8	29 1/2	25 1/8
4½	14 5/8	26 1/2	21 3/4	24 7/8	24 23/32	19 9/16	30 27/32	21 1/2	35	27	17 1/2	14	16	19	18 7/8	24 1/2	20 1/8	30 1/2	26 1/8
5	15 5/8	28 1/4	23 1/4	26 1/2	26 1/4	20 25/32	32 3/4	23	37	28 1/2	18 1/2	15	17	20	19 3/4	25 3/8	21	31 3/4	27 3/8
5½A 5½	16 7/8	30 5/8	25 1/4	28 5/8	28 15/32	22 1/2	35 17/32	24 1/2	40	30 1/2	20	16	19	22	21 1/8	25 7/8	21 1/2	33	28 5/8
6A 6	18 3/8	33 3/8	27 1/2	31 1/4	31 1/32	24 9/16	38 23/32	26 1/2	43	33	21 1/2	18	21	25	22 3/4	27 1/8	22 3/4	35	30 5/8
6½A 6½	19 7/8	36 1/8	29 13/16	33 7/8	33 21/32	26 5/8	42 1/32	28 1/2	46 1/2	35 1/2	23	19	23	27	24 3/8	28 1/4	23 1/2	38	33 1/4
7	21 3/8	38 7/8	32 3/16	36 1/2	36 1/4	28 21/32	45 1/4	30 1/2	49 1/2	38 1/2	24 1/2	20	25	29	26	29 3/8	24 3/8	39 1/4	34 1/2
7½	23	41 3/4	34 5/8	39 1/8	38 13/16	30 23/32	48 7/16	33	52 1/2	41	26	21	27	31	27 5/8	30 1/2	25 1/8	43	37 5/8
8	24 1/2	44 1/2	37	41 3/4	41 7/16	32 25/32	51 1/16	35	56	43 1/2	28	23	29	33	29 1/4	33 1/4	27 3/4	44 1/2	39 1/8
8½	26	47 1/4	39 5/16	44 3/8	44	34 27/32	54 5/16	37	59	46	29 1/2	24	30	35	30 7/8	34 3/8	29	47	41 5/8
9	27 1/2	50	41 11/16	46 7/8	46 21/32	36 29/32	58 7/32	39	63	48 1/2	31	25	32	38	32 1/2	36 3/4	31 3/8	48 3/4	43 3/8
10	30 1/2	55 1/2	45 15/16	52 1/8	51 13/16	41	64 11/16	43 1/2	70	54	34 1/2	27	36	48	36 3/8	39 3/8	33 1/2	52 1/4	46 3/8
11	33 3/4	61 1/8	50 3/4	57 1/4	57	45 1/8	71 3/8	47 1/2	76 1/2	59 1/2	38	30	40	52	39 5/8	41 3/8	35 1/4	56 1/2	50 3/8
12	36 3/4	66 3/4	55 1/2	62 1/2	62 3/16	49 3/16	77 5/8	51 1/2	83	64 1/2	41	32	43	56	42 7/8	43 1/8	37	59 1/2	53 3/8
13	39 3/4	72 1/4	59 7/8	67 3/4	67 5/16	53 1/4	84	56	90	70	44 1/2	35	46	60	46 3/8	46 3/8	39 1/8	65 3/4	58 1/2
14	42 3/4	77 3/4	64 1/2	73	72 1/2	57 3/8	90 1/2	60	96 1/2	75	48	37	50	65	49 5/8	48 3/8	40 3/4	69	61
15	46	83 1/2	69 9/16	78	77 5/8	61 7/16	96 7/8	64	102 1/2	80	51	40	54	70	52 3/4	50 1/4	42 1/4	72	64

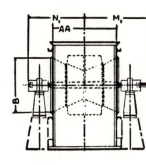

Double Width

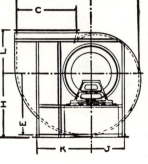

Up Discharge

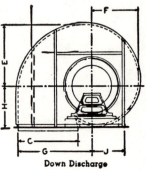

Down Discharge

PLATE 11

BOILER FEED PUMPS DIMENSIONS

PUMP SIZE — NUMBER OF STAGES MOTOR HP RATING (16 ft minimum *NPSH*) (3,500 rpm)								
Developed Head Psi	Gallons per Minute							
	50	75	100	125	150	200	250	300
200			#2-2 30	#2-2 30	#2-2 40	#2-2 50	#2-2 50	#2-2 60
250	#1½-4 30	#1½-4 30	#2-2 40	#2-2 40	#2-2 50	#2-2 50	#2-2 60	#2-2 75
300	#1½-4 30	#1½-4 40	#1½-4 40	#1½-4 40	#1½-4 50	#2-4 60	#2-4 75	#2-4 100
400	#1½-6 40	#1½-6 50	#1½-6 50	#1½-6 60	#1½-6 60	#2-6 75	#2-6 100	#2-4 125
500	#1½-6 50	#1½-6 60	#1½-6 60	#2-6 75	#2-6 100	#2-6 100	#2-6 125	#2-6 150
600			#2-6 100	#2-6 100	#2-6 100	#2-6 125	#2-6 150	#2½-6 200
700			#2-8 100	#2-8 125	#2-8 125	#2-8 150	#2-8 200	#2-8 200
800			#2-8 125	#2-8 125	#2-8 150	#2-8 200	#2-8 200	#2-8 250

DIMENSIONS

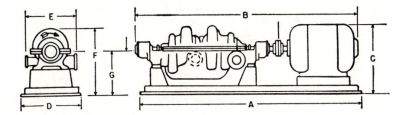

Pump Size	Number Stages	Diameter"		A	B	C	D	E	F	G
		Suction	Discharge							
1½	4	2½	1½	91"	95"	30"	33"	18"	29"	19"
1½	6	2½	1½	99"	103"	30"	33"	18"	29"	19"
2	2	3	2	72"	74"	30"	33"	22"	26"	19"
2	4	4	2	105"	109"	35"	38"	24"	30"	20"
2	6	4	2	114"	119"	35"	38"	24"	30"	20"
2	8	4	2	128"	130"	35"	38"	24"	30"	20"
2½	6	4	2½	122"	126"	36"	38"	26"	31"	21"

PLATE 12

DIMENSIONS OF DIRECT-ACTING PUMPS

Double-acting

(All dimensions are in inches.)

Size	Strokes per Minute, Boiler Feed Service	Strokes per Minute, General Service	Simplex Direct-Acting Steam	Simplex Direct-Acting Exhaust	Simplex Direct-Acting Suction	Simplex Direct-Acting Discharge	Duplex Direct-Acting Steam	Duplex Direct-Acting Exhaust	Duplex Direct-Acting Suction	Duplex Direct-Acting Discharge	Floor Space* Simplex, Piston Type	Floor Space* Duplex, Piston Type	Floor Space* Duplex, Plunger Type Center Packed	Floor Space* Duplex, Plunger Type Outside End Packed
3 × 2 × 4	55	150	3/8	1/2	1¼	1	3/8	1/2	1¼	1	30 × 6	29 × 10		
4 × 2⅝ × 5	50	120	3/4	1	2	1½	1/2	3/4	2	1½		35 × 12		
5½ × 3¾ × 7	45	100			3		1	1½	3	2	45 × 8			
6 × 4 × 6	50	100	1	1¼	4	2½			5	4		55 × 27	64 × 16	75 × 22
6 × 4 × 10	40	80	1	1½		3	1½	2	5	4	56 × 9			
8 × 5 × 12	40	60					2	2½	5	4	64 × 10	82 × 32	94 × 22	
10 × 6 × 12	40	60					2½	3	7	6		84 × 40		114 × 27
12 × 6 × 12	40	60					2½	3				84 × 40		117 × 33
12 × 8 × 12	35	50	2	2½	6	5					72 × 15	84 × 40		117 × 35
14 × 10 × 20			2	3	8	7					101 × 19			

* These dimensions will depend on the design pressure and arrangement of water valves.

PLATE 13

Boiler Dimensions

(Total surface includes superheater surface; maximum design pressure 300 psi)
(Babcock & Wilcox Co.)

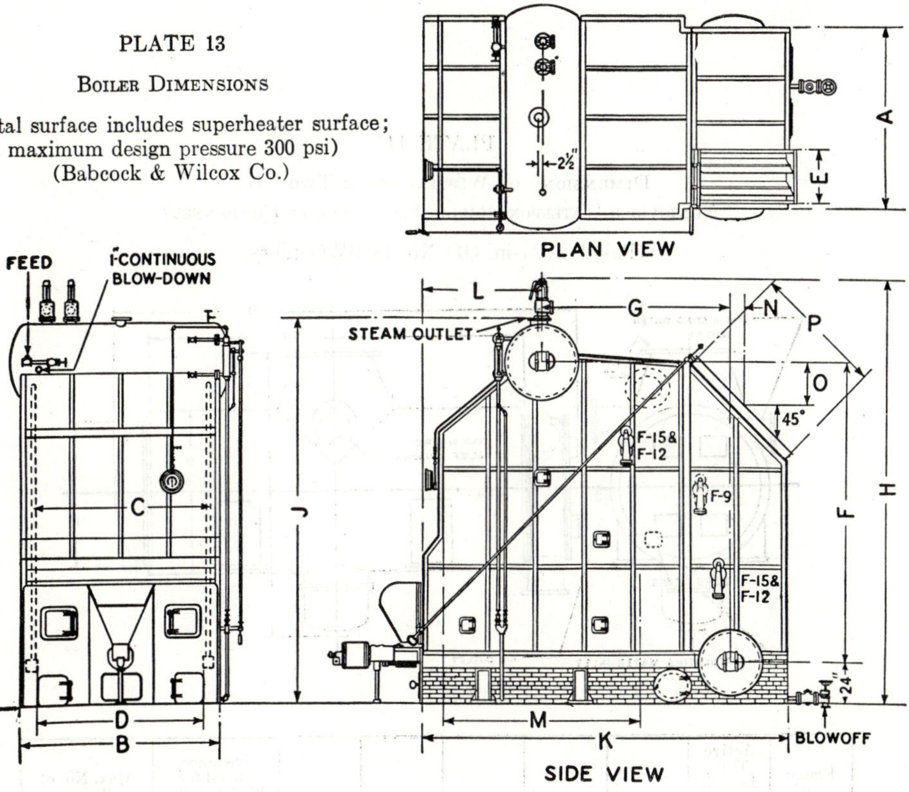

PRINCIPAL DIMENSIONS

Size of Boiler		Tubes Wide: 16	20	24	28	32	36	40	44
F-9	Heating Surface	1353	1654	1956	2258	2560	2863	3164	
	Steam Outlet Size	4"	4"	4"	4"	4"	4"	5"	
	Feed Size	1½"	1½"	1½"	1½"	1½"	2"	2"	
F-12	Heating Surface	1984	2414	2843	3273	3703	4134	4563	
	Steam Outlet Size	4"	4"	4"	5"	5"	5"	5"	
	Feed Size	1½"	1½"	2"	2"	2"	2"	2"	
F-15	Heating Surface		3160	3717	4274	4832	5392	5949	6506
	Steam Outlet Size		5"	5"	5"	6"	6"	6"	6"
	Feed Size		2"	2"	2"	2½"	2½"	2½"	2½"
A	Boiler Width—Outside Casing	5'- 6¾"	6'- 9"	7'-11¼"	9'- 1½"	10'- 3¾"	11'- 6"	12'- 8¼"	13'-10½"
B	Furnace Width—Outside Casing	6'- 8¼"	7'-10½"	9'- 0¾"	10'- 3"	11'- 5¼"	12'- 7½"	13'- 9¾"	15'- 0"
C	℄ to ℄ Furnace Wall Tubes	5'- 5"	6'- 7¼"	7'- 9½"	8'-11¾"	10'- 2"	11'- 4¼"	12'- 6½"	13'- 8¾"
D	Face to Face 7¼" Square Headers	4'-10¾"	6'- 1"	7'- 3¼"	8'- 5½"	9'- 7¾"	10'-10"	12'- 0¼"	13'- 2½"
E	Width of Dampers	13¾"	20⅞"	20⅞"	2'- 4"	2'- 4"	2'-11⅛"	3'- 6¼"	3'- 6¼"

		Size of Boiler: F-9	F-12	F-15
F	℄ to ℄ Drums	9'- 2"	12'- 2"	15'- 2"
G	℄ to ℄ Drums	5'- 4"	7'- 6½"	9'- 9"
H	Overall Height	15'- 0"	18'- 0"	21'- 0"
J	Height over Steam Outlet (West. Det. Oil)	13'- 4"	16'- 4"	19'- 4"
K	Overall Length	14'-11⅜"	16'- 7⅞"	18'-10⅜"
L	℄ Drum to Outside Casing Front Wall	6'- 8"	6'- 2"	6'- 2"
M	Inside Face Front Wall to Inside Face Bridgewall	6'- 0"	7'- 0"	9'- 0"
N	Damper Location, Horizontal	2'- 1"	17⅛"	9¹¹⁄₁₆"
O	Damper Location, Vertical	8½"	16¼"	24⅞"
P	Length of Damper	4'- 0"	5'- 3"	6'- 6"

PLATE 14

Dimensions of Westinghouse Two-Pass,
Nondivided Waterbox, Radial-Flow Surface Condensers

(Based on ¾-in. OD No. 18 BWG tubes.)

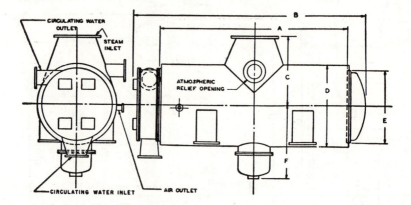

Frame No.	Active Tube Length						Distance Req'd for Withdrawing Tubes	Max. No. of Tubes
	A	B	C	D	E	F		
900	10'-0''	13'-3''	5'-8''	3'-8''	3'-3''	3'-4''	11'-8''	494
1250	12'-0''	15'-5''	5'-11''	4'-3''	3'-9''	4'-0''	13'-9''	720
2000	14'-0''	17'-1''	6'-1''	5'-1''	4'-2''	4'-3''	15'-7''	878
3000	16'-0''	18'-9''	6'-3''	5'-5''	4'-6''	4'-9''	17'-5''	1,052
4000	16'-0''	19'-1''	6'-6''	6'-1''	5'-1''	5'-0''	17'-7''	1,398
5000	16'-0''	19'-6''	6'-10''	6'-7''	5'-7''	5'-6''	17'-9''	1,762
6000	18'-0''	22'-4''	7'-2''	7'-3''	6'-3''	6'-0''	20'-2''	2,230
7000	18'-0''	23'-8''	7'-4''	7'-8''	6'-8''	6'-6''	20'-10''	2,580
8500	18'-0''	24'-3''	7'-8''	8'-3''	7'-4''	7'-0''	21'-1''	3,106

NOTE: Steam inlet to be of size and shape to match the turbine exhaust. Circulating water connections to be sized for maximum water velocity of seven feet per second.

Dimension (A) represents what is normally considered a balanced condenser design. Any increase or decrease in dimension (A) will correspondingly increase or decrease dimensions (B) and distance required for withdrawing tubes.

PLATE 15

DIMENSIONS OF WESTINGHOUSE, SINGLE AUTOMATIC EXTRACTION, CONDENSING, 3600-RPM TURBINE GENERATOR UNITS

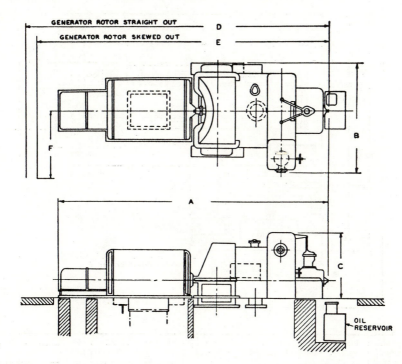

Rating Kw	A	B	C	D	E	F	Exhaust Size	Min Height Floor to Crane Hook
1,000	23'-4''	8'-2''	6'-2''	24'-9''	22'-9''	4'-0''	2'-0''	10'-0''
1,500	23'-11''	8'-3''	6'-8''	27'-2''	25'-4''	4'-3''	3'-0''	12'-0''
2,000	24'-1''	9'-0''	6'-8''	26'-10''	25'-0''	4'-4''	3'-0''	12'-0''
2,500	25'-1''	9'-8''	7'-0''	28'-4''	26'-3''	4'-10''	3'-0''	12'-0''
3,000	25'-7''	9'-8''	7'-0''	29'-5''	28'-4''	6'-2''	3'-6''	12'-0''
4,000	28'-9''	10'-4''	7'-3''	33'-5''	30'-11''	6'-8''	4'-0''	12'-0''
5,000	28'-11''	10'-4''	7'-3''	34'-0''	31'-0''	7'-1''	4'-0''	12'-0''
6,000	30'-5''	10'-9''	7'-4''	36'-4''	33'-5''	7'-1''	3'-4'' × 6'-4''	12'-0''
7,500	31'-6''	12'-0''	8'-0''	37'-8''	34'-8''	6'-4''	4'-0'' × 7'-6''	12'-0''

PLATE 16

Uniflow Engine Dimensions

(Skinner Engine Co.)

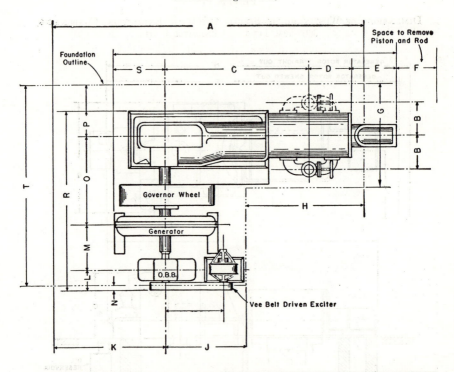

Cyl Size Rpm Gov Wh } Size	11 × 13 250–277 62½ × 11	13 × 15 250–260 63½ × 12	16 × 16 225–260 69¼ × 15	18 × 20 200–225 85 × 20	21 × 21 200 88 × 20	22 × 24 150–200 90 × 20	26 × 32 150–164 120 × 20	29 × 36 150 126 × 24
A	13′-3″	15′-10″	16′-8″	20′-11″	22′-0″	25′-1″	31′-10″	34′-9″
B	19″	20″	23″	29″	31″	32″	34″	42″
C	6′-10″	7′-9″	7′-10″	10′-2″	10′-4″	11′-6″	14′-7″	16′-10″
D	23″	30″	31″	3′-0″	3′-0″	3′-4″	4′-2″	4′-12″
E	6″	0	0	0	39″	43″	4′-9″	5′-0″
F	36″	36″	36″	36″	36″	39″	4′-0″	5′-0″
G	5′-0″	6′-0″	6′-6″	8′-0″	8′-0″	8′-0″	8′-6″	9′-6″
H	5′-1″	5′-7″	5′-3″	7′-10	8′-9″	9′-7″	15′-4″	16′-10″
J	4′-6″	5′-9″	6′-6″	6′-0″	6′-1″	6′-6″	6′-6″	8′-0″
K	3′-8″	4′-6″	5′-0″	7′-1″	7′-3″	9′-0″	10′-0″	10′-0″
L	12″	16″	15″	16″	21″	19″	23″	28″
M	30″	33″	37″	35″	37″	40″	48″	47″
N	0″	3″	0″	0″	5″	4″	3″	0″
O	5′	4′-6″	5′-7″	6′-4″	6′-8″	6′-12″	7′-10″	8′-5″
P	16″	18″	16″	18″	19″	2′-0″	23″	27″
R	11′	11′-3″	13′-2″	13′-8″	14′-8″	15′-3″	17′-10″	20′-2″
S	31″	37″	43″	45″	4′-1″	4′-6″	5′-5″	5′-6″
T	11′-0″	11′-3″	13′-2″	14′-0″	14′-7″	15′-4″	17′-10″	19′-7″

PLATE 17

Dimensions of Counterflow Steam Engines

(Erie City Iron Works)

SIDE CRANK HIGH SPEED ENGINES

Standard R. P. M.	Diam. Cyl.	Stroke	Diam. Steam pipe	Diam. Exhaust	Main Bearing		Out-B. Bearing		Main Shaft		Crank Pin		X-head Pin		Diam. Piston Rod	X-head Shoe		Fly-wheel			Floor Space		Approximate Shipping Weights	
					Diam.	Length	Diam.	Length	Max. Diam.	Approx. Length	Diam.	Length	Diam.	Length		Width	Length	Diam.	Face	Approx. Weight	Width	Length	Belted	D.C.
300	10	10	3½	4½	5	11	4½	10	6	TO SUIT CONDITIONS	4¾	3¼	3	3½	1¾	6½	11¾	54	12½	2025	85″	99″	6800	7000
300	11	10	3½	4½	5	11	4½	10	6		4¾	3¼	3	3½	1¾	6½	11¾	54	12½	2025	91″	99″	7000	7200
275	12	12	4	5	5½	13	5	10	6½		5¼	4	3½	4	2⅛	7	13	54	12½	2025	106″	117″	10200	11000
275	13	12	4½	5	5½	13	5	10	6½		5¼	4	3½	4	2⅛	7	13	54	12½	2025	106″	117″	10500	11300
250	14	14	5	6	6	15	5½	11	7		5¾	4¾	4	4½	2⅜	8	15	60	14½	3000	125″	135″	14000	14400
250	15	14	5	6	6	15	5½	11	7		5¾	4¾	4	4½	2⅜	8	15	60	14½	3000	125″	135″	14400	14800
225	16	16	6	8	7½	17	7	14½	8½		7½	6	4½	5½	2¾	9½	17	72	16½	4500	127″	150″	20400	21200
225	17	16	6	8	7½	17	7	14½	8½		7½	6	4½	5½	2¾	9½	17	72	16½	4500	127″	150″	20900	21700
200	18	18	7	8	9	19	8½	18	10		8½	8	5¾	6	3	10	19	72	18½	5200	139″	167″	32000	32600
200	19	18	7	8	9	19	8½	18	10		8½	8	5¾	6	3	10	19	72	18½	5200	139″	167″	32600	33200

Main Shaft Diameters are for direct connected Engines.
Floor Space Dimensions are for D C. Engines and are Approximate.

CENTER CRANK HIGH SPEED ENGINES

Standard R. P. M.	Diam. Cyl.	Stroke	Diam. Steam pipe	Diam. Exhaust	Main Bearing		Out-B. Bearing		Main Shaft		Crank Pin		X-head Pin		Diam. Piston Rod	X-head Shoe		Fly-wheel			Floor Space		Approximate Shipping Weights	
					Diam.	Length	Diam.	Length	Max. Diam.	Approx Length	Diam.	Length	Diam.	Length		Width	Length	Diam.	Face	Approx. Weight	Width	Length	Belted	D.C
300	10	10	3½	4½	4	9	3½	8	4	TO SUIT CONDITIONS	4¾	3¼	3	3½	1¾	6½	11¾	48	9½	1350	85″	99″	6800	7000
300	11	10	3½	4½	4	9	3½	8	4		4¾	3¼	3	3½	1¾	6½	11¾	48	9½	1350	91″	99″	7000	7200
275	12	12	4	5	4¾	11	4¼	10	4¾		5¼	4	3½	4	2⅛	7	13	54	12½	2000	108″	117″	10200	11000
275	13	12	4½	5	4¾	11	4¼	10	4¾		5¼	4	3½	4	2⅛	7	13	54	12½	2000	108″	117″	10500	11300
250	14	14	5	6	5½	13	5	10	5½		5¾	4¾	4	4½	2⅜	8	15	60	14½	3000	130″	135″	14000	14400
250	15	14	5	6	5½	13	5	10	5½		5¾	4¾	4	4½	2⅜	8	15	60	14½	3000	130″	135″	14400	14800
225	16	16	6	8	7	15	7	14½	7		7½	6	4½	5½	2¾	9½	17	72	16½	4500	134″	150″	20400	21200
225	17	16	6	8	7	15	7	14½	7		7½	6	4½	5½	2¾	9½	17	72	16½	4500	134″	150″	20900	21700
200	18	18	7	8	8½	17	8	18	8½		8½	8	5¾	6	3	10	19	72	18½	5200	148″	167″	32000	32600
200	19	18	7	8	8½	17	8	18	8½		8½	8	5¾	6	3	10	19	72	18½	5200	148″	167″	32800	33200

Out-board bearing furnished only with direct connected engines.
Two Fly-wheels on Engines for belt drive.
Floor Space Dimensions are for D C. and are Approximate.

PLATE 18

TURBINE GENERATOR ENGINE EFFICIENCIES

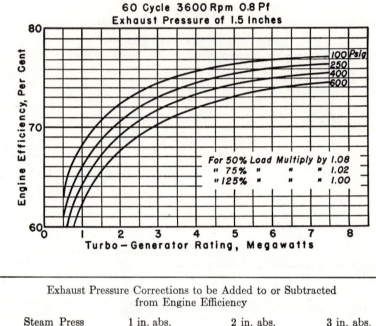

60 Cycle 3600 Rpm 0.8 Pf
Exhaust Pressure of 1.5 Inches

For 50% Load Multiply by 1.08
" 75% " " " 1.02
" 125% " " " 1.00

Exhaust Pressure Corrections to be Added to or Subtracted from Engine Efficiency			
Steam Press psig	1 in. abs. (subtract)	2 in. abs. (add)	3 in. abs. (add)
100	1.0%	0.6%	1.2%
250	0.8%	0.5%	1.0%
400	0.7%	0.4%	0.8%
600	0.6%	0.3%	0.7%

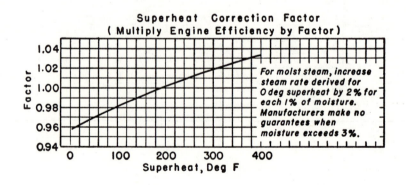

Superheat Correction Factor
(Multiply Engine Efficiency by Factor)

For moist steam, increase steam rate derived for 0 deg superheat by 2% for each 1% of moisture. Manufacturers make no guarantees when moisture exceeds 3%.

PLATE 19

DIESEL ENGINE GENERATORS

(Nordberg Manufacturing Company)

Engine Bore and Stroke, In.	No. of Cyl.	Speed Range, RPM	Range of Brake Horsepower Ratings			
			Nonsuper-charged	Super-charged	Supercharged and Intercooled (See Note)	Supair-thermal* (See Note)
9 x 11½	5	600–720	220–265	330–400	360–440	
9 x 11½	6	600–1000	265–440	400–665	440–735	535–890
9 x 11½	7	600–720	310–370	465–555	510–615	620–745
9 x 11½	8	600–1000	355–590	530–885	580–965	710–1180
13 x 16½	5	450–514	500–570	750–850	825–940	
13 x 16½	6	450–600	600–800	900–1200	990–1320	1200–1600
13 x 16½	7	450–514	700–800	1050–1200	1150–1310	1400–1595
13 x 16½	8	450–600	800–1065	1200–1600	1320–1760	1600–2130
13 x 20	5	360–400	480–535	725–805	795–885	
13 x 20	6	360–450	580–725	870–1090	955–1200	1160–1455
13 x 20	7	360–400	680–755	1020–1130	1120–1245	1360–1510
13 x 20	8	360–450	775–970	1165–1450	1280–1600	1555–1940

NOTE: Based on 90 F water to the intercooler.

*Increased turbocharger capacity and pressure over engine listed in preceding column.

PLATE 19 (*Continued*)

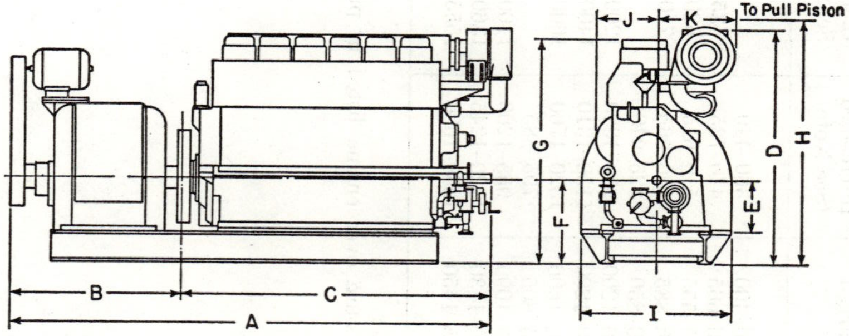

Approximate Dimensions FS-9 Engines (9-in. bore)

Engine Size, In.	No. Cyl.	A	B	C	D	E	F	G	H	I	J	K
9 x 11½	5	13′-10⅞″	4′-11⅝″	8′-11¼″	8′-0⅝″	1′-8⅛″	2′-9⅞″	7′-8¼″	8′-3⅜″	4′-2¾″	1′-11⅝″	2′-6″
9 x 11½	6	15′-1″	5′-0¼″	10′-0¾″	8′-0⅝″	1′-8⅛″	2′-9⅞″	7′-8¼″	8′-3⅜″	4′-2¾″	1′-11⅝″	2′-6″
9 x 11½	7	16′-2½″	5′-0¼″	11′-2¼″	8′-0⅝″	1′-8⅛″	2′-9⅞″	7′-8¼″	8′-3⅜″	4′-2¾″	1′-11⅝″	2′-6″
9 x 11½	8	17′-7″	5′-3¼″	12′-3¾″	8′-0⅝″	1′-8⅛″	2′-9⅞″	7′-8¼″	8′-3⅜″	4′-2¾″	1′-11⅝″	2′-6″

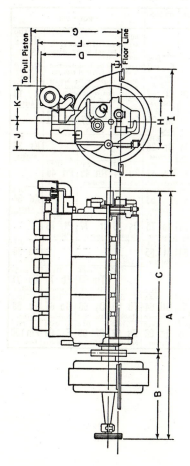

Approximate Dimensions FS-13 Engines (13-in. bore)

Engine Size, In.	No. Cyl.	A	B	C	D	E	F	G	H	I	J	K
13 x 16½	5	19'-2"	6'-7"	12'-7"	7'-11"	10"	7'-8½"	9'-4"	4'-6"	7'-6½"	2'-7½"	3'-1"
13 x 16½	6	21'-8"	7'-5"	14'-3"	7'-11"	10"	7'-8½"	9'-4"	4'-6"	8'-6"	2'-7½"	3'-1"
13 x 16½	7	24'-2"	8'-3"	15'-11"	7'-11"	10"	7'-8½"	9'-4"	4'-6"	8'-8"	2'-7½"	3'-1"
13 x 16½	8	26'-5"	8'-10"	17'-7"	7'-11"	10"	7'-8½"	9'-4"	4'-6"	9'-0"	2'-7½"	3'-1"
13 x 20	5	19'-2"	6'-7"	12'-7"	8'-8"	10"	8'-5¼"	10'-1"	4'-6"	9'-6½"	2'-7½"	3'-1"
13 x 20	6	21'-8"	7'-5"	14'-3"	8'-8"	10"	8'-5¼"	10'-1"	4'-6"	10'-6"	2'-7½"	3'-1"
13 x 20	7	24'-2"	8'-3"	15'-11"	8'-8"	10"	8'-5¼"	10'-1"	4'-6"	10'-8"	2'-7½"	3'-1"
13 x 20	8	26'-5"	8'-10"	17'-7"	8'-8"	10"	8'-5¼"	10'-1"	4'-6"	11'-0"	2'-7½"	3'-1"

PLATE 20

International Atomic Symbols and Weights
(Westinghouse Electric Corp.)

	GROUP I	GROUP II	GROUP III	GROUP IV	GROUP V	GROUP VI	GROUP VII	GROUP VIII			
PERIOD 1	1.008 H 1 Hydrogen							NOTE: At top of each box is shown atomic weight, followed by symbol and atomic number. The name of the chemical element is shown at bottom of each box. Atomic weight is the average weight of an atom compared to an average atom of ordinary terrestrial oxygen as 16.			4.003 He 2 Helium
PERIOD 2	6.940 Li 3 Lithium	9.013 Be 4 Beryllium	10.82 B 5 Boron	12.010 C 6 Carbon	14.008 N 7 Nitrogen	16.000 O 8 Oxygen	19.00 F 9 Fluorine				20.183 Ne 10 Neon
PERIOD 3	22.997 Na 11 Sodium	24.32 Mg 12 Magnesium	26.97 Al 13 Aluminum	28.06 Si 14 Silicon	30.98 P 15 Phosphorus	32.066 S 16 Sulfur	35.457 Cl 17 Chlorine				39.944 A 18 Argon
PERIOD 4	39.096 K 19 Potassium	40.08 Ca 20 Calcium	45.10 Sc 21 Scandium	47.90 Ti 22 Titanium	50.95 V 23 Vanadium	52.01 Cr 24 Chromium	54.93 Mn 25 Manganese	55.85 Fe 26 Iron	58.94 Co 27 Cobalt	58.69 Ni 28 Nickel	
	63.54 Cu 29 Copper	65.38 Zn 30 Zinc	69.72 Ga 31 Gallium	72.60 Ge 32 Germanium	74.91 As 33 Arsenic	78.96 Se 34 Selenium	79.916 Br 35 Bromine				83.7 Kr 36 Krypton
PERIOD 5	85.48 Rb 37 Rubidium	87.63 Sr 38 Strontium	88.92 Y 39 Yttrium	91.22 Zr 40 Zirconium	92.91 Nb 41 Niobium	95.95 Mo 42 Molybdenum	99.0 Tc 43 Technetium	101.7 Ru 44 Ruthenium	102.91 Rh 45 Rhodium	106.7 Pd 46 Palladium	
	107.88 Ag 47 Silver	112.41 Cd 48 Cadmium	114.76 In 49 Indium	118.70 Sn 50 Tin	121.76 Sb 51 Antimony	127.61 Te 52 Tellurium	126.91 I 53 Iodine				131.3 Xe 54 Xenon
PERIOD 6	132.91 Cs 55 Cesium	137.36 Ba 56 Barium	138.92 La 57 Lanthanum	178.6 Hf 72 Hafnium	180.88 Ta 73 Tantalum	183.92 W 74 Wolfram	186.31 Re 75 Rhenium	190.2 Os 76 Osmium	193.1 Ir 77 Iridium	195.23 Pt 78 Platinum	
	197.2 Au 79 Gold	200.61 Hg 80 Mercury	204.39 Tl 81 Thallium	207.21 Pb 82 Lead	209.0 Bi 83 Bismuth	210. Po 84 Polonium	210. At 85 Astatine				222. Ra 86 Radon
PERIOD 7	223. Fr 87 Francium	226.05 Ra 88 Radium	227.0 Ac 89 Actinium						169.4 Tm 69 Thulium	173.04 Yb 70 Ytterbium	174.99 Lu 71 Lutecium
PERIOD 6 Rare Earths	140.13 Ce 58 Cerium	140.92 Pr 59 Praseodymium	144.27 Nd 60 Neodymium	147.0 Pm 61 Promethium	150.43 Sm 62 Samarium	152.0 Eu 63 Europium	156.9 Gd 64 Gadolinium	159.2 Tb 65 Terbium	162.46 Dy 66 Dysprosium	164.94 Ho 67 Holmium	167.2 Er 68 Erbium
PERIOD 7 Actinides	232.12 Th 90 Thorium	231.0 Pa 91 Protoactinium	238.07 U 92 Uranium	237.0 Np 93 Neptunium	239.0 Pu 94 Plutonium	241.0 Am 95 Americium	242.0 Cm 96 Curium	243.0 Bk 97 Berkelium	244.0 Cf 98 Californium	253.0 E 99 Einsteinium	254.0 Fm 100 Fermium

DESIGN AND LAYOUT PROBLEMS

DESIGN AND LAYOUT PROBLEM "A" FOR AN INDUSTRIAL POWER PLANT

GENERAL

The power plant will contain two identical, controlled-extraction, 4000-kw turbine generators. Steam will be extracted from each turbine for process use and for a deaerator. Steam also will be extracted at a lower pressure for use in a surface type of feedwater heater. A by-pass, pressure-reducing station will provide process steam when the turbines are not in use. All process steam becomes contaminated and is wasted.

Four identical boilers are to supply the plant steam requirements (two boilers to carry maximum requirements of one turbine). The boiler room will be separate from the turbine hall, and in addition to the boilers, will contain one fan (forced draft) per boiler, coal-handling equipment and ducts.

Equipment to be located in the turbine hall: turbine generators, deaerators, low-pressure heaters, boiler feed pumps, condensers, circulating pumps, air ejectors, condensate pumps, turbine control board, overhead crane.

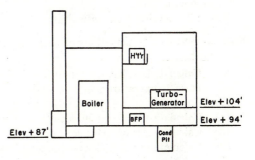

An end view of the plant with elevations is shown. The plant is not isolated.

DATA

Turbine generators: 96% mechanical efficiency, 95% generator efficiency, straight-line condition curve, design point is 5/4 load, 40-psia automatic extraction and 10-psia nonautomatic extraction pressures, steam conditions 400 psig, 550 F, and 2 in. Hg.

Process Steam: 30,000 lb per hr for the plant. Show connection for safety valve after PRV.

Boilers: Single settings. Minimum distances, 10 ft between boilers, 18-ft aisle in front of boiler setting for control board and coal-handling equipment, 6-ft aisle behind boiler. Air enters at 80 F and gases leave at 450 F. Ash removal from pit below aisle in back of boilers. Boiler rating, 250%. Franklin County, Illinois, coal with 35% excess air.

Forced-Draft Fans: 2.5-in. water static pressure.

Condensers: Circulating water enters at 80 F, surface type, 2-pass, 18-gage ¾ in. tubes, 6-fps water velocity, 75% cleanliness factor. Design condensers for 2 in. Hg abs, at 4000 kw, with steam extraction from the turbines for feed heating only. Condenser axis will be perpendicular to turbine axis.

Air Ejectors: 200 lb per hr of steam required for each unit; 1000 Btu per lb available for feed heating in air-ejector condensers.

Condensate Pumps: Two per condenser, 100% spare, centrifugal, motor-driven, 150-ft total head. Allow 10% extra capacity based on the flow obtained for the condenser.

Boiler Feed Pumps: Four pumps for the plant, 100% spare, centrifugal, motor-driven, 325-psig discharge pressure. Allow 10% extra flow to take care of emergencies.

Deaerators: Two for entire plant, 100% spare. Make-up water enters de-aerators at 60 F.

Surface Heaters: One per turbine generator, no spare, 5 F TD, drains cascaded to condenser. Design on basis of heat-balance data.

Turbine Control Board: One for both turbines, 8 ft high by 14 ft long, and 5 ft deep (to allow for switchgear) with 4-ft clearance.

Piping: Of sufficient strength for pressures and temperatures involved, flanged. Water piping on pump discharges designed for velocity of about 450 fpm and less than 200 fpm on suction. 10% pressure drop in low-pressure bleed line and 5% in high-pressure bleed line. Use a hand by-pass around automatic PRV.

Stack: To develop 1-in. static draft.

REQUIREMENTS (All calculations to be bound in a folder, in order, labeled, and neat.)

(a) Calculate heat balance for turbine hall and make schematic diagram with flows, temperatures, and enthalpies.

(b) Calculate boiler size and select boiler.

(c) Calculate surface requirements of condenser and select condenser.

(d) Calculate size of, and select, deaerators and surface heaters.

(e) Calculate size of boiler feed pumps and select pumps.

(f) Calculate size of condensate pumps.

(g) Make a plan layout of equipment first by templates and then by pencil on tracing paper. Show major dimension of equipment and building, and show centerline dimensions.

(h) Lay out main steam pipe (single line) on drawing and show *all* fittings.

(i) Calculate pipe sizes for main steam lines.

(j) Write a bill of material for the main steam piping.

(k) Calculate the maximum friction loss in the steam piping from boiler to turbine.

(l) Calculate the fan horsepower and select a fan.

(m) Determine the diameter and height of the stack.

(n) Draw an end view of the plant showing equipment and piping.

(o) Lay out the water piping from the deaerator to the boiler, showing all fittings, and include an automatic valve for feedwater regulation.

(p) Determine the friction loss of the water piping.

(q) Determine cost of building based on 50¢ per cu ft.

DESIGN AND LAYOUT PROBLEM "B" FOR AN INDUSTRIAL POWER PLANT

GENERAL

The power plant will consist of two identical boilers supplying steam to two identical controlled-extraction turbine generators of 2000 kw, each operating at 250 psig, 550 F and 1½ Hg abs. Process steam, which is contaminated and wasted, is to be supplied from the controlled extraction points at 20 psig (at the turbine) in the total amount of 20,000 lb per hr for the plant. A by-pass, pressure-reducing valve station is to connect the process steam line and the boiler header to supply process steam when the turbines are operating at low load and cannot supply sufficient process steam. Steam for a contact type of deaerating heater is also supplied from the extraction points. The heaters will operate at 25 psia.

An end view of the plant is shown with some dimensions. Other dimensions are to be determined by the layout. The boiler room will be separate from the turbine hall. The boiler room will contain the two boilers, two fans,

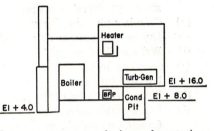

coal-handling equipment, ducts, and the by-pass, pressure-reducing valve station. The turbine hall will contain the two turbines, two deaerators, 3 boiler feed pumps, two condensers, four circulating pumps, two air ejectors, two condensate pumps for each condenser. Make-up enters deaerator at 60 F. The plant is not isolated.

DATA

Turbine generators: 94.4% generator efficiency, 95.5% mechanical efficiency, straight-line condition curve.

Boilers: Fayette County, Pennsylvania, bituminous coal, 40% excess air, cross-drum 250% of rating, 8 ft between boilers, 15-ft aisle in front of boilers for control board and coal-handling equipment, 8-ft aisle in back of boilers, stack to supply 0.9-in. water draft with temperatures of 80 F and 450 F, ash removal from aisle beside chimney.

Condensers: Circulating water enters at 70 F, surface type, 2-pass, 18-gage ¾ in. tubes, 7-fps water velocity, 70% cleanliness factor, condenser axis perpendicular to turbine axis.

Condensate Pumps: Two per condenser, centrifugal, motor-driven, 100% spare, 125-ft head.

Boiler Feed Pumps: Three total, 200% spare, centrifugal, motor-driven, 400-psig discharge pressure.

Deaerators: Two for entire plant, 100% spare.

Piping: Of sufficient strength for pressure and temperatures involved; flanged. Water piping on jump discharge to have a velocity of 400 to 500 fpm, on pump suctions not more than 200 fpm, automatic pressure-reducing valve with hand by-pass.

Air Ejectors: 200 lb of steam per hr each.

Forced-Draft Fans: 3.5-in. static pressure.

REQUIREMENTS

(a) Calculate heat balance for turbine hall.

(b) Calculate boiler size and select boiler.

(c) Calculate surface requirements of condenser and select condenser.

(d) Calculate size of boiler feed pumps and select pumps.

(e) Calculate size of condensate pumps.

(f) Make a plan layout of equipment, first by templates and then by pencil on tracing paper. Instructor to approve template layout before pencil drawing is made. Show major dimension of equipment and building, and show centerline dimensions.

(g) Calculate pipe sizes for main steam lines.

(h) Lay out main steam pipe (single line) on drawing and show *all* fittings.

(i) Write a bill of material for the steam piping.

(j) Calculate to the maximum the friction loss in the steam piping from boiler to turbine.

(k) Draw an end view of the plant showing equipment and piping.

(l) Lay out the water piping from the heater to the boiler, showing all fittings, and include an automatic valve for feedwater regulation.

(m) Determine the friction loss of the water piping.

(n) Calculate the fan horsepower and select a forced-draft fan.

(o) Determine the diameter and height of the stack.

(p) Determine cost of building based on 55¢ per cu ft.

INDEX

Air
  flow of
    economical velocities in ducts, 77
    high pressure, 73
    low pressure, 75
  leakage in condensers, 362
  primary, 278
  secondary, 278
  tempering, 279
  thermal properties, 680
  velocity pressure, 75
  viscosity, 49
Air ejectors (evactors), 340, 362
  condensate recirculation, 365
  heat balance calculations, 521
  location in the cycle, 365, 519
Air preheaters, 10, 246
  heat transfer, 270
  primary, 279, 291
  temperature from, 247
  tube sizes, 270
Amortization, 17
Arithmetic mean temperature difference, 150
Ash softening temperatures, 276
Aspect ratio, 80
Atomic mass units, 596
Atomic symbols and weights, 698
Attemperation, 242
Automatic extraction turbines, 458
Auxiliary power, 511, 560

Babcock and Wilcox friction factor, 58, 63
Before-and-after pump, 587
Benson boilers, 251
Bernoulli, 28
Binary cycles, 14, 533
Binding energy, 600
Bleeder Heater Mfr's. Association, 373, 374
Blow-out discs, 356
Blowdown, boiler, 7
  safety valves, 47
Boiler horsepower, 255
Boilers (see Generators, steam)
Brayton cycle, 557
Break-even point, 21

Burners, 294
  cyclone, 227, 300
  tilting, 241, 300

Calorimeters
  oxygen bomb, 177
  peroxide, 175, 178
  Sargent, 149
Campbell's classification of coal, 182
Capacity factor, 624
Capitalized cost, 18
Carpenter's friction factor, 63
Cavitation
  centrifugal pumps, 110, 252
  hydraulic turbines, 660
Chemical cleaning, 254
Chimneys, 302
  draft, 305
  gas temperature variation, 305
  gas velocities, 306
  resistance to flow in, 307
Circulating pumps, 10, 122, 359
  costs, 358
Coal
  air for combustion, 188, 193
  analyses, table, 172
  ash softening temperatures, 276
  burners, 294
  classification, 181
    by Campbell, 182
    by Frazer, 183
    by Ralston, 183
  geographical distribution, 186
  grindability, 276
    table, 277
  heating value, 174
    table, 172
  proximate analysis, 170
  ultimate analysis, 171
Combustion
  computations, 188
  incomplete, 197
  products of, 194
Condensate pump control, 541
Condensate pumps, 104, 122, 365, 539, 545
Condensers
  barometric, 339
  circulating water for, 341, 356, 359

Condensers (*Continued*)
  low-level jet, 339
  surface, 10, 341
    air leakage, 363
    costs, 358
    dimensions, 690
    economic selection, 357
    expansion, 344
    friction, 355
    heat transfer, 350
    laning, 346
    nomenclature, 341
    reheating hotwell, 347
    rolled and packed tubes, 343
    supports, 348
    tube area, 351, 360
    tube length, recommended, 354
    tube properties and composition, 343,
      352; table, 353
Conductivity
  electrical, 325
  thermal, 140
    of insulations, 141
    of metals, 141
Connected load, 626
Contact type condensers, 340
  heaters, 377, 508, 519, 529, 541
Convection coefficient, 146
  table, 145
Cooling towers, 587
Curtis, C. G., 402
Curtis stages, 405, 427
Curve ratio, 79
Cycles
  binary, 14, 533
  Brayton, 557
  extraction pressures, 525
  gas, open and closed, 558
  heat balances for, 518
  heat flow diagram, 534
  heat rates, 15, 510
    estimates, 512
  heater arrangements, 506
  open and closed, 529
  Rankine, 505
  reheat, 509
Cyclone burners, 227, 300

Dalton's law, 365
Dampers, 242, 310, 318
Day tank, 590
de Laval, Gustaf, 402
Deaerators, 6, 377
  dimensions, 683
Demand factor, 623
Demineralization, 320, 509

Depreciation, 17, 631
Desuperheaters, 242
Desuperheating zones, 375
Diagram factor, 499
Diesel engines (*see* Engines, diesel)
Diffusors, 104, 363, 412
Dimensions
  centrifugal pumps, 687
  condensers, 690
  counterflow steam engines, 693
  deaerators, 683
  diesel engines, 695
  direct-acting pumps, 688
  fans, 684
  pipe, 674
  steam generators, 689
  surface heaters, 682
  turbine generators, 691
  uniflow steam engines, 692
Direct-acting pumps (*see* Pumps, reciprocating)
Direct contact
  condensers, 339
  heaters, 377
Diversity factors, 623
Draft in chimneys, 304
Drain coolers, 392, 507, 519
Drainers, 6, 507
Dry bottom furnaces, 227, 275
Ducts, 75
  splitters, 79
  turning vanes, 79
Dulong's equation, 175
Dump power, 626

Economic selection of equipment, 16
  centrifugal pump speeds, 131
  condensers, surface, 357
  evaporators, 389
  pipe sizes, 20
Economizers, 10, 246, 270
  design pressures, 545
  extended surface, 246
  heat transfer, 270
  steaming, 247
Efficiency
  diesel cycle, 580
  engine, 442
  fans, 312
  gas turbine cycle, 564
  pumps
    centrifugal, 127
    reciprocating, 99
  steam cycle, 15
  steam engines, 498
  steam generators, 199, 256

syphon, 361
turbine, hydro, 650, 657
turbine generators, 442, 445, 694
Ejectors (evactors), air, 340, 362
priming, 362, 366
Electric rates (*see* Rates, electric)
Electrons, 596
Emissivity
definition, 154
gases, 265
tables, 156
Emmet, Dr. W. LeR., 538
Engine efficiency, 442
Engines, 3
diesel, 573
before-and-after pump, 587
buildings, 590
descriptions, 573
dimensions, 695
governing, 583
heat rates, 581
mean effective pressure, 582
performance, 580
pilot oil, 579
steam
clearance volume, 484, 486, 493
Corliss, 488
cut-off, 484, 494, 500
cylinder condensation, 493
D slide valves, 487
diagram factor, 499
dimensions, 692, 693
efficiencies of, 498
governing systems, 498
indicator card, 485, 496
lead and lap, 487
nomenclature, 485
piston displacement, 486
poppet valves, 488
slide valves, 487
speed regulation, 494
uniflow, 488, 491
Equivalent lengths of pipe fittings, 68
Euler, 28
Evaporation
equivalent, 256
factor of, 255
Evaporative coolers, 587
Evaporators, 326, 383
costs, 389
cracking, 385
heat balance, 388, 521
heat head, 388
heat transfer, 390
relieving rate (velocity), 385
Evasé, 82, 310

Excess air, 191, 202, 244, 290, 546, 565
Expansion joints for condensers, 345, 350
Extraction factors, 457, 523

Fans, 308
control of, 318
dimensions, 684
economical selection, 18
forced draft, 7, 289, 308, 546
horsepower, 312, 322
induced draft, 10, 275, 308, 546
inlet vanes, 311, 319
parallel operation, 316
performance curves, 314, 317, 322, 684
selection of capacities and pressures, 546
vane shapes, 309
variable speed, 314, 319
variable speed drives, 128, 308
Feedwater hardness, 325
Feedwater heaters, 11, 366, 507, 525, 545
contact type, 377
dimensions, 683
heat balance, 382, 508, 519
surface type, 366
desuperheating zone, 375
dimensions, 682
heat balance, 371, 506, 519
heat transfer, 372
selection of design pressures, 545
Feedwater treatment, 6, 323, 383
conductivity, 325
demineralization, 330
lime-soda process, 327
pH, 325
phosphates, 329
zeolite, 326
Filters, 329
Firm power, 626, 646
Fission and fusion, 599
Fixed costs, 630
Flanges, 35
Flash tank, 7
Flow duration curves, 664
Flue gas analysis, 193
weight of actual air from, 198
weight of dry gas from, 196
Flue gas recirculation, 244
Flue gas specific heats, 262
Fly-ash, 275, 290, 300, 309
Foster, Dean E., 69
Fourier's equation, 139
Francis, James B., 651
Friction factor, 56, 58
Fuels (see specific kind)

Furnaces (*see* Generators, steam)
Fusion, and fission, 599

Gas
  artificial and natural, 557
  burners, 294
  table, 189
Gas turbines (*see* Turbines, gas)
General energy equation, 26, 28, 90, 408
Generators
  efficiency, 445
  electric, hydrogen cooled, 7, 445, 527
Generators, steam, 4, 208, 509, 545
  baffles, 217
  bent-tube, 10, 218
  chemical cleaning, 254
  circulation, 212, 222
  controlled circulation, 251
  cross-drum, 217
  cyclone furnaces, 253
  dimensions, 691
  drum, 233
    concentrations, 234
  dry bottom furnace, 227, 275
  dry pipe, 234
  dryness fraction, 213
  equivalent evaporation, 256
  factor of evaporation, 255
  fire tube, 208
  foaming, 234
  forced circulation, 10, 251
  furnace calculations, 258
  heat balance, 198
  heat release, 212, 228
  horizontal return tubular (HRT), 209
  insulation, 157, 229
  internal furnace, 209
  longitudinal drum, 216
  manufacturer's rating, 256
  mercury, 539
  natural circulation, 10, 212
  packaged, 210, 222
  performance, 255
  pressurized, 13, 232, 308, 569
  priming, 234
  selection of capacities, 545
  slag screen, 227
  slagging bottom (wet bottom) furnaces, 228
  Stirling, 218
  straight tube, 216
  surface blowdown, 234
  three-drum, 221, 225
  two-drum, 10, 222
  water tube, 212, 216
Governors
  centrifugal pumps, 124

diesel engines, 583
fans, 318
steam engines, 498
steam turbines, 447
  effect on performance, 450
  overspeed, 463
Grindability, 276
Gutermuth friction factor, 63

Half-life, 599
Hawksley friction factor, 63
Heat balances
  station, 518
  steam generator, 198
Heat Exchange Institute, 350, 355, 356, 366
Heat flow diagram, 534
Heat head, 149, 388
Heat rates, 15, 510, 523, 581, 619
  estimates, 512
  typical, 518
Heat release, 212, 228
Heat transfer
  air preheaters, 270
  arithmetic mean temperature difference, 150
  composite
    cylinders, 142
    walls, 143
  condensers, 350, 359
  conduction, 139, 146, 161
  conductivity values, 141
  convection, 139, 144, 146, 161, 237, 263
  cylindrical walls, 143
  economizers, 270
  emissivity, 154, 265
  evaporators, 390
  feedwater heaters, 372
  film coefficients, table, 145
  heat head, 149
  insulation, 155
    efficiency, 164
  log mean temperature difference, 150, 264
  plain walls, 139
  radiation, 139, 154, 161, 237, 261
  reheaters, 270
  resistance, 143
  scale, 152
  superheaters, 270
  temperature-surface diagrams, 149
  terminal difference, 149
Heaters (*see* Feedwater heaters)
Heating values
  determination for fuels, 174
  empirical equations, 176

tables for
coal, 172
gas, 189
oil, 187
Heating values of fuels, 15
Horizontal return tubular boilers, 209
Hydraulic couplings, 128, 308
economics of, 131, 319
Hydraulic radius, 56
Hydrogen cooled generators, 445, 527
Hydrogen ion concentration, pH, 325
Hydrograph, 665

Impulse blades, 404, 419, 448
Insulation
conductivity, 141
descriptions, 155
efficiency, 164
recommended thickness, 162
steam generators, 229, 233
Investment charges, 17
Investors' profit, 630
Isochronous governors, 584

Jennison Station, 8
Jet deflector, 648

Kaplan turbines, 651, 657

Laminar flow, 52
LaMont boiler, 251
Lantern gland, 115
Lime-soda treatment, 327
Load curve, 624, 627
Load duration curve, 624
Load factor, 624
Log mean temperature difference, 150, 265, 350, 359, 374

Make-up water, plant, 383
Mass defect, 600
Mercury, thermal properties, 679
Mercury cycles, 14, 537
Mercury steam generators, 539
Moody, Lewis F., 60

Napier's formula, 411
Net positive suction head, 90, 105, 109, 124, 543, 545
Neutrons, 596
Nikuradse, J., 59
Nozzles
critical pressures, 409
efficiency, 413
flow, 409

forms of, 417
overexpansion and underexpansion, 413, 430
supersaturation, 417
Nuclear plants, 557
classification of reactors, 605
economics, 595, 617
reactor materials, 602
Nuclear power plants, 14, 595, 605
Nuclear reactor materials, 602

Off-peak electric rates, 639
Oil
burners, 294
gas turbine fuel, 557
geographical location, 187
in feedwater, 324
ultimate analysis, table, 187
Operating costs, 632
Operation factor, 626
Orsat analyzer, 193
Output factor, 624
Overfire air, 10, 293
Oxygen, 65, 332, 362, 377

Package plants, 11
Packaged boilers, 212, 222
Parsons, Sir Charles A., 402
Peak load, 626
Pelton, Lester A., 649
Penstock, 652
Petroleum (see Oil)
Phosphate treatment, 329
Pipe
allowable stress, 31
cast iron, 31
economical sizing, 20
expansion, 26
expansion joints, 348
fittings, 42
equivalent lengths, 68
flanges, 34
friction factors, 58
gaskets, 40
graphitization, 32
heat transfer from, 161
insulation, 155
recommended thickness, 162
physical properties, 674
pressure-seal joints, 39
recommended velocities, 72
roughness, 60
schedule number, 34, 674
thickness, 34, 674
welding, 40
wrought iron, 31
wrought steel, 31

Plant factor, 624
Preferred standards for steam turbine
  generators, 447, 527
Present worth, 18
Pressure head, 27
Prime power, 626
Prony brake, 498
Protons, 596
Proximate analysis of coal, 170
Pulverizers, 275, 631
  air temperatures for, 279
  ball-race mills, 284
  bin and unit systems, 281
  bowl mill, 285
  coal fineness, 276
  impact, 286
  tube (ball) mills, 282
  use of, 275
Pumps
  centrifugal, 90, 102, 313
    balancing drum or disc, 116
    boiler feed, 107, 545
    capacity ranges, 94
    cavitation, 110, 543
    characteristics, 105, 112
    circulating, 10, 122, 359
    condensate, 104, 122, 343; control, 541
    descriptions, 114
    dimensions, 687
    efficiency, 127
    hydraulic balance, 115
    impeller contours, 111
    instability, 108
    parallel operation, 316
    specific speed, 113
    vane shapes, 105
    variable-speed calculations, 126
    variable-speed drives, 128
  classifications, 90
  control of, 125
  developed head, 91
  direct acting (see reciprocating, below)
  dry-vacuum, 363
  dynamic head, 90
  hydraulic horsepower, 92
  jet, 90
  location in the cycle, 6, 540
  reciprocating, 90, 94
    capacity ranges, 94
    cylinder dimensions, 95, 99, 688
    dimensions, 688
    efficiencies, 99
    liquid valves, 96
    slip, 100
    steam valves, 97
    volumetric efficiency, 100
  rotary, 90, 101

selection of heads and capacities, 545
shaft (brake) horsepower, 93
suction lift, 90
system head curves, 124
wet-vacuum, 362

Radiation losses, steam generators, 198
Radioactivity, 598
Radius ratio, 79
Ralston's classification of coal, 183
Rankine cycle, 505
Rankine cycle ratio, 424, 505
Rateau stages, 404, 419, 429
Rates, electric, 628, 634
  block Hopkinson demand, 637
  block meter, 636
  off-peak, 639
  riders, 640
  step meter, 636
  straight line, 636
  Wright demand, 636
Reaction stages, 404, 429, 434, 465
Refractories, 160, 229, 259
Regenerative cycle, 506, 561
Reheaters, 235
  temperature control, 240
Relieving rate (velocity), evaporators,
  385
Replacement factors, 457, 523
Reserve equipment, 627
Reynolds, O., 52
Reynolds' number, 53, 59, 75, 76, 215
Run-of-river turbines, 627, 662

Scale, formations, 153, 329, 385
Semi-outdoor plants, 11, 631
Sigma function, 660
Slide valves, 97, 487
Soot blowers, 510
Spark ignition engines, 3, 573, 583
Specific gravity
  API and Baumé, 188
  water, 678
Specific speed
  centrifugal pumps, 113
  hydraulic turbines, 648
Speed regulation, 451, 496, 568, 584, 650
Splitters, ducts, 79
Stacks (see Chimneys)
Static head (pressure), 27, 90, 313
Station heat rates, 15, 510, 523
Steam, viscosity, 49
Steam engines (see Engines, steam)
Steam generators (see Generators, steam)
Steam rates, 442, 454
Steam turbines (see Turbines)
Stokers, 7, 275, 288

air temperatures, 289, 294, 558
chain grate, 288, 290
combustion rates, 290, 291, 294
overfeed, 288
spreader, 288
traveling grate, 7, 288, 293
underfeed, 288, 293
use of, 288
Superheaters, 10, 235
design pressures, 545
heat transfer, 270
location, 237
radiant and convection characteristics, 240
temperature control, 240
tube sizes, 270
Supersaturation, 417
Syphon efficiency, 361

Taxes, 4, 17, 630
Temperature control, superheaters and reheaters, 240
Tempering air, 279
Terminal difference, 149
Thermal conductivity, 140
Thermal efficiency, 15, 505, 540
Tubes
air preheater, 270
condenser, 343, 351, 358
economizer, 270
feedwater heater, 373
superheater, 270
Turbines
gas, 3, 13, 556
cycles, 557
descriptions, 564
efficiencies, 560
performance, 564
regenerator effectiveness, 562
regenerators, 559
hydraulic, 3, 645
cavitation, 660
descriptions, 649
efficiencies, 650, 657
Kaplan, 651, 657
power, 647
run-of-river, 662
sigma function, 660
specific speed, 648
speed calculations, 658
steam, 402, 512, 518, 523
automatic or controlled extraction, 458
blade shields, 431
classifications, 404
condition curve, 442, 448, 519
constructional details, 461

controlled extraction steam chests, 458
Curtis staging, 404, 419, 426, 428, 432
diagram efficiency, 429
dimensions, 691
dummy pistons, 438
emergency governor, 463
engine efficiency, 442, 694
extraction or replacement factors, 457, 523
governing, 447
grid valves, 458
impulse blade theory, 419
labyrinth seals, 439
leaving or exhaust loss, 425, 433, 441
multi-valve governing, 447
nozzles (see Nozzles)
poppet valves, 458
prepared standards for, 447, 527
pressure and velocity compounding, 404, 426
Rateau stages, 404, 419, 427
reaction blade theory, 424
selection of extraction pressures, 525
stage efficiency, 430
stage pressures, 440
steam rates, 442, 454
throttle governing, 446
topping or superposed, 406
vector diagrams, 420, 424, 428, 435
warped blades, 423
Willans lines, 454, 525
Turbulent flow, 52, 54
Turning vanes, ducts, 79

Ultimate analysis, 171
coal, table, 172
oil, table, 187
Uniflow engines, 488, 491
dimensions, 692
Unwin's friction factor, 63
Utilization factor, 626

Valves, 42
atmospheric relief, 356
blowdown for safety valves, 47
equivalent length, 68
safety, 46, 545
stop, 4, 440
stop-check, 46
throttle, 4, 440
Velocity head (pressure), 90, 312
Velocity pressure, 75
Viscosity, 47
air, 48
oil, 50

Viscosity (*Continued*)
  steam, 49
  units, 50
  water, 48

Waste heat boilers, 585
Water
  concentration, 234, 325, 385
  hardness, 325
  impurities, 234, 323, 509
  specific gravity, 678
  treatment, 323
  viscosity, 48

Water-tube steam generators, 212, 216
Water wheels, 645
Waterwalls
  heat transfer, 259
  insulation, 229
Welded joints, 40
Wet bottom furnaces, 228, 275
Weymouth, T. R., 75
Willans lines, 454, 525
Williams and Hazen formula, 63
Wright demand rate, 636

Zeolite treatment, 326, 386